SURFACE CHEMISTRY

Theory and Industrial Applications

Lloyd I. Osipow

Director, Surface Chemistry Department
Foster D. Snell, Inc.

ROBERT E. KRIEGER PUBLISHING COMPANY
Huntington, New York
1977

ORIGINAL EDITION 1962

REPRINT 1972, 1977

Printed and Published by

ROBERT E. KRIEGER PUBLISHING CO., INC.
BOX 542, HUNTINGTON, NEW YORK. 11743

Reprinted by arrangement with

VAN NOSTRAND REINHOLD COMPANY

Library of Congress Card Number: 62-20782

SBN 0-88275-076-3

Printed in the United States of America

To my wife,
whose patience and cooperation
made this work possible

General Introduction

American Chemical Society's Series of Chemical Monographs

By arrangement with the Interallied Conference of Pure and Applied Chemistry, which met in London and Brussels in July, 1919, the American Chemical Society was to undertake the production and publication of Scientific and Technologic Monographs on chemical subjects. At the same time it was agreed that the National Research Council, in cooperation with the American Chemical Society and the American Physical Society, should undertake the production and publication of Critical Tables of Chemical and Physical Constants. The American Chemical Society and the National Research Council mutually agreed to care for these two fields of chemical progress. The American Chemical Society named as Trustees, to make the necessary arrangements for the publication of the Monographs, Charles L. Parsons, secretary of the Society, Washington, D. C.; the late John E. Teeple, then treasurer of the Society, New York; and the late Professor Gellert Alleman of Swarthmore College. The Trustees arranged for the publication of the ACS Series of (a) Scientific and (b) Technological Monographs by the Chemical Catalog Company, Inc. (Reinhold Publishing Corporation, successor) of New York.

The Council of the American Chemical Society, acting through its Committee on National Policy, appointed editors (the present list of whom appears at the close of this sketch) to select authors of competent authority in their respective fields and to consider critically the manuscripts submitted.

The first Monograph of the Series appeared in 1921. After twenty-three years of experience certain modifications of general policy were indicated. In the beginning there still remained from the preceding five decades a distinct though arbitrary differentiation between so-called "pure science" publications and technologic or applied science literature. By 1944 this differentiation was fast becoming nebulous. Research in private enterprises had grown apace and not a little of it was pursued on the frontiers of knowledge. Furthermore, most workers in the sciences were coming to see the artificiality of the separation. The methods of both groups of workers are the same. They employ the same instrumentalities, and frankly recognize that their objectives are common, namely, the search for new knowledge for the service of man. The officers of the Society therefore combined the two editorial Boards in a single Board of twelve representative members.

Also in the beginning of the Series, it seemed expedient to construe rather broadly the definition of a Monograph. Needs of workers had to be recognized. Consequently among the first hundred Monographs appeared works in the form of treatises covering in some instances rather broad areas. Because such necessary works do not now want for publishers, it is considered advisable to hew more strictly to the line of the Monograph character, which means more complete and critical treatment of relatively restricted areas, and, where a broader field needs coverage, to subdivide it into logical subareas. The prodigious expansion of new knowledge makes such a change desirable.

These Monographs are intended to serve two principal purposes: first, to make available to chemists a thorough treatment of a selected area in form usable by persons working in more or less unrelated fields to the end that they may correlate their own work with a larger area of physical science discipline; second, to stimulate further research in the specific field treated. To implement this purpose the authors of Monographs are expected to give extended references to the literature. Where the literature is of such volume that a complete bibliography is impracticable, the authors are expected to append a list of references critically selected on the basis of their relative importance and significance.

Preface

In his review of progress in surface chemistry from 1917–57, the eminent A. S. C. Lawrence[1] wrote: "No one has dared to write a book on the science of soaps; I don't think that anybody could. There are plenty of books on parts of the subject, which are useful compilations for reference purposes and most valuable to Ph.D. *aspirants* writing theses, but we still don't know how soaps wash; we don't understand frothing and foam stability, nor do we understand emulsification. We can give lectures on all these, but we don't really understand them."

I have devoted most of the past fifteen years to industrial problems concerned with these phenomena. It is true that no one *really* knows how soaps clean or how deflocculants function. But we have some pretty good ideas. We know how to modify a solution so that it will clean under practical conditions. We can form or prevent foam. We can separate a large variety of materials by selective flotation. We can form stable emulsions or break them. We can deposit or remove oils. We can prevent corrosion and reduce friction.

My object has been to review those theories and supporting data that appear to provide the best current explanations of these phenomena. This book was written primarily for industrial workers, with ideas and tabulations of data intended as tools for solving practical technological problems.

I gratefully acknowledge the inspiration of Dr. Foster Dee Snell, who led me into this field of chemical endeavor, provided ideas for many of my own experimental efforts, and spent many long hours reading and correcting this manuscript.

LLOYD I. OSIPOW

New York, N.Y.

[1] Lawrence, A. S. C., "Surface Phenomena In Chemistry and Biology," ed. by J. F. Danielli, K. G. A. Pankhurst, and A. C. Riddiford, New York, Pergamon Press, Inc., 1958.

Contents

CHAPTER 1

Introduction

PLAN OF BOOK

This book is not intended as an exhaustive treatment of all aspects of surface chemistry, but is limited to those industrial applications in which the author has a significant experience. Applications covered are wetting, dispersion and flocculation, emulsions, foams, detergency, ore flotation, lubrication, and corrosion inhibition. In these areas, technology has far outpaced our theoretical understanding. Even some of the most fundamental concepts remain the subjects of controversy.

The presentation of many conflicting points of view, with their supporting data, is more likely to confuse than enlighten the reader. Instead, the author has arbitrarily selected for discussion those concepts that appear most credible on the basis of the supporting evidence. These ideas will also be found useful in the design of experiments directed toward the solution of technological problems. Figures and tabulations of data were selected either to illustrate concepts, or because of their practical importance.

The earlier chapters provide background for the discussions on practical applications in Chapters 10 through 16. Chapter 5 on electrical phenomena also includes a discussion of flocculation and dispersion. Chapters 6 and 7 on monolayers, and Chapter 9 on the properties of solutions containing surfactants provide more thorough discussions of these subjects than will be found elsewhere. Because excellent reviews are already available, the treatment of the subject matter in Chapters 2, 3, 4, and 9 is scanty. The reader is referred to Adamson (1) for a detailed discussion of the subjects covered in Chapters 2, 3, and 4, and to the works of Schwartz and coworkers (2, 3) for more detailed information about the classes of surfactants.

Glossary

In some instances, terms specific to the field of surface chemistry have been used without definition, or a discussion of the term does not appear until later in the book. This glossary may be found helpful to the reader.

Adsorption. The adhesion of a thin film of molecules to a solid or liquid surface.

Amphoteric Surfactant. A surface-active material that ionizes in aqueous solution. The surface-active ion may bear a negative or a positive charge, depending upon the pH of the solution.

Amphipathic Compound. A substance containing one or more polar groups segregated from a relatively large nonpolar group.

Anionic surfactant. A surface-active material that ionizes in aqueous.solution. The ion that bears a negative charge has a pronounced tendency to concentrate at the interface between two phases.

Anisotropic. The condition of having different properties in different directions, invariably due to an orientation of the elements making up the structure.

Autophobic Liquid. One that is unable to spread on its own adsorbed film.

Cationic Surfactant. A surface-active agent that ionizes in aqueous solution, with the surface-active ion bearing a positive charge.

Cation Exchange Capacity. The capability of a solid to exchange cations initially present on the surface for those in the contacting solution.

Chemisorption. An adsorption process in which the forces involved are of the same magnitude as in chemical reactions.

Cloud Point. The temperature at which turbidity appears when an aqueous solution of a nonionic surfactant is heated.

Coalesce. The act of combining to form a single body, as in the case of two oil droplets in an emulsion coalescing to form a larger droplet.

Collector. An agent used in ore flotation to promote attachment of solid particles to air bubbles.

Contact Angle. The angle formed by a droplet in contact with a solid surface, measured from within the droplet. An advancing contact angle is formed when the droplet advances onto a fresh surface. A receding contact angle is formed when the droplet is withdrawn from a portion of the surface on which it has been in contact.

Coupling Agent. A material that increases the miscibility of two liquids or a liquid and a solid.

Counter-ion. An ion with electrical charge opposite to the charge on the surface of an aggregate.

Critical Micelle Concentration — (cmc). A narrow concentration range in which surfactant ions or molecules begin to aggregate and form micelles.

Critical Surface Tension. That value of the liquid surface tension below which liquids will spread on a given solid.

Cryoscopic Forces. Forces associated with the freezing of liquids and the crystallization of solids.

Detergency. The removal of contaminants from the surface of a solid.

Dispersion. The breaking up of solid aggregates, and maintaining the individual particles in suspension.

Electric Double Layer. The excess of ions of one charge type present at an interface and the equivalent amount of ions of opposite charge present in one liquid phase, generally water. In the diffuse double layer, it is assumed that the charges in the liquid phase are distributed in accordance with a Boltzmann relation.

Electrocapillary Effect. The change in interfacial tension that occurs with the charging of an interface.

Electrochemical Potential. The sum of the chemical potential and a term whose magnitude depends upon the internal potential of the phase.

Electrokinetic Phenomena. The relative motion of a charged surface with reference to the bulk solution.

Electro-osmosis. The movement of a liquid with respect to a solid wall as the result of an applied potential gradient.

Electrophoretic. The movement of colloidal particles in an electric field.

Emulsion. A system consisting of two immiscible liquids, with one dispersed as small droplets in the other. In certain complex emulsions, a portion of the liquid constituting the external phase may also be found dispersed within droplets of the second liquid.

Film Balance. An instrument for measuring the difference in surface tension between a pure liquid and one covered with a surface film.

Film Elasticity. The tendency of a foam film to resist distortion by changing its surface tension to oppose expansion or compression.

Flocculation. The agglomeration or sticking together of solid particles or liquid droplets.

Foam. Bubbles of gas whose walls are thin liquid films.

Foam Fractionation. The separation of solutes by frothing.

Foam Transition Temperature. The temperature below which the foam is slow draining, and above which it is fast draining.

Free Surface Energy. A two-dimensional free energy term applied to a hypothetical surface phase. Used interchangeably with surface free energy.

Frothing Agent. In ore flotation, an agent used to create a froth which entraps particles brought to the surface after bubble attachment.

Hofmeister Series. A tabulation of ions of a given charge and valence according to the hydrated radius of the ions.

Hydrophilic. Having an affinity for water. A hydrophilic surface is one that is wet by water. A hydrophilic emulsifier is soluble in water and promotes the formation of an oil-in-water emulsion.

Hydrophobic. The opposite of hydrophilic.

Hydrophile-lipophile Balance (HLB). The number and type of hydrophilic and lipophilic groups present in an emulsifier or in a combination of emulsifiers.

Hydrotropic Agent. Same as coupling agent.

Interface. The region between two contacting phases, generally two condensed phases.

Interfacial Tension. A force with the dimensions of dynes/cm. A measure of the work required to enlarge the interface by one sq. cm.

Krafft Temperature. A temperature above which there is a rapid increase in the solubility of an ionic surfactant. Also, the temperature at which the solubility of a surfactant coincides with its critical micelle concentration.

Lipophilic. Having an affinity for oil. A lipophilic surface is wet by oils. A lipophilic emulsifier promotes the formation of a water-in-oil emulsion.

Lipophobic. The opposite of lipophilic.

Liquid-crystalline Phase. A viscous liquid exhibiting optical anisotropy.

Lyotropic Series. Same as Hofmeister series.

Micelle. An oriented aggregation of surfactant ions or molecules.

Migration Potential. The development of an electrical potential as the result of the movement of small particles suspended in a liquid.

Monomolecular Layer; Monolayer. A film one molecule thick on the surface of a solid or liquid.

Multimolecular Layer; Multilayer. A film two or more molecules thick on the surface of a solid or liquid.

Nonionic Surfactant. A surface-active agent that does not ionize in water.

Nonpolar Molecule. One containing an equal number of positive and negative charges with coinciding centers of gravity. The term nonpolar group is applied to a portion of a large molecule with nonpolar characteristics.

Oil-in-Water Emulsion (O/W). An emulsion in which oil droplets are dispersed in a continuous water phase.

Oleophilic. Same as lipophilic.

Oleophobic. Same as lipophobic.

Polar Molecule. An uncharged molecule in which the centers of gravity of positive and negative charges do not coincide. The term polar group is applied to a portion of a molecule with polar characteristics.

Saturation Adsorption. As determined by the Gibbs adsorption isotherm, the maximum amount of surfactant adsorbed from its solution.

Secondary Minimum. A shallow minimum observed in a plot of interaction *versus* distance between particles. At this minimum, forces of attraction are only slightly greater than those of repulsion, and flocculated particles are readily redispersed.

Sedimentation. The settling of particles due to gravity.

Selective Adsorption. The tendency for one adsorbable species to concentrate at a surface or interface in preference to another.

Solubilization. The increased solubility of a substance in a surfactant solution relative to its solubility in the pure solvent.

Spreading Coefficient. The change in surface and interfacial tensions caused by the spreading of one liquid on another.

Spreading Pressure. The decrease in surface tension caused by the spreading of an insoluble substance as a monolayer.

Streaming Potential. The potential difference that arises from forcing a liquid through a porous plug or capillary.

Substrate. A substance acted upon, as by adsorption.

Surface. The region between two contacting phases, generally a condensed and a gaseous phase.

Surface-active Agent. A substance that exhibits a marked tendency to adsorb at a surface or interface.

Surface Charge Density. The excess of ions of one charge type per unit area of surface.

Surface Chemistry. A study of phenomena arising primarily from surface forces.

Surface Excess. The difference between the concentration of solute in the surface region and in the interior of the solution.

Surface Phases. The physical states of monolayers, analogous to bulk phases.

Surface Potential. The change in the potential difference between a liquid and air arising from the presence of a surface film.

Surface Pressure. The decrease in the surface tension of a liquid due to the presence of a surface film.

Surface Tension. A force with the dimensions of dynes/cm that is a measure of the work required to increase the area of a surface by one sq cm.

Surfactant. A surface-active agent.

Syndet. A synthetic detergent.

Water-in-Oil Emulsion (W/O). An emulsion consisting of water droplets dispersed in a continuous oil phase.

Work Function. A measure of the heat of evaporation of electrons.

Work of Adhesion. The work per unit area required to separate two liquids, equal to the sum of the surface tensions of the two liquids less the interfacial tension of the liquid-liquid interface.

Work of Cohesion. The work required to form two sq cm of liquid surface, equal to twice the surface tension of the liquid.

References

1. Adamson, A. W., "Physical Chemistry of Surfaces", New York, Interscience Publishers, 1960.
2. Schwartz, A. M., and Perry, J. W., "Surface Active Agents", New York, Interscience Publishers, 1949.
3. Schwartz, A. M., Perry, J. W., and Berch, J., "Surface Active Agents and Detergents", Vol. II, New York, Interscience Publishers, 1958.

CHAPTER 2

Surface Energy and Surface Tension

SURFACE PHENOMENA

Surface Free Energy

For any system consisting of two phases, there is a surface of separation between the two phases. While the precise thickness of this interfacial region is not definitely known, it includes all parts of the system that are influenced by surface forces. The two bulk phases, which may be two liquids or a liquid and vapor, can be considered as being separated by a surface phase. While the surface phase is not a true phase in the usual physical sense, the description is convenient.

The total free energy F of a system comprising two bulk phases and an interface is

$$F = F^{\alpha} + F^{\beta} + F^{s} \tag{2.1}$$

where the superscripts α and β refer to the bulk phases, and s to the surface phase. The free energies of the bulk phases are calculated on the assumption that they both remain homogeneous right up to a hypothetical geometric surface.

If a small, reversible change occurs in the system, the free energy change dF is expressed

$$dF = dF^{\alpha} + dF^{\beta} + dF^{s} \tag{2.2}$$

For the homogeneous bulk phases, the free energy charges are given by

$$dF^{\alpha} = -S^{\alpha}dT + V^{\alpha}dP^{\alpha} + \mu_1^{\alpha}dn_1^{\alpha} + \mu_2^{\alpha}dn_2^{\alpha} + \ldots \tag{2.3}$$

$$dF^{\beta} = -S^{\beta}dT + V^{\beta}dP^{\beta} + \mu_1^{\beta}dn_1^{\beta} + \mu_2^{\beta}dn_2^{\beta} + \ldots \tag{2.4}$$

where S, T, V and P are the usual notations referring to entropy, temperature, volume and pressure, respectively. The term n refers to the number of moles of a component whose chemical potential is μ. The subscripts refer to components 1, 2, ...

The surface free-energy change must include a term for the work required to increase the area of the surface by an infinitesimal amount dA, at constant temperature, pressure and composition. The work of surface expansion is done against a tension γ, referred to as the surface tension. This reversible work is equal to γdA. Since the surface contribution to the volume is negligible, the quantity VdP can be omitted. Then

$$dF^s = -S^s dT + \gamma dA + \mu_1^s dn_1^s + \mu_2^s dn_2^s + \ldots \tag{2.5}$$

where μ_1^s, $\mu_2^s \ldots$, are the surface chemical potentials of the various components of the system.

Summing dF^α, dF^β and dF^s gives the total free energy change for the system,

$$dF = -SdT + V^\alpha dP^\alpha + V^\beta dP^\beta + \gamma dA + \sum \mu_i^\alpha dn_i^\alpha + \sum \mu_i^\beta dn_i^\beta + \sum \mu_i^s dn_i^s \tag{2.6}$$

where $S = S^\alpha + S^\beta + S^s$ is the total entropy of the system. At constant temperature, pressure, and composition,

$$dF = \gamma dA \tag{2.7}$$

or

$$\gamma = \left(\frac{\partial F}{\partial A}\right)_{T,P,n} = F_s \tag{2.8}$$

Equations 2.7 and 2.8 express a fundamental relation of surface chemistry. The surface tension γ has the dimension dynes/cm. It is a measure of the work required to increase the surface by unit area, at constant temperature, pressure, and composition. The term F_s is the surface free energy per unit area, expressed in ergs cm^{-2}. Since it is numerically equal to the surface tension for a pure liquid, γ is commonly used to express either surface tension or surface free energy per unit area.

Surface Entropy and Energy

The entropy of any system at constant pressure, surface area and composition is

$$-S = \left(\frac{\partial F}{\partial T}\right)_{P,A,n} \tag{2.9}$$

For a pure liquid, the surface entropy per square centimeter S_s is

$$-S_s = \frac{d\gamma}{dT} \tag{2.10}$$

The total surface energy per square centimeter E_s for a pure liquid is

$$E_s = F_s + TS_s \tag{2.11}$$

or, as usually expressed

$$E_s = \gamma - T\frac{d\gamma}{dT} \tag{2.12}$$

The total surface energy of a pure liquid is generally larger than the surface free energy.

The surface specific heat C_s is given by

$$C_s = \frac{dE_s}{dT} \tag{2.13}$$

Effect of Temperature on Surface Tension

The work required to increase the area of a surface is the work required to bring additional molecules from the interior to the surface. This work must be done against the attraction of surrounding molecules. Consequently, surface tension is a measure of the attraction between molecules.

As the temperature is raised, the kinetic energy of the molecules increases, and the attraction between molecules is partially overcome. Consequently the surface tension invariably decreases with a rise in temperature. As the critical temperature of the liquid is approached, the surface tension diminishes and finally vanishes altogether.

The equation of Ramsay and Shields (1) for the variation of surface tension with temperature is

$$\gamma\left(\frac{M}{\rho}\right)^{2/3} = k\,(T_c - T - 6) \tag{2.14}$$

where M is the molecular weight and ρ is the density of the liquid. The ratio M/ρ is the molecular volume. T_c is the critical temperature of the liquid. The quantity $\gamma\,(M/\rho)^{2/3}$ is called the molecular free surface energy. A less accurate equation had previously been given by Eotvos (2) in which the constant 6° was not subtracted from the critical temperature.

The constant k has a value of about 2.1 ergs/degree for hydrocarbon liquids. For alcohols k varies from about 0.95 to 1.5, and for acids from 0.90 to 1.7, decreasing as the hydrocarbon chain is lengthened to five carbons. The value of k for water increases from 0.87 at 0° to 1.21 at 140°.

It was thought for a long time that k the "constant of Eotvos" was a measure of the degree of association of liquids. Adam (3) has pointed out that not only is k frequently not constant, but its actual numerical value is without significance with regard to determining the molecular complexity of liquids.

There are a number of empirical equations relating surface tension to temperature. The simplest of these expresses the nearly linear variation of surface tension with temperature

$$\gamma = \gamma_0 \, (1 - bT) \tag{2.15}$$

where b is a constant.

McLeod's empirical equation (4) relates density with surface tension

$$\frac{\gamma}{(\rho_L - \rho_V)^4} = C \tag{2.16}$$

where ρ_L and ρ_V refer to the density of the liquid and vapor, respectively. The constant C is different for each liquid. McLeod's equation holds accurately for the majority of organic liquids over a broad range of temperature. It does not appear to hold for liquid metals.

Parachor. The molecular volume of organic liquids depends on their chemical constitution. It would be possible from molecular volumes to decide between different possible arrangements of atoms in a molecule, if it were not for the complicating effect of temperature. Sugden (5) pointed out that McLeod's relation afforded this basis for comparison, because the effect of temperature was nullified by the surface tension. He defined a parachor P,

$$P = \frac{M}{\rho_L - \rho_V} \cdot \gamma^{\frac{1}{4}} \tag{2.17}$$

This is the fourth root of the McLeod constant multiplied by the molecular weight. The parachor is also a molecular volume multiplied by the fourth root of the surface tension, and does not vary with temperature. The parachors of organic compounds have been reviewed by Quayle (6).

Effect of Pressure on Surface Tension

An increase in the pressure of vapor over the surface of a liquid has the effect of bringing more gaseous molecules into contact with the surface. The attraction of these molecules neutralizes to some extent the inward attraction on the surface molecules. The net effect would be expected to be a decrease in surface tension with increasing pressure. This has been confirmed. With some liquids, the surface tension decreases 50 per cent at about 150 atmospheres.

Effect of Surface Curvature

Due to surface forces, the pressure within a curved surface is greater than the external pressure. This is conveniently illustrated by considering a soap bubble. In the absence of other fields of force, such as gravitational, a soap bubble is spherical. The surface free-energy of a soap bubble of radius r is $4\pi r^2\gamma$. If the radius is decreased by dr, then the change in free surface-energy is $8\pi r\gamma dr$.

The tendency for the bubble to shrink must be balanced by a pressure difference ΔP across the film. The work against this pressure difference is $\Delta P 4\pi r^2 dr$. Equating this work term with the decrease in surface free-energy gives

$$\Delta P 4\pi r^2 dr = 8\pi r\gamma dr \tag{2.18}$$

or

$$\Delta P = 2\frac{\gamma}{r} \tag{2.19}$$

The smaller the bubble, the greater is the pressure of air inside as compared to that outside. Equation 2.19 applies to a spherical liquid having a single surface. However, a soap bubble is composed of a film of liquid with two surfaces. Consequently, the surface area, the surface free-energy and the change in surface free-energy upon shrinkage must be multiplied by 2. Then the difference in pressure is

$$\Delta P = 4\frac{\gamma}{r} \tag{2.20}$$

The curved surface of a liquid is best described by two radii of curvature r_1 and r_2. The pressure difference across this surface is given by the equation of Young and Laplace,

$$\Delta P = \gamma\left(\frac{1}{r_1} + \frac{1}{r_2}\right) \tag{2.21}$$

If both radii are equal, the liquid is in the form of a sphere, and Equation 2.21 reduces to Equation 2.19. For a plane surface, the radii are infinite and $\Delta P = 0$. There is no pressure difference across a plane surface.

The vapor pressure over a convex surface is greater than that over a plane surface, while over a concave surface the vapor pressure is lower. This difference results from the fact that the condensation of vapor on a small convex drop of a liquid will raise its surface area, and consequently, its surface free-energy. Condensation does not change the surface area of a plane surface, while it diminishes that of a concave surface.

The increase in the vapor pressure Δp over a spherical drop of liquid of radius r, as compared with the vapor pressure p over a plane surface of the liquid is

$$RTln\left(1 + \frac{\Delta p}{p}\right) = \frac{2\gamma M}{r\rho} \tag{2.22}$$

where M is the molecular weight, ρ is the density and γ is the surface tension of the liquid. For water, Δp is 0.1 per cent of p if r is 10^{-4} cm, but 11 per cent if r is 10^{-6} cm.

One consequence of the increase in vapor pressure, due to the curvature of small droplets, is that water vapor will not condense in a dust-free atmosphere. Nuclei providing a less curved surface must be present in order for the vapor to condense near the usual saturation vapor pressure.

Work of Cohesion and Adhesion

The work of cohesion was defined by Harkins (7) as the work required to separate a column of liquid 1 cm² in cross section. The surface tension of the liquid is γ_A. The net effect is to form two new surfaces whose total area is 2 cm². The cohesional work W_c is equal to the change in surface free-energy,

$$W_c = 2\gamma_A \tag{2.23}$$

Consider instead two immiscible liquids, A and B in a column of cross-sectional area equal to 1 cm². Separation of the two liquids results in the formation of two new surfaces of surface free-energy γ_A and γ_B. There is a simultaneous loss of surface free-energy γ_{AB}, corresponding to the interfacial tension between the two liquid phases. The work of adhesion W_a is equal to the change in surface free-energy

$$W_a = \gamma_A + \gamma_B - \gamma_{AB} \tag{2.24}$$

Comparison of the cohesional work with the work of adhesion to a second liquid gives an indication of the relative attraction between the molecules of one liquid and that between the molecules of the two different liquids.

Surface-Tension Values

The term surface tension of a liquid, when used without qualification, refers to the equilibrium surface tension in the boundary between the liquid

TABLE 2.1. SURFACE-TENSION VALUES FOR VARIOUS TYPES OF LIQUIDS.

Liquid	Surface Tension (dynes/cm)
Water	72.6
Fluorocarbons	8–15
Hydrocarbons	18–30
Polar Organic Compounds	22–50
Aqueous Detergent Solutions	24–40
Molten Glasses	200–400
Molten Metals	350–1800

and its vapor. Surface-tension values for a number of liquids are given in the chapter on Wetting. More complete data for highly-purified liquids can be found in various tabulations of physical data, such as International Critical Tables I, 103 and IV, 439, Annual Tables of Physical Constants 1942, No 700, and Landolt-Bornstein Tabellen.

Table 2.1 shows surface-tension values for various types of liquids. Fluorocarbon liquids have the lowest surface tensions, and molten metals the highest.

Surface Tension of Solutions

The simplest relation expressing the surface tension of binary mixtures is an additive rule (8)

$$\gamma = \gamma_1 X + \gamma_2(1 - X) \tag{2.25}$$

where γ is the surface tension of a solution composed of X mole fraction of

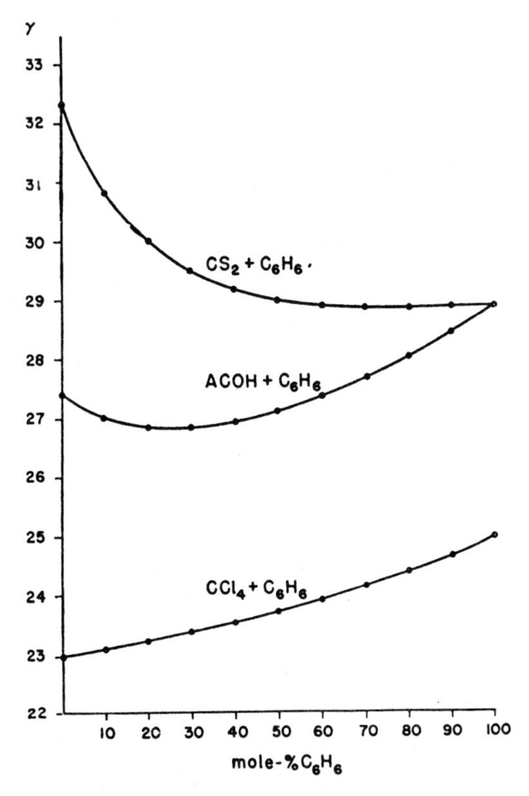

Figure 2.1. Surface tension of binary mixtures of benzene with carbon disulfide at 20°, acetic acid at 20°, and carbon tetrachloride at 50°C (8, 9).

a component of surface tension γ_1, and $(1-X)$ mole fraction of a second component of surface tension γ_2.

This relation is sometimes observed where the vapor pressure of the solution is a linear function of the composition. However, most binary mixtures show positive or negative deviations, with the latter more common. Figure 2.1 illustrates the surface tension behavior of mixtures of organic liquids, showing negative deviation from the simple additive rule.

The addition of inorganic electrolytes to water invariably results in an increase in surface tension as shown in Figure 2.2. The behavior of binary

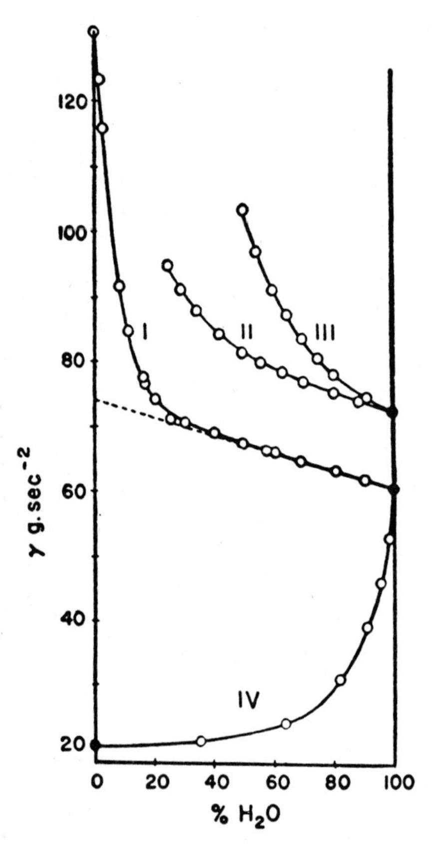

Figure 2.2. Surface tension of aqueous solutions of $AgTl(NO_3)_2$ at 90° (curve I), KNO_2 at 20° (curve II), K_2CO_3 at 20° (curve III), and *n*-butyric acid at 90° C (curve IV) (8, 10).

mixtures of water and sulfuric acid is given in Figure 2.3. In general, when the heat of mixing has a large, positive value, the deviation from the additivity rule is positive.

Surface-active agents lower the surface tension of water considerably, even in very dilute solutions. Other types of organic compounds, such as acetic acid or ethanol, lower the surface tension of water only slightly. Inorganic electrolytes raise the surface tension of water to only a small extent in dilute solution. The effect of ions on increasing the surface tension of water is in the same order as the lyotropic or Hofmeister series: Li > Na > K, and F > Cl > Br > I.

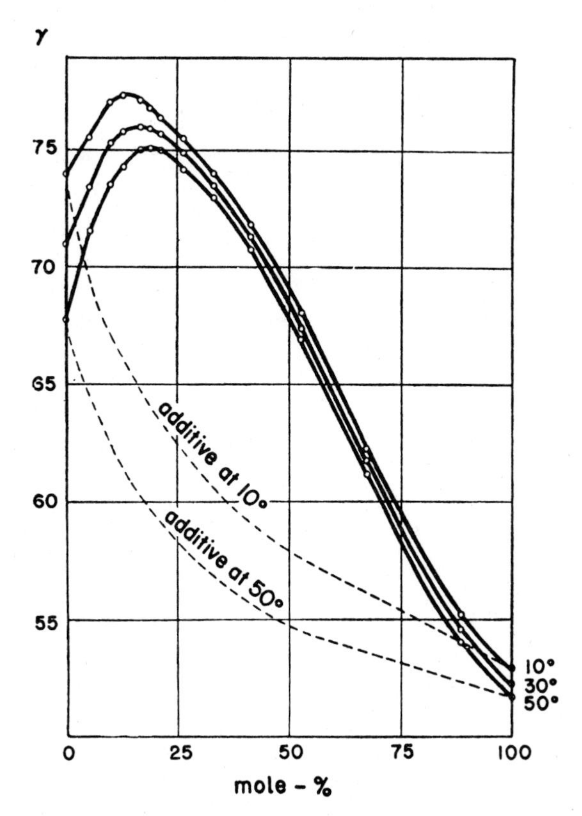

Figure 2.3. Surface tension of sulfuric acid-water mixtures. The continuous curves are drawn through experimental points determined at 10°, 30°, and 50°C. The discontinuous curves show the values calculated by means of an additivity rule (8, 11).

The surface tension of fresh solutions differs from that at equilibrium. The rate of change of surface tension, surface aging, depends upon a number of factors. These include the rate of diffusion of the solute to or from the surface, agitation, and other external circumstances. The surface aging effect is shown in Figure 2.4.

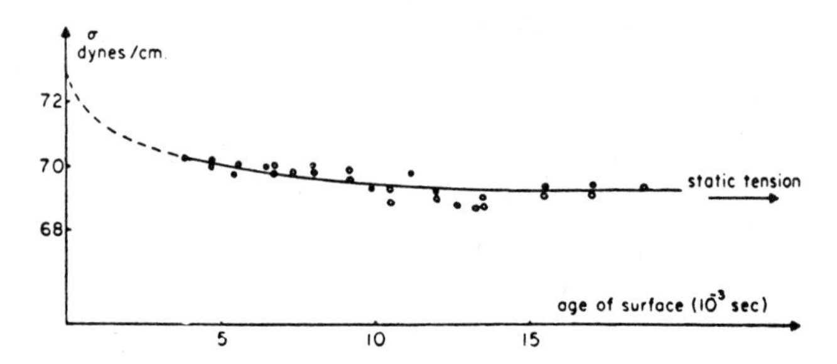

Figure 2.4. Dynamic surface tensions of 0.1 N adipic acid solution at 20°C (12).

Gibbs Adsorption Equation (13, 14). Consider a system composed of two phases, each of which contains one or more components. At equilibrium, the chemical potential of any one component is the same in each phase and in the surface. This can be seen from Equation 2.6, since at equilibrium

$$dF_{T,P,A} = 0 \tag{2.26}$$

It then follows that

$$\sum \mu_i^\alpha dn_i^\alpha + \sum \mu_i^\beta dn_i^\beta + \sum \mu_i^s dn_i^s = 0 \tag{2.27}$$

If matter can pass freely between the two bulk phases and the surface phase, and no material is lost, the sum of the variations for each component is zero. Therefore, it follows from Equation 2.27 that for each component

$$\mu_i^\alpha = \mu_i^\beta = \mu_i^s \tag{2.28}$$

Thus, the surface chemical-potential of any component of a system is equal to its chemical potential in the bulk phases at equilibrium.

The surface free-energy is given by the equation

$$F^s = \gamma A + n_1^s\mu_1^s + n_2^s\mu_2^s + \ldots \tag{2.29}$$

Since F^s is a definite property of the surface, depending only upon the thermodynamic state of the system, F^s is a complete differential. Differentiation of Equation 2.29 gives

$$dF^s = \gamma dA + Ad\gamma + \sum n_i^s d\mu_i^s + \sum \mu_i^s dn_i^s \tag{2.30}$$

Comparison with Equation 2.5 shows that

$$S^s dT + Ad\gamma + \sum n_i^s d\mu_i^s = 0 \tag{2.31}$$

At constant temperature this becomes

$$Ad\gamma + \sum n_i^s d\mu_i^s = 0 \tag{2.32}$$

Dividing through by the area of the surface A

$$d\gamma + \frac{n_1^s}{A} d\mu_1^s + \frac{n_2^s}{A} d\mu_2^s + \ldots = 0 \tag{2.33}$$

If

$$\Gamma_1 = n_1^s/A, \Gamma_2 = n_2^s/A, \ldots$$

$$d\gamma + \Gamma_1 d\mu_1^s + \Gamma_2 d\mu_2^s + \ldots = 0 \tag{2.34}$$

For a system at equilibrium, at constant temperature, pressure, surface area, and composition, the chemical potential of any component is the same at the surface as it is in the bulk phase. Then

$$d\gamma + \Gamma_1 d\mu_1 + \Gamma_2 d\mu_2 + \ldots = 0 \tag{2.35}$$

where μ_1, μ_2, . . . refer to the chemical potential of the component in the bulk phase, that is in the solution itself.

The terms Γ_1, Γ_2, . . . , are the excess surface concentrations of the various components of the system. Surface concentrations are expressed as amounts per unit area of surface. The actual numerical values depend upon the arbitrary position chosen for the geometrical surface. In the study of dilute solutions, it is convenient to choose the surface so as to make the surface excess of the solvent equal to zero.

For a system composed of a single solute dissolved in a solvent, and with the surface chosen to make the surface excess of the solvent Γ_1, equal zero, then

$$d\gamma + \Gamma_2 d\mu_2 = 0 \tag{2.36}$$

or

$$\Gamma_2 = -\left(\frac{\partial \gamma}{\partial \mu_2}\right)_T \tag{2.37}$$

Since, for any component

$$\mu_i = \mu_i^0 + RT \ln a_i \tag{2.38}$$

where μ_i^0 is the chemical potential of the component in a standard state, and a_i is the activity of the component. Then

$$\Gamma_2 = -\frac{1}{RT}\left(\frac{\partial \gamma}{\partial \ln a_2}\right)_T \tag{2.39}$$

For dilute solutions, the activity of the solute can be replaced by its concentration c, so that

$$\Gamma_2 = -\frac{1}{RT}\left(\frac{\partial \gamma}{\partial \ln c}\right) \tag{2.40}$$

Equation 2.35 and the simpler Equations 2.39 and 2.40 are the fundamental thermodynamic equations expressing the adsorption of solutes from solution. Their importance in surface chemistry will be shown in many of the chapters that follow.

SURFACE TENSION MEASUREMENTS

The various methods that are available for the accurate measurement of surface tension have been reviewed elsewhere (3, 8, 15, 16). Consequently, the present treatment is much abbreviated.

Capillary Rise

The capillary rise method for determining surface tension is the classic one. A liquid that wets the walls of a capillary tube will rise in the capillary, otherwise it will fall, as illustrated in Figure 2.5. If the capillary is circular

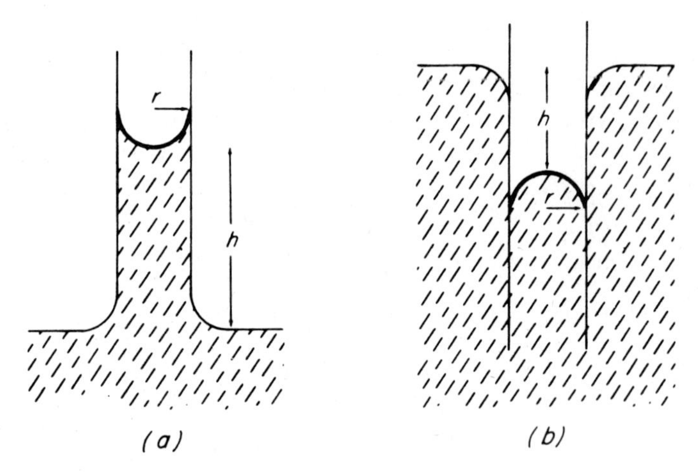

Figure 2.5. (*a*) Capillary rise where the liquid wets the walls of the capillary. (*b*) Capillary depression, where the walls are not wet by the liquid.

in cross section, the pressure difference across the interface ΔP is that of Equation 2.19,

$$\Delta P = \frac{2\gamma}{r} \tag{2.41}$$

where γ is the surface tension of the liquid and r is the radius of the capillary. The pressure difference must also equal the hydrostatic-pressure drop in the column of liquid in the capillary. This is given by

$$\Delta P = \Delta \rho g h \tag{2.42}$$

where $\Delta\rho$ is the difference in density between the liquid and the gas phases, g is the acceleration due to gravity, and h is the height of the meniscus above a flat liquid surface. Equating the two values for ΔP gives

$$\gamma = \frac{rhg\Delta\rho}{2} \tag{2.43}$$

This equation is applicable if the liquid completely wets the walls of the capillary, giving a zero contact angle. If the liquid completely fails to wet the walls of the capillary, the contact angle is 180°. Then the meniscus is convex, and h is the depth of the depression. If the contact angle θ is not 0° or 180°, then the expression is

$$\gamma \cos\theta = \tfrac{1}{2}\, grh\Delta\rho \tag{2.44}$$

The experimental details necessary to obtain high accuracy, and the correction terms that must be applied, are discussed by Richards and Carver (17), and Harkins and Brown (18).

Maximum Bubble Pressure Method

A tube is projected vertically beneath the surface of a liquid and curves upward. A bubble is blown at the tip of the tube. The pressure in the bubble increases at first, as the bubble grows and the radius of curvature diminishes. The smallest radius of curvature and the maximum pressure occur when the bubble is a hemisphere. Further growth causes a decrease of pressure, so that air rushes in and bursts the bubble. The maximum pressure is

$$P + P_1 = gh\Delta\rho + \frac{2\gamma}{r} \tag{2.45}$$

where $P_1 = gh\Delta\rho$ is the part of the measured pressure that is required to force the liquid down the tube to the level h at the end of the tube below the plane liquid surface. In the case of a liquid that wets the tube moderately well, r is the internal radius of the tube. Experimental details and correction terms are given by Sugden (19, 20).

Drop Weight Method

One method that is in frequent use is to determine the surface tension by measuring the weight or volume of drops which fall slowly from the tip of a vertical tube. The method can be used to measure the surface tension at either a liquid-air or a liquid-liquid interface.

The behavior of a falling drop is quite complex, which explains the absence of a simple formula. As a drop grows at the end of a tip, it becomes hemispherical, then lengthens and develops a waist. At about the point of instability the waist becomes longer and narrower and finally two drops break away. The second drop is extremely small, and is not ordinarily observed. Considerable liquid remains attached to the tip.

The equation relating the surface tension of the liquid to the weight of the falling drop W and the radius of the tube r is

$$\gamma = \frac{W}{r}\mathbf{F} \tag{2.46}$$

where **F** is the correction factor. Tables of **F** values are given by Harkins and Brown (18).

Wilhelmy Slide Method

A thin platinum plate or slide is suspended from one arm of a balance and dipped into a liquid (21). The slide is then gradually raised from the liquid until the point of detachment. If the slide is completely wetted by the liquid, so that the contact angle is zero, then the total weight W_{total} at the point of detachment is

$$W_{\text{total}} = W_{\text{slide}} + 2\,(w + t)\,\gamma \tag{2.47}$$

where w is the width and t is the thickness of the slide.

Ring Method

This method is most widely used in commercial laboratories, because it is capable of rapid and accurate measurement of the interfacial tension between two liquids or the surface tension between liquid and vapor. It involves the determination of the force required to detach a loop of wire from the surface of the liquid. It is necessary that the wire be completely wetted by the liquid.

Elementary theory suggests that at the point of detachment of the wire ring, the total weight W_{total} is

$$W_{\text{total}} = W_{\text{ring}} + 4\pi R\gamma \tag{2.48}$$

where R is the radius of the ring. Harkins and Jordan (22) showed that this equation is in serious error and developed the necessary correction factors. These factors have been extended for liquids of high density and low surface tension (23).

Static Drop Methods

Methods based on the shape of static drops or bubbles are useful for studying slow changes of surface tension. More recently they have been used to determine the surface tension of molten metals.

The sessile drop method consists of placing a drop of liquid on a flat plate and measuring the dimensions of the drop. When the diameter of the drop is sufficiently large that the curvature at the apex can be neglected, the surface tension is related to the height h as follows

$$h^2 = \frac{2\gamma}{g\Delta\rho} \qquad (2.49)$$

Where this curvature cannot be neglected, Porter's results (24) can be used. These are based on Bashforth and Adams tables (25).

The pendant drop method depends upon the shape of a drop hanging from a tip. The surface tension can be calculated with Bashforth and Adams tables (25, 26, 27).

Dynamic Methods

If a jet of liquid is forced from an orifice that is not circular, the surface tension tends to correct the noncircular cross-section of the jet. However, the momentum of this motion causes the jet to pass through the circular form and become unsymmetrical again. The result is an oscillation about the preferred circular cross-section. The surface tension can be calculated from the distance apart of the nodes and swellings that appear periodically on the jet, when viewed from one side. The theory has been given by Rayleigh (28). For the ideal case

$$\gamma = (k^2\rho/\pi^2 r)\,(f/\lambda)^2 \qquad (2.50)$$

where λ is the wavelength of the oscillation, f is the flow rate, and r is the radius of the jet where the cross-section is circular. The method has been discussed by Addison (29).

A dynamic method based on the shape of a falling column of liquid was developed by Addison and Elliott (30). These dynamic methods are used to study rapid changes in surface tension.

References

1. Ramsay, W., and Shields, J., *J. Chem. Soc.* **1893**, 1089.
2. Eotvos, R., *Wied. Ann.* **27**, 456 (1886).
3. Adam, N. K., "The Physics and Chemistry of Surfaces", London, Oxford University Press, 1941.
4. McLeod, D. B., *Trans. Faraday Soc.* **19**, 38 (1923).
5. Garner, F. B., and Sugden, S., *J. Chem. Soc.*, **1929**, 1298.
6. Quayle, O. R., *Chem. Revs.* **53**, 439 (1953).
7. Harkins, W. D., and Cheng, V. C., *J. Am. Chem. Soc.* **43**, 36 (1921).
8. Bikerman, J. J., "Surface Chemistry: Theory and Applications", New York, Academic Press, 1958.
9. Belton, J. W., *Trans. Faraday Soc.* **31**, 1642 (1935).
10. Rehbinder, P., *Z. physik Chem.* **121**, 103 (1926).
11. Sabinina, L., and Terpugov, L., *Z. physik Chem.* **A173**, 237 (1935).
12. Defay, R., and Hommelen, J. R., *J. Colloid Sci.* **13**, 553 (1958).

13. Gibbs, J. W., "Collected Works", Vol. I, New Haven, Yale University Press, 1948.
14. Glasstone, S., "Thermodynamics For Chemists", New York, D. Van Nostrand Company, 1947.
15. Harkins, W. D., "The Physical Chemistry of Surface Films", New York, Reinhold Publishing Corp., 1952.
16. Adamson, A. W., "Physical Chemistry of Surfaces", New York, Interscience Publishers Inc., 1960.
17. Richards, T. W., and Carver, E. K., *J. Am. Chem. Soc.* **43**, 827 (1921).
18. Harkins, W. D., and Brown, F. E., *J. Am. Chem. Soc.* **41**, 499 (1919).
19. Sugden, S., *J. Chem. Soc.*, **1922**, 858.
20. Sugden, S., *J. Chem. Soc.*, **1924**, 27.
21. Wilhelmy, L., *Ann. Physik* **119**, 117 (1863).
22. Harkins, W. D., and Jordan, H. F., *J. Am. Chem. Soc.* **52**, 1751 (1930).
23. Fox, H. W., and Chrisman, C. H., Jr., *J. Phys. Chem.* **56**, 284 (1952).
24. Porter, A. W., *Phil. Mag.* **15**, 163 (1933).
25. Bashforth, F., and Adams, J. C., "An Attempt to Test the Theories of Capillary Action", Cambridge, University Press, 1883.
26. Andreas, J. M., Hauser, E. A., and Tucker, W. B., *J. Phys. Chem.* **42**, 1001 (1938).
27. Niederhauser, D. O., and Bartell, F. E., "Report of Progress—Fundamental Research on the Occurrence and Recovery of Petroleum", Publication of the American Petroleum Institute, Baltimore. The Lord Baltimore Press, 1950.
28. Rayleigh, L., *Proc. Roy. Soc.* **29**, 71 (1879).
29. Addison, C. C., *Phil. Mag.* **36**, 73 (1945).
30. Addison, C. C., and Elliott, T. A., *J. Chem. Soc.* **1949**, 2789.

CHAPTER 3

Physical Adsorption by Solids

ADSORPTION OF GASES

The process by which atoms or molecules of one material become attached to the surface of another is called adsorption. The contiguous area between two phases, such as a solid and a gaseous phase, is referred to as an interface. The material that is adsorbed becomes concentrated at an interface.

Adsorption is a means of neutralizing or satisfying the forces of attraction that exist at a surface. These forces at the surface of a solid or a liquid are merely extensions of the forces within the body of the material. At the surface, the average number of atomic or molecular neighbors is only half as great as underneath the surface. Thus, there is an unbalance of forces at the surface, and a marked attraction of the surface toward atoms and molecules in its environment.

There is also a greater attraction of surface atoms toward neighboring atoms in the liquid or solid. This results in stronger bonds between surface atoms and closer distances as compared with atoms underneath the surface. This tendency for the atoms to compress gives rise to a surface tension.

The unfilled forces at the surface can be satisfied by the adsorption of atoms or molecules of another species. This reduces the attraction of the surface atoms or molecules of the solid or liquid toward its neighbors of the same kind, and thus reduces surface tension. Thus, adsorption is always accompanied by a decrease in surface tension. The process of adsorption continues until the free surface energy of the system, due to the imbalance of surface forces, has reached a minimum value.

Adsorption of a gas by a solid may be a relatively weak physical adsorption. If it is a strong interaction of a chemical nature, such as the formation of the oxide of a metal, it is known as activated adsorption or chemisorption.

Van der Waals forces are considered to be responsible for physical adsorption. These forces are associated with liquefaction and with condensation in capillaries. Thus, the energy change accompanying physical adsorption is of the same order of magnitude as that of the liquefaction of a gas. Similarly, physical adsorption occurs rapidly and is readily reversible. Chemisorption, on the other hand, involves an energy of activation and heats of adsorption of the order of chemical reactions. Chemisorption is irreversible or reversible with great difficulty.

Applications of the physical adsorption of gases on solids include the measurement of the surface area of catalysts, chromatographic analysis, gas purification, desiccation, and corrosion protection.

Gibbs Adsorption Isotherm

The general form of the Gibbs adsorption isotherm for the adsorption of a soluble substance at an interface is:

$$\Gamma = -\frac{1}{RT}\left(\frac{\partial\gamma}{\partial lna}\right)_T \tag{3.1}$$

or

$$\Gamma = -\frac{a}{RT}\left(\frac{\partial\gamma}{\partial a}\right)_T \tag{3.2}$$

where Γ is the surface excess of solute in moles/cm², γ is the surface tension in dynes/cm, a is the activity of the solute in moles, T is the absolute temperature, and R is the molar gas content.

Bangham (1, 2, 3) was the first to show that this equation could be applied to the adsorption of a gas on a solid. At low gas pressures, p the equilibrium pressure of the gas can be substituted for a, the activity of the solute. The amount of gas adsorbed v/V is equivalent to the surface excess, with v equal to the volume of gas adsorbed per gram of solid and V the molar volume of gas. The total free energy change at constant temperature is $\sum d\gamma$ where $\sum$ is the area per gram of solid. Then

$$\frac{v}{V} = -\frac{p}{RT}\cdot\Sigma\left(\frac{\partial\gamma}{\partial p}\right)_T \tag{3.3}$$

The integration of this equation is expressed by the relation:

$$\gamma_s - \gamma_{sf} = \frac{RT}{V\Sigma}\int_0^p \frac{v}{p}dp \tag{3.4}$$

where γ_s and γ_{sf} are the surface tensions of a clean solid surface and the solid with an adsorbed gas film, respectively.

While no experimental procedure has been devised for the determination of the surface tension of a clean solid surface, the equation can be used to calculate the decrease of the free surface energy caused by the adsorption of a gas or vapor. This can be done by plotting v/p versus p as shown in Figure 3.1 and calculating the area under the curve. Values for the decrease in free surface energy of anatase obtained by extrapolation to the saturation vapor pressure are 190, 56, 43 and 46 erg cm^{-2} for water, nitrogen, n-butane and n-heptane, respectively (4).

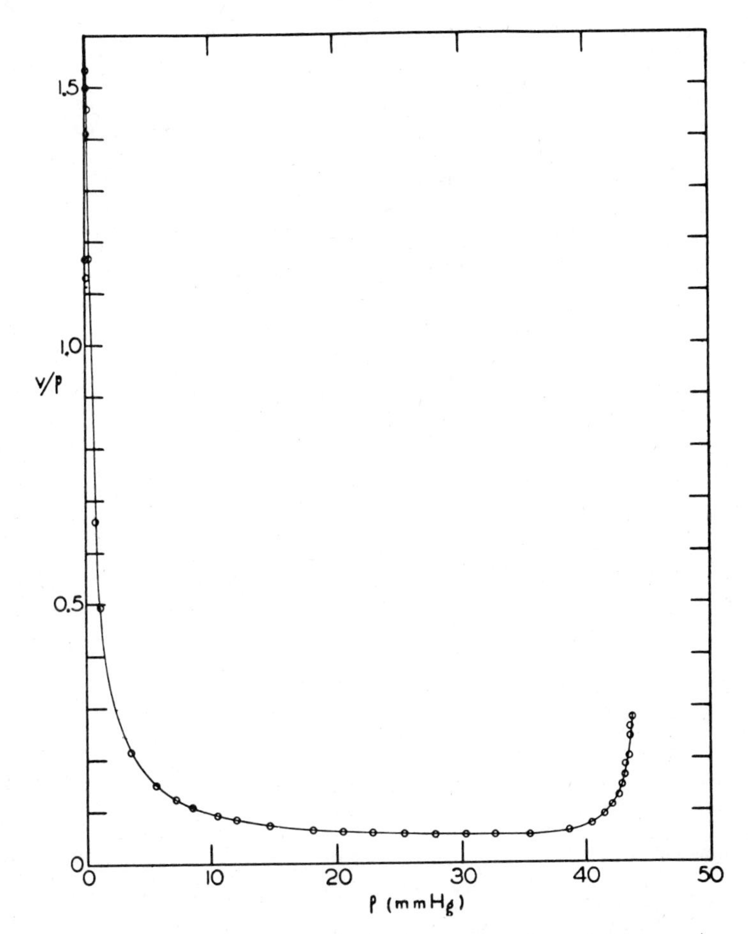

Figure 3.1. Adsorption of n-heptane on anatase plotted as v/p versus p. The total area under the curve can be used to calculate the decrease of the free surface energy caused by the adsorption of the vapor (4).

Langmuir Equation

The adsorption equation developed by Langmuir (5, 6) was one of the first and most important equations based on theory. Langmuir postulated that adsorption occurred as a monomolecular film. He envisioned a dynamic equilibrium, such that the rate of adsorption equaled the rate of desorption. Thus, S_0 represents the fraction of the surface that is bare and S_1, the fraction of the surface covered by a monolayer of adsorbed molecules. The rate of adsorption is proportional to the pressure of the gas and the fraction of the surface that is available,

$$\frac{dS_1}{dt} = k_1 p S_0 = k_1 p (1 - S_1) \tag{3.5}$$

The rate at which molecules leave the surface is proportional to the fraction of the surface that is covered,

$$-\frac{dS_1}{dt} = k_2 S_1 \tag{3.6}$$

At equilibrium, the rates of evaporation and condensation are equal.

$$S_1 = \frac{k_1 p}{k_2 + k_1 p} \tag{3.7}$$

Many simplifying assumptions were made in the derivation of this adsorption isotherm. Thus, it was assumed that the heat of adsorption is independent of the fraction of the surface that is covered and that only elastic collision — monomolecular adsorption — occurs on the covered surface. While Langmuir's equation fits experimental data in only a limited number of cases, it is important in the further development of the theory.

Types of Adsorption Isotherms.

Adsorption is most generally described in terms of isotherms which show the relationship between the pressure of the adsorbate gas and the amount of gas adsorbed at a constant temperature. Three phenomena may be involved in physical adsorption: monomolecular adsorption, multimolecular adsorption, and condensation in capillaries or pores. Since there is frequently considerable overlapping of these phenomena, the interpretation of adsorption isotherms can be quite complicated.

The contour of a complete isotherm, from zero pressure to the saturation pressure, where $p = p_0$, depends upon the gas, the nature of the substrate, and its pore structure. Brunauer, Deming, Deming, and Teller have distinguished five types of contours involving physical adsorption as shown in Figure 3.2.

Type 1 approximates monomolecular adsorption and is frequently referred to as the Langmuir type. This type of curve is obtained by low temperature adsorption of oxygen or nitrogen on certain charcoals and silica xerogels, because a monolayer saturates the surface or fills the pores. Chemisorption phenomena frequently produce a curve of this shape.

Type II is the most frequently encountered adsorption isotherm and is referred to as the sigmoid or S-shaped isotherm. The first portion of the curve

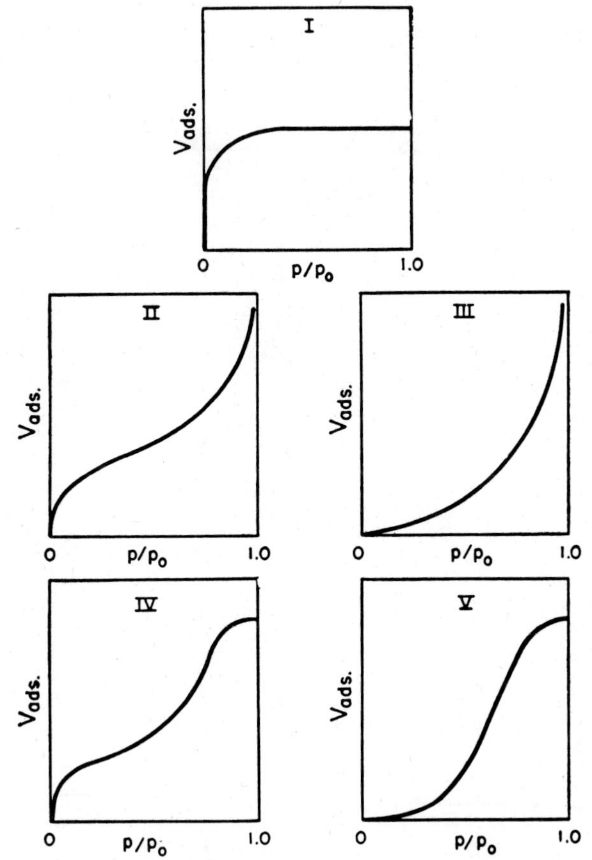

Figure 3.2. Five types of isotherms for physical adsorption (7).

up to a relative pressure of about 0.1 corresponds to monomolecular adsorption. This is followed by a multilayer region. Capillary condensation occurs in the region above about 0.4 relative pressure.

Type IV is similar to Type II, except in that portion of the isotherm approaching 1.0 relative pressure. The sharp approach to the line corresponding to the saturation pressure indicates a limited pore volume, because the diameter of the pores is only a small multiple of the diameter of the adsorbate molecules. This contour is common for many kinds of porous substrates.

Types III and V occur only when the forces of monomolecular adsorption are small. They are rarely encountered. A limited pore volume distinguishes the Type V isotherm, as compared with Type III. The adsorption of water vapor on graphite and charcoal are examples of Type III and V isotherms, respectively (8, 9).

BET Adsorption Theory

The theory of multimolecular adsorption developed by Brunauer, Emmett, and Teller (10) is known as the BET theory. It is widely used for determining the surface area of solids. The BET theory is an extension of Langmuir's treatment of monomolecular adsorption. The derivation is based on the same kinetic picture and the assumption that in physical adsorption, the forces of condensation are the predominant forces.

In deriving the BET isotherm, S_0, S_1, S_2, ... S_i represent the fraction of the total surface area that is covered with 0, 1, 2, ... i layers of adsorbed molecules, respectively.

At equilibrium for the monolayer, when the rate of condensation on the bare surface is equal to the rate of evaporation from the first layer, Equations 3.5 and 3.6 are equivalent. With the rate constant k_1 now expressed as a_1, we have Langmuir's equation for unimolecular adsorption,

$$a_1 p S_0 = k_2 S_1 \tag{3.8}$$

Since k_2 actually gives the fraction of the adsorbed molecules which possess sufficient energy to leave the surface, it must vary with temperature according to the Boltzmann exponential function,

$$k_2 = b_1 \mathbf{e}^{-E_1/RT} \tag{3.9}$$

Then

$$a_1 p S_0 = b_1 S_1 \mathbf{e}^{-E_1/RT}$$

For the second and successive layers the same relationship can be shown to apply, so that we have

$$a_i p S_{i-1} = b_i S_i \, \mathbf{e}^{-Ei/RT} \tag{3.10}$$

The total surface area of the solid is

$$A = \sum_{i=0}^{\infty} S_i \tag{3.11}$$

The total volume of gas adsorbed is

$$v = v_0 \sum_{i=0}^{\infty} S_i \tag{3.12}$$

where v_0 is the volume of gas adsorbed on one cm² of the adsorbent surface when it is covered with a complete unimolecular layer of adsorbed gas. Then

$$\frac{v}{Av_0} = \frac{v}{v_m} = \frac{\sum_{i=0}^{\infty} iS_i}{\sum_{i+0}^{\infty} S_i} \tag{3.13}$$

where v_m is the volume of gas adsorbed when the entire adsorbent surface A is covered with a complete monolayer.

This summation can be carried out if the simplifying assumptions are made that

$$E_2 = E_3 = E_i = E_L \tag{3.14}$$

and

$$\frac{b_2}{a_2} = \frac{b_3}{a_3} = \frac{b_i}{a_i} = a \tag{3.15}$$

where E_L is the heat of liquefaction of the gas. This is equivalent to saying that the evaporation-condensation properties of the second and higher adsorbed layers are the same as those in the liquid state.

The final result of the summation is

$$\frac{p}{v\,(p_0 - p)} = \frac{1}{v_m c} + \frac{c-1}{v_m c}\,\frac{p}{p_0} \tag{3.16}$$

where p_0 is the liquefaction or saturation pressure,

$$c = \frac{b_2 a_1}{b_1 a_2}\,\mathrm{e}^{E_1 - E_L/RT} \tag{3.17}$$

and

$$\frac{b_2\,a_1}{b_1\,a_2} \approx 1 \tag{3.18}$$

In the form of the equation

$$v = \frac{v_m\,c\,p}{(p_0 - p)\,[1 + (c-1)\,(p/p_0)]} \tag{3.19}$$

a plot of the volume of gas adsorbed v against the relative pressure p/p_0 results in a Type II curve, such as that of Figure 3.3.

Subsequent extensions of the theory have resulted in more complicated expressions corresponding to other types of adsorption isotherms. Equation 3.16 is the form in which the isotherm is used for the determination of the surface area of solids. Typical BET plots of $\frac{p}{v(p_0 - p)}$ against p/p_0 are shown in Figure 3.4. From the slope and the intercept, values for v_m and c can be calculated.

BET plots of volume adsorbed *versus* relative pressure are usually linear from 0.05 to 0.35 relative pressure. In general, the linear portion of the plot will occur on both sides of the relative pressure corresponding to a monolayer of adsorbed gas v_m. The location of the linear portion will depend upon the value of C and therefore on the heat of adsorption. When $C=100$, the mono-

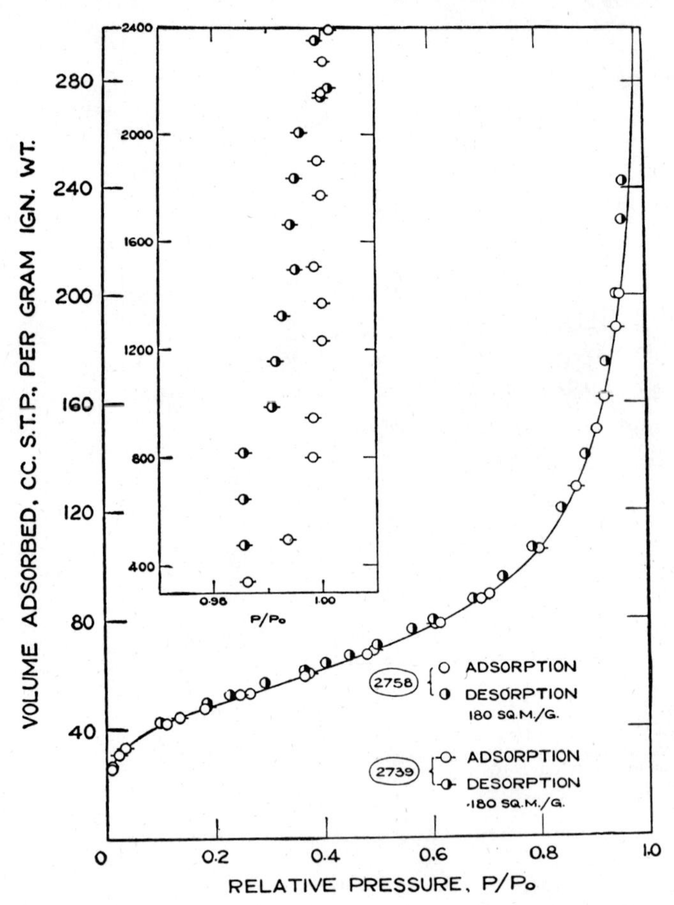

Figure 3.3. Nitrogen adsorption and desorption isotherms for Linde silica (11).

layer point occurs at a relative pressure of about 0.1; when $C=1$, a monolayer is reached at a relative pressure of 0.5.

In the absence of more reliable data the area of each adsorbed molecule can be calculated from the density of the liquefied or solidified adsorbate according to the equation

$$\text{Area per adsorbate molecule} = 4(.866)\left[\frac{M}{4\sqrt{2}\,A\rho}\right]^{2/3} \tag{3.20}$$

where M is the molecular weight of the gas, A is Avogadro's number and ρ is the density of the liquefied or solidified adsorbate. In the derivation of this equation it is assumed that the adsorbate molecules are held in two-dimensional close packing on the surface and that the area occupied by each

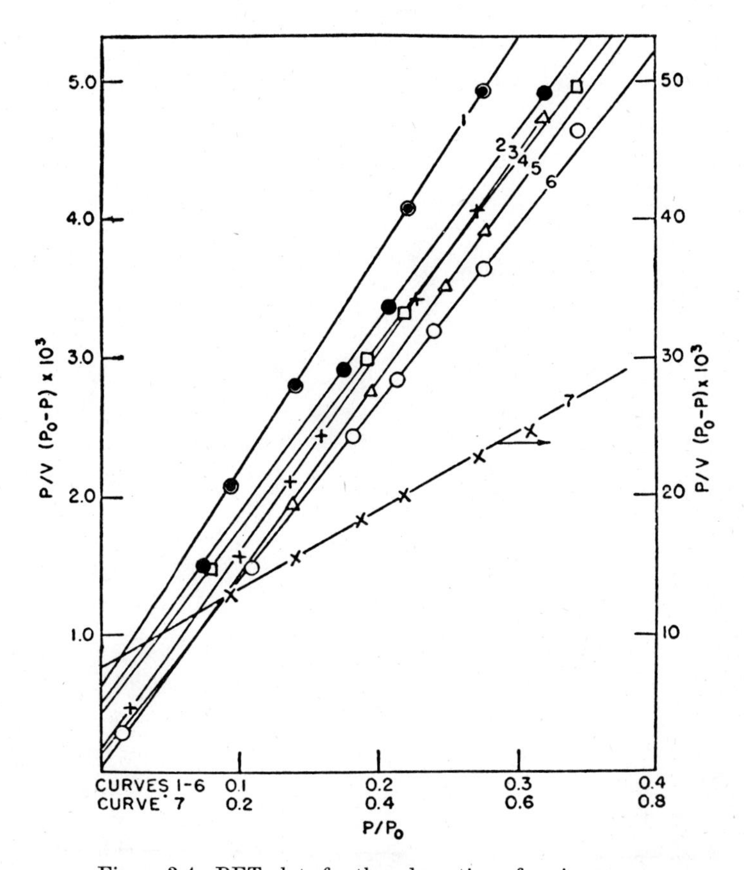

Figure 3.4. BET plots for the adsorption of various gases on 0.606 g of silica gel as follows:
Curve 1, carbon dioxide at − 78°; Curve 2, argon at − 183°; Curve 3, nitrogen at − 183°; Curve 4, oxygen at − 183°; Curve 5, carbon monoxide at − 183°; Curve 6, nitrogen at − 195.8°; and Curve 7, n-butane at 0°C (11).

molecule is the projected cross-section of the molecular volume calculated from the density of the condensed adsorbate. A list of molecular areas calculated in this manner is given in Table 3.1. The value for nitrogen of 16.2Å² is most commonly used in surface area measurements.

When the solid has pores of very small diameter, the usual form of the BET isotherm is replaced with the more general equation

$$v = \frac{v_m cx}{1-x} \cdot \frac{1-(n+1)x^n + nx^{n+1}}{1+(c-1)x - cx^{n+1}} \tag{3.21}$$

where X is the relative pressure, p/p_0, and n is the maximum number of layers of adsorbate that can be retained on the narrow walls of the pores (12).

TABLE 3.1. CALCULATED CROSS-SECTIONAL MOLECULAR AREAS OF TYPICAL ADSORBENTS (12).

Gas	Density of Solidified Gas	Temperature (°C)	Calculated Molecular Area (Å²)	Density of Liquefied Gas	Temperature (°C)	Calculated Molecular Area (Å²)
N_2	1.126	−253	13.8	0.751	−183	17.0
				0.808	−195.8	16.2
O_2	1.426	−253	12.1	1.14	−183	14.1
A	1.65	−233	12.8	1.374	−183	14.4
CO		−253	13.7	0.763	−183	16.8
CO_2	1.565	− 80	14.1	1.179	− 56.6	17.0
CH_4		−253	15.0	0.392	−140	18.1
n-C_4H_{10}			32.0	0.601	0	32.1
NH_3		− 80	11.7	0.688	− 36	12.9

When $n=1$ the equation reduces to the Langmuir isotherm. Under conditions where the surface is not limited, $n = \infty$, the usual form of the Equation 3.19 is applicable.

While the assumptions employed in the derivation of the BET isotherm have been criticized by many, it is important to note that the surface areas obtained with the BET equation are in agreement with areas obtained by other methods. Emmett (12) has observed that for nonporous solids there is good agreement between the physical adsorption method and (*1*) the heat of immersion method of Harkins and Jura on titanium dioxide (13), (*2*) the electron microscope photographs for carbon black (14), (*3*) direct microscopic observations on glass spheres (15), (*4*) liquid permeability measurements on zinc oxide pigments (16), (*5*) geometric measurements on single crystals of copper (17, 18), and others.

For porous solids, there are few independent methods of confirming surface area measurements. These include low angle x-ray scattering (19) and the adsorption of fatty acids from benzene (20). Both methods give surface areas in reasonable agreement with BET plots.

All adsorption methods require thorough degassing of the sample. Desorption of contaminents is more thorough the higher the temperature of the adsorbent, the better the vacuum, and the longer the pumping time. However, care must be taken that sintering or other changes in the adsorbent do not occur at elevated temperatures.

The most widely used method for measuring surface areas is the BET method using nitrogen at −195°C with an assigned area of 16.2Å² per molecule. The equation is in good agreement with actual values within the range p/p_0 equals 0.05 to 0.35. Outside of this range, deviation from experimental values can be considerable. Fortunately, the completion of a monomolecular layer generally takes place within the valid range of the equation. The deposition of a completed nitrogen monolayer is generally considered to occur at a relative pressure of about 0.1. The major theoretical criticisms of the BET isotherm are concerned with (1) the assumption that the heat of vaporization is the same for all layers following the first and is equal to the heat of vaporization for the bulk liquid, and (2) failure to consider interaction between adsorbed molecules. Statistical and thermodynamic analyses of the theory have been made by Cassie (21), Everett (22), Halsey (23), Harkins and Jura (24), and Hill (25). Other discussions and modifications of the BET theory have been made by Frenkel (26), Anderson and Hall (27), Keenan (28), and Cook (29). Detailed experimental techniques and procedures have been described (30, 31).

Harkins and Jura Method

Two methods have been developed by Harkins and Jura for the determination of the area of finely divided solids (13, 24). One method, which is experimentally the most difficult and which cannot be used with either porous or coarse particles, they refer to as an absolute method. It can be used to standardize the BET method or the relative method of Harkins and Jura.

Thus, the BET method does not give the area of the solid. Instead the number of molecules in what is supposed to be a completed monolayer is obtained. The area of the solid is equal to the product of the number of molecules in the monolayer and the mean area occupied per molecule in the complete monolayer. The latter value is not known, but is calculated from Equation 3.20, which assumes that all molecules are spherical. This assumption appears to be valid for small molecules, like nitrogen, but not for larger molecules.

The relative method of Harkins and Jura depends upon the fact that a condensed film on a liquid or a solid substrate exhibits a linear relationship between surface pressure (π) and the mean area (σ) occupied per molecule, or

$$\pi = b - a\sigma \tag{3.22}$$

where a and b are constants.

Surface pressure is the decrease in surface tension which results from the adsorption of molecules onto a substrate,

$$\pi = \gamma - \gamma_f \tag{3.23}$$

where γ is the surface tension of the pure solid or liquid and γ_f is the surface tension of the same substrate with an adsorbed film.

Equation 3.22 can be transformed to

$$\log (f/f_0) = B^1 - \frac{A^1}{v^2} \tag{3.24}$$

or with sufficient accuracy to

$$\log (p/p_0) = B - \frac{A}{v^2} \tag{3.25}$$

A plot of log (p/p_0) against the reciprocal of the square of the amount — volume at STP, mass, moles or molecules — adsorbed will result in a linear relationship if the film is condensed. Otherwise, it is necessary to either lower the temperature or change to another adsorbate which gives a condensed phase at the temperature of measurement.

Harkins and Jura found that the area $\sum$ of the solid is proportional to the square root $A^{\frac{1}{2}}$ of the slope.

$$\sum = kA^{\frac{1}{2}} \tag{3.26}$$

The constant k is evaluated by the absolute method. Thus, they exposed a finely divided solid, such as titanium dioxide to a sufficient pressure of water vapor to form four or five statistical adsorbed layers. The sample was then immersed in liquid water in a sensitive calorimeter. The heat of immersion obtained in this manner was then divided by 118.5 ergs, the value for the normal surface energy per sq cm of liquid water, to obtain a value for the number of sq cm of area in the sample. Using this surface area for the powder, they were then able to evaluate the constant k of Equation 3.26. By assuming that the value of k depends only on the adsorbate, they were able to calculate surface area values for different solids.

Harkins and Jura also observed that in some of the films studied, two or even three condensed phases are present. When this situation exists, the slope of the appropriate phase must be selected.

In Table 3.2 is shown the area σ occupied by various molecules in a completed monolayer as calculated by the absolute method of Harkins and Jura and the BET method of Equation 3.20. The method of calculating molecular areas by the latter method becomes more in error as the molecules become larger.

TABLE 3.2. COMPARISON OF VALUES FOR THE AREA PER MOLECULE OF ADSORBENT IN A COMPLETED MONOLAYER BY THE HARKINS-JURA AND BET METHODS (4).

Gas	Temperature (°C)	Area Per Molecule in a Completed Monolayer ($Å^2$) Harkins-Jura Absolute Method	BET Method Eq. 3.20
Nitrogen	−195.8	16.2	16.2
Argon	−195.6	16.0	14.4
n-Pentane	20	52.4	36.2
Pentene-1	20	49.3	34.9
n-Heptane	20	64.0	42.5

Table 3.3 shows a comparison of surface area results on various solids obtained by the BET method and the relative method of Harkins and Jura. Emmett (12) has observed that this excellent agreement between the two methods as shown in the table is surprising. Surface areas deduced by the two methods may differ by as much as 25 to 30 per cent. This deviation appears to be largely a function of the heat of adsorption, upon which the C value in the BET isotherm depends. Thus, as C increases from 50 to 250, it is necessary to assign values of 13.6 to 18.6 $Å^2$ to the mean cross-sectional area of the nitrogen molecule in order to obtain areas that are equal to those obtained from the Harkins and Jura plots.

TABLE 3.3. SURFACE AREAS IN SQUARE METERS PER GRAM OF SIX ADSORBENTS CALCULATED BY THE HARKINS-JURA AND BET METHODS. FOR THE BET METHOD, THE CROSS-SECTIONAL AREA FOR THE NITROGEN MOLECULE WAS TAKEN TO BE 16.2$Å^2$. THE MOLECULAR AREA VALUES FOR THE OTHER ADSORBED GASES WERE SELECTED TO MAKE THE AREA FOR THE STANDARD TiO_2 AGREE WITH THE ONE CALCULATED BY THE NITROGEN ISOTHERM. THESE VALUES ARE: WATER, 14.8$Å^2$; N-BUTANE, 56.6$Å^2$; N-HEPTANE, 64.9$Å^2$ (12).

Solid	Surface Area (Square Meters per Gram) Harkins-Jura Method: Nitrogen	Water	*n*-Butane	*n*-Heptane	BET Method: Nitrogen	Water	*n*-Butane	*n*-Heptane
TiO_2 (standard)	13.8	13.8	13.8	13.8	13.8	13.8	13.8	13.8
TiO_2 II	8.7	8.4		8.7	8.6	11.7		8.7
SiO_2 (quartz)	3.2	3.3		3.3	3.2	4.2		3.6
$BaSO_4$	2.4	2.3	2.2	2.3	2.4	2.8	2.7	2.4
$ZrSiO_4$	2.9	2.7			2.8	3.5		
$TiO_2+Al_2O_3$	9.6	11.8			9.5	12.5		

Absolute Method of Harkins and Jura. The fundamental method of Harkins and Jura for the determination of the area of solids will be discussed in some detail, because of the general usefulness of the concepts involved. The data obtained by immersing a finely divided solid in the liquid contained in an extremely sensitive calorimeter can be used to compute the following types of information:

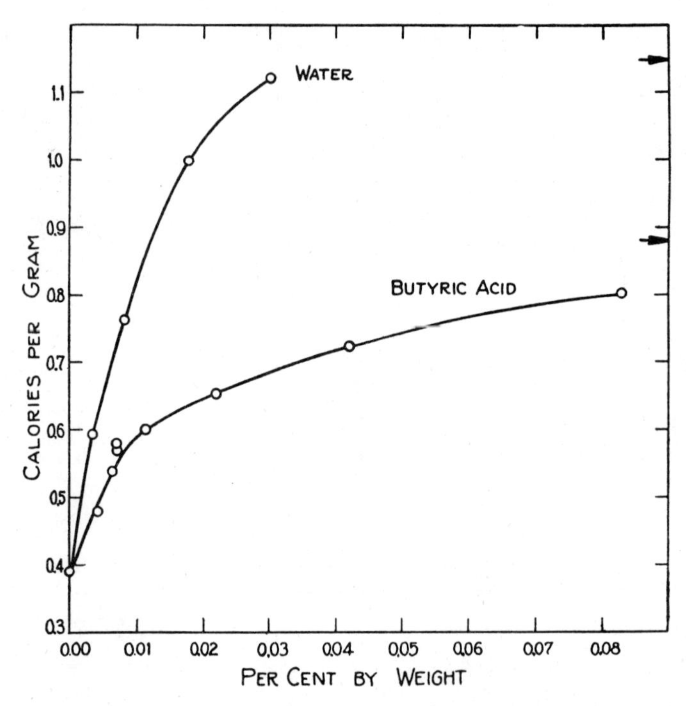

Figure 3.5. Energy of immersion of titanium dioxide in benzene which contains water or butyric acid. The arrows give the values for pure water or pure butyric acid (4).

(a) The energy of adhesion of the solid for the liquid can be calculated from the heat of immersion of the clean solid in the pure liquid.

(b) The integral and differential heats of adsorption can be calculated from the heat of adsorption as a function of the amount of vapor adsorbed on the solid.

(c) The decrease in the total surface energy caused by the adsorption of a gas or liquid on the surface of a solid can also be determined from heat of immersion data, with the solid containing known quantities of adsorbate.

(d) The area of the solid can be obtained from the heat of immersion of the solid saturated with the vapor of the same liquid into which the solid is to be immersed.

(e) The free energy at any temperature can be obtained from the temperature variation of the heat of immersion, if the free energy at one of the temperatures is known.

If the heat of immersion of a clean surface is desired, the solid should be degassed at as high a temperature as can be employed without affecting the surface. Titanium dioxide, for example, was heated in a high vacuum, 10^{-5} mm of Hg, for 24 hours at 500°C. Impurities in the liquid can have a very large effect on the values obtained. Harkins and Dohlstrom (32) found the heat of immersion of a sample of anatase to be 1.15 cal g^{-1} in water and 0.390 cal g^{-1} in thoroughly dried benzene. With as little as 0.03 per cent of water by weight initially present in the benzene, the value rose to 1.12 cal g^{-1}, almost the same as that of water. The effect of traces of water and butyric acid on the energy of immersion of titanium dioxide in benzene is shown in Figure 3.5.

The energy of adhesion h_A and the work of adhesion W_A are defined as the increase in internal energy and free energy, respectively, involved in the separation of a liquid from a solid, when the area of the interface separating the two phases is unity. According to the definition, separation is complete with no trace of adsorbed film remaining on the surface of the solid. These values are determined by carrying out the experiment in the reverse direction, beginning with a degassed, clean solid surface.

The total surface energy E_S or enthalpy H_S of a clean, dry solid is given by the relation

$$H_S = \Sigma_S\left[\gamma_S - T\left(\frac{\partial \gamma_S}{\partial T}\right)_P\right] \tag{3.27}$$

If the nonporous solid is immersed in a liquid, the solid surface is replaced by a solid-liquid interface with an enthalpy

$$H_{SL} = \Sigma_{SL}\left[\gamma_{SL} - T\left(\frac{\partial \gamma_{SL}}{\partial T}\right)_P\right] \tag{3.28}$$

Since the area Σ_S of the nonporous solid and that of the interface Σ_{SL} are equal, the enthalpy of emersion is

$$H_{E(SL)} = H_S - H_{SL} = \Sigma\left[\gamma_S - \gamma_{SL} - T\left(\frac{\partial \gamma_S}{\partial T} - \frac{\partial \gamma_{SL}}{\partial T}\right)_P\right] \tag{3.29}$$

or per unit area

$$h_{E(SL)} = \frac{H_{E(SL)}}{\Sigma} = \gamma_S - \gamma_{SL} - T\left(\frac{\partial \gamma_S}{\partial T} - \frac{\partial \gamma_{SL}}{\partial T}\right)_{P,\Sigma} \tag{3.30}$$

At constant pressure p the relationship between the change in enthalpy and the change in internal energy is

$$\Delta h = \Delta \varepsilon + p\Delta v \tag{3.31}$$

Since the volume change Δv is very small, it can be neglected. Then

$$h \cong \epsilon \tag{3.32}$$

and the change in enthalpy is equal to the change in internal energy.

The energy of adhesion is equal to the sum of the energies of the clean solid and liquid surfaces less the energy at the interface of the two phases

$$\varepsilon_{A(SL)} = \varepsilon_S - \varepsilon_{SL} + \varepsilon_L = \varepsilon_E + \varepsilon_L \tag{3.33}$$

$$\varepsilon_{A(SL)} = h_{E(SL)} + \gamma_L - T\left(\frac{\partial \gamma_L}{\partial T}\right) \tag{3.34}$$

Some values for the heat of emersion and energy of adhesion of solids in various liquids are shown in Table 3.4. It will be observed that the relative order of the energy of separation of the solids is the same against any of the

TABLE 3.4. ENERGY OF SEPARATION OF VARIOUS LIQUIDS FROM THE SURFACE OF CRYSTALLINE SOLIDS AT 25°C. ENERGY IN ERGS CM^{-2} (4).

Liquid	Energy of Emersion (h_E or ε_E)			Energy of Adhesion (h_A or ε_A)		
	$BaSO_4$	TiO_2	$ZiSiO_4$	$BaSO_4$	TiO_2	$ZiSiO_4$
Water	490	520	850	610	640	970
Ethyl alcohol		500			550	
Ethyl acetate	370	360		430	420	
Butyl alcohol	360	350		410	400	
Nitrobenzene		280	430		360	510
Carbon tetrachloride	220	240	410	280	300	470
Benzene	140	150	260	210	220	330
Isooctane		105	190		155	240

liquids. Thus zirconium silicate has a higher energy of separation than titanium dioxide, regardless of the liquid used. Similarly, the order of decreasing energy of separation of the liquids for one solid is the same for all of the other solids.

Harkins (4) explained the relationship between the heat of adsorption and the heat of emersion in the following manner. If a solid is emersed from a liquid into the vapor of the liquid at the pressure p, and the film on the solid is the correct thickness to give it equilibrium with the vapor at this pressure, then

$$\Delta h_1 = h_{E(SfL)} \tag{3.35}$$

If the solid is instead emersed from the liquid into a vacuum, so that the solid emerges without an adsorbed film, then

$$\Delta h_2 = h_{E(SL)} \tag{3.36}$$

Liquid is then evaporated from the liquid phase equal to the number of moles n of adsorbed liquid on the solid in the first case

$$\Delta h_3 = n\lambda \tag{3.37}$$

where λ is the molar heat of evaporation.

The physical difference between the two cases is the conversion of the adsorbed film to vapor. Then this difference is equal to the heat of desorption, which has the same absolute value as the heat of adsorption.

$$h_{D(VSf)} = \Delta h_2 + \Delta h_3 - \Delta h_1 \tag{3.38}$$

$$h_{D(VSf)} = h_{E(SL)} + n\lambda - h_{E(SfL)} \tag{3.39}$$

It is customary to express the heat of adsorption in cal g^{-1}, where $h_{E(SLf)}$ and $h_{E(SL)}$ are the heats of emersion in cal g^{-1} of adsorbent and n is the number of moles of vapor adsorbed g^{-1}. Figure 3.6 shows the decrease in the energy of desorption of water from a titanium dioxide surface as a function of the amount of water adsorbed.

The heat of emersion is employed by Harkins and Jura as an absolute method for the determination of the area of a finely divided crystalline solid. Thus, if a solid is exposed to vapor in contact with its liquid phase, the solid will continue to adsorb the vapor until the vapor pressure of the adsorbed material is equal to that of the liquid. If a duplex film is formed, the total surface energy as well as the free surface energy of the film is equal to that of the liquid phase of the adsorbed material. If the solid is then immersed in the liquid, the heat liberated is the same as that occasioned by the disappearance of a liquid surface of the same area. Thus, the heat of emersion $h_{E(SeL)}$ of a solid containing an adsorbed film in equilibrium with its saturated vapor can be found from the relation analogous to Equation 3.30

$$h_{E(SeL)} = \gamma_{Se} - \gamma_{SL} - T\left(\frac{\partial \gamma_{Se}}{\partial T} - \frac{\partial \gamma_{SL}}{\partial T}\right)_{P,\Sigma} \tag{3.40}$$

where γ_{Se} is the surface tension of the solid containing an adsorbed film in equilibrium with its saturated vapor and γ_{SL} is the interfacial tension at the solid-liquid interface.

Equation 3.40 may be simplified by the use of the relation

$$\gamma_{Se} = \gamma_{SL} + \gamma_L \cos\theta \tag{3.41}$$

to give

$$h_{E(SeL)} = \left(\gamma_L - T\frac{\partial\gamma_L}{\partial T}\right)\cos\theta + T\gamma_L \sin\theta\frac{\partial\theta}{\partial T} \tag{3.42}$$

where θ is the contact angle of the liquid on the solid. If it is assumed that $\theta = 0$, then

$$h_{E(SeL)} = \gamma_L - T\left(\frac{\partial\gamma_L}{\partial T}\right) = h_L \tag{3.43}$$

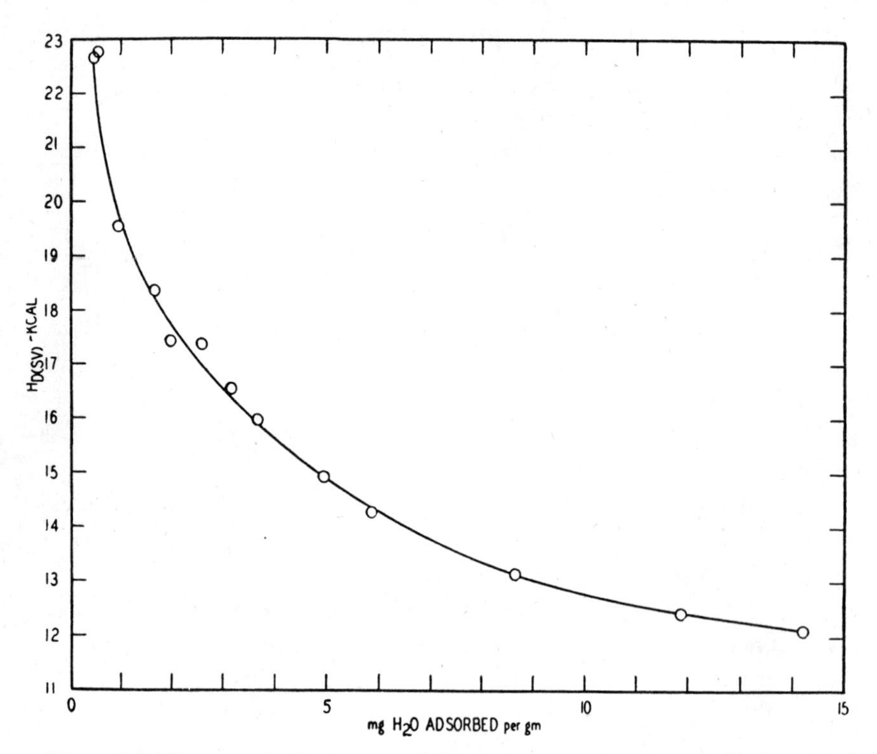

Figure 3.6. Decrease in the energy of desorption, in calories per mole of water, from the surface of TiO_2 (anatase) as a function of the weight of water absorbed (4).

Thus, the heat of emersion of each sq cm of the film-covered powder is equal to the total surface energy or the enthalpy of the surface of the liquid used. The outer area of the film Σ' is

$$\Sigma' = 4.185 \times 10^7 H_{E(SeL)}/h_L \tag{3.44}$$

where $H_{E(SeL)}$ is the heat of emersion in cal g^{-1}, h_L is expressed in ergs cm^{-2} and Σ' in $cm^2\, g^{-1}$.

The outer area of the film $\sum'$ would be equal to the specific area of the solid $\sum$, if the solid were a single plane surface. If the solid is porous, $\sum'$ is much less than $\sum$, and the latter cannot be determined. However, if the solid is nonporous and crystalline, the area of the solid can be calculated by correcting for the thickness of the water film, which is obtained by dividing $\sum'$ by the total volume of liquid water adsorbed.

If water is the liquid used for the heat of emersion measurements, $\sum'$ is obtained by converting the heat evolved in cal g^{-1} to ergs and dividing this by 118.5, the total surface energy of water.

Equations of State of Adsorbed Films (4)

In the case of insoluble films on liquid subphases, at least five different phases have been identified. These phases are gas, liquid expanded, liquid

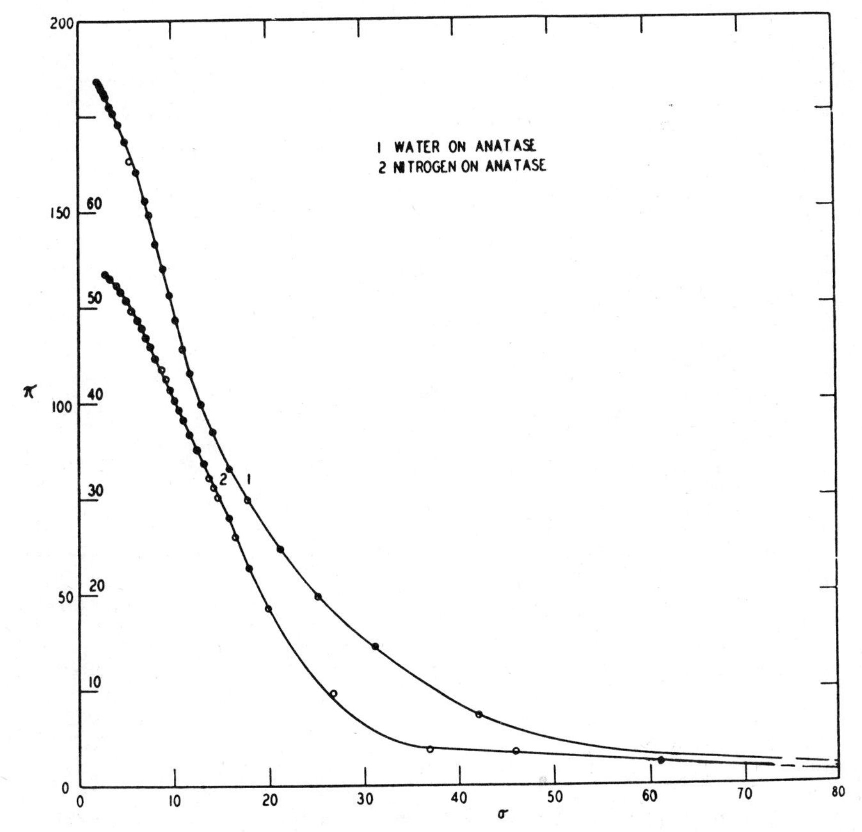

Figures 3.7. The pressure (π) — area (σ) isotherms of nitrogen and water films on anatase at − 195.6°C and 25.0°C. Both films exhibit a long condensed region (4).

intermediate, liquid condensed, and solid. The identification and verification of these phases follow from experimental work on surface pressure *versus* surface area, ($\pi-\sigma$), viscosity, and surface potential relationships.

In the study of films on solids, the pressure-area relationship is the only one of these methods that can be applied. The surface pressure π can be obtained by the use of Gibbs adsorption equation, while the area of the surface σ available per molecule is obtained by dividing the total surface area

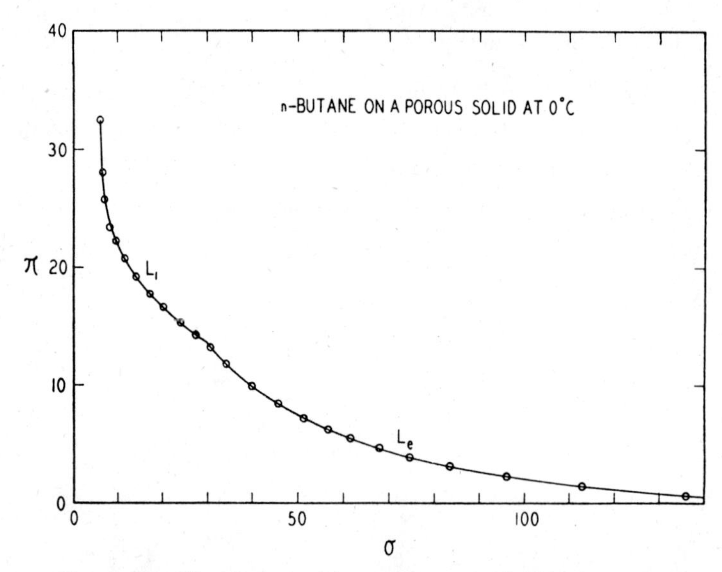

Figure 3.8. The pressure (π) — area (σ) isotherm of an n-butane film on an alumina-silica gel. The isotherm exhibits the liquid expanded and liquid intermediate states with a second-order phase change between the two (4).

by the number of molecules adsorbed. Figures 3.7 and 3.8 are typical $\pi-\sigma$ isotherms of adsorbed films on solid substrates.

If the adsorbed film behaves as a perfect two-dimensional gas, the equation of state is

$$\pi\sigma = kT \tag{3.45}$$

In the low pressure area, all films behave as gases, though not as perfect gases. The deviation from the behavior of a perfect gas is similar to that of an ordinary gas in three dimensions. If the molecular area is not too low — below about 200–300 Å² per molecule — the following equation is usually applicable,

$$\pi(\sigma - \sigma_0) = kT \tag{3.46}$$

where σ_0 is a constant.

For condensed phases, π is a linear function of σ,

$$\pi = a - b\sigma \tag{3.47}$$

where the constants a and b assume different values for each of the possible condensed phases. Empirical equations of state have been found for the liquid-expanded and the liquid-intermediate states on liquid subphases. Thus, the following equation is valid for a liquid expanded phase on water

$$\pi = c + \frac{\sigma}{a} + \frac{b}{a} ln\ \sigma \tag{3.48}$$

The liquid intermediate phase is represented by

$$\pi = c + \frac{1}{b} ln\ (b\ \sigma + a) \tag{3.49}$$

Figure 3.8 shows that these equations are applicable to adsorption on solids.

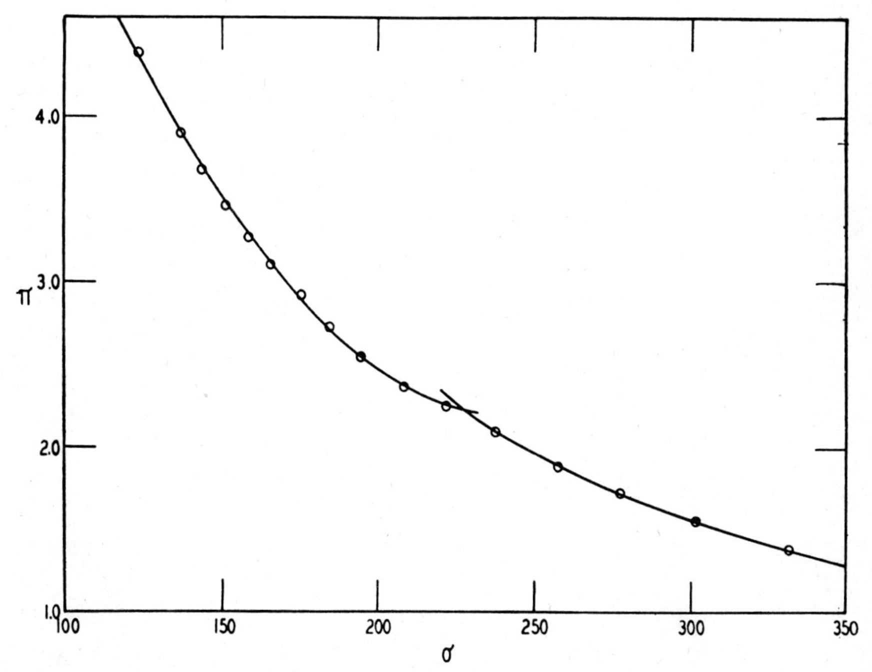

Figures 3.9. The pressure (π) — area (σ) isotherm for n-butane on an alumina-silica gel at 25°C. A second-order phase change from the gaseous phase occurs at the kink (4).

The circles are experimental points while the solid lines are calculated values from these equations. The slight kink between the curves for liquid-expanded L_e and liquid-intermediate L_i films is characteristic of second-order phase

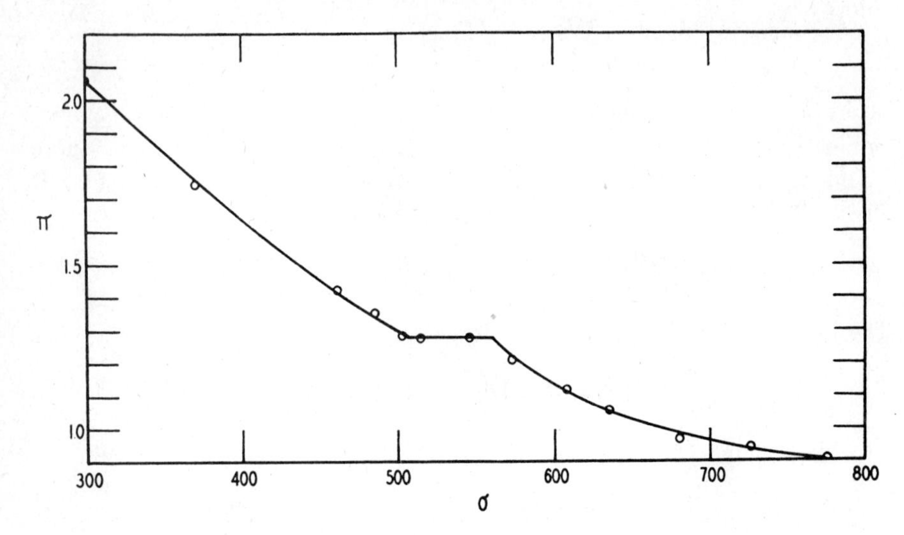

Figure 3.10. The pressure (π) — area (σ) isotherm of water on a sample of graphite containing 0.46% ash. The flat portion of the curve is characteristic of a first-order phase change (4).

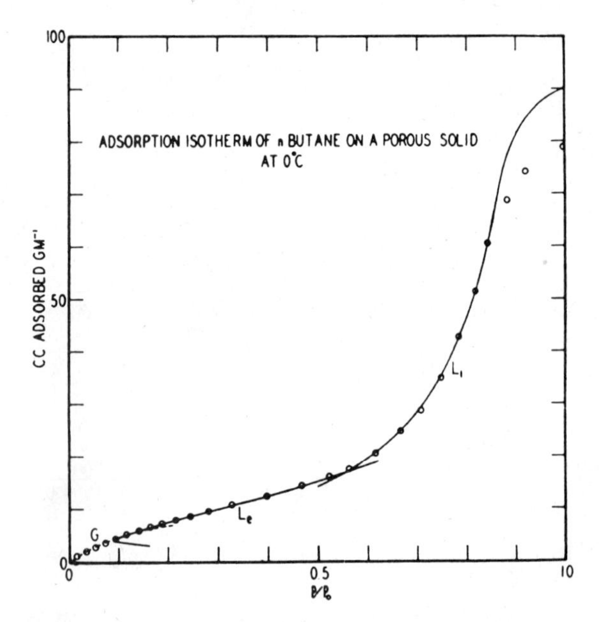

Figure 3.11. A pressure-volume isotherm for the adsorption of n-butane on a porous solid at 0°C, showing gaseous, liquid expanded, and liquid intermediate films (4).

changes. This is also shown in Figure 3.9 which illustrates a second-order phase change from the gaseous phase at the kink.

A first-order phase change is illustrated in Figure 3.10 for the isotherm of water on a sample of graphite. The flat portion of the curve is typical of first-order changes. Phase changes on a solid subphase are generally second-order, except for the transition from a gaseous phase which can be first- or second-order.

The two-dimensional equations of state can be transformed from $\pi-\sigma$ equations to $p-v$ equations, in which p is the vapor pressure of the adsorbed material and v is the volume of gas adsorbed. This is done by the use of the Gibbs equation. First- and second-order transitions with $p-v$ isotherms are shown in Figures 3.11 and 3.12.

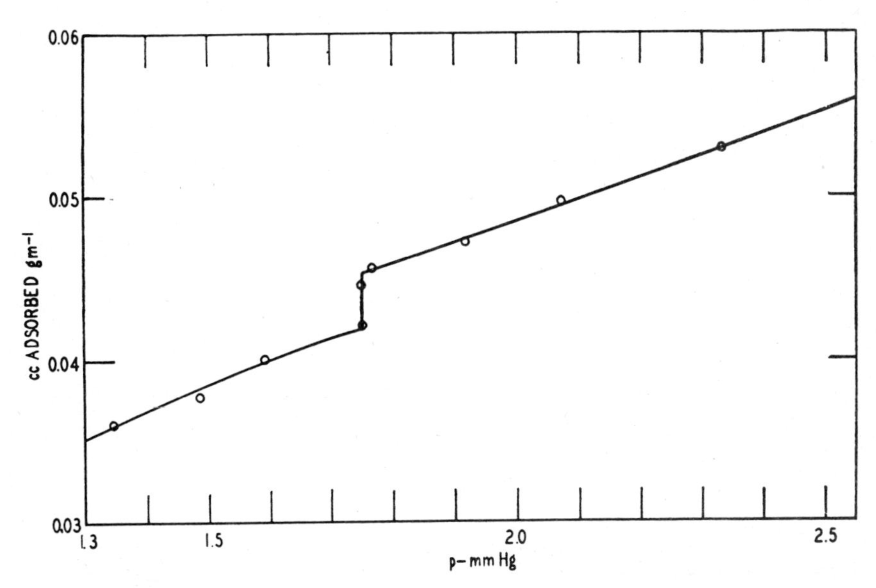

Figure 3.12. The pressure-volume isotherm of water on a sample of graphite. At the first-order transition, the volume adsorbed increases without any change in pressure (4).

ADSORPTION FROM SOLUTION

Adsorption Isotherms

The form of isotherms of physical and chemical adsorption by solids from dilute solutions has received little attention, as contrasted with the study of isotherms of vapor-phase adsorption by solids. Giles and MacEwan (33) classified adsorption isotherms from dilute solution into four main types, as shown in Figure 3.13. The *L* or Langmuir type is the same as Brunauer's type I and is the one usually encountered in adsorption from dilution solutions. Types *S* (Brunauer's type *V*), *ln* (linear) and *HA* (high affinity) are much less common.

In the *S* type isotherm, the convex shape of the curve relative to the solution-concentration axis means that adsorption is difficult at low concentrations, but becomes progressively easier as the concentration is raised. This isotherm is observed only occasionally and generally in systems involving low affinity solutes, where a high concentration of the solute is required to achieve measurable adsorption.

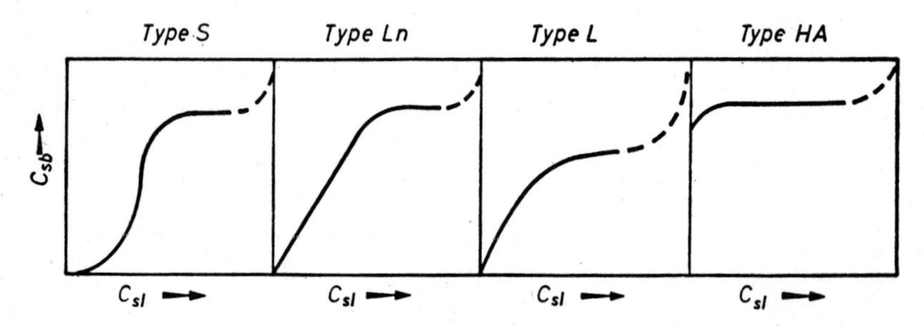

Figure 3.13. Forms of isotherms for the adsorption by solids from dilute solution. C_{sb} is the equilibrium concentration of the adsorbed solute on the substrate, while C_{sl} is the equilibrium concentration of the solute in solution (33).

The *ln* isotherm has been observed in the dyeing of cellulose acetate (34) and polyethylene terephthalate (35) with nonionic polar dyes. It appears to represent a condition where the number of sites for adsorption remains constant even though the amount of solute adsorbed increases. This will occur when the solute, but not the solvent alone, can swell and penetrate the substrate structure. Once the amorphous regions of the substrate have been penetrated, while the crystalline regions remain impenetrable to solute and solvent, the process of sorption ends abruptly (36).

The *L* isotherm is usually encountered in adsorption from solution. It is characterized by the condition that all of the sites at which adsorption can occur are available to the solute molecules even in the most dilute solutions. Further, the solute has high affinity and can readily displace solvent at the substrate surface.

The *HA* isotherm is a special form of the *L* curve in which a solute of very high affinity for the solid, in a solvent of low affinity, is completely adsorbed from dilute solutions. It has been observed in the adsorption from water of anionic dyes by graphite, and of phenol by charcoal (33).

Kipling (37) has emphasized the necessity for distinguishing the term "adsorption" as applied in studies with gases and with solutions. The isotherm of adsorption of a gas refers to the actual amount of one substance adsorbed by the unit mass of a solid. However, the adsorption isotherm which refers to adsorption from a solution refers to a change in concentration in the solution and is a measure of the difference in the amounts of solute and solvent adsorbed.

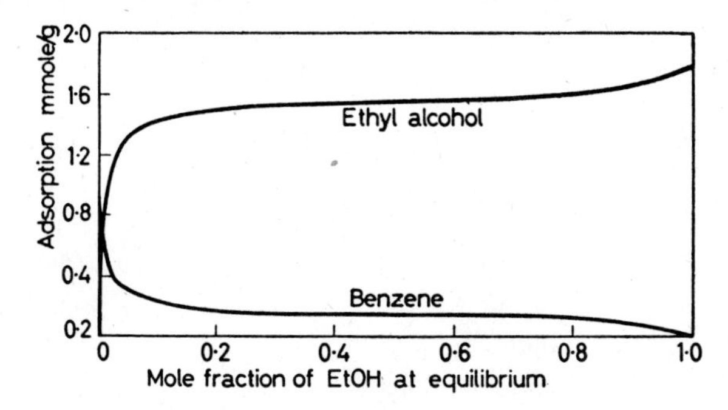

Figure 3.14. Adsorption of each component from mixtures of ethyl alcohol and benzene on a Gibbsite-type surface (37).

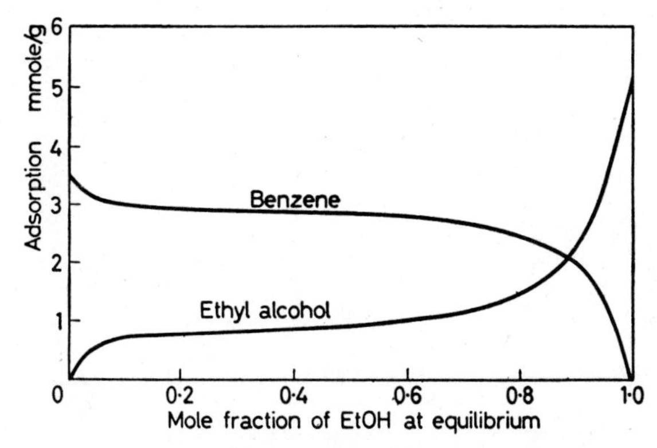

Figure 3.15. Adsorption of each component from mixtures of ethyl alcohol and benzene on charcoal (37)

It is a composite isotherm, referring to the behavior of the whole system, rather than to a single component. Kipling notes that a great deal more can be learned about the system if the isotherm can be resolved to show the separate adsorption of each component.

In Figures 3.14 and 3.15 are plotted individual adsorption isotherms for binary mixtures of ethyl alcohol and benzene on a gibbsite-type surface,

$\gamma - Al(OH)_3$, and on charcoal. All of the isotherms shown are of the type associated with monomolecular adsorption. The polar gibbsite-type surface shows an overwhelming preference for ethyl alcohol as compared with benzene, probably due to hydrogen bonding of the hydroxide with the alcohol. On charcoal, benzene is adsorbed preferentially to ethyl alcohol. Charcoal, which is a condensed aromatic system, may be regarded as interacting more strongly with the aromatic benzene molecules than with the aliphatic ethyl group of the alcohol molecule (37).

Forces of Adsorption

The forces that are responsible for adsorption from solution have been discussed by Giles (38, 39, 40). They may be classified as (*a*) non-polar van der Waals attraction, (*b*) formation of hydrogen bonds, (*c*) ion exchange, and (*d*) covalent bond formation.

The importance of van der Waals forces of attraction between hydrocarbon chains in the adsorption of large molecules will become apparent in subsequent chapters. While other forces are frequently more effective in the dyeing of fabrics, where the aromatic dye molecules are sufficiently large, intermolecular, van der Waals forces can be important. This is shown in Figure 3.16 for vat dyes, which give a linear relation between cellulose affinity and the logarithm of the number of carbon atoms in the longest axis of the conjugated chain of the molecule (41).

Ion exchange or electrostatic attraction promotes the adsorption of organic ions by solids with ionogenic groups. Solids which are normally negatively charged in water, such as silica and carbon, readily adsorb cationic dyes and surfactants. Alumina and protein fibers in acid solution are positive and readily adsorb anionic dyes and surfactants. Since van der Waals and other forces may assist adsorption, it sometimes happens that ions carrying the same charge as the surface are adsorbed. Ion-exchange adsorption is virtually independent of temperature, since surface charge varies very little with temperature.

The strongest forces involved in the adsorption of small molecules from dilute solutions are ion-exchange and hydrogen-bonding. Adsorption of a solute by a solid substrate will usually occur if either contains hydrogen-bond donor groups and the other contains acceptor groups. But there are exceptions. Hydrogen-bond adsorption does not occur in aqueous solutions when one of the bonding groups has too strong an affinity for water. Thus, the lack of reactivity of the hydroxy-groups of simple hydrocarbons and cellulose for hydrogen-bonding is due to their powerful solvation by water preventing a close approach by any other solute (40). Solvation of strongly ionized groups may screen hydrogen-bonding groups.

Phenols are adsorbed by hydrogen bonding from non-aqueous solutions by practically all polar solid substrates, including alumina, silica, cellulose, cellulose acetate, chitin, nylon, silk, and wool. In water, adsorption of phenolic

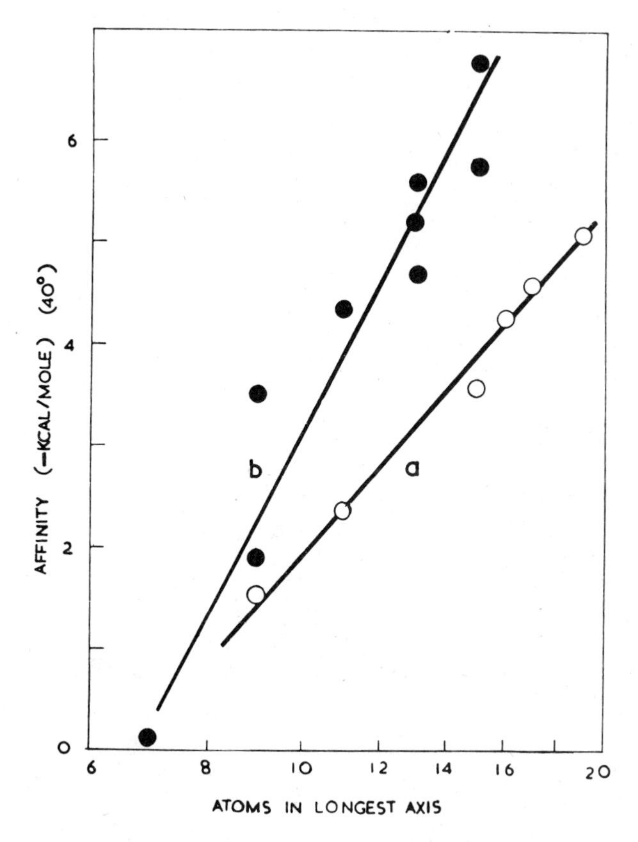

Figure 3.16. Relation between the affinity of vat dyes for cuprammonium rayon and the length of the longest conjugated chain of the dye molecules. *a* dyes with, and *b* dyes without the NHCO group (39).

groups occurs with all of these substrates, except cellulose. Hydrogen bonding between proton-acceptors and ester groups appear to occur by the formation of weak CH . . . bonds under the activating influence of the adjacent $C = O$ double bond. A simplified guide showing the tendency for hydrogen bonding by different groups is presented in Table 3.5.

TABLE 3.5. INTERACTION OF GROUP A WITH GROUP B IN AQUEOUS AND NON-AQUEOUS MEDIA. W = WATER, N = NON-AQUEOUS SOLVENT, + = INTERACTION DETECTED BETWEEN A AND B, − = WEAK OR NO INTERACTION (40).

Group A	Group B									
	OH (carbohydrate)	Alk—OH	Ar—OH	Alk_3N	Ar—NH_2	Alk—C = O	O—CO—Alk	—N = N—	—O—	—CONH—
OH (carbohydrate)	W−	W−	W−	W−	W−	W−	W−	W−		W−
Alk—OH	W−N+	W−N+	W+N+	W+N+		W−		W−	W+	W+N+
Ar—OH	W−N+	W+N+	W+N+	W+N+	W+N+	W−	W+N+	W+N+	W+	W+N+
Alk_3N	W−	W+N+	W+N+			W−	W+N+	W−N−		W+N+
Ar—NH_2	W−		W+N+			W−		N+		N+
Alk—C = O	W−		W−	W−				W−		W−
—O—CO—Alk	W−		W+	W+N+				N+		W+
—N = N—	W−	W−	W+	N−	N+	W−	N+			W+N+
—O—		W+	W+							
—CONH—	W−	N+	W+N+			W−		W+N+		

Measurement of Surface Area by Fatty Acid Adsorption

Orr and Dallavalle (42) have reviewed the literature concerning the measurement of the surface area of solids by the adsorption of solutes. The method employing the adsorption of straight-chain fatty acids from non-aqueous solutions is based on studies with insoluble monolayers. It was shown that properly compressed films of these straight-chain fatty acids on a water surface are monomolecular and occupy the same area of the interface per molecule regardless of the number of carbon atoms in the molecule. The evidence indicated that the molecules were oriented perpendicular to the interface,

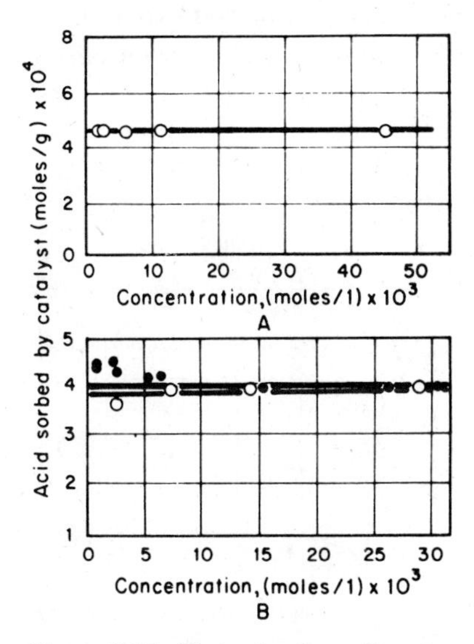

Figure 3.17. Plot of adsorption isotherms for palmitic acid on Adams-platinum catalyst (A) and lauric and palmitic acids on Raney-nickel catalyst (B) (45).

with the area of the carboxylic acid group determining the interfacial area per molecule. According to Adam (43), each molecule occupies an area of 20.5Å². Harkins and Gans (44) found that oleic acid adsorbed as a monomolecular layer on titanium dioxide and calculated a surface area for the solid which was in good agreement with results obtained by measurement under the microscope.

Irreversible chemisorption of fatty acids occurs on such active metals as Adams-platinum and Raney-nickel catalysts. Smith and Fuzek (45) found that on these metals the quantity of acid adsorbed was independent of the concentration of fatty acid in solution. If the quantity of acid in solution was insufficient to completely cover the solid with a monomolecular layer, all of the acid was removed from solution. As shown in Figure 3.17 the adsorption isotherm is of the *HA* type.

However, fatty acids are physically adsorbed rather than chemisorbed by most solids at normal temperatures. The isotherms obtained are of the Langmuir or *L* type, with the horizontal portion of the curve corresponding to the formation of a monomolecular layer. Figure 3.18 is typical of adsorption isotherms of fatty acids from non-aqueous solutions. Here stearic acid was adsorbed from benzene. It is important that the solution and the solid be freed from traces of water. Hirst and Lancaster (46) found that the presence of water reduced the amount of acid adsorbed on SiC, SiO_2, TiC, and TiO_2, while water initiated chemisorption with reactive materials such as Cu, Cu_2O, CuO, Zn, and ZnO.

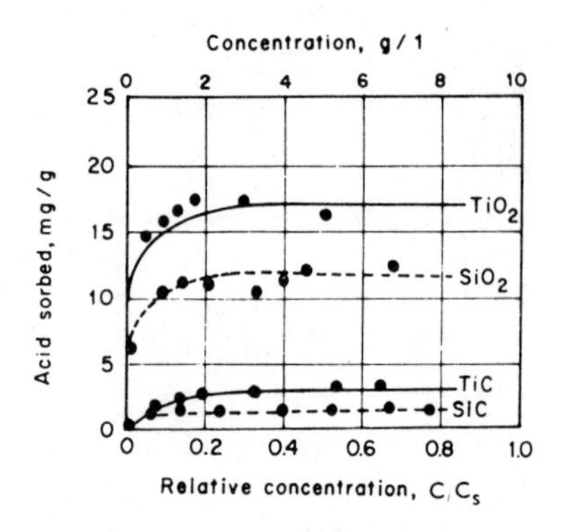

Figure 3.18. Plot of adsorption isotherms for stearic acid on various solids at 18°C (46).

In studying the adsorption by graphite of fatty acids from aqueous solution, Fu, Hansen, and Bartell (47) found evidence of multimolecular layers of adsorbed acid at high concentrations. At the completion of the first molecular layer a high proportion of water molecules were present in the layer, indicating competition between solute and solvent. By plotting the relative concentration *versus* apparent adsorption, they obtained curves characteristic of a BET type of isotherm. The relative concentration C/C_s, which is the ratio of actual concentration to the saturation concentration of the solute, is analogous to the relative pressure p/p_0 used in BET plots of gas adsorption. The data are presented graphically in Figure 3.19 (48).

Measurement of the surface area of solids by fatty acid adsorption is very simple in principle. A sample of the solid is place in contact with a fatty-acid solution of known concentration. After adsorption, the concentration

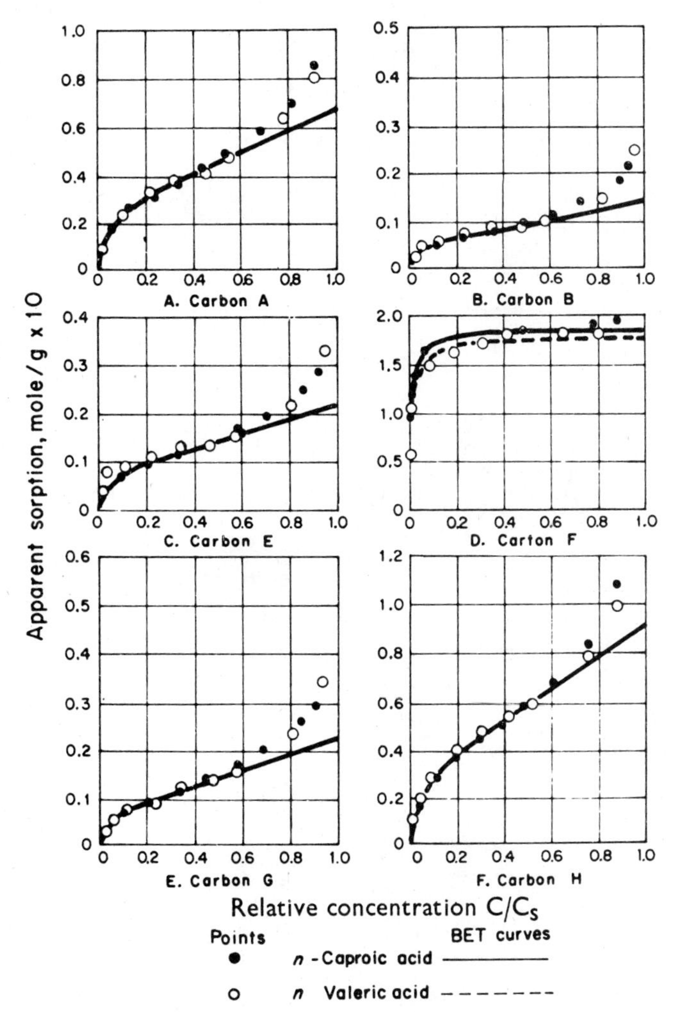

Figure 3.19. Plot of adsorption isotherms for acids on carbon from aqueous solution (48).

of fatty acid is again determined by analysis. The difference between initial and final quantities of fatty acid in solution is equal to the amount adsorbed. If the area occupied by each molecule of fatty acid is known, the total area of the solid can be readily calculated. Water cannot be used as the solvent, since the adsorbed films are multimolecular and contain water molecules. While a BET type isotherm is obtained, the surface area of the solid cannot be calculated by the BET method, because of the presence of adsorbed water molecules. However, condensed monomolecular films of adsorbed fatty acid are obtained if an anhydrous solvent such as benzene is used which is not adsorbed in preference to the fatty acid.

Measurement of Surface Area by Dye Adsorption

Perhaps the most convenient method for measuring surface area is by the adsorption of dye from aqueous solution. Dye concentrations can be determined readily by spectrophotometric methods. Since aqueous solutions are

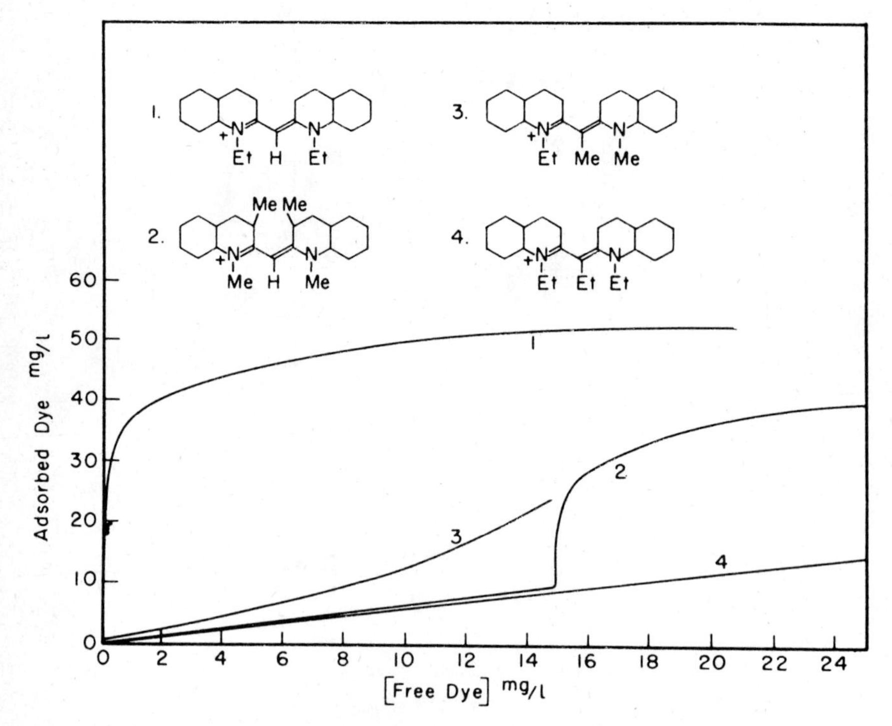

Figure 3.20. Adsorption isotherns for cyanine dyes on silver halide grains, showing the effect of nonplanarity (51).

used, exhaustive drying of solid and solution is unnecessary. Unfortunately, the method is not of general validity at present, can only be used in specific instances. Generally, monomolecular layer adsorption is not obtained due to competition between the solvent molecules, and the dissolved dye. The results of this competition are largely unpredictable at the present time. For example, crystals do not adsorb equally on different faces (49). Some crystals adsorb one dye on only one set of faces, but may adsorb a second dye on a different set of faces. Ewing and Lui (50) studied the adsorption of the ionic dyes, crystal violet chloride and Orange II, on samples of anatase and rutile and concluded that the dyes adsorbed to form a bimolecular film, with the dyes lying flat on the surface.

A detailed study of the adsorption of cyanine dyes on silver halide grains was made by West, Carroll, and Whitcomb (51). A Langmuir type adsorption isotherm was obtained with 1, 1′—dimethyl—2, 2′ cyanine iodide on silver bromide. Calculation of the area occupied per dye molecule indicated edge-on orientation of the dye cations. However, more complicated adsorption behavior was also observed, as graphed in Figure 3.20, depending upon the configuration of the dye molecules. Curve 2 of Figure 3.20 is the *S*-type of Giles. It shows a region of low adsorbability at low concentrations of added dye and a region of strong adsorption at higher concentrations. The methyl groups substituted for the hydrogen atoms in the 3- and 3′-positions force the two nuclei from coplanarity. Poor adsorbability is often associated with nonplanarity of the molecule. Comparison can be made between curve 2 for the tetramethyl dye and that of its planar counterpart, curve 1. Curves 3 and 4 show the low adsorption of 2, 2′-cyanines forced from planarity by bulky substituents in the methine bridge.

Soap Titration Method (52)

The method of determining the surface area and particle size of synthetic latices by soap titration was employed in the Government Synthetic Rubber Program (53, 54, 55, 56, 57). The principles of the method have been discussed by Maron and his associates (58). Further improvements in the method, particularly with regard to determining the effective surface area of the soaps, were made by Brodnyan and Brown (52).

The latex is titrated with approximately $0.01M$ soap solution, with the end point of the titration corresponding to the cmc of the soap in the latex. Any convenient method for determining the cmc can be used to follow the progress of the soap titration. A typical plot of surface tension versus milliliters of soap solution, obtained by using a Du Nouy ring tensiometer, is shown in Figure 3.21. The concentration of soap at the cmc is obtained at several concentrations of the emulsion, and plotted as shown in Figure 3.22, where c is the concentration of soap at the cmc and m is the concentration of the emulsion titrated.

The slope of this line is S_a, the grams of adsorbed soap per gram of emulsion solid, and the intercept equals the cmc of the soap in the continuous phase of the emulsion. If S_i is the grams of soap initially on the surface of one gram of emulsion solid, and the soap used in the titration is the same as the original

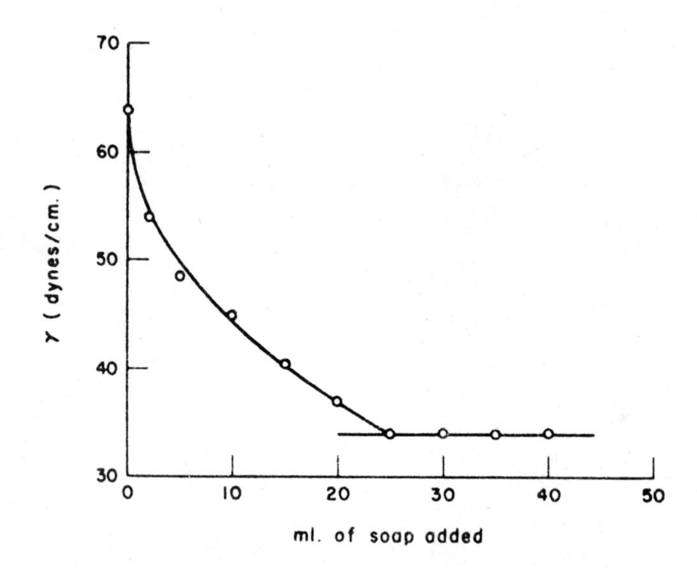

Figure 3.21. Typical curve of surface tension versus milliliters of soap (52).

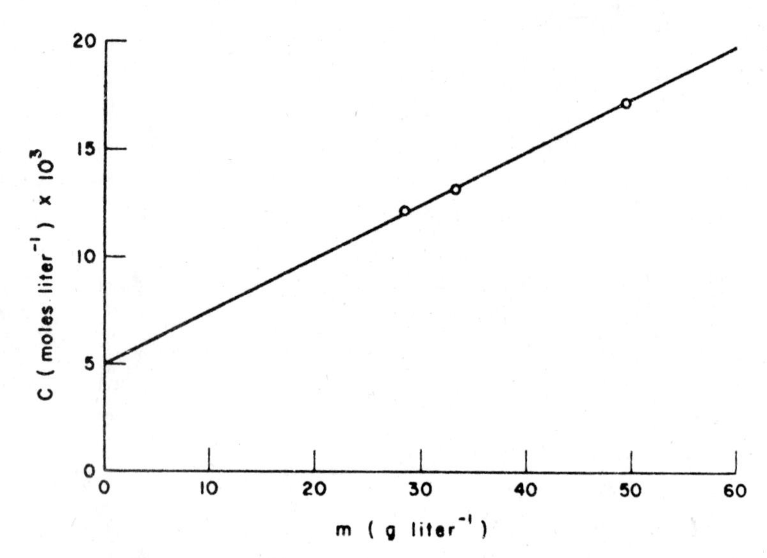

Figure 3.22. Typical curve of c versus m (52).

soap, then the surface of one gram of emulsion solid is $a\ (S_a + S_i)$, where a is the effective area of one gram of soap.

This value can be obtained using the Gibbs adsorption isotherm

$$\frac{-\,d\gamma}{d\log c} = 2.303\ RT\,\Gamma\,n \tag{3.50}$$

where γ is the interfacial tension at the soap concentration c, Γ is the surface concentration, R is the gas constant 8.3×10^7, T is the absolute temperature, and $1 \leqq n \leqq 2$, depending upon the concentration of ions in addition to the surfactant ions present.

The area A is calculated from the relation

$$A = \frac{10^{16}}{\Gamma N} \tag{3.51}$$

where N is Avogadro's number.

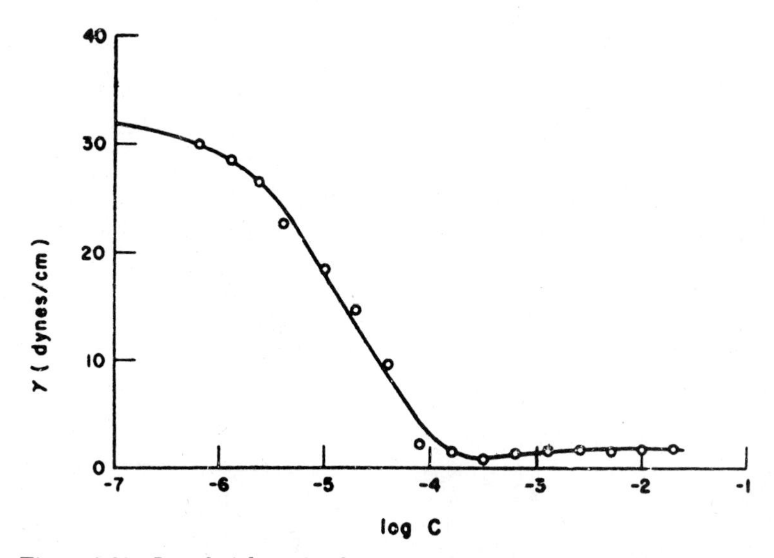

Figure 3.23. Interfacial tension between n-hexane and soap solutions (52).

The value of $n = 1$, in the Gibbs adsorption equation, when the surfactant is of the nonionic type, or when there is an excess of salt present. The determination of the effective surface area of the soap should be made in an environment similar to that of the polymer particles. Thus, Brodnyan and Brown (52) determined the effective surface area of the soap by measuring the interfacial tension between n-hexane and the soap solution, rather than that at the soap solution-air interface. The soap solutions were made $0.2M$ with respect to sodium chloride, so that $n = 1$. A plot of interfacial tension versus the logarithm of the soap concentration is shown in Figure 3.23. For sodium lauryl sulfate, an effective surface area of 61Å^2 was found, as compared with an area of 38Å^2 at the solution-air interface. An octylphenoxy polyethoxy ethanol was found to have an effective area of 88.5Å^2.

The surface area of the particles in any emulsion can be determined by titrating with the same soap that is initially on the surface, provided that there is a reasonable differential between the surface tension of the emulsion and that of the surfactant solution beyond its cmc.

References

1. Bangham, D. H., *Trans. Faraday Soc.* **33**, 805 (1937).
2. Bangham, D. H., and Razouk, R. I., *Trans. Faraday Soc.* **33**, 1463 (1937).
3. Bangham, D. H., and Razouk, R. I., *Proc. Roy. Soc.* (London) **A166**, 572 (1938).
4. Harkins, W. D., "The Physical Chemistry of Surface Films", New York, Reinhold Publishing Corp., 1952.
5. Langmuir, I., *J. Am. Chem. Soc.* **40**, 1361 (1918).
6. Langmuir, I., *J. Chem. Soc.* **1940**, 511.
7. Brunauer, S., Deming, L. S., Deming, W. E., and Teller, E., *J. Am. Chem. Soc.* **62**, 1723 (1940).
8. Harkins, W. D., Jura, G., and Loeser, E. H., *J. Am. Chem. Soc.* **68**, 554 (1946).
9. Coolidge, A. S., *J. Am. Chem. Soc.* **49**, 708 (1927).
10. Brunauer, S., Emmett, P. H., and Teller, E., *J. Am. Chem. Soc.* **60**, 309 (1938).
11. Ries, H. E., Jr., in "Catalysis", edited by P. H. Emmett, Vol. 1, p. 1, New York, Reinhold Publishing Corp., 1954.
12. Emmett, P. H., in "Catalysis", edited by P. H. Emmett, Vol. 1, p. 31, New York, Reinhold Publishing Corp., 1954.
13. Harkins, W. D., and Jura, G., *J. Am. Chem. Soc.* **66**, 919 (1944).
14. Anderson, R. B., and Emmett, P. H., *J. App. Phys.* **19**, 367 (1948).
15. Emmett, P. H., in "Advances in Colloid Sciences", Vol. 1, p. 1, New York, Interscience Publishers, Inc., 1942.
16. Emmett, P. H., and DeWitt, T. W., *Ind. Eng. Chem., Anal. Ed.* **13**, 28 (1941).
17. Rhodin, T. N., *J. Am. Chem. Soc.* **72**, 5691 (1950); ibid. **72**, 4343 (1950).
18. Rhodin, T. N., *J. Phys. Chem.* **57**, 143 (1953).
19. Elkin, P. B., Shull, C. G., and Roess, L. C., *Ind. Eng. Chem.* **37**, 327 (1945).
20. Smith, H. A., and Fuzek, J. F., *J. Am. Chem. Soc.* **68**, 229 (1946).
21. Cassie, A. B. D., *Trans. Faraday Soc.* **41**, 450 (1945).
22. Everett, D. H., *Trans. Faraday Soc.* **46**, 453 (1950).
23. Halsey, G. D., "Advances in Catalysis", edited by W. G. Frankenburg, V. I. Komarewsky, and E. K. Rideal, Vol. 4, p. 259, New York, Academic Press, Inc., 1952.
24. Harkins, W. D., and Jura, G., *J. Am. Chem. Soc.* **66**, 1366 (1944).
25. Hill, T. L., *J. Chem. Phys.* **14**, 263 (1946).
26. Frenkel, J., "Kinetic Theory of Liquids", Oxford, Oxford University Press, 1946.
27. Anderson, R. B., and Hall, W. K., *J. Am. Chem. Soc.* **70**, 1727 (1948).
28. Keenan, A. G., *J. Am. Chem. Soc.* **70**, 3947 (1948).

29. Cook, M. A., *J. Am. Chem. Soc.* **70**, 2925 (1948).
30. Loebenstein, W. V., and Deitz, V. R., *J. Chem. Phys.* **15**, 687 (1947).
31. Barr, W. E., and Anhorn, V. J., "Scientific and Industrial Glass Blowing and Laboratory Techniques", p. 267, Pittsburgh, Pa., Instrument Publishing Co., 1949.
32. Harkins, W. D., and Kahlstrom, R., *Ind. Eng. Chem.* **22**, 897 (1930).
33. Giles, C. H., and MacEwan, T. H., Proc. Intern. Congr. Surface Activity, 2nd, Vol. 3, p. 457, London, 1957.
34. Bird, C. L., and Manchester, P. *J. Soc. Dyers and Colourists* **71**, 604 (1955).
35. Schuler, M. J., and Remington, W. R., *Disc. Faraday Soc.* **16**, 201 (1954).
36. Giles, C. H., *Disc. Faraday Soc.* **16**, 112 (1954).
37. Kipling, J. J., Proc. Intern. Congr. Surface Activity, 2nd, London, 1957.
38. Giles, C. H., Compt. rend. 27e cong. international de chim. ind., Brussels, p. 610, 1954.
39. Giles, C. H., Compt. rend. 31e cong. international de chim. ind., 1958.
40. Giles, C. H., "Hydrogen Bonding", edited by D. Hadzi, New York, Pergamon Press, 1959.
41. Peters, R. H., and Summer, H. H., *J. Soc. Dyers and Colourists* **71**, 130 (1955).
42. Orr, C., Jr., and Dallavalle, J. M., "Fine Particle Measurement", Chapt. 8, New York, The MacMillan Co., 1959.
43. Adam, N. K., "Physics and Chemistry of Surfaces", London, Oxford University Press, 1941.
44. Harkins, W. D., and Gans, D. M., *J. Am. Chem. Soc.* **53**, 2804 (1931).
45. Smith, H. A., and Fuzek, J. K., *J. Am. Chem. Soc.* **68**, 229 (1946).
46. Hirst, W., and Lancaster, J. K., *Trans. Faraday Soc.* **47**, 318 (1951).
47. Fu, Y., Hansen, R. S., and Bartell, F. E., *J. Phys. and Colloid Chem.* **52**, 374 (1948).
48. Hansen, R. S., Fu, Y., and Bartell, F. E., *J. Phys. and Colloid Chem.* **53**, 769 (1949).
49. Gwathmey, A. T., *Record Chem. Progr.* **14**, 117 (1953).
50. Ewing, W. W., and Liu, F. W., Jr., *J. Colloid Sci.* **8**, 204 (1953).
51. West, W., Carroll, B. H., and Whitcomb, D. L., *Ann. N.Y. Acad. Sci.* **58**, 893 (1954).
52. Brodnyan, J. G., and Brown, G. L., *J. Colloid Sci.* **15**, 76 (1960).
53. Klevens, N. B., *J. Colloid Sci.* **2**, 365 (1947).
54. Borders, A. M., and Pierson, R. M., *Ind. Eng. Chem.* **40**, 1473 (1948).
55. Willson, E. A., Miller, J. R., and Rowe, E. H., *J. Phys. and Colloid Chem.* **53**, 357 (1949).
56. Maron, S. H., Madow, B. P., and Krieger, I. M., *J. Colloid Sci.* **6**, 584 (1951).
57. Morton, M., Cala, J. A., and Altier, M. W., *J. Polymer Sci.* **19**, 547 (1956).
58. Maron, S. H., Elder, M. E., and Ulevitch, I. N., *J. Colloid Sci.* **9**, 89 (1954).

CHAPTER 4

Chemisorption

All adsorption processes, whether physical or chemical in character, are accompanied by a decrease in free surface-energy. Physical adsorption, frequently referred to as van der Waals adsorption, occurs where there are relatively weak adhesional forces between adsorbate and adsorbent. The heat evolved when a gas is physically adsorbed is usually similar to the heat of liquefaction of the gas, a few hundred cal per mole. On the other hand, chemical adsorption arises from the actual formation of a chemical bond with the surface. The heats evolved are of the same order as those liberated in chemical reactions, from about 10 to 100 kcal per mole.

Physical adsorption also differs from chemical adsorption in that the former requires little if any activation energy. With chemisorption, the activation energy can be very considerable. The difference between physical and chemical adsorption is illustrated by the potential energy curves shown in Figure 4.1 after Lennard-Jones (1). Curve I represents the van der Waals interaction. As the molecule is brought close to the surface, there is a small attraction and a corresponding exothermic heat of adsorption, H_w. At closer distances this attraction changes over to repulsion. If adsorption is accompanied by chemical bond formation, the potential energy curve is that of type II, with the heat of adsorption indicated by H_A. The point where the two curves cross represents the change over from physical to chemical adsorption, and the activation energy for chemisorption is indicated by E.

It is evident that the activation energy will depend upon where the two curves cross. If the dashed line of Figure 4.1 corresponds to curve I,. the curves will cross at point A and there will not be any activation energy. In the absence of an activation energy, chemisorption like physical adsorption can occur rapidly even at liquid air temperatures.

Typical examples of chemisorption are the oxide films on copper, aluminum and other metal surfaces, including liquid mercury. Chemisorbed fatty acids are important in lubrication, as discussed Chapter 15. The most important and most thoroughly studied applications of chemisorption are in catalysis and electrode processes. Catalytic reactions do not occur on surfaces unless at least one of the reactants is chemisorbed.

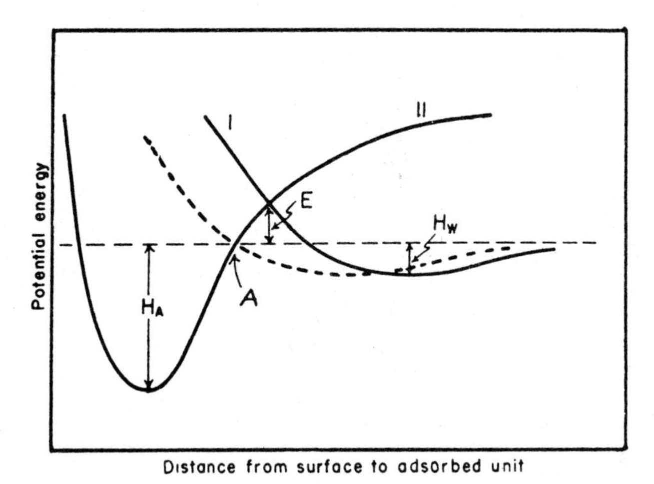

Figure 4.1. Potential energy versus distance from the surface to the adsorbed molecule (1).

Nature of Chemisorption

Langmuir (2) emphasized that chemical adsorption cannot extend beyond one molecular layer. Studies by other investigators have also shown that the amount of chemisorbed gas required to saturate a surface corresponds to a unimolecular layer (3).

The concept that an activation energy is frequently required for chemisorption, which originated with Taylor (4), provides a satisfactory explanation for many adsorption phenomena. Thus, the quantity of gas adsorbed sometimes varies with temperature as shown in Figure 4.2. At the lowest temperatures van der Waals adsorption prevails, with a small heat of adsorption. Since the process is exothermic, it follows that the extent of adsorption decreases with an increase in temperature. However, as the temperature is raised, the rate of activated adsorption increases as does the quantity of gas adsorbed. At still higher temperatures the extent of adsorption again decreases, since chemisorption is also an exothermic process.

Chemisorption on oxides and metals that have not been especially cleaned, and therefore contain an oxide surface, involves appreciable activation energies. With carefully cleaned metal surfaces, there is usually a small but significant activation-energy (5).

Another characteristic of chemisorption is that the surfaces involved appear to be heterogeneous and show variations in adsorptive power. Taylor (6) suggested that lone surface atoms at peaks and edges are partially unsaturated and constitute "active centers" at which chemisorption and catalytic activity predominate. More recent work (7) suggests that these active centers are lattice defects. Boudart (8) emphasized that they do not occupy fixed positions, but are constantly being created and destroyed by the movement of excited electrons in the lattice. One type of phenomenon that is frequently cited as evidence for active centers is that poisons can almost completely inactive a catalyst when present at such low concentrations that they have only a slight effect on the degree of adsorption. Thus, Pease and Stewart (9) found that mercury vapor reduced the rate of hydrogenation with a copper catalyst to 0.5 per cent, while decreasing the adsorption of ethylene to 80 per cent and of hydrogen to 5 per cent of normal.

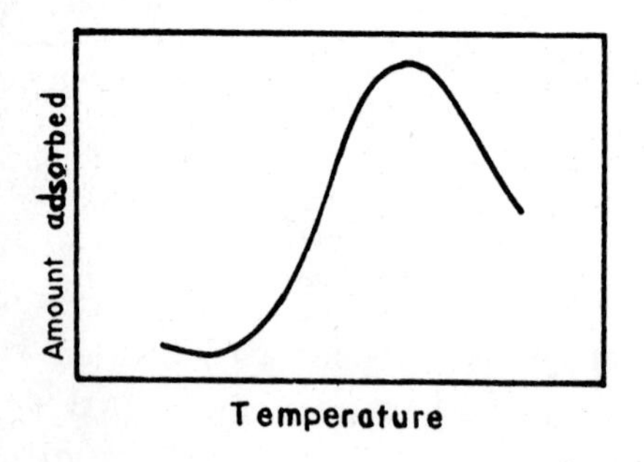

Figure 4.2. Amount of gas adsorbed as a function of temperature.

There appears also to be a repulsive interaction between atoms or molecules chemisorbed on a surface. Differential heats of adsorption as well as desorption rates are cited as evidence of repulsive interaction. However, it is frequently difficult to distinguish between repulsion and surface heterogeneity.

Ordinary repulsive forces existing between neighboring atoms would be expected to be of the same order of magnitude as van der Waals forces, and therefore would not exert much effect on chemisorption. However, Boudart (10) points out that a covalent bond formed between an adsorbed atom and the surface will have an ionic character depending upon the relative electronegativities of the two types of atoms. This bond has a dipole moment and an electric double layer at the surface, the effect of which is to reduce the strength of bonds subsequently formed and thus to reduce the heat liberated on further chemisorption.

The Langmuir adsorption isotherm, derived in Chapter 3, is considered to represent ideal monomolecular adsorption. However, few if any actual systems give results that are in agreement with the Langmuir equation. The reasons for these deviations are based on assumptions inherent in the derivation. Thus, Langmuir assumed a uniform surface, free from heterogeneity, and the absence of interaction between adsorbed atoms or molecules. Parlin, et al (11) show several derivations of the Langmuir equation from statistical mechanics which emphasize the implied restrictions. Thus, it is necessary to assume localized adsorption, in which the adsorbed molecules are held in fixed positions on the uniform surface. In addition to the absence of interaction between adsorbed molecules, the surface itself is assumed not to change properties when a molecule is adsorbed. In actual chemisorption, all of these assumptions may well be invalidated.

In experimental studies on catalytic surfaces, it is generally difficult to distinguish between differences in the activity of different parts of a surface towards adsorption and the effect of lateral interaction of adsorbed molecules on subsequent adsorption on neighboring sites. An additional complication is that the surface may be initially heterogeneous, or heterogeneity may be induced by the interaction between adsorbed molecules and the surface. Initial or *a priori* heterogeneity arises in the case of corners or edges. Different crystal faces have been shown to differ in catalytic activity (11). The surface of a multiphase solid will be heterogeneous because of the variation in composition of the phases and the presence of phase boundaries. The introduction of impurities in a crystal will make the surface non-uniform.

Taylor and Liang (12) investigated the adsorption of hydrogen on zinc oxide in a temperature range at which physical adsorption could be neglected. By measuring the isobaric rate of sorption with an instantaneous increase in temperature, they found that at the higher temperature rapid desorption occurred followed by a slow readsorption of a larger volume of gas. This result was shown to be general for oxide catalysts as well as certain metals, such as iron and nickel. These results could be explained by assuming the surface to be composed of regions of differing activity, such that an increase in temperature leads to rapid desorption in one region followed by slow adsorption in areas where the activation energy is high. Thus the area of the surface that is active in chemisorption would appear to vary, depending upon the temperature. Kubokawa and Toyama (13) showed a further distinction between the two types of chemisorption. Below 110°C, hydrogen chemisorption did not appreciably affect the conductivity of zinc oxide. At higher temperatures, hydrogen adsorption was accompanied by a substantial increase in conductivity.

It appears likely that chemisorption on nonideal surfaces involves three distinct processes. (*1*) At low coverage, adsorption occurs at the most favorable sites, as determined by temperature and other experimental conditions. The extent of binding will depend upon the heat of adsorption and the energy of activation. (*2*) In addition, the process of adsorption will alter the energetics of the surface and affect the tendency for adsorption on neighboring sites. (*3*) Lateral interactions, while of cardinal concern to catalytic reactions, can only be important in adsorption and desorption processes when the surface is almost completely covered (11).

Experimental Methods

In a review, Emmett (14) predicted that, as a result of the powerful new tools that are being brought to bear to elucidate the mechanisms of catalysis, more progress would be made in putting catalysis on a truly scientific basis

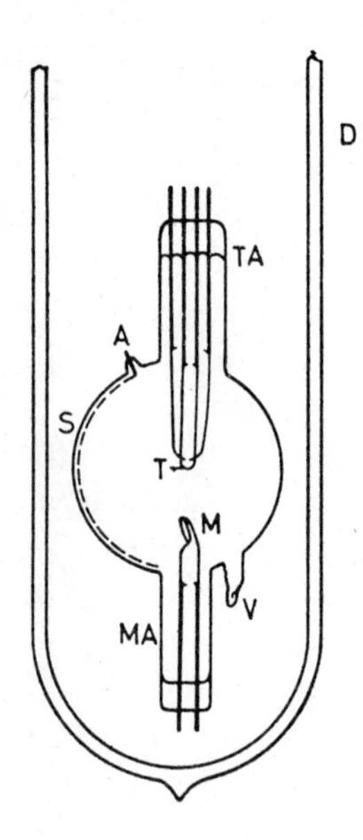

Figure 4.3. Diagram of field emission tube for mobility studies.
TA, tip assembly; I, tip; M, platinum crucible; MA, crucible assembly; S, fluorescent screen; A, anode lead; V, seal-off; D, cryostat (schematic) (15).

during the succeeding ten years than had been made in all work up to that time. The measurements referred to by Emmett were magnetic, conductivity, infrared, nuclear magnetic resonance, contact potential, x-ray and electron microscope, radioactivity, and gas chromatography. A number of these measurements are illustrated below, as well as typical chemisorption data.

The question of whether chemisorbed films are mobile is important in setting up theoretical models. Gomer (15) described one method for studying the diffusion of sorbed gases on metal substrates. Thus, at 4.2° K the vapor pressures of all gases except helium are negligibly small (10^{-15} mm of mercury), while their sticking coefficients on all surfaces are very large. A sealed-off field emission tube was immersed in a bath of liquid helium. Gas was then evaporated from an electrically heated source, using CuO or ZrH_2 as the source for oxygen or hydrogen respectively, and allowed to adsorb onto the tungsten field-emission tip. Since chemisorbates generally raise the work function and thus decrease emission, these areas appear dark when the emission tip is electrically heated. With the aid of a fluorescent screen, it is possible to follow the migration of the chemisorbed gas. The apparatus is depicted in Figure 4.3.

Gomer found qualitatively similar results in the diffusion of hydrogen and oxygen on tungsten. However, activation energies and temperatures required for a given phenomenon with oxygen were at least double those with hydrogen. At least three distinct types of diffusion were found to occur. At

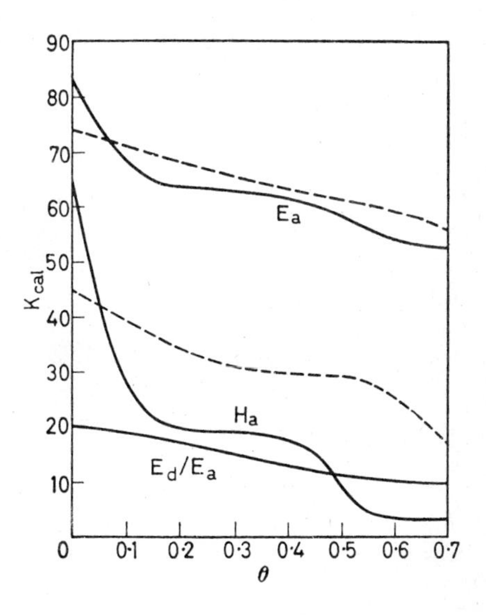

Figure 4.4. Heat of adsorption of hydrogen on tungsten (H_a), and heat of binding of hydrogen atoms (E_a) as a function of surface coverage (θ). E_d/E_a is the ratio of activation energy for surface diffusion to binding energy of hydrogen atoms. Ordinate is in parts per hundred (15).

very low temperatures, below 70° K for oxygen and very much less for hydrogen, the chemisorbed monolayers were immobile. At higher temperatures — 180–240° K for hydrogen and 500–530° K for oxygen — corresponding to initial deposits of about 0.8 to 1.0 monolayer, boundary diffusion was observed with activation energies of 5.9 $\pm$ 1 kcal for hydrogen and about 19 kcal for oxygen. At very low coverage, the activation energy of diffusion for hydrogen increased from about 9.5 kcal to about 16 kcal with decreasing coverage. These values were about double for oxygen. Data obtained by Gomer on the heat of adsorption and the heat of binding of hydrogen atoms on tungsten parallel the results of earlier investigators, as shown in Figure 4.4. The heat of adsorption decreases sharply with increasing coverage.

Gray and McCain (16) studied the chemisorption of oxygen and hydrogen on platinum by changes in magnetic susceptibility determined under dynamic conditions. Measurements were made using a sensitive Gouy balance and uniform fields up to 10,000 gauss. The extent to which platinum forms an oxide coating and the difficulty in its removal is not generally realized. Langmuir (17) observed that platinum was activated by successive oxidations and reductions. Roginskii and Rozing (18) indicated that pretreatment of platinum with an oxygen-hydrogen mixture activated the surface, while pretreatment with hydrogen alone rendered platinum inactive. Pretreatment

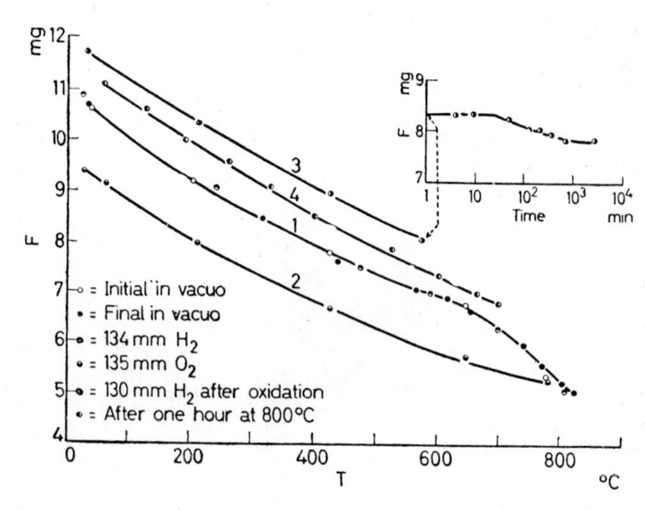

Figure 4.5. The magnetic attractive force on a platinum gauze cylinder as a function of temperature (16).

with oxygen alone above 430° C activates the metal, while removal of the oxygen with hydrogen destroys its activity.

The magnetic attractive force on a platinum gauze cylinder as a function of temperature is shown in Figure 4.5. Curve 1 is for the platinum in a completely reduced state, while curve 2 is after adsorption of oxygen at 783° C and 135 mm. The effect of chemisorbed oxygen is to diminish the magnetic attraction. Subsequent treatment of the oxidized platinum with hydrogen at 569° C and 130 mm increased the magnetic attraction to that of curve 3, with a slow decrease with time as shown in the insert. After heating the platinum gauze of curve 3 for one hour at 800° C, the magnetic attraction corresponded to curve 4. After additional heating and pumping at 824° C for one hour, the points again lay on curve 1, showing that the platinum had returned to its original magnetic state. Other experiments showed that in the absence of hydrogen, the magnetic effect of the adsorbed oxygen could not be removed by pumping for 12 hours at 805° C. Hydrogen did not readily reduce oxidized platinum below 200 to 300° C. Gray and McCain also compared theoretical and experimental curves and concluded that the reduction in magnetic attraction by adsorption of oxygen on platinum corresponded to a reduction in the *d*-band from 0.6 to 0.59 holes per atom. They deduced that each chemisorbed oxygen atom immobilizes electrons in 10 to 50 platinum atoms in its vicinity.

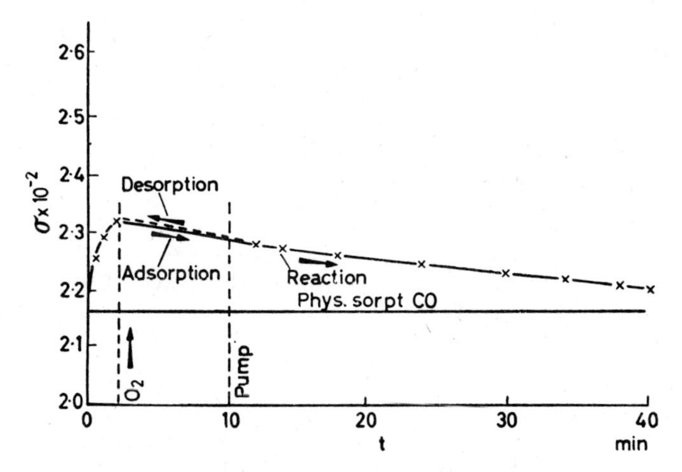

Figure 4.6. Electrical conductivity of manganese dioxide containing adsorbed carbon monoxide (20).

Elovitch (19) studied the chemisorption of carbon monoxide on manganese dioxide by means of electrical conductivity measurements. This adsorption is an important stage in the low-temperature oxidation of carbon monoxide. Previous studies showed that below −10 to −20° C physical adsorption takes place on manganese dioxide. At higher temperatures, an activation energy of about 6 kcal per mole is required and adsorption is accompanied by reaction with the surface atoms of oxygen of the manganese dioxide with the formation of carbon dioxide and a decrease in the oxygen to manganese ratio of the oxide. This is a self-poisoning reaction and the rate of reaction decreases exponentially with the number of molecules reacting in a given time interval.

It has been shown (20) that chemisorption, but not physical adsorption.

results in a detectable change in the electrical conductivity of manganese dioxide. Referring to Figure 4.6, it can be observed that the initial adsorption of carbon monoxide at 20° C results in an increase in electrical conductivity indicating that the adsorption of carbon monoxide bears an acceptor nature. The adsorption of carbon dioxide is of the donor type and leads to a decrease in the electrical conductivity. Thus, the subsequent decrease in conductivity testifies to the formation of carbon dioxide on the surface of manganese dioxide.

Stephens (21) studied surface reactions on evaporated palladium films at 0° C. A monolayer of adsorbed gas was first formed and allowed to react with a second gas, added in small doses. The extent of reaction was followed volumetrically and by gas analysis. Extensive reaction occurred in all of the systems used: $Pd-O_2+CO$, $Pd-CO+O_2$, $Pd-O_2+H_2$, $Pd-H_2+O_2$, $Pd-O_2+C_2H_4$ and $Pd-C_2H_4+O_2$.

References

1. Lennard-Jones, J. E., *Trans. Faraday Soc.* **28**, 334 (1932).
2. Langmuir, I., *J. Am. Chem. Soc.* **38**, 221 (1916).
3. Roberts, J. K., *Proc. Roy. Soc.* (London) **A152**, 445 (1935).
4. Taylor, H. S., *J. Am. Chem. Soc.* **53**, 578 (1931).
5. Laidler, K. J., in Catalysis, edited by Emmett, P. H., Vol. 1, Chapt. 3, New York, Reinhold Publishing Corp., 1954.
6. Taylor, H. S., *Proc. Roy. Soc.* (London) **A108**, 105 (1925); *J. Phys. Chem.* **30**, 145 (1926).
7. Volkenshtein, F. F., *Zhur. Fiz. Khim.* **23**, 917 (1949).
8. Boudart, M., Thesis, Princeton, 1950.
9. Pease, R. N., and Stewart, R., *J. Am. Chem. Soc.* **47**, 1235 (1925).
10. Boudart, M., *J. Am. Chem. Soc.* **74**, 3556 (1952).
11. Parlin, R. B., Wallenstein, M. B., Zwolinski, B. J., and Eyring, H., in "Catalysis", edited by Emmett, P. H., Vol. 2, Chapt. 5, New York, Reinhold Publishing Corp., 1955.
12. Taylor, H. S., and Liang, S. C., *J. Am. Chem. Soc.* **69**, 1306 (1947).
13. Kubokawa, Y., and Toyama, O., *Bull. Naniwa Univ.* **A2**, 103 (1954).
14. Emmett, P. H., *J. Phys. Chem.* **63**, 449 (1959).
15. Gomer, R., Proc. Intern. Congr. Surface Activity, 2nd, Vol. 2, p. 236, London, Butterworth, 1957.
16. Gray, T. J., and McCain, C. C., ibid., p. 260.
17. Langmuir, I., *J. Am. Chem. Soc.* **40**, 1361 (1918).
18. Roginskii, S. Z., and Rozing, V. S., *Uchenze Zapiski Leningrad Gosudarst Univ.*, Ser. Fiz. Nauk **5**, 67 (1939).
19. Elovitch, S. J., Proc. Intern. Congr. Surface Activity, 2nd, Vol. 2, p. 252, London, Butterworth, 1957.
20. Elovitch, S., and Margolis, L., *Dokl. Akad. Sci.* Nauk **107**, 112 (1956).
21. Stephens, S. J., *J. Phys. Chem.* **63**, 188 (1959).

CHAPTER 5

Electrical Phenomena at Interfaces

The electrical properties of small particles are treated in standard books on colloid chemistry, including that of Kruyt (1). The electrical phenomena that occur at interfaces are factors in adsorption, detergency, emulsion stability, and foam stability, and are discussed further in subsequent chapters. These electrical phenomena are basic to our understanding of flocculation and the dispersion of solids.

ELECTRIC DOUBLE LAYER

Potential Difference Across an Interface

The potential difference across an interface can be obtained by applying an external potential difference to a system containing an electrode with a completely polarizable electrode, represented schematically in Figure 5.1.

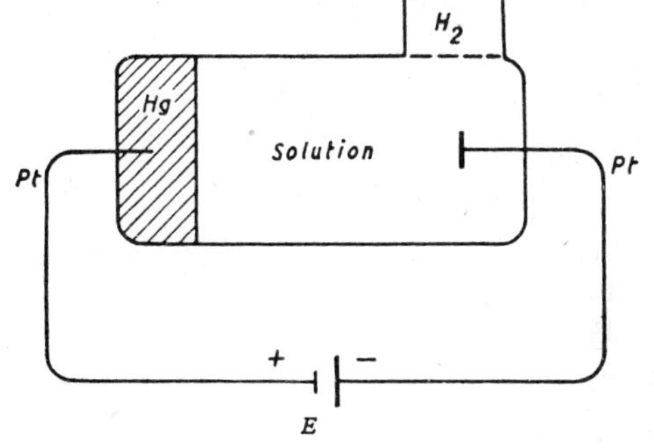

Figure 5.1. Potential difference across a mercury-solution interface. The mercury electrode is assumed to be completely polarizable (1).

The cell consists of a solution, a mercury electrode, and a hydrogen electrode saturated with one atmosphere of hydrogen. The mercury electrode is assumed to be completely polarizable, that is, no ions from the solution are discharged at the electrode. Any charge carried to it from the applied potential difference remains on the mercury. The differential of the Gibbs free energy is represented by

$$dF = -SdT + VdP + \gamma dA + \sum_{a} \sum_{i=1}^{r} \eta_i^a dn_i^a + EdQ \tag{5.1}$$

where γ is the free surface-energy per cm² and A the interfacial area of the mercury-solution interface. E is the potential difference and Q is the charge on the mercury. Then η_i^a is the electrochemical potential of component i in phase a, defined by the relation

$$\eta_i^a = \mu_i^a + z_i e\phi^a \tag{5.2}$$

where μ_i^a is the chemical potential of component i in phase a, z_i is the valence of component i and ϕ is the internal potential of phase a.

The important result relates the change in free surface-energy to the charging of the interface

$$d\gamma = -\sum_{i=3}^{r} \Gamma_i d\mu_i - \sigma dE_s \tag{5.3}$$

where σ is the charge on the mercury per cm² and E_s is the potential difference between mercury and the solution.

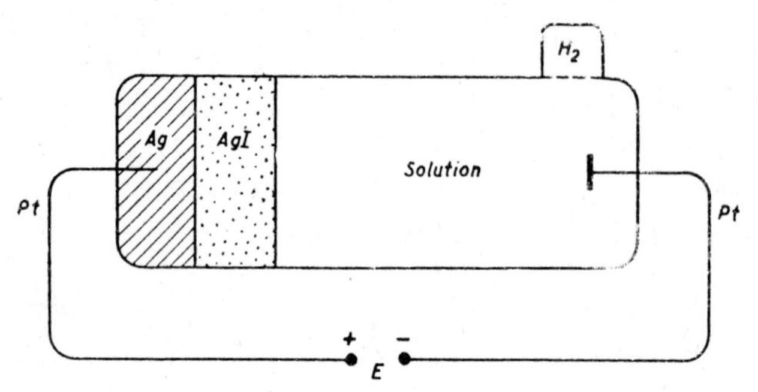

Figure 5.2. Reversible cell containing a silver iodide-solution interface (1).

The other type of system studied is the completely reversible galvanic cell. The reversible cell shown in Figure 5.2 consists of a silver-silver iodide electrode, a solution, and a platinum electrode saturated with hydrogen at one atmosphere pressure. The EMF of the cell is determined by the composition of the phases present. For this case also,

$$d\gamma = -\sum_{i=3}^{r} \Gamma_i d\mu_i - \sigma dE_s \tag{5.4}$$

for the AgI-solution interface.

Diffuse Double Layer

If a solid or water-immiscible liquid is placed in contact with water containing an electrolyte, an excess of ions of one type will generally be present on the surface of the non-aqueous phase and an equivalent amount of ions of opposite charge will be distributed in the aqueous phase near the interface. The distribution of excess charges on the surface and in the solution constitute an electric double-layer.

The charge on the solid is treated as a surface charge spread uniformly over the surface. The charge in solution is considered to be composed of an unequal distribution of point-like ions. The solvent influences the double layer only through its dielectric constant.

According to the Gouy-Chapman treatment, the distribution of the ions in the solution is governed by a Boltzmann relation. Negative ions are concentrated at places of positive potential and repelled at places of negative potential. The reverse occurs for positive ions,

$$n_i = n_{i0}\, \mathbf{e}^{-ze\psi/kT} \tag{5.5}$$

where n_i is the concentration of ions of kind i at a point where the potential is ψ, n_{i0} is the concentration in the bulk of the solution, and z is the valence with the sign of the charge included.

The coulombic interaction between the charges present in the system is expressed by Poisson's equation,

$$\nabla\, \psi = -\frac{4\pi\rho}{D} \tag{5.6}$$

where ψ is the potential which varies from ψ_0 at the interface to zero in the bulk of the solution, ρ is the charge density, D is the dieletric constant, and ∇ is the Laplace operator.

The space charge density ρ is the sum of the ionic charges per unit volume,

$$\rho = \sum z_i e n_i \tag{5.7}$$

The combination of Equations 5.5. and 5.6 leads to the differential equation

$$\nabla\, \psi = \frac{-4\pi}{D} \sum z_i e n_{i0}\, \mathbf{e}^{-z_i e\psi/kT} \tag{5.8}$$

This equation can only be solved with certain simplifying assumptions.

If we assume an infinitely large plane surface and a single binary electrolyte of valency z in the solution, then the change in potential with distance from the surface is

$$\frac{d\psi}{dx} = -\sqrt{\frac{8\pi nkT}{D}}\,(\mathbf{e}^{ze\psi/2kT} - \mathbf{e}^{-ze\psi 2kT}) \tag{5.9}$$

and the surface charge density σ is expressed by

$$\sigma = \sqrt{\frac{DnkT}{2\pi}}\,(\mathbf{e}^{ze\psi_0/2kT} - \mathbf{e}^{-ze\psi_0/2kT}) \tag{5.10}$$

When the potential is so small that the exponentials can be expanded,

$$\sigma = \sqrt{\frac{DnkT}{2\pi}}\ \frac{ze\psi_0}{kT} = \frac{D\kappa}{4\pi}\psi_0 \tag{5.11}$$

where κ is the Debye-Huckel function

$$\kappa = \sqrt{\frac{4\pi e^2 \sum n_{i0} z_i^2}{DkT}} \tag{5.12}$$

Similarly, for small values of ψ,

$$\psi = \psi_0 \mathbf{e}^{-\kappa x} \tag{5.13}$$

The potential drops exponentially to zero as the distance approximates $1/\kappa$.

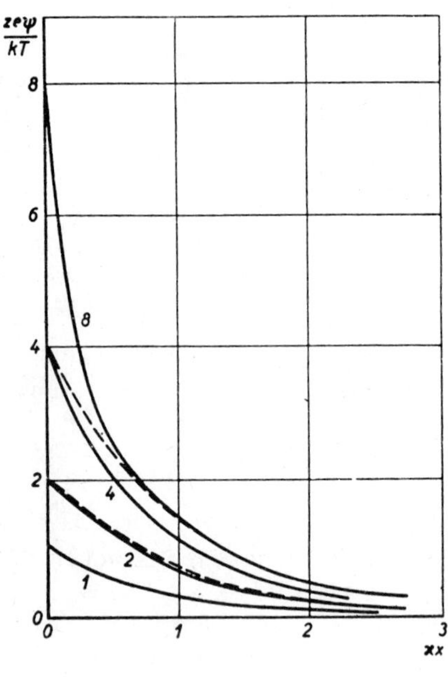

Figure 5.3. Decrease in the electrical potential in the double layer with increasing κX, according to the Gouy-Chapman theory (1).

Thus, the thickness of the double layer is said to equal $1/\kappa$. Representations of the change in potential with distance from the surface are shown in Figure 5.3.

Stern Treatment of the Double Layer

The Gouy-Chapman theory of the diffuse double-layer is an over-simplification that suffers from a number of defects. Stern introduced two important corrections. One is the correction for the finite dimensions of the ions in the first ionic layer adjacent to the wall. The second correction was the consideration of the possibility of the specific adsorption of ions and the location of these ions in a plane δ distance from the surface. This layer of adsorbed ions is called the Stern layer.

The total potential drop ψ_0 is divided into two portions, the potential ψ_δ over the diffuse part of the double layer, and $\psi_0 - \psi_\delta$ over the molecular condenser. Figure 5.4 is a schematic representation of the double layer according to Stern. The treatment of the diffuse layer is the same as in the Gouy-Chapman theory, except that the diffuse layer starts with potential ψ_δ at δ distance from the wall.

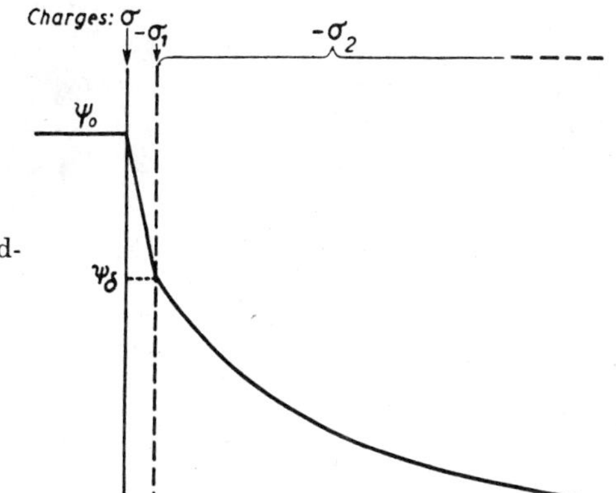

Figure 5.4. The double layer according to the Stern theory (1).

The charge on the wall σ is opposite and equal to the sum of the charge of the ions in the liquid part of the molecular condenser σ_1 and the charges of the diffuse part of the double layer σ_2. The diffuse double layer charge σ_2 is related to ψ_δ according to the Gouy-Chapman theory. The calculation of σ_1 is discussed in the chapter on emulsions.

While Stern's theory is an improvement over that of Gouy, it is not in complete accord with experimental findings. Other useful approaches to the problem include those of Frumkin (2), Bikerman (3), and Graham (4).

ELECTROCAPILLARY EFFECTS

The interface between mercury and an aqueous solution is very favorable for studying the properties of the double layer. The potential difference between the two phases can be readily altered and conveniently measured, and the interfacial tension can be measured directly.

A sketch of the apparatus used by Lippmann (5) to determine the relation between surface tension and polarizing potential is shown in Figure 5.5. A polarizing potential E is applied between the mercury and the calomel electrode. The interface between mercury and the aqueous solution is maintained at a predetermined position in the capillary by applying a suitable pressure h. The interfacial tension is determined from the height h and the diameter of the capillary at the interface.

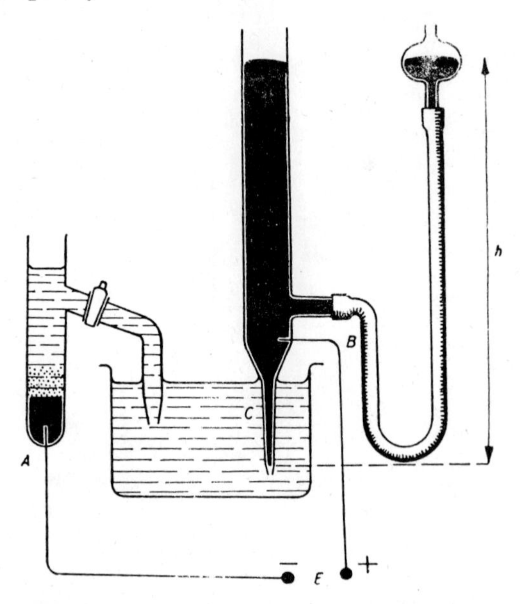

Figure 5.5. Capillary electrometer after Lippmann. The polarizing potential E is applied between a calomel electrode A and a mercury electrode B. The interface between mercury and the solution is in the capillary c. The surface tension is determined from the height h (1).

The general relation between the interfacial tension and the polarizing potential is shown in Figure 5.6. The mercury surface is initially positively charged. On reducing this charge by an applied potential, the interfacial tension increases. At the electrocapillary maximum, the charge on the mercury is zero. On further increasing the polarising potential, the charge on the mercury becomes negative and the interfacial tension decreases. The charge on the mercury surface follows from the equation

$$\left(\frac{\partial \gamma}{\partial E}\right)_{T,P,\mu} = -\sigma \qquad (5.14)$$

At the interfacial tension maximum, $\sigma = 0$.

Figure 5.7 shows the effect of various potassium salts on the electrocapillary curve. In the anodic branches, the more highly polarizable anions are strongly adsorbed, reducing the interfacial tension and displacing the maximum to lower values at more negative potentials. All of the salts show the same cathodic branch of the electrocapillary curve, since the anions are electrostatically repelled from the surface and only the cations are adsorbed.

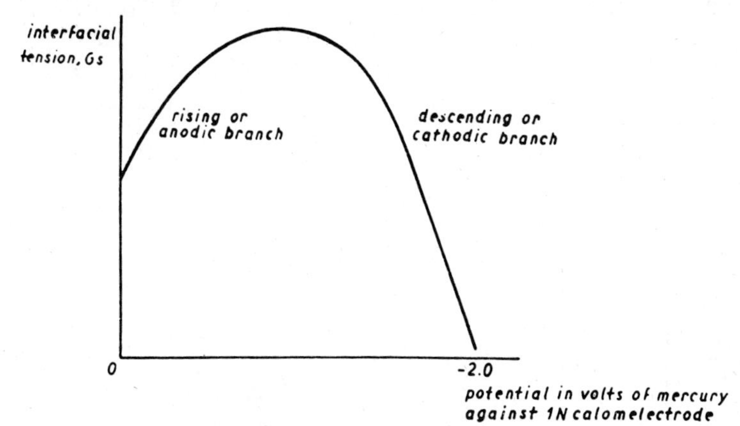

Figure 5.6. General form of the electrocapillary curve (1).

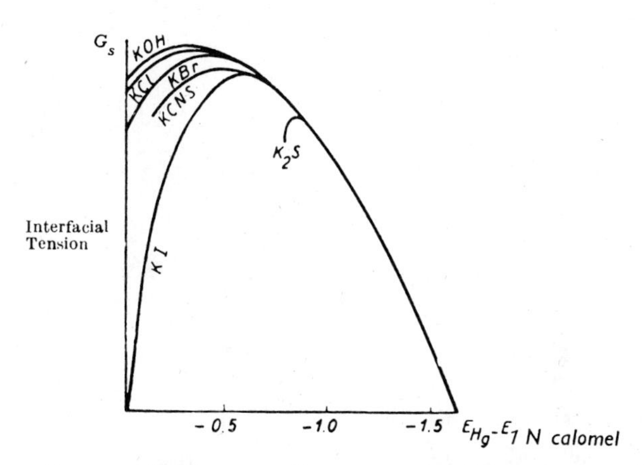

Figures 5.7. Electrocapillary curves for different anions (1).

Figure 5.8 shows the effect of different cations. While the curves for different inorganic cations are almost identical, organic cations are more strongly adsorbed and depress the cathodic branch of the curve. The surface charge calculated by differentiation of the electrocapillary curve can also be determined directly. The two methods are in satisfactory agreement.

The capacity C of the electrical double layer can be calculated by a second differentiation of the electrocapillary curve,

$$\left(\frac{\partial^2 \gamma}{\partial E^2}\right)_{T,P,\mu} = C_{\text{differential}} \tag{5.15}$$

The capacity of the double layer is most readily compared with different models of the structure of the double layer. It can be determined directly by using the cell containing the mercury-solution interface as one arm of a

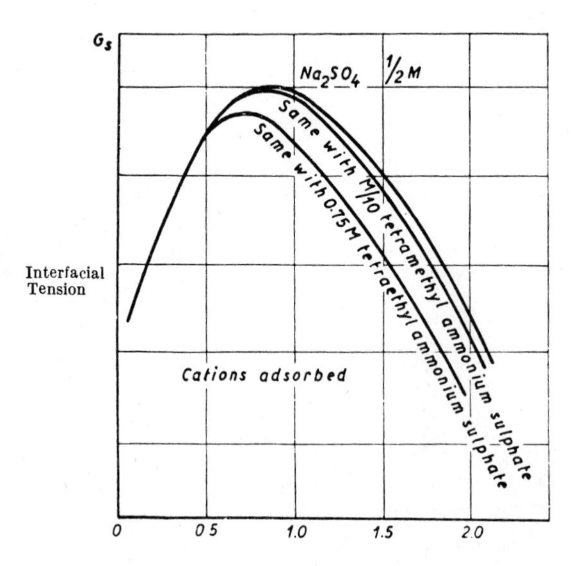

Figure 5.8. Electrocapillary curves for different cations (1).

capacitance bridge, constructed so that a variable dc potential can be applied to the cell. Results obtained agree with those calculated from electrocapillary curves.

In addition, the separate contributions of negative and positive ions to the solution part of the double layer can be calculated from the equations

$$\left(\frac{d\gamma}{d\mu_c}\right)_{E_A} = -\Gamma_c \tag{5.16}$$

and

$$\left(\frac{d\gamma}{d\mu_A}\right)_{E_C} = -\Gamma_A \tag{5.17}$$

The change in surface tension with composition at constant potential leads

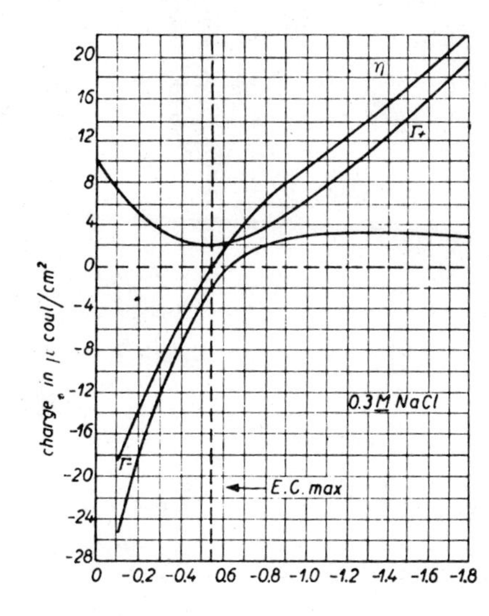

Figure 5.9. Charge η and adsorption of cations Γ^+ and anions Γ^- at the mercury – $0.3M$ NaCl interface. Potentials are measured relative to $0.3M$ $NaCl/Hg_2Cl_2/Hg$ (1).

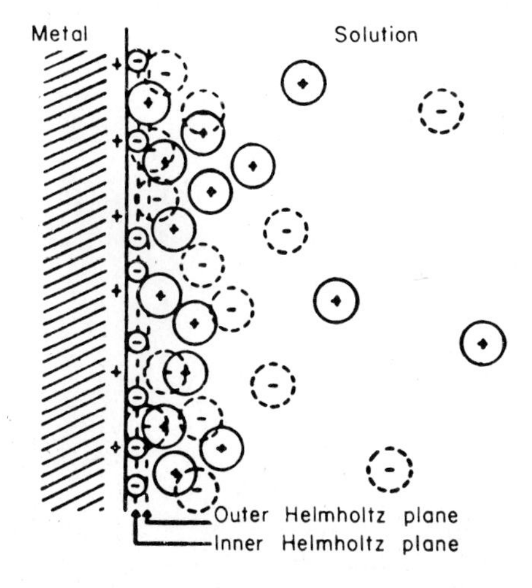

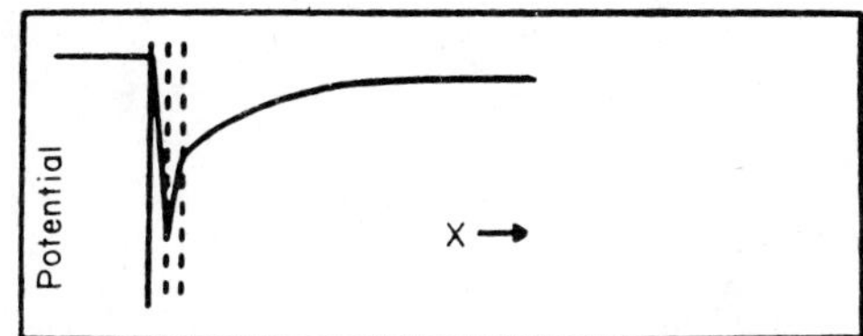

Figure 5.10. Schematic representation of the electrical double layer with positive polarization (4).

to the separate adsorption of cations and anions. The constant potential E_A or E_C means that the second electrode is reversible to the anion or the cation.

The contribution of Na^+ and Cl^- to the charge is shown in Figure 5.9. The sum of charges is shown by the curve η. At the electrocapillary maximum, where $\eta = 0$, there is a slight adsorption of cations and an equal adsorption of anions. With increasing negative polarization, adsorption of Na^+ increases while there is a slight desorption of Cl^-. Positive polarization is compensated by a large adsorption of Cl^-, while Na^+ adsorption also increases, but to a lesser extent. Grahame (4) concluded that there is a strong non-electrostatic adsorption of anions at the mercury-solution interface, while specific cation adsorption was probably absent.

According to Grahame (4), there are two kinds of molecular condenser. The first type involves the polarized mercury surface and the anions, which are strongly adsorbed. The potential in the plane of anions is called the "inner Helmholtz plane" or "Stern plane". The distance of closest approach of the cations to the interface is called the "outer Helmholtz plane" or "limiting Gouy plane". This is shown diagramatically in Figure 5.10.

ELECTROKINETIC PHENOMENA

Much of our information concerning the charge on small particles is deduced from electrokinetic phenomena, that is, the relative motion of a charged surface with reference to the bulk solution. Electro-osmosis and electrophoresis

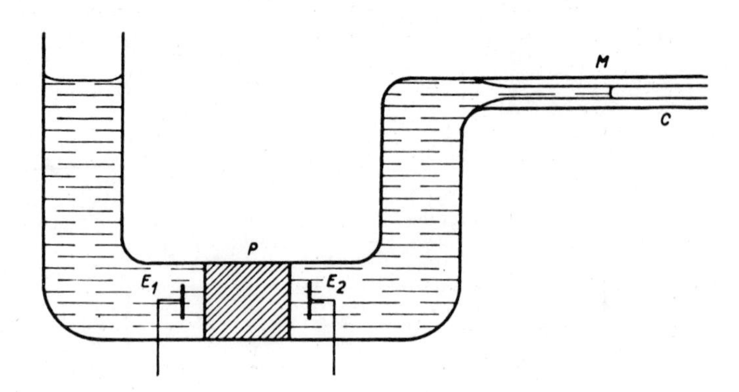

Figure 5.11. Apparatus for measuring electro-osmosis (1).

result from the movement of charged particles upon the application of an external electric field. Streaming potential and migration potential are examples of the transport of electricity arising from the movement of phases.

Electro-osmosis

The movement of a liquid with respect to a solid wall as the result of an applied potential gradient is called electro-osmosis or electro-endosmosis. The arrangement shown in Figure 5.11 can be used to observe electro-osmosis. P is a porous plug through which water flows when a potential difference is applied between the electrodes E_1 and E_2. The extent of flow of the liquid is measured by the displacement of the meniscus M in the capillary tube. The direction and velocity of the liquid flow depends upon the properties of the liquid and the plug.

To avoid gas development at the electrodes, it is usually necessary to use reversible electrodes, such as Ag—Ag Cl or Zn—Zn SO_4. A single capillary may be used in place of the plug. However, only a few materials, such as quartz and glass can be used as capillaries, while most solids can be fabricated as porous plugs.

Electrophoresis

Electrophoresis is the movement of colloidal particles in an electric field. It is comparable to the movement of ions and can be measured in the same

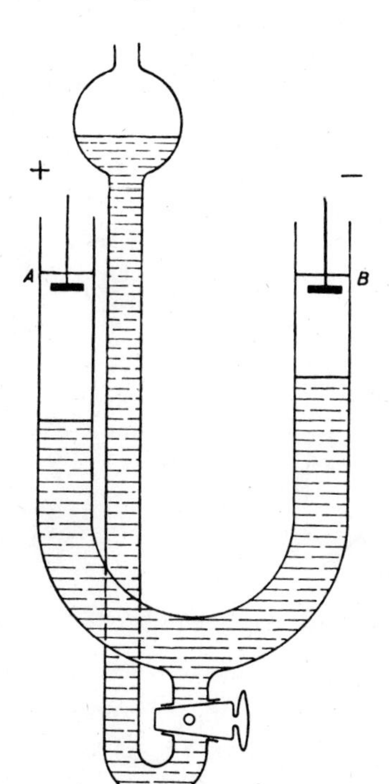

Figure 5.12. Moving boundary method for the determination of the electrophoretic velocity (1).

manner as ionic mobilities. The visibility of colloidal particles permits the direct measurement of electrophoretic mobility under the ultramiscroscope or light microscope.

The moving boundary method is frequently used to measure electrophoretic velocity. A simple form of the apparatus is shown in Figure 5.12.

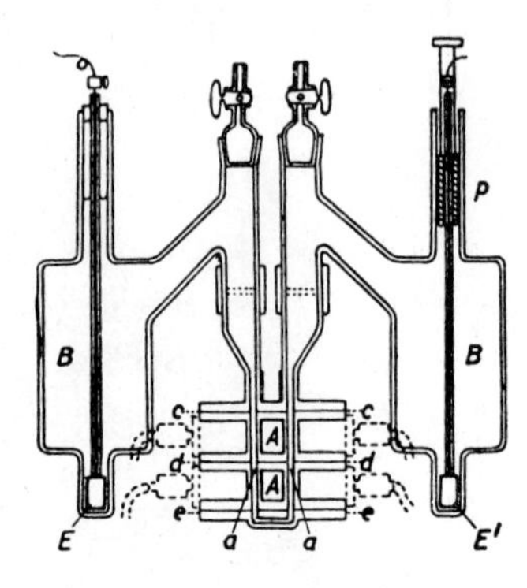

Figure 5.13. Tiselius electrophoresis apparatus (1).

The lower portion of the U-shaped tube is filled with the colloidal solution, while the upper part contains the pure medium without colloid. A much improved unit is the Tiselius apparatus shown in Figure 5.13.

Streaming Potential

The measurement of streaming potential is similar to that of electro-osmosis, except that the liquid is forced through the porous plug or capillary and the

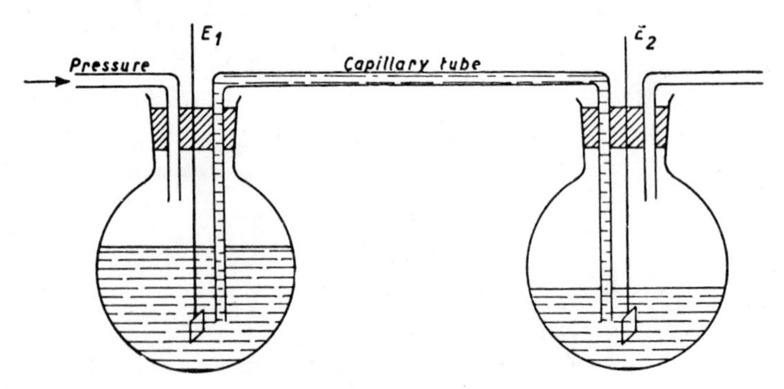

Figure 5.14. Apparatus for observing streaming-potentials (1).

potential difference developed between the electrodes is measured. Larger quantities of liquid are displaced and the resulting potential difference is much smaller than in the case of electro-osmosis. An apparatus due to Kruyt (6) is shown in Figure 5.14.

Migration Potential

The migration potential is the reverse of electrophoresis. When small particles suspended in a liquid are forced to move as the result of a gravitational or centrifugal field an electric field is generated in the direction of the movement and the potential difference can be measured. The development of an electric field in this manner is called the Dorn-effect. The experimental techniques are difficult.

The Zeta Potential

The principal object of most electrophoretic measurements is to determine the zeta potential ζ, and all four methods discussed result in essentially the same value for ζ. The zeta potential is the electrical potential in the slipping plane between the fixed and flowing liquid. The slipping plane is located somewhere in the liquid and not exactly at the solid-liquid phase boundary.

Since electrophoresis is the more usual procedure for determining the zeta potential, the discussion which follows is directed to this method. The relation between electrophoretic velocity v and the zeta potential is

$$v = \frac{D\zeta E}{4\pi\eta} \tag{5.18}$$

where the particle is considered to be exposed to a homogeneous electric field of strength E in a medium of dielectric strength D and viscosity η. While the shape of the particle is not important, it is necessary that the double layer be thin as compared with the dimensions of the particle and the particle must be insulating, with the surface conductance at the interface small.

Henry (7) showed that the factor 1/4 in the equation is applicable when κa is large, where κ is the reciprocal of the thickness of the double layer and a is the radius of the particle. When κa is small, the factor is 1/6.

A number of authors (8, 9, 10) have considered possible errors in the calculation of the zeta potential due to neglect of the effect of relaxation. This effect originates in the deformation of the double layer in a direction opposite to the movement of the particle and thus has a retarding effect on electrophoresis. The conclusions are that the relaxation effect may be neglected if $\zeta \ll 25$ mv, or if κa is very small or very large. However, if ζ is not very small and κa has an intermediate value (0.1–100), the relaxation effect can be very important. Thus conclusions concerning the value of the zeta potential, as determined by electrophoretic measurements, are often open to question.

FLOCCULATION AND DISPERSION

The tendency of colloidal particles to flocculate depends upon the interaction of forces of attraction and repulsion between the particles. If repulsion is of sufficient strength, the particles remain dispersed. Otherwise, the particles coagulate.

The repulsive factor is said to originate in the electrochemical double-layer. London-van der Waals interaction is responsible for the attraction of particles. The rate of flocculation can vary from a few seconds for complete flocculation to many years for the first evidence of coagulation.

Rapid Coagulation

The theory of rapid coagulation was developed by von Smoluchowski. The rate of coagulation depends upon the Brownian motion of the particles and their interaction when they are close together. In the simplest case, a swamping amount of electrolyte is added to the sol so that the electrical repulsion vanishes. The attraction can be represented as a sphere of action surrounding each particle. If a second particle enters this sphere of action, the two particles coalesce irreversibly.

In this form of coagulation, the rate depends entirely upon Brownian motion. Every contact between particles is permanent. The time $t_{\frac{1}{2}}$ in which the number of particles is just halved is given by the relation

$$t_{\frac{1}{2}} = \frac{1}{8\pi D a v_0} \tag{5.19}$$

where D is the diffusion constant, a is the effective radius of the particles, and v_0 the number of particles initially present.

If the diffusion constant $D = kT/6\pi\eta a$,
then

$$t_{\frac{1}{2}} = \frac{3\eta}{4\pi T v_0} \tag{5.20}$$

where η is the viscosity of the medium and T is the absolute temperature. If water is the dispersion medium and $T = 298°$,

$$t_{\frac{1}{2}} \sim 2{\cdot}10^{11}/v_0 \tag{5.21}$$

Slow Coagulation

According to Smoluchowski's theory (11), for rapid coagulation every encounter between particles results in a permanent adhesion of the particles. For slow coagulation, only a fraction a of the collisions lead to permanent sticking of the particles. Then

$$t_{\frac{1}{2}} = \frac{1}{8\pi D a v_0 a} \tag{5.22}$$

The theory of Fuchs (12), originally developed for smokes or mists, has been extended to slow coagulation in liquid dispersion media. According to this theory there is an energy barrier to the number of collisions between

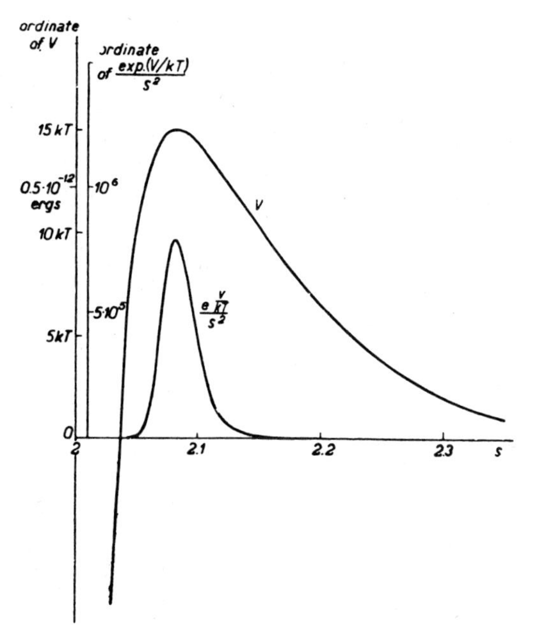

Figure 5.15. Curves of potential energy of interaction V and $\frac{\exp(V/kT)}{s^2}$ versus s for the interaction of two spherical particles of radius $a = 10^{-5}$ cm, with $\kappa = 10^{-6}$, $A = 10^{-12}$ erg, and $\psi_0 = 28.2$ mv (1).

particles. In the presence of a potential energy barrier V, the coagulation is slowed down by a factor W, where

$$W = 2\int_{2}^{\infty} e^{V/kT} \cdot \frac{ds}{s^2} \tag{5.23}$$

with $s = r/a$, where r is the distance between particles. As a useful approximation

$$W \sim \frac{1}{2\kappa a} \cdot e^{(V_{max}/kT)} \tag{5.24}$$

where V_{max} is the maximum in the potential energy curve. A typical curve relating the potential energy V to s is shown in Figure 5.15. The dependence of the stability ratio W on the electrolyte concentration and the valency is shown in Figure 5.16. The potential energy V can be evaluated from equations such as 5.40 and 5.42.

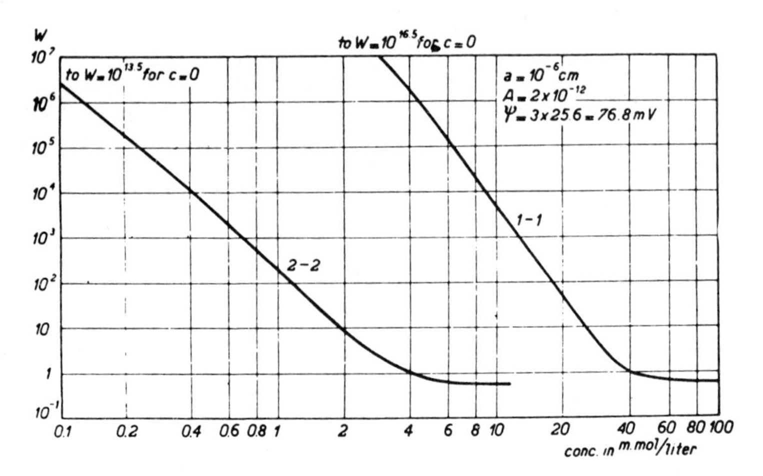

Figure 5.16. Stability ratio W as a function of the concentration of 1–1 and 2–2 valent electrolytes (1).

Theory of Verwey and Overbeek

In their development of the theory of the stability of lyophobic colloids, Verwey and Overbeek (13) treated the problem as an energy of interaction between particles. The total energy is not involved, but rather the free energy is a measure of the work that can be performed.

The free energy of the double layer, or of a system of double layers, is equal to

$$F_{\text{double layer}} = -\int_0^{\psi_0} Q\,d\psi \tag{5.25}$$

where Q is the charge of the surface, ψ is the potential difference between the two phases, arbitrarily assuming that the potential difference is zero at the zero point of charge, and ψ_0 is the double layer potential after equilibrium has been established. The equilibrium value of the surface potential ψ_0 is dependent upon the difference in chemical potential μ_i of the potential determining ions in the two phases,

$$\psi_0 = C + \frac{\Delta\mu_i}{z_i e} \tag{5.26}$$

where C is a constant, e is the electronic charge, and z_i is the valence.

The distribution of charge and potential in one double layer is shown in Figure 5.17. In the case of two parallel double layers that have approached each other until the diffuse parts of the double layers overlap is shown in Figure 5.17 and 5.18. This occurs when the distance between the surfaces is of the order of the thickness of a double layer, $1/\kappa$.

Where the two diffuse double layers bearing the same sign overlap each other, neither can develop fully. Consequently, the potential drop between the two surfaces is never as great as the potential between one surface and a large distance from the surface, where the potential is taken as zero. The

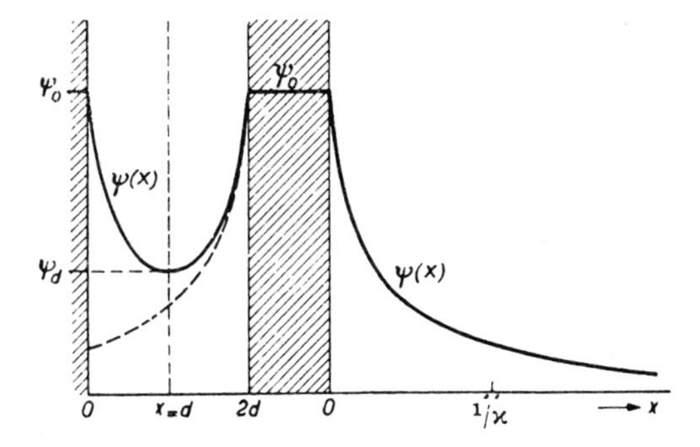

Figure 5.17. Comparison of the double layer potential between two plates with that of a single double layer (1).

potential on the two surfaces is the same as on an independent surface, since this potential is completely determined by the thermodynamic equilibrium expressed by Equation 5.26. Symmetry considerations require that the potential minimum occur just halfway between the two surfaces.

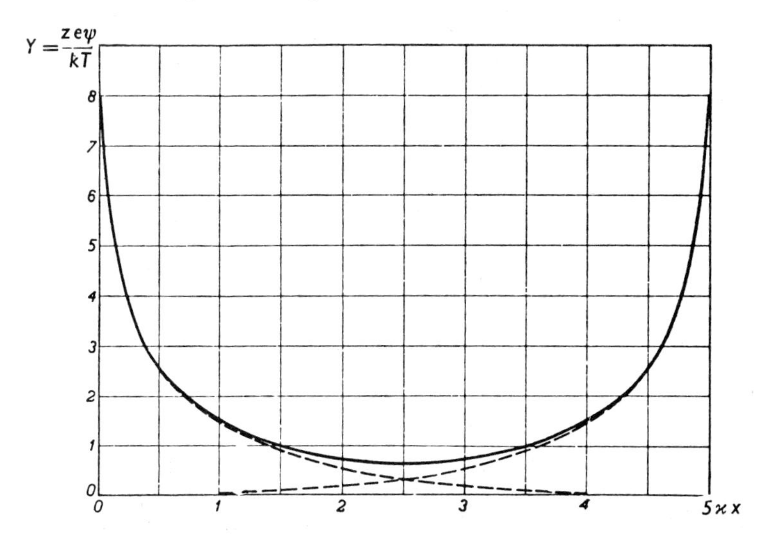

Figure 5.18. Electrical potential between two plates with small interaction, showing that the potential can be built up additively from that of two single double layers (dotted lines) (1).

For double layers of the Gouy-Chapman type, the potential between two surfaces can be described by the combination of the Poisson and Boltzmann equations. The equation is the same as that for a single double layer, except for the boundary conditions. For the single double layer, the boundary conditions are that the potential and its derivatives are zero at distances far from the surface. For two overlapping double layers, halfway between the surfaces the potential has a minimum value so that its first derivative is zero.

$$\frac{d^2\psi}{dX^2} = \frac{4\pi e}{D}\left(n_- z_- \mathbf{e}^{(z_- e\psi/kT)} - n_+ z_+ \mathbf{e}^{(-z_+ e\psi/kT)}\right) \tag{5.27}$$

For the interacting double layers, the boundary conditions are

$\psi = \psi_0$ for $x = 0$,

$\psi = \psi_d$ for $x = d$

$\frac{d\psi}{dx} = 0$ for $x = d$

The distance between the flat surfaces is $2d$.

The first integration gives

$$\frac{d\psi}{dx} = -\sqrt{\frac{8\pi kT}{D}}\sqrt{n_- \mathbf{e}^{z_- e\psi/kT} + n_+ \mathbf{e}^{-z_+ e\psi/kT} - n_- \mathbf{e}^{z_- e\psi_d/kT} - n_+ \mathbf{e}^{z_+ e\psi_d/kT}} \tag{5.28}$$

The negative sign is used for positive values of ψ when $0 \leqslant x < d$.

When the potential is small everywhere,

$$\frac{d\psi}{dx} = -\kappa\sqrt{\psi^2 - \psi_d^2} \tag{5.29}$$

Integration leads to

$$\psi = \psi_0 \frac{\cosh \kappa (d - x)}{\cosh \kappa d} \tag{5.30}$$

with

$$\psi_d = \psi_0 \frac{1}{\cosh \kappa d} \tag{5.31}$$

and

$$\kappa = \sqrt{\frac{4\pi e^2(n_+ z_+^2 - n_- z_-^2)}{Dkt}} \tag{5.32}$$

However, the approximation of small potentials does not lead to a result that shows the important feature of the stability of hydrophobic colloids. This is shown with high values of ψ. Here the value of Equation 5.28 is almost entirely determined by the positive powers of **e**. For simplicity the discussion is restricted to symmetrical electrolytes of valency z and concentration n. Then Equation 5.28 simplifies to

$$\frac{d\psi}{dx} = -\sqrt{\frac{8\pi nkT}{D}}\sqrt{2\cosh\frac{ze\psi}{kT} - 2\cosh\frac{ze\psi_d}{kT}} \tag{5.33}$$

When the surfaces are still far apart, there is small interaction between the double layers. Then the potential between the two surfaces is to a good approximation equal to the summation of two undisturbed double layers, as illustrated in Figure 5.18. For large values of κd,

$$\frac{ze\psi_d}{kT} = 8\,\frac{\mathrm{e}^{z/2} - 1}{\mathrm{e}^{z/2} + 1}\cdot \mathrm{e}^{-\kappa d} \tag{5.34}$$

The potential ψ_d halfway between the surfaces as a function of plate distance $2d$ is illustrated in Figure 5.19.

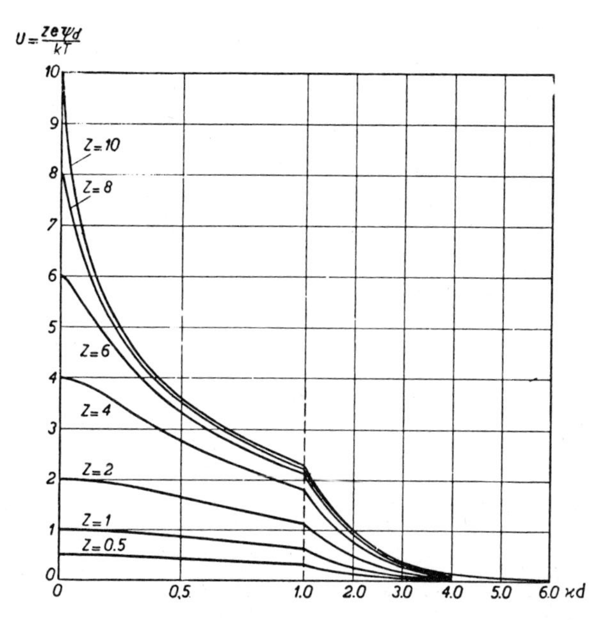

Figure 5.19. Potential ψ_d halfway between two plates as a function of the plate distance $2d$. $U = \dfrac{ze\psi_d}{kT}$, $Z = \dfrac{ze\psi_0}{kT}$

Examination of Figure 5.19 shows that the greater the value of ψ_d, the smaller the slope of the potential curve. The slope is also directly proportional to the surface charge. Thus the surface charge diminishes with increasing interaction and goes to zero when the two parallel surfaces come into contact. The expression for the double-layer charge is

$$\sigma = -\frac{D}{4\pi}\left(\frac{d\psi}{dx}\right)_{x=0} \tag{5.35}$$

and

$$\sigma = \sqrt{\frac{nDkT}{2\pi}}\sqrt{2\cosh\frac{ze\psi_0}{kT} - 2\cosh\frac{ze\psi_d}{kT}} \tag{5.36}$$

The potential energy of repulsion V_R per cm^2 cross section of the parallel surfaces is the amount of work that has to be performed against the forces arising from the interaction of double layers. This is the change in free energy when the two surfaces approach each other from an infinite distance of separation. The derivation leads to the result

$$V_R = \frac{n}{\kappa kT}(ze\psi_d)^2 \tag{5.37}$$

or

$$V_R = \frac{64nkT}{\kappa}\left(\frac{e^{z/2} - 1}{e^{z/2} + 1}\right)^2 e^{-2\kappa d} \tag{5.38}$$

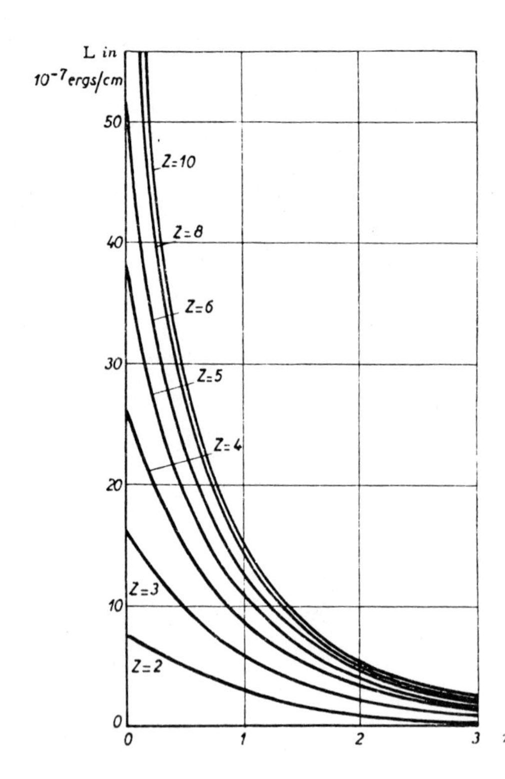

Figure 5.20. The repulsive potential V_R between two spherical particles. $Z = \frac{ze\psi_0}{kT}$. (1).

The solution to the problem for two spherical double layers is mathematically more difficult. For the case where the radius a of the spheres is very much greater than the thickness of the double layer, Derjaguin found

$$V_R = \frac{Da\psi_0^2}{2} \ln\ (1 + \mathbf{e}^{-\kappa H_0}) \tag{5.39}$$

where H_0 is the distance between the centers of the interacting spheres.

Figure 5.20 shows repulsive potential curves for two spherical particles obtained with the more exact expression of Verwey and Overbeek,

$$V_R = \frac{a}{z^2} L \tag{5.40}$$

where L is a function of $\frac{ze\psi_0}{kT}$ and κH_0.

The above treatment has been based on the Gouy-Chapman theory of the double layer. To apply the Stern correction, it is necessary to replace the surface potential ψ_0 with the Stern potential ψ_δ. This eliminates very high potentials in the double layer where it is questionable to apply the Bolzmann equation in its simple form.

Attractive Forces Between Particles

According to the theory of Verwey and Overbeek, in addition to the double layer repulsion there is an attraction between small particles. This attraction must be of a very general kind, since all lyophobic colloids are known to flocculate when the double layer repulsion is removed. London-van der Waals forces are considered to be responsible for the attraction between particles. Due to the additivity of dispersion forces, the range of the attraction is said to be of the order of colloidal dimensions.

London-van der Waals attraction is due to the interaction of dipole moments, the polarizing action of a dipole in one molecule on the other molecule, and a quantum mechanical effect leading to attraction between nonpolar atoms. The latter attraction, called the London force, results from a rapidly fluctuating dipole in a neutral atom at the zero-point energy of the electrons. The frequency of the fluctuation is of the order of 10^{15} or 10^{16} per second.

The three components of the attraction give rise to an attractive energy that varies inversely as the sixth power of the distance between the two atoms. For a large number of atoms, the London energy is approximately equal to the sum of the attraction between all atomic pairs.

The energy of attraction V_A per cm^2 between infinitely thick parallel flat plates is

$$V_A = -\frac{A}{48\pi d^2} \tag{5.41}$$

when the distance between the plates $2d$ is small compared with their thickness. Then $A = \pi^2 q^2 \beta$, with q the number of atoms per cm^3, and β the attraction between two atoms 1 cm apart.

The attraction decays slowly with increasing distance. For very small distances it assumes an infinitely large negative value. If a retardation correction to the London attraction is introduced, the attraction between two infinitely thick plates is found to decay as $1/d^3$ instead of $1/d^2$.

The problem of the attraction between spheres is more complicated. Neglecting the retardation correction, for small distances between two equal spheres of radius a

$$V_A = -\frac{Aa}{12} \cdot \frac{1}{H} \tag{5.42}$$

where H is the shortest distance between the spheres. For small distances the attraction between spheres decays still slower than between flat plates.

Total Interaction Between Particles

The total interaction between colloidal particles is obtained by the summation of the attraction and the repulsion curves. This is illustrated in

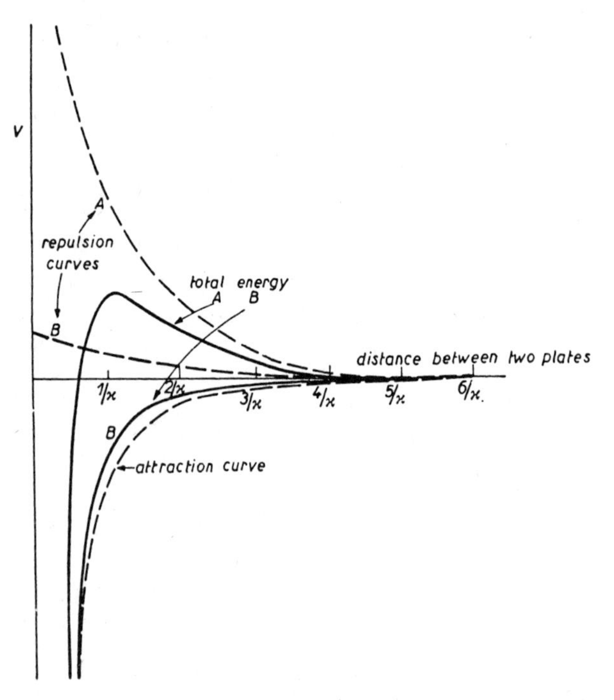

Figure 5.21. Total energy of interaction curves obtained by the summation of one attraction curve with two repulsion curves of different heights (1).

Figure 5.21 for the interaction between flat plates. The character of the total interaction curve can be deduced from the properties of the two forces. The repulsion is an exponential function with a range of the order of the thickness of the double layer. It remains finite for all values of the distance between the plates. The attraction decreases as an inverse power of the distance. For very small distances it has very large negative values. The attraction force will predominate at very small and large distances. At intermediate distances the repulsion may predominate, depending upon the actual values of the two forces.

Figure 5.21 shows the two general types of curves for total energetic interaction. Where repulsion is greater than attraction at an intermediate distance, there is a maximum at these distances and a minimum at larger

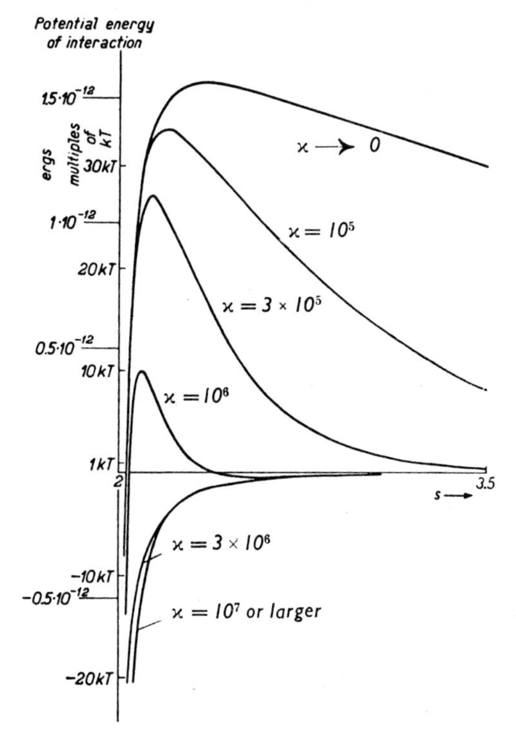

Figure 5.22. The effect of the concentration of electrolyte κ on the total potential energy of interaction of two spherical particles of radius $a = 10^{-5}$ cm; $A = 10^{-12}$ erg; $\psi_0 = 25.6$ mv. (1).

distances. The other curve shows a monotonic decrease of the energy with decreasing distance. The maxima and minima occur at distances of the order of the thickness of the double layer. With decreasing concentrations of electrolyte, or increasing values of $1/\kappa$, the maximum is displaced to larger distances.

Repulsion and attractive curves can also be combined for spherical particles to give the total interaction. The interaction energy per pair of particles can be compared with the energy of Brownian motion. When the energy maximum is large compared with kT, the system should be stable. A maximum of kT or lower should be easily overcome by Brownian motion, leading to flocculation.

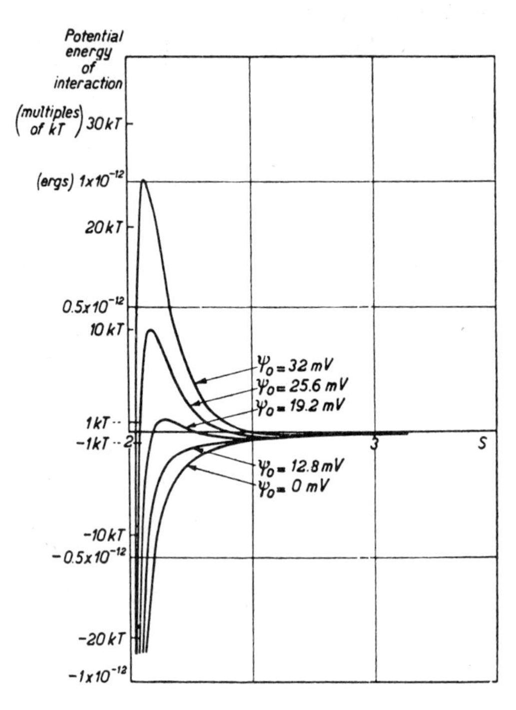

Figure 5.23. The influence of the surface potential ψ_0 on the total potential energy of the interaction of two spherical particles of radius $a = 10^{-5}$ cm; $A = 10^{-12}$ erg; $\kappa = 10^6$ cm^{-1} (1).

Figure 5.22 shows interaction curves for various values of κ. As the electrolyte concentration κ is increased, the height of the potential energy maximum is reduced and shifted to closer distances between the spheres. Increase in surface potential increases the potential maximum, as shown in Figure 5.23.

VALIDITY OF VERWEY-OVERBEEK THEORY

Schulze-Hardy Rule

An important test of the validity of a theory dealing with the stability of hydrophobic colloids is that it accounts for the rule of Schulze and Hardy. This rule is based on observations concerning the flocculating values of various electrolytes for colloidal dispersions. The flocculation value of arsenic trisulfide sols is determined by mixing the sol in a series of test tubes with various concentrations of electrolyte. After two hours the mixtures are again agitated, and a half hour later the flocculation value is taken as that concentration of electrolyte for which the supernatant liquid is just clear and colorless.

A gold sol is red when stable, but turns blue at a certain degree of flocculation. The flocculation value is that concentration of electrolyte that produces the first indication of a blue color five minutes after mixing.

Flocculation values are strongly dependent upon the valency of the ions that are charged oppositely to the sol. In general, these values are independent of the specific nature of the ions and depend only slightly on the concentration and nature of the sol.

The rule of Schulze and Hardy is that flocculation values for ions of opposite charge to the sol range from 25 to 150 millimolls/1 for monovalent ions, 0.5 to 2 millimolls/1 for divalent ions, and 0.01 to 0.1 millimoll/1 for trivalent ions.

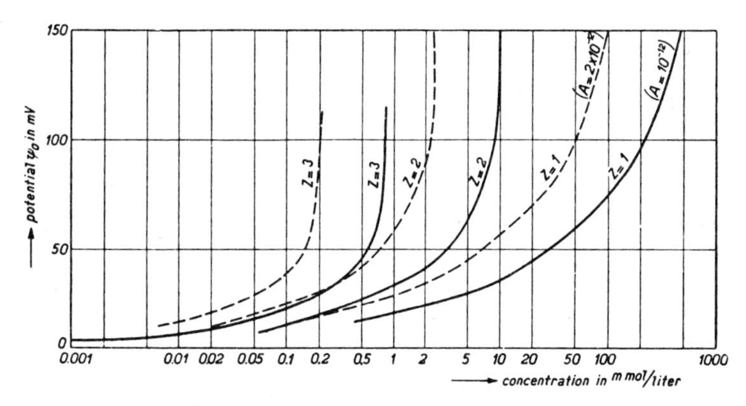

Figure 5.24. Conditions of limiting stability for two values of the London-van der Waals constant A and the valences z = 1, 2, and 3. The system is stable above and to the left of each curve, and flocculated to the right and below (1).

The curves drawn in Figure 5.24 are based on the theory of Verwey and Overbeek. All values above and to the left of each curve represent a stable system, while the sols flocculate for values below and to the right of each curve. The curves represent flocculation values for mono-, di- and tri-valent electrolytes with sols having different surface potentials. In all cases, the London-van der Waals constant A is taken as 2×10^{-12}. Thus, for $\psi_0 = 100$ mv, the flocculation values are 50, 2 and 0.2 millimoll/1 for mono-, di- and trivalent electrolytes, respectively. Both the ratios and the actual values are in good agreement with the rule of Schulze and Hardy.

Surface Potential

The Verwey-Overbeek theory predicts that if the surface potential is low, the colloidal dispersion is unstable. This has been observed experimentally. When the surface potential is high, the flocculation value does not appear to depend upon the exact value of the potential. Values obtained are frequently in reverse of the expected direction.

Specific Effects of Counter-ions

While the valency of the counter-ion is the predominant factor in flocculation, different ions of the same valency are frequently found to have different flocculation values. To account for the specific effect of counter-ions, it is necessary to introduce a refinement like that of Stern. This theory accounts for the dimensions of the ions and for specific adsorption. One effect is the lowering of the potential in the diffuse double-layer. This potential, otherwise dependent only on the amount of potential-determining ions, is lowered with an increase in the electrolyte concentration.

The potential ψ_0 depends inversely on the thickness of the molecular condenser. If the counter-ions are large, the molecular condenser will be thick, giving rise to a lower ψ_δ. Consequently, larger counter-ions should have smaller flocculation values. Further, the larger counter-ions have greater polarizability and should be more strongly adsorbed. This is observed for negative sols. The usual order of flocculating values is

$$\mathrm{Li} > \mathrm{Na} > \mathrm{K} > \mathrm{Rb} > \mathrm{Cs}$$

and

$$\mathrm{Mg} > \mathrm{Ca} > \mathrm{Sr} > \mathrm{Ba}$$

For many positive sols, however, the order is reversed, with the larger ions having the greater flocculating value:

$$1 > \mathrm{Br} > \mathrm{Cl}$$

Rate of Flocculation

The theory of the stability of colloidal dispersions can be tested by examining the effect of electrolyte concentration on the rate of coagulation. The retardation factor W, which gives the ratio between the rates of rapid and slow flocculation, was shown in equation 5.23 as

$$W = 2 \int_2^\infty \mathrm{e}^{V/kT} \cdot \frac{ds}{s^2} \tag{5.43}$$

In Figure 5.16, it is shown that in the major region of slow flocculation log

W is a linear function of log c, where c is the electrolyte concentration. This linear relation has been observed by several investigators.

Secondary Minimum

A schematic curve for the potential energy of interaction is shown in Figure 5.25. A characteristic feature of such a curve is the presence of a secondary minimum at a relatively large distance between the particles. This minimum is due to the longer range of the van der Waals attractive force as compared with the repulsion.

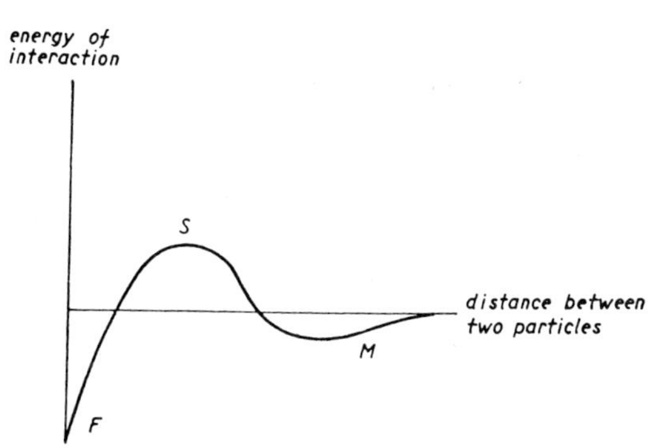

Figure 5.25. Schematic potential energy curve, showing the primary minimum F, the secondary minimum M and the maximum S (1).

If the minimum is several times kT deep, it should overcome the effect of Brownian motion and give rise to coagulation. The character of this coagulation would be different from that in the primary minimum. It would be completely reversible and equilibrium distances would be of the order of several times the thickness of the double layer.

For colloidal particles of radius 10^{-6} cm and smaller, the secondary minimum is never deep enough when the maximum is large enough to prevent normal flocculation. If the particles are larger, the secondary minimum may cause observable effects.

A number of observations can be explained on the basis of flocculation in a secondary minimum. One example is the reversible separation into two phases of colloidal systems containing anisodimensional particles. When an iron oxide sol is sufficiently concentrated and the electrolyte content is not so high as to cause irreversible flocculation, the sol separates into two phases. One is a dilute isotropic phase and the other is a more concentrated birefringent one. Examination of the concentrated phase by X-ray measurements shows that the particles are separated by the dispersion medium. The particles are aligned as parallel rods with the distance between the rods varying from 180 to 600Å. The distance varies with pH and salt content, showing that the repulsive force is of an electrical nature.

References

1. Kruyt, H. R., "Colloid Science", Vol. 1, New York, Elsevier Publishing Company, 1952.
2. Frumkin, A., *Trans. Faraday Soc.* **36**, 117 (1940).
3. Bikerman, J., *Phil. Mag.* (7) **33**, 384 (1942).
4. Grahame, D. C., *Chem. Revs.* **41**, 441 (1947).
5. Lippmann, G., *Ann. Physik* **149**, 546 (1873).
6. Kruyt, H. R., *Kolloid-Z.* **22**, 81 (1918).
7. Henry, D. C., *Proc. Roy. Soc.* (London) **133**, 106 (1931).
8. Paine, H. H., *Trans. Faraday Soc.* **24**, 412 (1928).
9. Bikerman, J. J., *Z. physik. Chem.* **A171**, 209 (1934).
10. Hermans, J. J., *Phil. Mag.* (7) **26**, 650 (1938).
11. Smoluchowski, M. von, *Physik Z.* **17**, 557, 585 (1916); *Z. physik. Chem.* **92**, 129 (1917).
12. Fuchs, N., *Z. Physik* **89**, 736 (1934).
13. Verwey, E. J. W., and Overbeek, J. Th. G., "Theory of the Stability of Lyophobic Colloids", New York, Elsevier Publishing Company, 1948.

CHAPTER 6

Insoluble Monolayers at Liquid Interfaces

In his review of modern film techniques, Trurnit (1) noted that the first record of the application of the methods by which insoluble mono-molecular layers are studied seems to be a 4000 year-old cuneiform inscription found in ancient Babylon. A priest dropped sesame oil in a flat wooden bowl filled with water and looked at the surface against the rising sun. The colors and movement of the oil film were thought to indicate the course of future events.

The fact that vegetable oils are effective in calming waves was known to the Egyptians. Benjamin Franklin described experiments he conducted on a pond at Clapham near London. He was greatly impressed by the visible speed and the force of the spreading oil. He observed that on a windy day a teaspoonful of oil was sufficient to smooth a half acre of water. Trurnit noted that this corresponds to 20Å for the thickness of the oil film, which is the correct order of magnitude for a monolayer.

Laboratory studies of thin layers at liquid surfaces date from the experiments made by Agnes Pockels (2). Rayleigh (3, 4) extended this work and demonstrated that the size of molecules could be determined by film techniques. He showed that they were in accord with the atomistic theory of matter. Langmuir (5) devised the well-known film balance that has been used, with modifications, for large numbers of monolayer experiments. Various improvements in the instrument have been made by Adam (6), Harkins (7), Dervickian (8), and others.

Film Balance

The film trough is a long rectangular vessel. Harkins used a shallow trough, about 6 mm deep, in which the water rises about 6 mm above the level of its edges. The trough is in an air thermostat and is hollow so that thermostated water can be circulated through it. Usually, the trough is covered with a film of paraffin. The surface of the water in the trough is cleaned by "sweeping" with a transverse barrier of paraffined glass or stainless steel.

Any device that measures directly the difference between two surface tensions is considered a film balance. This difference is usually between the surface tension of pure water γ_w and that of the film-covered surface γ_f, as illustrated in Figure 6.1. It is defined as the film pressure

$$\pi = \gamma_w - \gamma_f \tag{6.1}$$

The vertical type of film balance is due to Wilhelmy (9). The film-covered surface of the water in the trough exerts a vertical downward pull on a solid sheet partially immersed in the water. The contact angle made by the liquid on the solid must be zero. Since the sensitivity of the method varies inversely with the thickness of the solid, an extremely thin sheet of platinum is frequently used. The sheet is hung from one arm of an analytical balance. The decrease in the downward pull on the sheet is measured by the deflection of the beam of the balance.

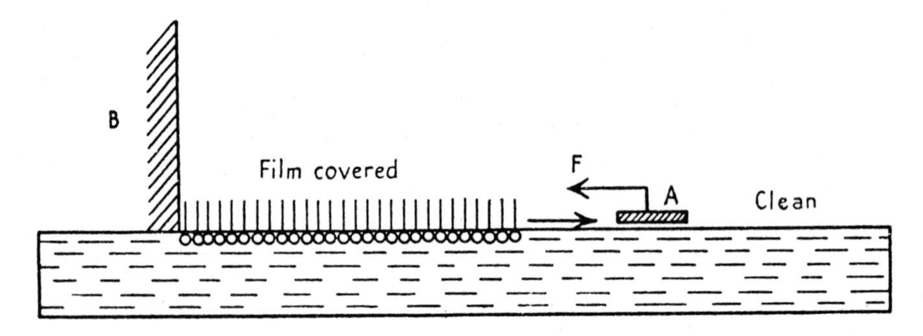

Figure 6.1. Diagramatic representation of film pressure. B is a fixed barrier, A is a floating barrier. Force F is the surface pressure (6).

The horizontal-float film-balance is more commonly used for the study of insoluble monolayers. The clean surface of the water is separated from the film-covered surface by a floating barrier of paraffined metal, mica or glass. In practice, the trough is first filled with the pure liquid. The surface on both sides of the float is swept clean. A known quantity of the filming material is applied from solvent to the surface in front of the float. After evaporation of the solvent, the sweep is moved forward in stages to obtain data concerning the film pressure and the area of the film.

Orientation of Molecules in Monolayers

Langmuir suggested that long-chain fatty acids, such as oleic acid, are adsorbed at an air-water interface with the polar group directed toward the water phase and the hydrocarbon chains toward the air. He showed that monolayers of saturated fatty acids can be compressed to a limiting area that was independent of the hydrocarbon chain length. This supported his view concerning the orientation of fatty acid monolayers as depicted in Figure 6.2. Subsequently, Lyons and Rideal (10) suggested that at the limiting surface area the molecules are not normal to the liquid surface but tilted.

When smaller concentrations of fatty acids are present on a water surface, it was thought that the molecules were lying flat on the surface. However, it has been shown that the limiting areas are not sufficiently great. Instead, the molecules are partially linked together (11, 12).

More recently, films of fatty acids on water were transferred from a film balance to a collodion support, shadowcast with chromium, and examined

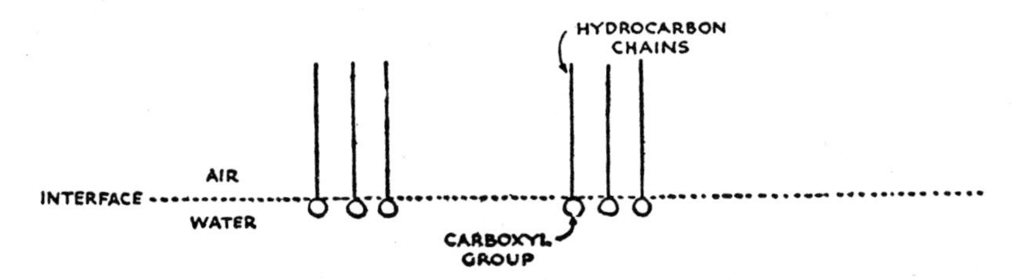

Figure 6.2. Diagram showing orientation of fatty acid molecules at air-water interface according to Langmuir (16).

in an electron microscope. Monolayers of *n*–hexatriacontanoic acid, which give pressure-area curves similar to stearic acid were examined. With this fatty acid the film collapses at 58 dynes/cm as compared with 42 dynes for stearic acid.

At a film pressure of 15 dynes/cm, the film thickness was close to 50Å. This would correspond to a vertically oriented monolayer of the 36 carbon acid. The electron micrograph showed that the partially compressed monolayer was composed of many islands or aggregates of irregular shape and size. When the film pressure was increased to 25 dynes per cm, the monolayer became the continuous phase with small patches of bare portions. It was suggested that the bare portions may contain film molecules that were less closely packed.

After further compression of the film beyond collapse, several long fiber-like structures appear. These are about 100Å or two molecules thick and rest on a continuous monolayer substrate (13).

Ryan and Shepard (14) studied monolayers of C^{14} tagged calcium stearate. Autoradiographs of the stearate monolayer deposited on aluminum at low film-pressures showed islands of an expanded liquid phase surrounded by areas of gaseous phase.

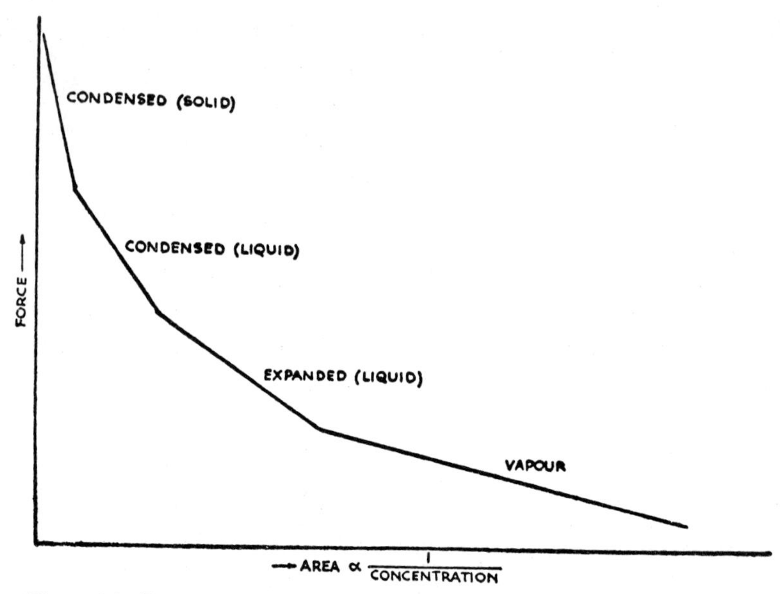

Figure 6.3. Diagrammatic force-area curve for long chain fatty acid films (16).

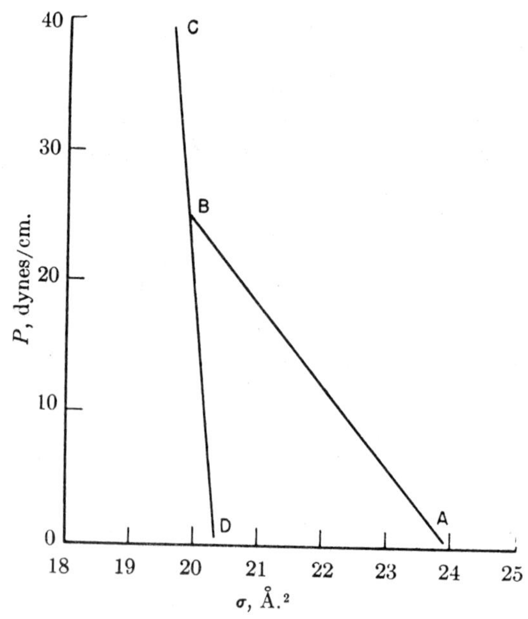

Figure 6.4. Typical pressure-molecular area isotherm for the two phases in which compressed fatty acid monolayers exist. *A–B* represents the liquid condensed phase and *D–C* the solid phase (17).

temperature at which the measurements were made is above the critical temperature for the liquid-expanded film.

Straight-chain hydrocarbon derivatives with polar end-groups, such as alcohols or acids, can be made to exhibit these various phases by a suitable selection of chain length and temperature. One CH_2 group is roughly equivalent to a 5° change in temperature. With large, bulky end-groups, a liquid-condensed rather than a solid film will be observed at high film pressures.

Where there is more than one polar group in the molecule, the film will tend to be of the liquid-expanded type. Considerable pressure is required to overcome the attraction of the second polar group and allow vertical orientation of the hydrocarbon chain. Molecules with more than one hydrocarbon chain also tend to give expanded films.

Film Potential

Surface potential measurements in combination with pressure-area data provide important information concerning the distribution and orientation of molecules at an interface. The difference in potential between the surface of a liquid and a metal sheet just above the surface can be determined if the metal is coated with an α emitter such as polonium. An electrode is inserted in the solution. A Lindemann or Compton electrometer and potentiometer are used to measure the potential difference. The apparatus is illustrated in Figure 6.7. An alternative method employs the vibrating electrode of Zisman (21).

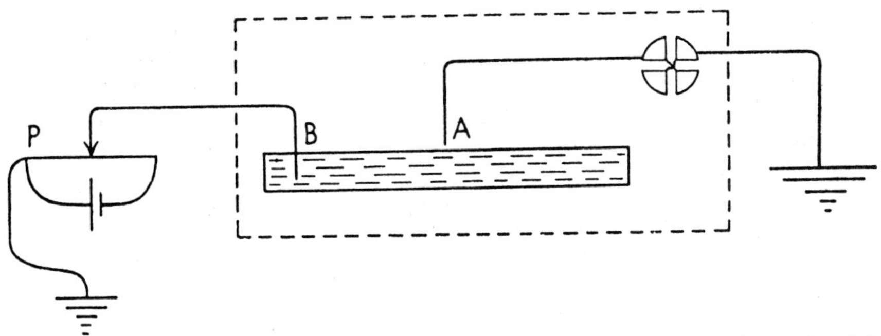

Figure 6.7. Schematic representation of apparatus for measuring the surface potential of a film. *A* is the "air electrode", an insulated metal wire held above the surface of the liquid. *B* is a reversible electrode, and *P* is a potentiometer (6).

The surface or film potential is the difference in potential obtained with the electrode over a clean water surface and over a surface covered with a monomolecular film. For a liquid condensed monolayer of stearic acid, the change of potential is close to 400 mv.

The potential difference ΔV between two parallel metal plates of an ordinary parallel-plate condenser is

$$\Delta V = \frac{4\pi\sigma d}{D} \tag{6.4}$$

where σ is the surface charge density, d is the distance between plates and D is the dielectric constant of the medium.

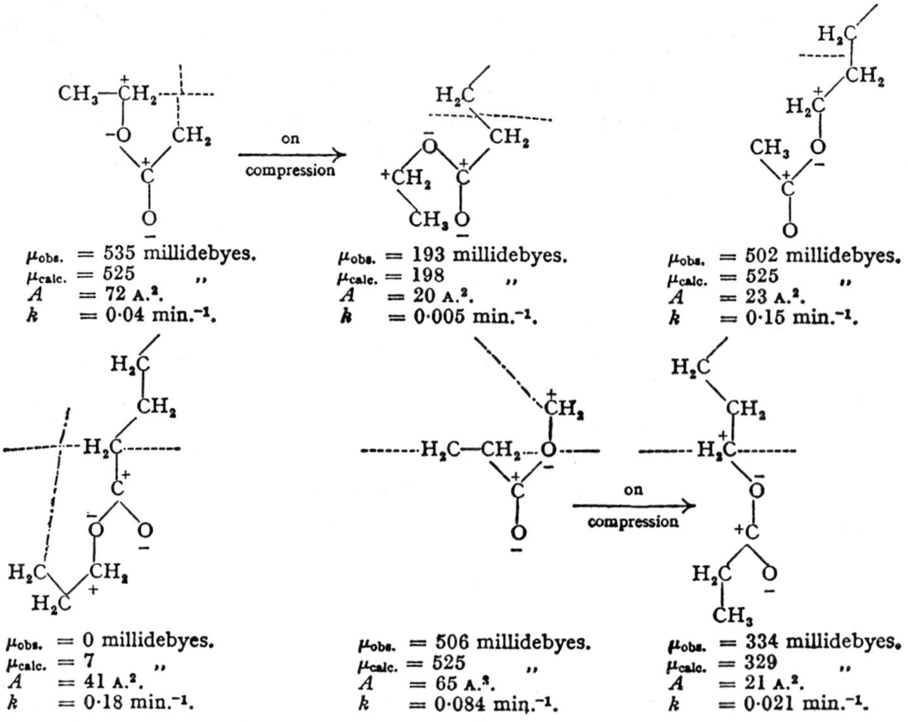

on compression

$\mu_{obs.}$ = 535 millidebyes.
$\mu_{calc.}$ = 525 ,,
A = 72 A.^{2}.
k = 0·04 min.$^{-1}$.

$\mu_{obs.}$ = 193 millidebyes.
$\mu_{calc.}$ = 198 ,,
A = 20 A.^{2}.
k = 0·005 min.$^{-1}$.

$\mu_{obs.}$ = 502 millidebyes.
$\mu_{calc.}$ = 525 ,,
A = 23 A.^{2}.
k = 0·15 min.$^{-1}$.

on compression

$\mu_{obs.}$ = 0 millidebyes.
$\mu_{calc.}$ = 7 ,,
A = 41 A.^{2}.
k = 0·18 min.$^{-1}$.

$\mu_{obs.}$ = 506 millidebyes.
$\mu_{calc.}$ = 525 ,,
A = 65 A.^{2}.
k = 0·084 min.$^{-1}$.

$\mu_{obs.}$ = 334 millidebyes.
$\mu_{calc.}$ = 329 ,,
A = 21 A.^{2}.
k = 0·021 min.$^{-1}$.

Figure 6.8. Orientation of long-chain ester molecules at a water surface. μ calc. is the dipole moment calculated for the orientation shown. A is the area per molecule. k is the hydrolysis-rate constant for the monolayer (16).

The dielectric constant of air is unity. Further, $\sigma = ne$, where n is the number of electronic charges per sq cm and e is the value for an electronic charge. Then,

$$\Delta V = 4\pi n\,(ed) = 4\pi nm \tag{6.5}$$

with m the product of the electronic charge and the distance between the plates.

The dipole moment of a molecule μ is also equal to the product of the electronic charge and the distance between the charges

$$\mu = ed \tag{6.6}$$

Thus, while m would appear to equal μ, values obtained by surface-potential measurements may be as little as one-tenth of those obtained by standard methods for determining the dipole moments of polar molecules. The explanation given is that the dipole does not lie perpendicular to the surface. Then,

$$m = \mu \cos\theta \tag{6.7}$$

where θ is the angle made by the dipole with the surface normal, as illustrated in Figure 6.8. Then

$$\Delta V = 4\pi n\mu \cos\theta \tag{6.8}$$

While this equation is not exact, it is useful in obtaining information concerning the relative orientation of dipoles at the surface and in studying surface reactions.

Surface Viscosity

A number of methods are available for measuring the viscosity of surface films. One method employs a canal viscometer, where the film flows through a channel formed by parallel plates (22). Other methods include the determination of the damping of the oscillations of a torsion pendulum (23), and the measurement of the torque developed on a wire attached to a disc that is in contact with the surface of a liquid flowing at a constant rate (24).

The viscosity of a monolayer is dependent upon the closeness of packing of the molecules in the film. The viscosity of gaseous films is too low to measure. Liquid-expanded films have low viscosities, but higher than those of the gaseous films. Liquid-condensed and solid films have high viscosities. The viscosity of liquid films increases rapidly with the length of the hydrocarbon chain. Monolayers of long-chain alcohols are from 17 to 33 times more viscous than those of the corresponding acids (20).

Evaporation of Water Through Monolayers

There are many areas of the world where rainfall is sparse. Conservation of water can be accomplished economically by covering the surface of reservoir water with a suitable monolayer to retard evaporation. The permeability of monolayers is also important in problems of biology and physics. The cosmetic industry is concerned with the application of thin films to the skin to prevent drying and chafing.

Langmuir and Schaefer (25) employed a solid desiccant above the surface of the trough to determine the rate of water evaporation. They found that the resistance to evaporation through fatty-acid monolayers spread from benzene increased with film pressure. Similar findings had been reported by Sebba and Briscoe (26).

In contrast to these observations, Archer and La Mer (17) determined that when the fatty-acid monolayers were spread from solution in petroleum ether, the resistance to evaporation was independent of film pressure. However, to obtain results showing that resistance was independent of film pressure, it was necessary to have a scrupulously clean surface, and to apply the solution rapidly from an initial pressure of above 10 dynes/cm. Otherwise,

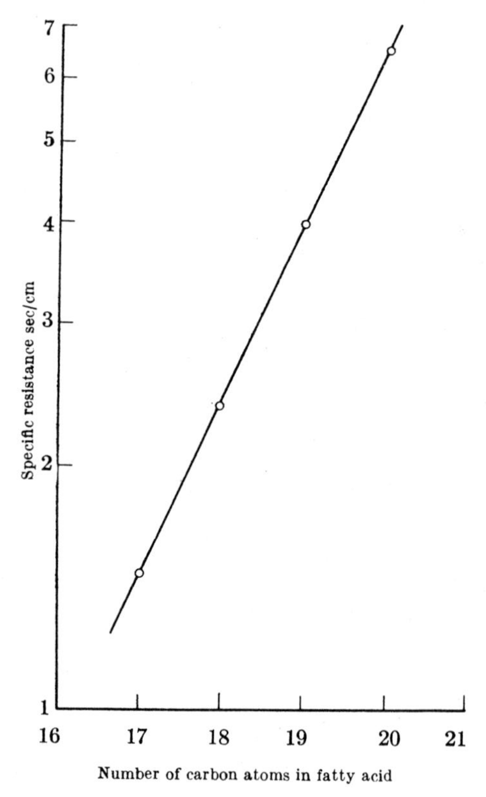

Figure 6.9. The logarithm of specific resistance in the liquid condensed phase as a function of the chain length of the monolayer molecules (17).

solvent molecules or small foreign molecules are occluded in the monolayer, offering a low-resistance path for water evaporation. These foreign molecules are expelled on compression of the film, thus explaining the effect of film pressure on the evaporation rate. Benzene would be occluded more strongly than petroleum ether, both because of its lower volatility and its greater interaction with water and the long-chain molecules.

Figure 6.9 shows that the specific resistance to evaporation r increases with the chain length of the fatty acid. All films were in the liquid-condensed phase. Where measurements were extended to the solid phase, the results indicated that resistance increased with pressure. The results were not reproducible due to instability of the films. These quickly collapsed soon after passing the transition point between the L_c and S phases. When the fatty acid film was spread on a subphase of $10^{-4}M$ $CaCl_2$ and $10^{-3}M$ $KHCO_3$ at pH 8, the resulting S phase showed a greater resistance than on water. The resistance increased with film pressure.

The surface film acts as an energy barrier E to the evaporation of water. Since resistance can be represented by the expression

$$r = \text{constant } \mathbf{e}^{E/RT} \qquad (6.9)$$

the log r should vary inversely with the absolute temperature. This has been confirmed. Fatty-acid monolayers decrease the rate of evaporation by a factor of about 10^4.

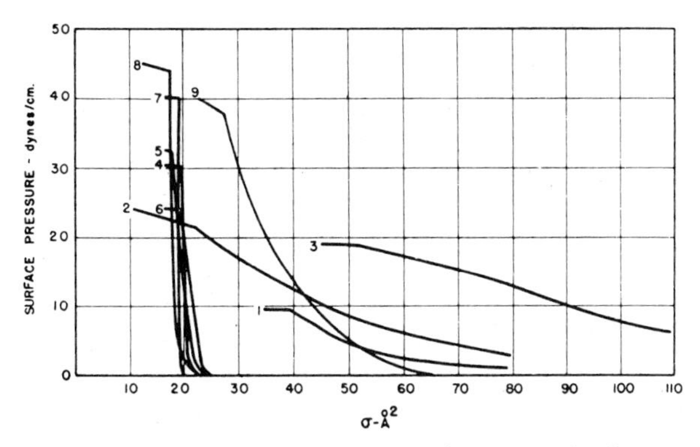

Figure 6.10. Surface pressure-surface area isotherms showing compressible (no. 1, 2, 3, 9) and non-compressible films (4-8) for substances of Table 6.1 (27).

Additional information relating the properties of monolayers to resistance to evaporation were obtained by Rosano and La Mer (27). Figures 6.10, 6.11 and 6.12 are from their work. The hydrocarbon-chain compounds that gave compressible films offered negligible resistance to evaporation. Ethyl palmitate, ethyl linoleate, and ethyl elaidate gave compressible films, while the fatty-acid and fatty-alcohol films were noncompressible.

High surface-viscosity films do not necessarily result in high resistance. However, the curves for resistance and viscosity have much the same slope. Discontinuities in the surface viscosity—surface pressure curve have been ascribed to changes in surface phases (28, 29), suggesting that a change in resistance reflects a change in the structure of the film.

The curves also show that at low film pressures, resistance does increase with pressure for all monolayers.

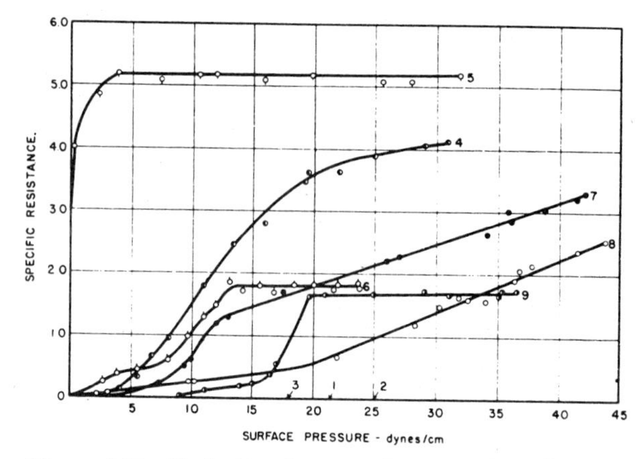

Figure 6.11. Reduction in rate of evaporation of water through a monolayer by substances listed in Table 6.1 in terms of the specific resistance to evaporation as a function of the surface pressure. Note the very low resistance of the esters 1, 2, 3 and the high resistance of saturated fatty acids—stearic and arachidic (27).

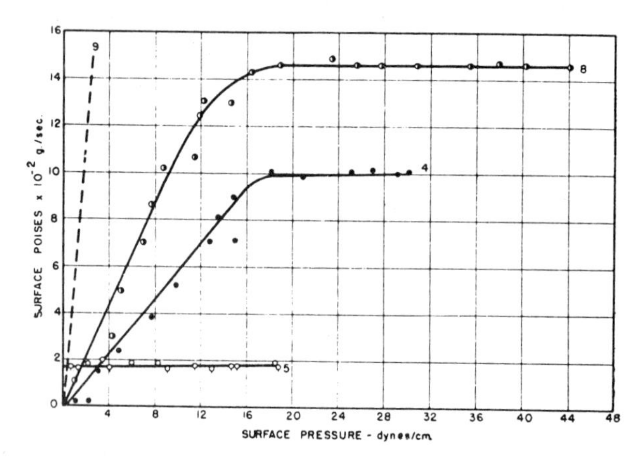

Figure 6.12. Surface viscosity (in 10^2 poises) as a function of surface pressure for substances of Table 6.1 (27).

The fluorinated alcohol was the only compound studied where resistance to evaporation was obtained with a compressible film. This surface film also had the highest viscosity. These results suggested to Rosano and La Mer that resistance to evaporation depends upon two effects.

(1) Cohesive forces in the monolayer. The compressible films are not cohesive. Increased cohesion in the monolayer of the fluorinated compound is evident from the high surface viscosity. The effect of chain length on resistance was shown earlier.

(2) Adhesive forces between the monolayer and the subphase.

TABLE 6.1.

1	Ethyl palmitate	$C_{15}H_{31}COOC_2H_5$	M.p. 24°
2	Ethyl linoleate	$C_{17}H_{31}COOC_2H_5$	n^{27}D 1.4573
3	Ethyl elaidate	$C_{17}H_{33}COOC_2H_5$	$n^{28.9}$D 1.4519
4	Ethyl stearate	$C_{17}H_{35}COOC_2H_5$	M.p. 34.4°
5	Arachidic acid	$CH_3(CH_2)_{18}COOH$	M.p. 76°
6	Stearic acid	$CH_3(CH_2)_{16}COOH$	M.p. 68.7°
7	1-Octadecanol	$CH_3(CH_2)_{16}CH_2OH$	M.p. 58.2°
8	Cetyl alcohol	$CH_3(CH_2)_{14}CH_2OH$	M.p. 49°
9	1, 1, 13-Trihydro-perfluorotridecyl alcohol	$H(CF_2)_{12}CH_2OH$	M.p. 110°

The products 1, 2, 3, 4, 5, 6, 7 and 8 were dissolved in redistilled Merck petroleum ether (b.p. 45–55°) and the product 9 in methyl alcohol (boiling range 64.2–64.7°) (27).

Energy Relations in Monolayer Formation

When a surface-active, very slightly-soluble substance in the form of a crystal or liquid drop is placed on a clean water surface, it will spread to form a monolayer. The equilibrium spreading pressure π_e is defined as the point where the change in free surface-energy of the system vanishes and spreading ceases.

$$\pi_e = \gamma_w - \gamma_f \tag{6.10}$$

where γ_w is the surface tension of pure water or of an aqueous solution and γ_f is the surface tension of the surface covered with an insoluble monolayer in equilibrium with the crystal or liquid lens.

The process of spreading is accompanied by a change in the heat content of the system in accordance with the Clapeyron equation

$$Q_s = T\,(\partial\pi_e/\partial T)_{Af}\ \ (A_e - A_s) \tag{6.11}$$

where Q_s is the molar latent heat of spreading, A_e is the molar area of the surface film, and A_s is the molar area occupied by the bulk phase, which may be made negligibly small relative to A_e (30).

The change in molar free energy ΔF_s and molar enthalpy ΔH_s on spreading is given by the relations

$$\Delta F_s = - \int_{A_s}^{A_e} \pi_e dA_f = - \pi_e(A_e - A_s) \tag{6.12}$$

and

$$\Delta H_s = \int_{A_s}^{A_e} h_s dA_f \tag{6.13}$$

where

$$h_s = - \left[\frac{\partial (\pi_e/T)}{\partial(1/T)} \right]_{A_f,\Sigma} \tag{6.14}$$

with $\sum$ denoting the total surface area of the system and A_f the film area. Also,

$$Q_s = \Delta H_s - \Delta F_s \tag{6.15}$$

Boyd and Schubert (18, 31) studied the variations of π_e and A_e with temperature and arrived at these conclusions.

(1) The general rule for adsorption is that all such processes are exothermic. However, monolayer formation by spreading from crystalline fatty acids is accompanied by the adsorption of a latent heat and by enthalpy and entropy increases.

(2) Spreading from the melted compound is accompanied by the evolution of heat and by decreases in entropy and enthalpy.

(3) Similar to heats of fusion, latent heats of spreading are smaller for odd-numbered than for even-numbered carbon-atom acids. The energy of binding in the crystal per carbon atom is less for the odd-numbered carbon acids.

(4) These results suggest that the molecules in the monolayer are more "ordered" than in the melt, but less than in the crystal. While X-ray diffraction measurements indicate closer packing of hydrocarbon chains in the melt than in the monolayer, the chains are arranged in a single plane in the monolayer, but randomly distributed in the melt. The result is a decrease in entropy on passing from the melt to the monolayer.

Transfer of Monolayers

In 1920 Langmuir (32) dipped a clean glass plate into water covered by a monolayer of oleic acid. Upon subsequent removal of the plate, he found that it was covered with a monolayer of oleic acid. While working in Langmuir's laboratory in 1934, Miss Blodgett (33, 34) found that any number of layers could be deposited on the glass plate by successive dippings.

During deposition, the monolayer on the aqueous subphase is preferably maintained under constant surface pressure. A piston oil behind a moving thread is frequently used to maintain constant pressure. The films may be transferred to the plate on entering the water only ("*X*" film), on both entering and leaving ("*Y*" film), or on leaving only ("*Z*" film). Generally, a substance that deposits as an "*X*" film can be made to deposit in "*Y*" fashion by increasing the surface pressure. Regardless of the method of deposition, the plate withdrawn from the water has a hydrophobic surface.

The area of the monolayer removed from the water is always equal to the apparent area of the plate. Holes and grooves in the plate are bridged by the film. A multilayer several thousand molecules thick can be formed by this technique.

Germer and Storks (35) described experiments where layers of stearic acid or barium stearate were deposited. Using a polished chromium plated surface, the block was always wetted by water on its first immersion and one layer of molecules was deposited as it was withdrawn from the surface. Two additional layers were deposited on each successive dip and withdrawal. With Resoglaz foil, which is not wetted as it passes into the water, two layers of molecules were deposited on the first and successive dips and withdrawals.

Electron diffraction studies showed that for the first layer, the axis of the hydrocarbon chains is normal to the surface for barium stearate. With stearic acid, the chains are slightly tilted from the normal (35).

Mixed Monolayers

Mixed surface films have been given considerable study because of their possible importance in biological systems. A number of investigators have presented evidence for the formation of "complexes" or molecular associations when the components of the monolayers were present in certain molecular proportions.

When a monolayer is spread from a mixture on water, the partial molecular areas are frequently found to be different from the molecular areas of the pure components. This situation is analogous to the variation of partial molar volumes with the composition in bulk phases, which is due to the interaction of the different molecules and to changes in their arrangement.

Dervichian (36) discussed two different cases in which changes in molecular areas were observed to occur at stoichiometric proportions of the constituents. The first case is due to a change of phase. An example is found in mixtures of myristic acid and trimyristin (37). At 32°C, both the trimyristin and the myristic acid monolayers are in the liquid state. The mean molecular area found for a mixture is almost precisely additive.

At 15° or 20°C, the pure trimyristin monolayer is solid while that of myristic acid is liquid. Qualitatively, the mixed monolayers are solid when the mole fraction of myristic acid in the monolayer is less than 0.5. The mean area per hydrocarbon chain — area of the mixed monolayer divided by the total number of hydrocarbon chains — is shown plotted against the mole fraction in Figure 6.13. Where the mole fraction of myristic acid is 0.5 or less, the monolayer is solid and the mean area per chain is constant at about 21 Å².

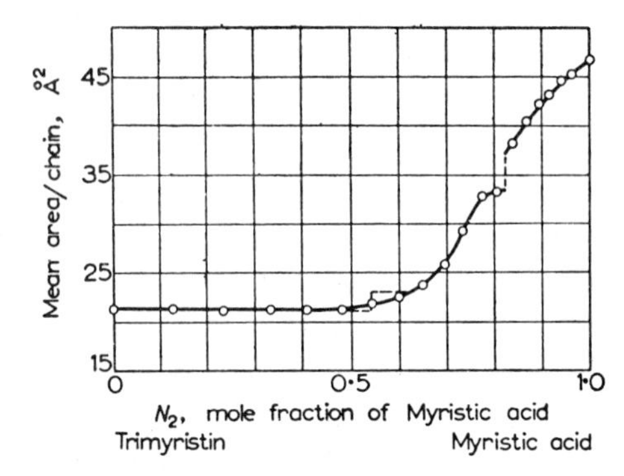

Figure 6.13. Variation of the mean area per chain at 15° C as a function of the composition of the spread mixture. (Mean area available per chain = area of monolayer divided by total number of chains, whether they belong to molecules of myristic acid or trimyristin) (36).

As the proportion of myristic acid is increased, the mean area per chain expands to 33 Å². Then there is a discontinuous expansion to 38 Å², at which point the mixed monolayer passes into an L_c state. Expansion continues with a further increase in the proportion of myristic acid.

The curve drawn in this way resembles the curve of Figure 6.14, showing the expansion of area with increase in temperature. The characteristic areas corresponding to the appearance of each physical state are found in both figures. Thus, the addition of myristic acid to trimyristin produces the same changes in phase as does an increase in temperature.

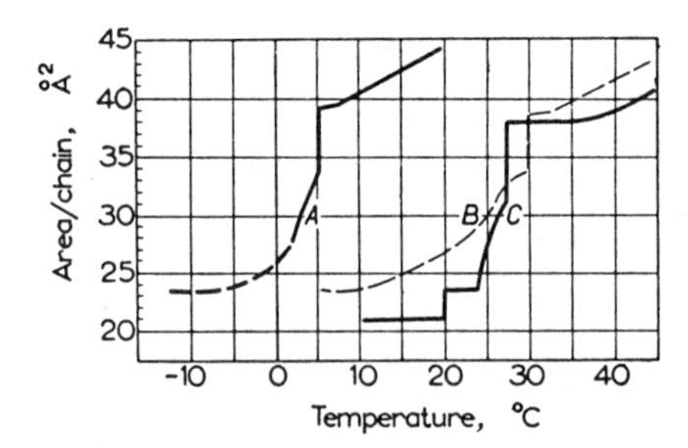

Figure 6.14. Variation with temperature of the area per chain of the condensed phase in equilibrium with the vapour phase (transition from the solid to the liquid state) for fatty acids and triglycerides: curve A, myristic acid; curve B, palmitic acid; curve C, trimyristin. Curve B for palmitic acid has been added to justify the extrapolation below 0° C of curve A for myristic acid (36).

The other extreme case examined by Dervichian is the "condensing effect" of cholesterol on a lecithin monolayer studied earlier by Adam and Jessop (38). Figure 6.15 shows the variation of the mean molecular area versus the mole fraction of lecithin. Measurements were made at a surface pressure of 5 dynes per cm. The curve shows that molecular interactions are modified in the molecular proportions of 1 to 3 and 3 to 1. The simple additivity rule is represented by the line CL, where $C=38$ Å^2, the molecular area for pure cholesterol. The point $L=96$ Å^2 corresponds to the molecular area of pure lecithin at 5 dynes per cm. The area of cholesterol undergoes very little change, either by compression or by change of temperature. Consequently, deviations from the CL line can be ascribed to a decrease in the molecular area occupied by lecithin.

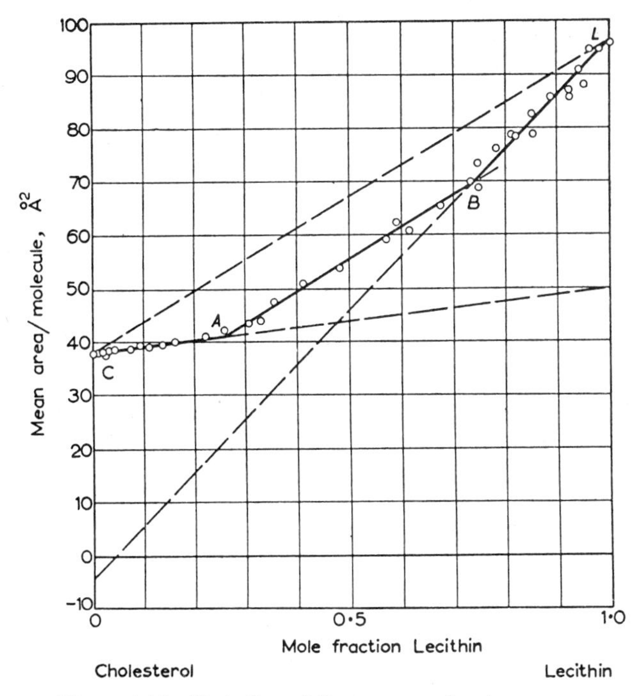

Figure 6.15. Variation of the mean molecular area, as a function of the composition, for monolayers spread from mixtures of cholesterol and egg lecithin. The pecked straight line CL represents the variation which would correspond to a simple additivity rule of the molecular areas of cholesterol (C) and lecithin (L) (36).

Extrapolation of the line CA to its intercept with the lecithin axis shows that lecithin occupies a molecular area of 50 Å^2 when the composition of the monolayer is within the range represented by the CA segment.

Extrapolation of the LB segment to its intercept with the cholesterol axis results in a value of -4 Å^2. The interpretation of this result is that every addition of a cholesterol molecule results in a decrease in the area of the monolayer of 38+4=42 Å². As long as there is less than one molecule of cholesterol for three of lecithin, it can be considered that three lecithin molecules associate with one cholesterol molecule which causes a decrease of 42/3=14 Å² in the area occupied by each lecithin molecule. It may be argued that points *A* and *B* correspond to "pure complexes", with both complexes present between *A* and *B*.

Dervishian's viewpoint is that there is no basis for chemical combination between lecithin and cholesterol. The area available per molecule depends upon the mutual arrangement and interactions of the molecules, as in crystallography. Each sudden change in the properties found in a mixed monolayer corresponds to a change in the association or structural arrangement of the molecules. Thus, points *A* and *B* in Figure 6.15 represent different structural arrangements of the molecules.

Reactions in Monolayers

It has been repeatedly observed that in biological systems, chemical reactions proceed with great rapidity and specificity. These biological systems are characterized by a high degree of order, with a well-defined orientation of molecules. Further, due to the microheterogeneity of the cell architecture, there are large interfacial-areas in the living cell. Since these features are characteristic of monolayers, study of such layers could provide insight concerning biochemical reactions. Factors influencing reaction rates in both types of systems are stereochemical configurations and electrical potential gradients (39).

The steric factor in the rate of reaction in monolayers can be shown by examination of the hydrolysis of an ethyl palmitate monolayer (16). The hydrolysis rate for the ester in a liquid expanded film is eight times greater than in a condensed film. The reason for this is shown diagramatically in the preceding Figure 6.8.

The geometric arrangement of the ester linkage was estimated by Rideal from the apparent dipole moment observed and that calculated on the basis of J. J. Thomson and Euken's principal of vectorial summation. In the expanded state, the ester group is unprotected and hydrolysis proceeds rapidly. When the film is compressed, the ethyl group is pushed below the surface of the water, as shown by the sharp drop in surface potential and the decrease in molecular area to 20 Å². In this position, the ester group is screened and the reaction rate is reduced.

An example of the effect of the electric potential on the rate of reaction is shown by the alkaline saponification of a monolayer of monocetyl succinate. The orientation and dipole moment of the ester group is shown in Figure 6.16. If the hydrolysis constant is calculated on the basis of the bulk concentration of sodium hydroxide, it shows a 300 per cent increase as the alkali content is increased from 0.4 *N* to 2 *N*. However, if the surface concentration of

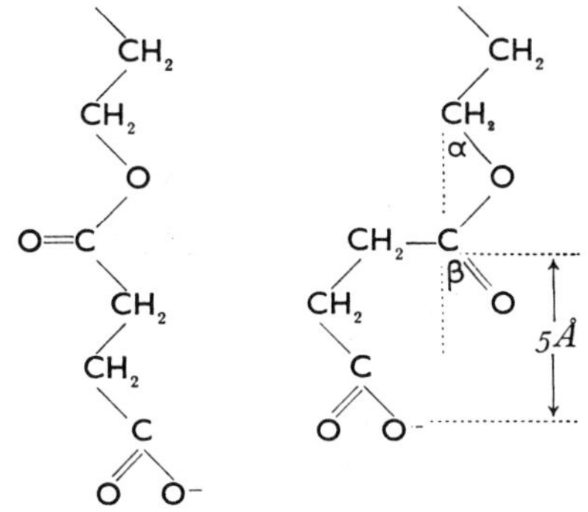

a

*p*H 0–13

$\mu_{obs} = 190$ mD
$\mu_{calc} = 186$ mD

b

*p*H 14

Expanded $\mu_{obs} = 430$ mD
$\mu_{calc} = 460$ mD
$\alpha = 30°$, $\beta = 30°$

Compressed $\mu_{obs} = 344$ mD
$\mu_{calc} = 330$ mD
$\alpha = 55°$, $\beta = 50°$

Figure 6.16. Effect of *p*H on orientation and moment of ester group of monocetyl succinate (40).

hydroxyl ions, calculated from the Donnan membrane equilibrate, is used in the rate equation, the variation in the hydrolysis constant is greatly reduced. Thus, the charge on the surface is shown to alter the concentration of alkali near the reacting group (40).

Monolayer Penetration

The phenomenon of penetration of an insoluble monolayer by a long-chain water-soluble compound was first discovered by Schulman and Hughes (41). The mixed films resulting from such penetration are of considerable technological interest, as shown in subsequent chapters.

Matalon (42) and Goddard and Schulman (43) examined the penetration of a cholesterol monolayer by sodium cetyl sulfate. Figure 6.17 shows the effect of injecting sodium cetyl sulfate under a monolayer of cholesterol maintained at a constant pressure. There is an initial rapid expansion of the film. The rate of expansion decreases with time and then becomes constant.

Extrapolation of the linear portion of the curve to zero time gave area values of 60 to 66 Å² per cholesterol molecule. If the initial value of the cholesterol monolayer is substracted, this leaves 20 to 27 Å² for the sodium

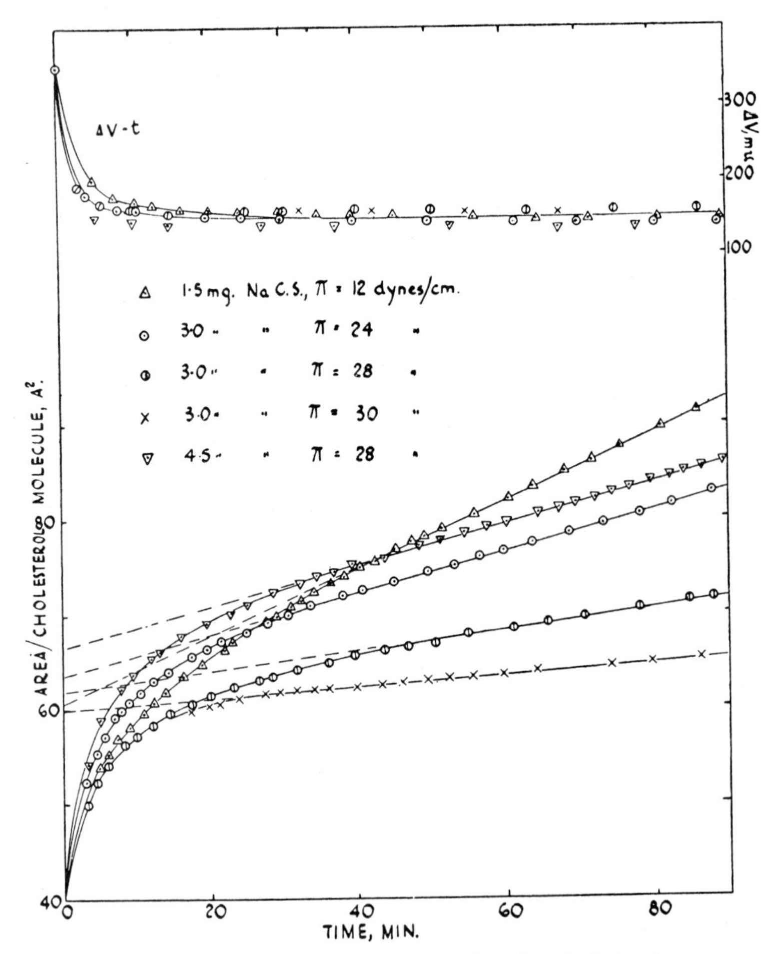

Figure 6.17. Sodium cetyl sulfate injected under cholesterol monolayers at constant pressure. Trough volume = 630 ml. (43).

cetyl sulfate. This corresponds approximately to the cross-sectional area of the penetrant molecule. Accordingly, the extrapolated area was identified as the area of a 1 : 1 complex.

When the cholesterol-sodium cetyl sulfate films were compressed, the surface remained fluid and the film collapsed at an area of 60 Å² per cholesterol molecule. This indicates that sodium cetyl sulfate present in the monolayer in excess of a 1 : 1 ratio is expelled during compression. The monolayer which collapses is the 1 : 1 ratio.

The change in the potential difference that occurred during penetration of the cholesterol monolayer by sodium cetyl sulfate, as compared with potential values obtained with sodium cetyl sulfate alone, suggested that the cetyl sulfate has a different orientation in the mixed film, or that the cetyl sulfate molecules are present in a sublayer (43).

Conditions for monolayer penetration have been defined by Schulman and Rideal (44). First, the penetrating molecule should contain a long hydrocarbon chain for strong van der Waals interaction with the hydrocarbon chain of the insoluble molecule. Second, there should be strong interaction between the polar groups of the insoluble and penetrating molecules. Such interaction can occur between ions of opposite charge, an ion and a dipole, or two dipoles. These conditions are met using monolayers of cetyl alcohol or 2-eicosanol and sodium cetyl sulfate. Penetration was observed, but the results did not suggest complex formation (45).

Dervichian (35) rejected the idea of complex formation by penetrating molecules for the reasons advanced previously with regard to monolayers containing two insoluble components. Changes in crystallogical structures are sufficient to explain the results, without the necessity for assuming the formation of stoichiometric molecular-complexes.

Insoluble Monolayers on Organic Liquids

A Teflon film balance has been described for the study of insoluble monolayers on organic liquids (46). Results obtained with an ethoxy end-blocked polymethylsiloxane of molecular weight 8250 spread on cetane are shown in Figure 6.18.

Aside from the extremely low pressures, the results are similar to those obtained with insoluble monolayers on water. At pressures below 0.32 dyne/cm and areas greater than 2200 Å^2 per molecule, the film is gaseous. The film is less compressible in the region between 0.32 and 0.57 dyne/cm. From pressures of 0.57 dyne/cm up to the collapse pressure, the film is even less compressible. Extrapolation of this third region to zero pressure yields an area per monomer of 17 Å^2.

According to Ellison and Zisman (46), the polysiloxane chains are fully extended in the first region. This allows a maximum number of the adsorbing groups to contact the surface. These are CH_3 groups on cetane. On water, they would be Si-O groups.

Near the boundary between the first and second regions the fully extended molecules become close-packed, with an area per monomer of 20 Å². The area per molecule is 2200 Å². The molecular chains buckle and form helices on further compression in the second region. Near the boundary of the second

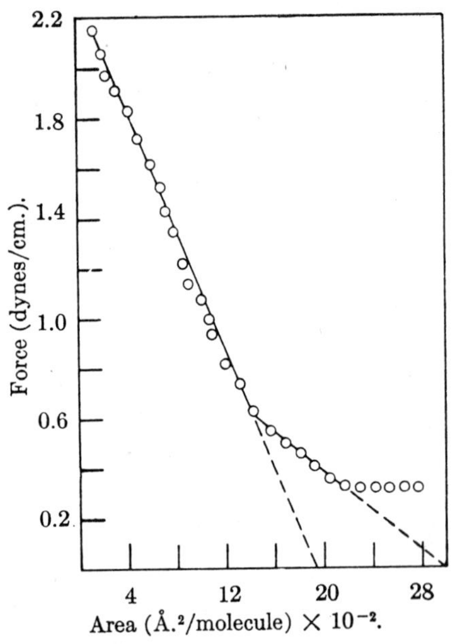

Figure 6.18. F–A curve for $C_2H_5O[Si(CH_3)_2O]_{110}C_2H_5$ on cetane (46).

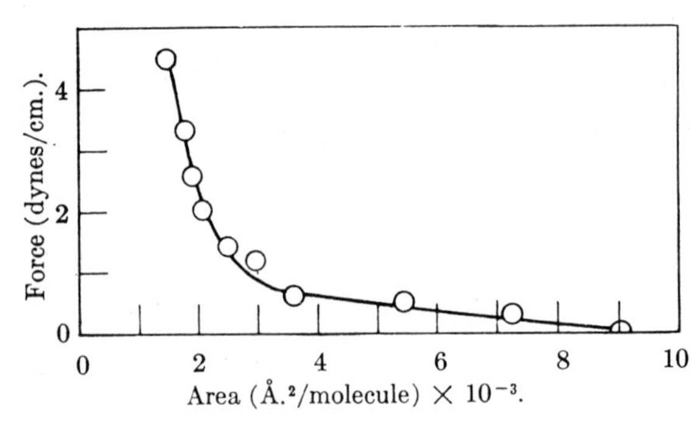

Figure 6.19. F — A curve for $C_2H_5O[Si(CH_3)_2O]_{110}$-$C_2H_5$ on tricresyl phosphate (46).

and third regions, the molecules reach a helical arrangement with six monomer units per turn. The third region corresponds to the packing together of the horizontally-orientated helical molecules.

When this polysiloxane is spread on tricresyl phosphate, a gaseous film is obtained, as illustrated in Figure 6.19. On this polar liquid the Si–O groups are adsorbed (46).

References

1. Trurnit, H. J., in "Monomolecular Layers", edited by H. Sobotka, Washington, D.C., American Association for the Advancement of Science, 1954.
2. Pockels, A., *Nature* **43**, 437 (1891).
3. Rayleigh, L., *Proc. Roy. Soc.* (London) **A47**, 364 (1890).
4. Rayleigh, L., *Phil. Mag.* **48**, 331 (1899).
5. Langmuir, I., *J. Am. Chem. Soc.* **39**, 1848 (1917).
6. Adam, N. K., and Jessop, G., *Proc. Roy. Soc.* (London) **A110**, 423 (1926).
7. Harkins, W. D., and Anderson, T. F., *J. Am. Chem. Soc.* **59**, 2189 (1937).
8. Dervichian, D. G., *J. Physique* (7), **6**, 221 (1935).
9. Wilhelmy, L., *Ann. Physik* **119**, 177 (1863).
10. Lyons, C. G., and Rideal, E. K., *Proc. Roy. Soc.* (London) **A124**, 323 (1929).
11. Schofield, R. K., and Rideal, E. K., *Proc. Roy. Soc.* (London) **A110**, 167 (1926).
12. Langmuir, I., 3rd Colloid Symposium Monograph, Chapter 6, 1925.
13. Ries, H. E., Jr., and Kimball, W. A., *J. Phys. Chem.* **59**, 94 (1955).
14. Ryan, J. P., and Shepard, J. W., *J. Phys. Chem.* **59**, 1181 (1955).
15. Harkins, W. D., and Boyd, E., *J. Phys. Chem.* **45**, 20 (1941).
16. Rideal, E. K., "Surface Chemistry", New York, Chemical Publishing Co., 1950.
17. Archer, R. J., and La Mer, V. K., *J. Phys. Chem.* **59**, 200 (1955).
18. Boyd, G. E., and Schubert, J., *J. Phys. Chem.* **61**, 1271 (1957).
19. Harkins, W. D., and Copeland, L. E., *J. Chem. Phys.* **10**, 272 (1942).
20. Harkins, W. D., "The Physical Chemistry of Surface Films", New York, Reinhold Publishing Corp. 1954.
21. Zisman, W. A., *Rev. Sci. Instr.* **3**, 7 (1932).
22. Nutting, G. C., and Harkins, W. D., *J. Am. Chem. Soc.* **62**, 3155 (1940).
23. Fourt, L., and Harkins, W. D., *J. Phys. Chem.* **42**, 897 (1938).
24. Brown, A. G., Thuman, W. C., and McBain, J. W., *J. Colloid Sci.* **8**, 491 (1953).
25. Langmuir, J., and Schaefer, V. J., *J. Franklin Inst.* **235**, 119 (1943).
26. Sebba, F., and Briscoe, H. V. A., *J. Chem. Soc.* **1940**, 106.
27. Rosano, H. L., and La Mer, V. K., *J. Phys. Chem.* **60**, 348 (1956).
28. Joly, M., *J. chim phys.* **44**, 213 (1947).
29. Joly, M., *Kolloid-Z.* **126**, 35 (1952).
30. Cary, A., and Rideal, E. K., *Proc. Roy. Soc.* (London) **109A**, 301 (1925).
31. Boyd, G. E., *J. Phys. Chem.* **62**, 536 (1958).
32. Longmuir, I., *Trans. Faraday Soc.* **15**, 68 (1920).
33. Blodgett, K. B., *J. Am. Chem. Soc.* **56**, 495 (1934).
34. Blodgett, K. B., *J. Am. Chem. Soc.* **57**, 1007 (1935).
35. Germer, L. H., and Storks, K. H., *J. Chem. Phys.* **6**, 280 (1938).
36. Dervichian, D. G., in "Surface Phenomena in Chemistry and Biology", edited by J. F. Danielli, K. G. A. Pankhurst and A. C. Riddiford, New York, Pergamon Press, 1958.
37. Dervichian, D.G., and DeBernard, L., *Bull. Soc. Chim. Biol.* **37**, 943 (1955).
38. Adam, N. K., and Jessop, G., *Proc. Roy. Soc.* (London) **120**, 473 (1928).
39. Havinga, E., in "Monomolecular Layers", edited by H. Sobotka, Washington, D.C., Am. Ass. Adv. Sci., 1954.

40. Davies, J. T., in "Surface Chemistry", New York, Interscience Publishers, 1949
41. Schulman, J. H., and Hughes, A. H., *Biochem. J.* **9**, 1243 (1935).
42. Matalon, R., *J. Colloid Sci.* **8**, 53 (1953).
43. Goddard, E. D., and Schulman, J. H., *J. Colloid Sci.* **8**, 309 (1953).
44. Schulman, J. H., and Rideal, E. K., *Proc. Roy. Soc.* (London) **B122**, 29, 46 (1937).
45. Goddard, E. D., and Schulman, J. H., *J. Colloid Sci.* **8**, 329 (1953).
46. Ellison, A. H., and Zisman, W. A., *J. Phys. Chem.* **60**, 416 (1956).

CHAPTER 7

Soluble Monolayers at Liquid Interfaces

When a surface-active substance is dissolved in water, the surface tension falls due to the formation of an adsorbed monolayer. These monolayers are in many respects similar to those formed by spreading insoluble long-chain polar molecules at the water-air interface. The surface tension of adsorbed monolayers can be measured by any of the methods described in Chapter 2. It can be expressed either as surface tension or as surface pressure, the difference in surface tension between the pure solvent and the solution.

ADSORBED NONIONIC COMPOUNDS

Surface Pressure

The surface concentration Γ of the adsorbed monolayer can be obtained from the Gibbs adsorption isotherm,

$$\Gamma = \frac{c}{kT} \cdot \frac{d\pi}{dc} \tag{7.1}$$

or

$$\Gamma = -\frac{c}{kT} \cdot \frac{d\gamma}{dc} \tag{7.2}$$

where the activity coefficient is assumed to be constant. In this, c is the concentration of the nonionic substance, while π and γ are surface pressure and surface tension, respectively.

A series of surface pressure *versus* concentration curves for *n*-heptyl alcohol at 12, 25 and 39°C is shown in Figure 7.1. The slope of the curve is constant from 0 to 10 dynes per cm, followed by a continuous decrease. The surface pressure decreases slightly with an increase in temperature.

Plots of π *versus* $-\log_{10} c$ for a series of normal aliphatic alcohols are shown in Figure 7.2. Above 10 dynes/cm surface pressure, the curves form a series of approximately equidistant parallel lines. There is considerable curvature at lower pressures. Surface pressure *versus* area curves can be constructed from the Gibbs equation for adsorbed monolayers, since $\Gamma=1/A$ and A is the area per molecule. Below 10 dynes/cm the films of these soluble alcohols are gaseous.

Above about 15 dynes/cm these films obey a modified form of the two-dimensional perfect gas law

$$\pi (A - A_0) = \times kT \tag{7.3}$$

where A_0 is approximately equal to the cross-sectional area of the polar end

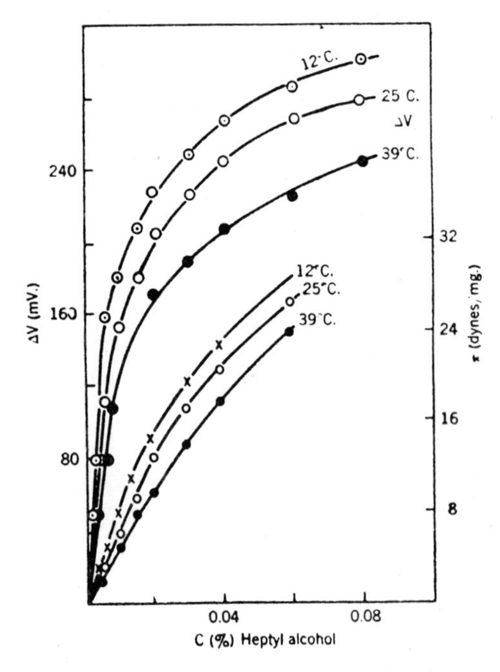

Figure 7.1. Surface pressure (II) plotted against bulk concentration (c in %) (1).

n-heptyl alcohol { × 12°C. ○ 25°C. ● 39°C.

Surface potential (ΔV) plotted against bulk concentration (c in %) for n-heptyl alcohol solution.

n-heptyl alcohol { ⊙ 12°C. ○ 25°C. ● 39°C.

group, while x depends upon the lateral adhesion of the molecules in the film. When there is no adhesion, $x=1$. It decreases as adhesion increases. At 25°C, A_0 is 16 Å^2 while x is about 0.88 for the normal aliphatic alcohols.

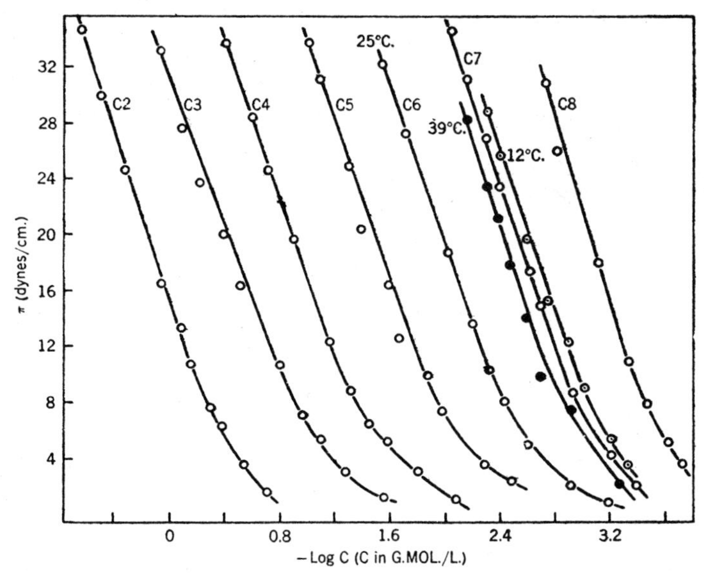

Figure 7.2. Surface pressure (II) plotted against $-\log_{10} c$ (c = concentration in g. moles/l.) for the normal alcohols C_2-C_8 inclusive). (1).

○ 25°C.
⊙ 12°C.
● 39°C.

Surface Potential

Plots of the change in potential difference ΔV with concentration for n–heptyl alcohol are shown in Figure 7.1. The curves show similar characteristics to the pressure *versus* concentration curves. Figure 7.1 shows that the curves for the change in potential *versus* surface excess are linear. This follows from the equation relating the electrical double layer at the surface to a parallel-plate condenser,

$$\Delta V = \frac{4\pi\mu_n \Gamma}{D} \tag{7.4}$$

$$\mu_n = \mu \cos\theta \tag{7.5}$$

where μ_n is the vertical component of the dipole moment μ of the film molecules.

For the normal aliphatic alcohols containing from 4 to 8 carbon atoms, μ_n equals about 0.22D. The Debye unit D$=10^{-18}$ e.s.v.

In the gaseous region of the film,

$$\pi = \Gamma kT \tag{7.6}$$

and

$$\pi = ac \tag{7.7}$$

Then

$$\Gamma = \frac{a}{kT} c \tag{7.8}$$

where a is the Traube constant, about 0.2×10^3 dynes/cm/mole/deg over the range 0 to 10 dynes per cm.

Substituting in Equation 7.4 gives

$$\Delta V = \frac{4\pi a \mu_n}{kT} c \tag{7.9}$$

This explains the linear relation between ΔV and c observed in the initial region of Figure 7.1.

Energy Relations

The standard free energy of adsorption from solution is expressible as

$$\Delta F^\circ = - RT \ln \frac{\Gamma}{c} \tag{7.10}$$

or from Equation 7.8,

$$\Delta F^\circ = - RT \ln \frac{a}{kT\delta} \tag{7.11}$$

TABLE 7.1. TRANSFER OF ALCOHOL FROM BULK TO SURFACE (1).

No. of Carbon Atoms	Temp. °C	δ^a A	$-10^{-2}\frac{a}{KT\delta}$	$-\Delta H^\circ$ (mean) kcal./mole	$-\Delta F^\circ$ kcal./mole	$-\Delta S^\circ$ (± 0.1) e.u./mole
	12		1.20			
4	25	8.8_9	0.90	2.6_8	2.66	0.1
	39		0.80			
5	25	9.5_3	2.99	4.4_1	3.36	3.5
	39		2.14			
	12		13.5			
6	25	9.9_2	9.5	3.8_8	4.05	−0.6
	39		7.5			
	12		39.5			
7	25	10.1_2	29.6	4.2_2	4.72	−1.7
	39		20.8			
	12		142.0			
8	25	10.2_4	97.7	4.8_7	5.42	−1.8
	39		67.4			

[a] Obtained from the most probable length of the hydrocarbon chain and taking the length of the OH group as 2.5 A.

where δ is the thickness of the surface layer, taken as the most probable length of the aliphatic alcohol. The standard state of 1 molecule per cc may be adopted for both the bulk of the solution and the surface.

Table 7.1 is a listing of ΔF° values for the adsorption of the alcohols calculated from Equation 7.11. ΔH° values were calculated from the temperature variations of the equilibrium constant $K = a/kT\delta$

$$\Delta H^\circ = RT^2 \cdot \frac{d \ln K}{dT} \tag{7.12}$$

The relation for the corresponding entropy change is

$$\Delta F^\circ = \Delta H^\circ - T\Delta S^\circ \tag{7.13}$$

Nonionic Surfactants

Surface tension — or surface pressure — *versus* concentration curves for nonionic surfactants at the water-air interface are similar to those obtained with the water-soluble alcohols. The one important difference, as shown in Figure 7.3., is that the surface tension drops to a limiting value in a narrow concentration range, called the critical micelle concentration (cmc). At this concentration, the molecules begin to associate to form micelles.

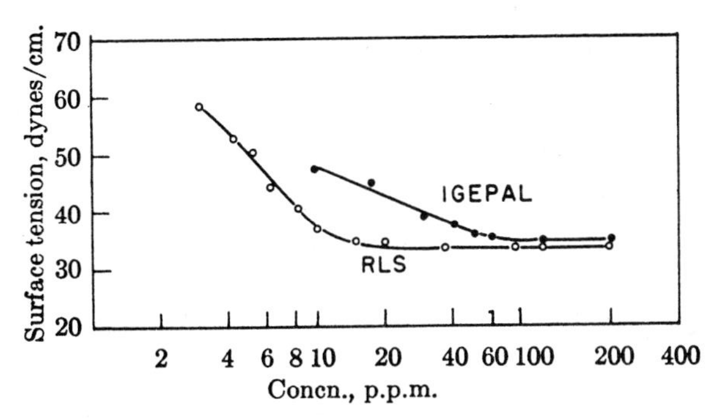

Figure 7.3. Surface tension versus concentration for Igepal CO — 710 and RLS, which contain more than a 10.5 molar ratio of ethylene oxide to alkylphenol (2).

At concentrations just below the cmc, the curves are linear, indicating the approach to a close-packed monomolecular layer. For Igepal CO–710, an octylphenol-ethylene oxide condensate with an average mole ratio of 10.5 moles of ethylene oxide per mole of hydrophobe, an area of 63 Å^2 per molecule was calculated with the Gibbs adsorption equation (2).

ADSORBED IONIC COMPOUNDS

Gibbs Adsorption Isotherm

Consider an aqueous solution of sodium dodecylsulfate (NaDS). With the usual convention that the surface plane of reference is selected so that the surface adsorption of the solvent is zero, then at constant temperature and pressure, the Gibbs adsorption equation takes the form

$$d\gamma = -RT\Gamma_{\mathrm{Na}^+}\, d \ln a_{\mathrm{Na}^+} - RT\Gamma_{\mathrm{DS}^-}\, d \ln a_{\mathrm{DS}^-} \tag{7.14}$$

The condition of electrical neutrality requires that the surface concentration of positive and negative charges be equal. Then

$$\Gamma_{\mathrm{Na}^+} = \Gamma_{\mathrm{DS}^-} \tag{7.15}$$

Further,

$$d \ln a_{\mathrm{Na}^+} = d \ln a_{\mathrm{DS}^-} \tag{7.16}$$

It then follows that

$$d\gamma = -\, 2RT\Gamma_{\mathrm{NaDS}} d \ln a_{\mathrm{NaDS}} \tag{7.17}$$

If a salt containing an ion in common with the surfactant, such as sodium chloride, is present in considerable excess, then

$$d \ln a_{\mathrm{Na}^+} \neq d \ln a_{\mathrm{DS}^-} \tag{7.18}$$

Instead, the concentration of the counter-ion remains sensibly constant with a variation in surfactant concentration. The adsorption equation is

$$d\gamma = -\, RT\, \Gamma_{\mathrm{DS}^-}\, d \ln a_{\mathrm{DS}^-} \tag{7.19}$$

Thus, in the absence of indifferent electrolyte, a factor 2 appears for the change in surface tension with a change in the concentration of ionic surfactant. In the presence of excess salt electrolyte the factor is unity.

When the added electrolyte is not in considerable excess, the factor X has the value (3)

$$X = 1 + \frac{c_{\mathrm{Na\,DS}}}{c_{\mathrm{NaDS}} + c_{\mathrm{NaCl}}} \tag{7.20}$$

where c_{NaDS} is the surfactant concentration and c_{NaCl} is the concentration of salt electrolyte having an ion in common with the surfactant.

The general form of the Gibbs adsorption equation is exact. However, in order to interpret surface tension *versus* activity or concentration data, it is necessary to make certain assumptions concerning the composition of the solution. The usual practice is to assume that the ions present in solution are unassociated at concentrations below the cmc. If ion pairs or micelles are present in solution, terms representing these species must also be included in the adsorption equation. There is considerable evidence that ion association occurs at concentrations below the cmc.

Adsorption Isotherms

Concentration c is frequently substituted for activity in the Gibbs adsorption equation. Surface tension *versus* log c curves for solutions of sodium dodecylsulfate with added sodium chloride are shown in Figure 7.4. The break points shown in the curves are interpreted as corresponding to the

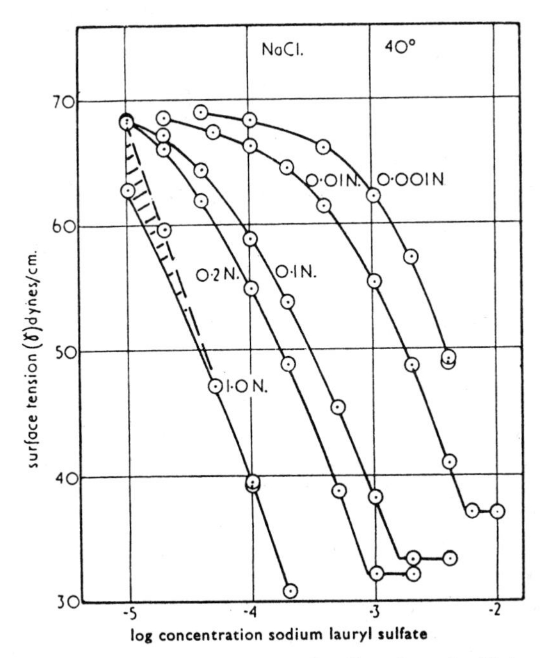

Figure 7.4. Surface tension of sodium lauryl sulfate at 40°C in the presence of 0.001, 0.01, 0.1, 0.2 and 1.0N NaCl (3).

maximum concentrations of unassociated surfactant ions. At slightly higher concentrations of surfactant, micelles form rapidly with a further increase in concentration. Thus, these break points correspond to the cmc values of the surfactant in the presence of 0.01N, 0.1N and 0.2N sodium chloride.

Figure 7.5 shows plots of surface pressure *versus* concentration of sodium dodecylsulfate in the presence of added electrolyte at 20°, 40° and 60°C. At low concentrations, π is proportional to c. This linear relation holds at higher surfactant concentrations, as the level of NaCl is increased.

Surface pressure *versus* area curves for sodium dodecyl sulfate on 0.2N

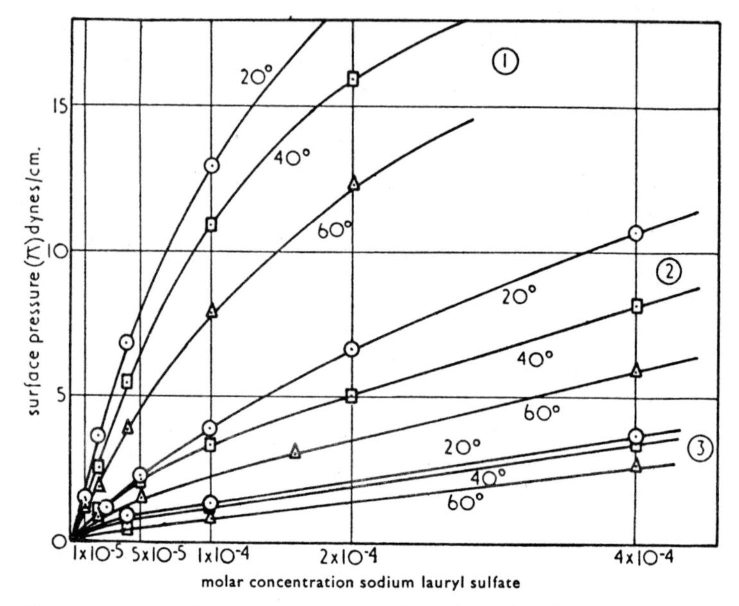

Figure 7.5. Surface pressure of sodium lauryl sulfate solution in (1) 0.1, (2) 0.01, and (3) 0.001 N NaCl at ○ 20° C, □ 40° C, and △ 60° C (3).

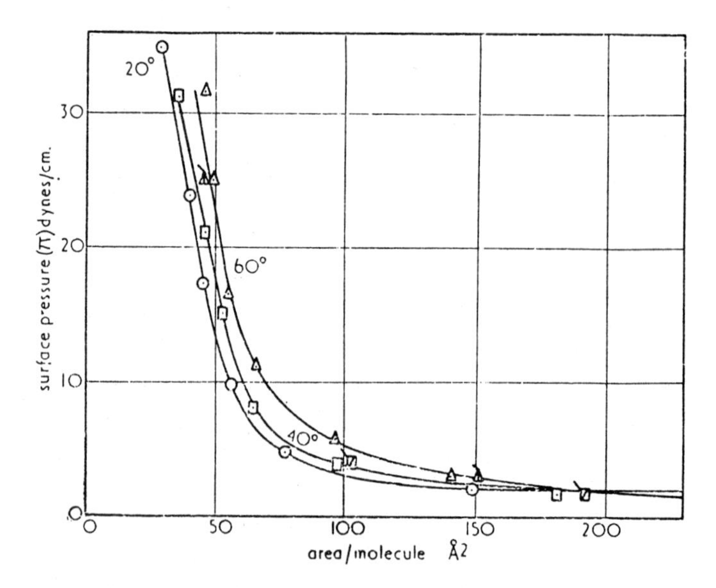

Figure 7.6. The force-area curves for sodium lauryl sulfate monolayers on 0.2 M NaCl solutions at ○ 20°C, □ 40°C, △ 60°C. Areas obtained from linear plots are noted with flags (3).

NaCl in Figure 7.6 were calculated from linear and log plots similar to those of the previous two figures. Expansion of the monolayers occurs at higher temperatures.

Saturation Adsorption (4)

It has been frequently observed that the relative adsorption of nonionic surfactants from their aqueous solutions becomes nearly constant at concentrations somewhat below the cmc. The same effect has been observed with solutions of anionic surfactants at constant cation concentrations. Generally, at surfactant concentrations ranging from the cmc to about a third of this value, the adsorption remains unchanged. This is evidenced by a linear relation between interfacial tension and the logarithm of the surfactant ion concentration. The constant value of the adsorption is referred to as the saturation adsorption of the surfactant.

TABLE 7.2. EFFECT OF THE ELECTROLYTE CONCENTRATION ON THE AREA PER SURFACTANT ION AT THE SATURATION ADSORPTION (4).

	$C_{11}H_{23}SO_3Na$ area/ion ($Å^2$) interface			$C_{11}H_{23}COONa$ area/ion ($Å^2$) interface			$C_{11}H_{23}COOK$ area/ion ($Å^2$) interface	
$[Na^+]$ (moles/l.)	air	*n*-heptane	$[Na^+]$ (moles/l.)	air	*n*-heptane	$[K^+]$ (moles/l.)	air	*n*-heptane
0.02	53	54	0.04	47	45	0.04	43	44
0.03	51	54	0.07	—	45	0.07	—	44
0.05	52	55	0.11	47	45	0.11	43	44
0.10	52	57	—	—	—	—	—	—

Table 7.2 gives the saturation adsorption of several anionic surfactants at a water-air and water-*n*-heptaine interface. The saturation adsorptions are essentially constant in the range 0.02 to 0.10N Na^+. Further, values at the air and *n*–heptane interfaces differ only slightly. Larger differences are found at interfaces formed with more polar organic solvents and the aqueous solution, as shown in Table 7.3. For sodium undecylsulfonate, the relation,

$$\Gamma_0 = K\gamma_0^{\frac{1}{3}} \tag{7.21}$$

is approximately valid for water-organic solvent interfaces. K is a constant, while Γ_0 is the saturation adsorption and γ_0 is the interfacial tension between pure water and the pure organic liquid. The effect of polar liquids in decreasing the saturation adsorption at the interface suggests that there is competition between the polar liquid and the surfactant for adsorption.

The saturation adsorption range for a number of ionic surfactants is given in Table 7.4. In general, chain length or branching has a negligible effect on the saturation adsorption value. The nature of the polar group has a somewhat greater effect.

TABLE 7.3. EFFECT OF THE NATURE OF THE INTERFACE ON THE SATURATION ADSORPTION OF SODIUM UNDECYL SULFONATE (4).

$[Na^+]$ (moles/l.)	Nature of the Interface							
	air	n-heptane	benzene	$C_6H_{13}Cl$	$C_7H_{13}Cl$ impure	$C_{12}H_{25}Cl$	methyl caprate	dibutyl-ether
	saturation adsorption in $Å^2$/detergent ion							
0.02	53	54	57	—	66	—	78	—
0.03	51	54	59	62	66	61	79	78

TABLE 7.4. SATURATION ADSORPTION RANGE FOR VARIOUS SURFACTANTS (4).

Surfactant	cmc (N)	Concentration at which surface excess becomes constant, as % of cmc
Na-dodecyl sulphate	1.4×10^{-3}	11
Na-nonyl sulphate	4.3×10^{-2}	23
Na-laurate	1.7×10^{-2}	18
Na-undecyl sulphonate	1.4×10^{-2}	15
Na-α-heptyl undecylsulphonate	2.9×10^{-4}	30
K-laurate	1.37×10^{-2}	15
Na-dodecyl sulphate	1.45×10^{-3}	20
dodecylamine hydrochloride	3.50×10^{-3}	25
Na-decylsulphate	3.2×10^{-2}	30
Na-dodecyl sulphate	0.8×10^{-2}	15
Na-myristyl sulphate	0.2×10^{-2}	30
Na-propylene tetramer benzene sulphonate	1.55×10^{-4}	15

Charged Monolayers (5)

In a charged monolayer, the potential in the surface ψ_0 as given by the Gouy equation is

$$\psi_0 = \frac{2kT}{e} \sinh^{-1} \left(\frac{134}{A c_i^{\frac{1}{2}}} \right) \tag{7.22}$$

The value of 134 in the bracket is for a monolayer at 20°C. This number, which involves the dielectric constant of water and the temperature, becomes 139 at 50°C.

In Equation 7.22, ψ_0 is expressed in millivolts. The term kT/e is 25.2 mv at 20°C. A is the area in Å² per charged group in the monolayer and c_i is the total concentration of uni-univalent electrolyte in the bulk of the solution.

For a charged monolayer, the surface potential ΔV, the difference in potential between a clean surface and a film-covered surface, can be expressed by the relation (6)

$$\Delta V = 4\pi n\mu_n + \psi_0 \tag{7.23}$$

where n is the number of long-chain ions per cm², equal to $10^{16}/A$. The quantity μ_n is the vertical component of the dipole moment of the polar group, expressed in millidebyes. For monolayers of $C_{16}H_{33}N(CH_3)_3^+$, the C–N dipoles result in a value for μ_n of 400 millidebyes (7).

From Equations 7.22 and 7.23, at a constant area, ΔV and ψ_0 vary together as c_i is changed. Consequently, $\Delta V - \psi_0$ is independent of c_i when the long-chain ions are held at a constant area. This can be seen from Figure 7.7. Thus, since μ_n is practically independent of ionic strength, changes in ψ_0 can be obtained from surface potential measurements.

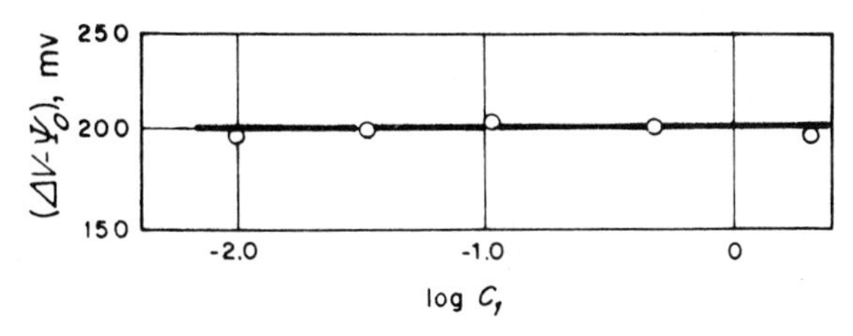

Figure 7.7. Plot of $(\Delta V - \psi_0)$ against log c_i, where c_i is the concentration of added sodium chloride. The monolayer is formed from $C_{18}H_{37}N(CH_3)_3^+$ ions spread to a constant area of 85 Å² per long chain at the air-water interface (5).

Further, Crisp (9) found that the surface potentials of films of α-bromopalmitic acid, on which the concentration of Na^+ was varied between 1 and 0.06N, obeyed approximately the relation

$$\left(\frac{\partial \Delta V}{\partial \log c_i}\right)_A = 2.303\, kT/e = +\, 59\ mv \tag{7.24}$$

For positively charged films, Davies (8) found

$$\left(\frac{\partial \Delta V}{\partial \log c_i}\right)_A = -\, 59\ mv \tag{7.25}$$

At low ionic strengths, when ψ_0 is large, differentiation of Equation 7.22 gives

$$\left(\frac{\partial \psi_0}{\partial \log c_i}\right)_A = -\, 59\ mv \tag{7.26}$$

Equations of State

Gaseous monolayers bearing no net charge, and formed at either the air-water or the oil-water interface closely obey the equation

$$(\pi^0 + B)\ (A - A_0) = kT \tag{7.27}$$

where π^0 is the surface pressure in dynes/cm. B is a measure of the cohesion in the surface, and is very low at the oil-water interface. A is the area, expressed in Å, occupied by each molecule in the surface. A_0 is the actual or limiting area occupied by each molecule when the film is highly compressed. At 20°C, kT has a value of 408×10^{-16} erg.

Davies (8) found that under certain conditions the exact equation of state for a gaseous film of ionized molecules is approximately

$$\pi^+ (A - A_0) = 3kT + 6.10 (A - A_0) \sqrt{c} - \frac{2kTA_0}{A} \tag{7.28}$$

where π^+ is the surface pressure of the charged monolayer, and c is the ionic strength of the uni-univalent electrolyte in the bulk phase, expressed in gram-ions per liter.

For very dilute substrates, where $c \to 0$, if $A \gg A_0$, then

$$\pi^+ (A - A_0) \to 3kT \tag{7.29}$$

However, if A is very large, the film tends to behave as though it were not charged and

$$\pi^+ (A - A_0) \to kT \tag{7.30}$$

Where A is small and approaches A_0, the term $(2kTA_0)/A$ becomes appreciable, and the product is lower than $3kT$ and tends to kT. For intermediate values of A, the product is closer to $3kT$, exhibiting a maximum.

Equation 7.28 can also be expressed in the form

$$\pi = \frac{kT}{A - A_0} + 6.1 \sqrt{c} \left[\sqrt{1 + \left(\frac{134}{A\sqrt{c}} \right)^2} - 1 \right] \tag{7.31}$$

A second equation proposed by Phillips and Rideal (10) differs from this by a factor of 2,

$$\pi = \frac{2kT}{\mathrm{A} - \mathrm{A}_0} + 6.1 \sqrt{c} \left[\sqrt{1 + \left(\frac{134}{A\sqrt{c}} \right)^2} - 1 \right] \tag{7.32}$$

Haydon (11) examined these equations on the basis of the Gibbs adsorption isotherm and concluded that Equation 7.31 is applicable in the presence of excess electrolyte. In the absence of added electrolyte, the derivation leads to Equation 7.32. Thus, if the factor 2 is applicable to the Gibbs equation, it should also appear in the equation of state.

RADIOTRACER STUDIES OF ADSORPTION

The Gibbs adsorption equation provides an indirect measure of the extent of adsorption. The reliability of conclusions deduced from surface tension *versus* concentration data is limited by the general lack of information concerning the precise state of the solute. Consequently, efforts have been aimed at the direct measurement of the extent of adsorption. Methods that have been used for the measurement of the adsorption of water-soluble surfactant at the air-water interface include the air-bubble method (12, 13) and the microtone method (14, 15). The radiotracer procedure (16, 17) provides the first convenient method for studying adsorption over a wide range of concentrations.

Radiotracer Method

The radioactivity of a solution containing a tracer is reduced due to the self-adsorption of the radiation from within the solution. However, the radioactive count from tracer materials adsorbed at the surface is not reduced by self-adsorption. Thus, if two solutions are identical in all respects including the geometry of the containing vessels, except that the tagged element is part of a surfactant electrolyte in one and a nonsurface-active electrolyte in the other, the difference in radioactivity is a measure of surface adsorption.

The method of Dixon and coworkers (16, 18) involves the direct measurement of the radioactivity of the solution with a correction for that due to the bulk of the solution. To minimize the correction, it is necessary to use tracer elements that emit soft beta radiation. These include S^{35}, C^{14}, Ca^{45}, and H^{3}.

TABLE 7.5. RATIO OF SURFACE TO BULK COUNTS FOR VARIOUS RADIOACTIVE ELEMENTS EMITTING BETA RADIATION. THE RATIO $I_{SURFACE}/I_{BULK}$ WAS CALCULATED FOR A MONOLAYER ASSUMED TO BE 2×10^{-10} MOLE CM^{-2} (18). RANGE IN WATER FROM (19).

Isotope	Energy (Mev)	Absorption Coefficient (cm^{-1})	Range in Water (10^{-3}mm)	$I_{surface}/I_{bulk}$ at Concentrations of $10^{-5}M$	$10^{-4}M$	$10^{-3}M$	$10^{-2}M$
H^3	0.0189	5600	6	100	10	1.0	0.1
C^{14}	0.154	290	300	6	0.6	0.06	0.006
S^{35}	0.167	320	340	6	0.6	0.06	0.006
Ca^{45}	0.254	90	650	2	0.2	0.02	0.002

Table 7.5 shows that tritium is particularly suited for surface adsorption studies, because the radiation is very soft. The other tracer elements shown in the table can not be used at high concentrations.

Hutchinson (17) used a platinum wire loop drawn through the solution. The count for the tagged species obtained from the film is compared with that of an equal weight of material taken from the interior of the solution. This method does not require the use of elements that emit soft radiation.

Aerosol OTN Adsorption. Adsorption isotherms for aqueous solutions of Aerosol OTN, di-*n*-octyl sodium sulfosuccinate, are shown in Figure 7.8. In

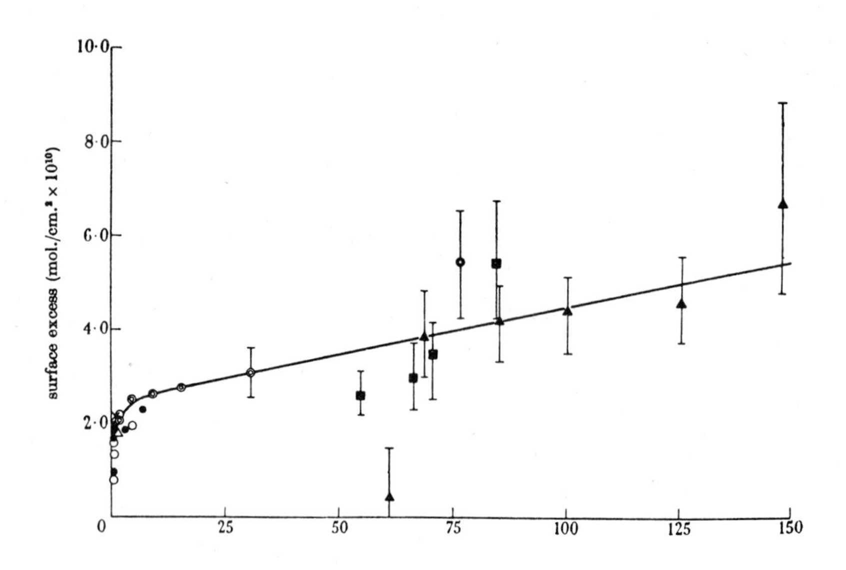

Figure 7.8. Adsorption of sulfosuccinate ion from Aerosol OTN solutions. Critical concentration, 65×10^{-5} mole per liter (20).

Figure 7.9, results obtained by the radioactivity method are compared with those calculated from the Gibbs adsorption isotherm in the form

$$\frac{d\gamma}{d \ln c_{\mathrm{NaOTN}}} = -RT\,\Gamma_{\mathrm{OTN^-}} \tag{7.33}$$

This equation results in a calculated surface adsorption of about 2.4×10^{-10} mole per cm² for the concentration range 0.2 to 65×10^{-5} mole per liter. The latter figure is the cmc of the surfactant. The calculated surface adsorption is in reasonable agreement with that obtained by the radiotracer method.

This is surprising, since in the absence of indifferent electrolyte, a factor of 2 should have been used in the Gibbs equation, corresponding to half the surface adsorption found. The explanation is that H^+ is adsorbed, rather than Na^+, along with surfactant anion. This is referred to as surface hydrolysis.

The reason for the factor 1 may be seen by writing the adsorption equation as

$$-d\gamma = RT\Gamma_{Na+} d \ln c_{Na+} + RT\Gamma_{OTN^-} d \ln c_{OTN^-} + RT\Gamma_{H+} d \ln c_{H+} \quad (7.34)$$

Since H^+ rather than Na^+ is adsorbed,

$$\Gamma_{H+} = \Gamma_{OTN^-} \quad (7.35)$$

$$\Gamma_{Na+} = 0 \quad (7.36)$$

Aerosol OTN is the sodium salt of a strong acid and does not hydrolyze significantly in dilute solution. Consequently, the pH of the solution does not

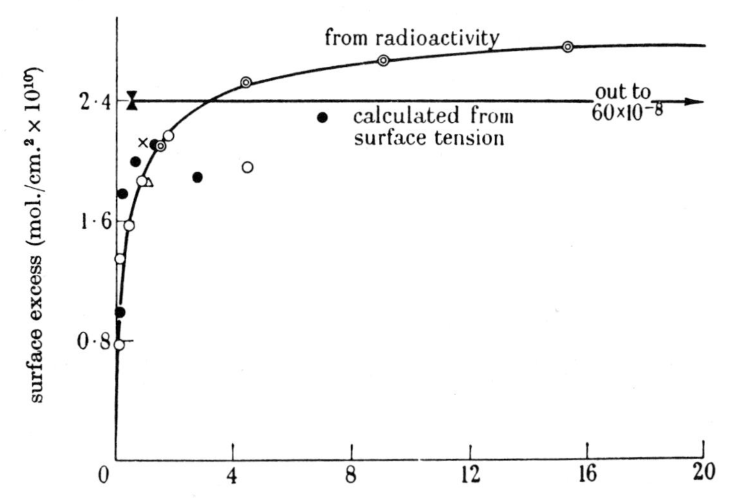

Figure 7.9. Enlargement of Aerosol OTN adsorption isotherm. Upper curve, adsorption measured by radioactivity method, data taken from Fig. 7.8; lower curve, adsorption calculated from surface tension using equation (7.33). Critical concentration, 65×10^{-5} mole per liter (20).

change appreciably with changes in concentration. Then $d \ln c_{H+} = 0$ and Equation 7.33 follows. Elimination of the factor 2 results in a calculated molecular area of 65 Å² per molecule or 32.5 Å² per alkyl chain.

The break in the curve of Figure 7.8, corresponding to the cmc, occurs at a surface concentration of about 2.4×10^{-10} moles cm^{-2}. With a further increase in bulk concentration, the surface concentration continues to increase. This is in marked contrast with calculations of surface concentration based on the Gibbs adsorption isotherm. These calculations show a constant surface concentration even before the cmc.

Adsorption data for both S^{35} tagged long-chain sulfosuccinate ions and Na^{22}

ions are shown in Figure 7.10. The low initial values for sodium ion adsorption substantiate the conclusion of surface hydrolysis at low concentrations. However, well above the knee of the adsorption isotherm, where the monolayer is complete, sodium-ion adsorption is increasing rapidly with concentration. At the cmc, sodium-ion and sulfosuccinate-ion adsorption are about equal.

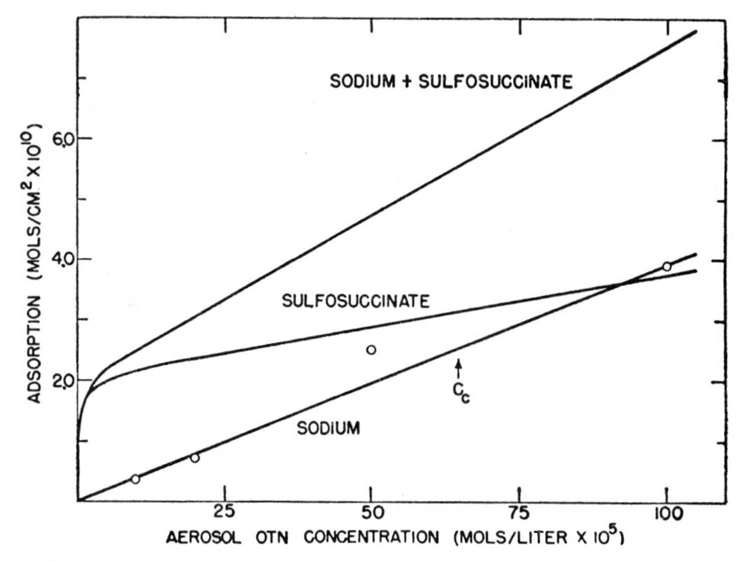

Figure 7.10. Adsorption of sodium and sulfosuccinate ion from Aerosol OTN solutions. Sodium ion adsorption from points shown; sulfosuccinate adsorption from Fig. 7.8. Critical concentration, 65×10^{-5} mole per liter (20).

These results as well as quite similar results obtained with the cationic agent Aerosol SE (stearamidopropyldimethyl-2-hydroxyethyl ammonium sulfate) suggest that the surfactant ion and counter-ion are adsorbed as multilayers. Adsorption data indicated that as many as 30 monolayers of Aerosol SE were adsorbed.

In considering possible explanations for these results, it should be noted that the count from a layer of S^{35} located 10,000 to 20,000 Å below the surface of the solution is experimentally indistinguishable from one at zero depth. Any increase in the concentration of tracer element in this entire depth will correspond to an increase in surface concentration. The most probable explanation for apparent multimolecular adsorption appears to be that micelles with their counter-ions tend to accumulate under the monolayer with its counter-ions.

Tritiated Sodium Dodecylsulfate. It can be expected that multilayer adsorption will not be observed with H^3 — tagged surfactant as compared with one containing S^{35}. This is shown to be the case in Figure 7.11. Curve *A* is for tritiated sodium dodecylsulfate in water, while curve *B* is for the surfactant in a buffer solution at pH 6.5 containing in addition an excess of neutral salt.

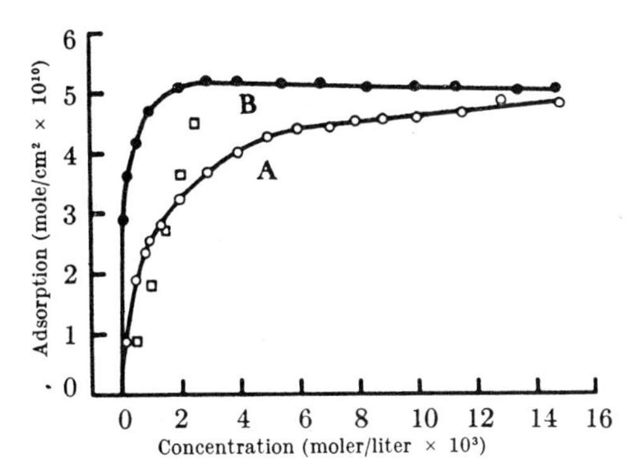

Figure 7.11. Adsorption isotherms of tritiated sodium dodecyl sulfate at 25° C in water, curve A, and in a pH 6.5 buffer solution and 0.1 mole per liter Na^+, curve B (21).

Examination of curve *A* shows that the isotherm changes and becomes linear at about 6×10^{-3} mole/liter. In comparison, the best value for the cmc of sodium dodecylsulfate obtained by other methods is 8.1×10^{-3} mole/liter. The lower value indicates that ion pair or micelle formation may take place below the cmc.

Curve *A* would probably show the same constant surface excess of about 5×10^{-10} mole cm^{-2} as curve *B*, but at higher concentrations. This value corresponds to a monolayer with a surface area of 33 Å^2 per molecule, in agreement with the limiting area of 33 Å^2 per molecule in Davies (22) equation of state. It is also in good agreement with the limiting value in Pethica's equation (23): $\pi(A-31)=1.2\ kT$ for the adsorption of sodium dodecylsulfate at the air-water interface.

A surface tension *versus* logarithm of concentration curve is shown in Figure 7.12. This curve was obtained by Roe and Brass (24) under the identical conditions as curve *B* in Figure 7.11. The surface tension curve shows a constant slope in a range below the cmc, which is 1.46×10^{-3} mole/liter. These results indicate that the surface concentration should be constant over the wide range, below the cmc, of constant slope. The direct measurement of surface excess with tritiated sodium dodecylsulfate shows that this is not the case. The surface concentration does not become constant below the cmc.

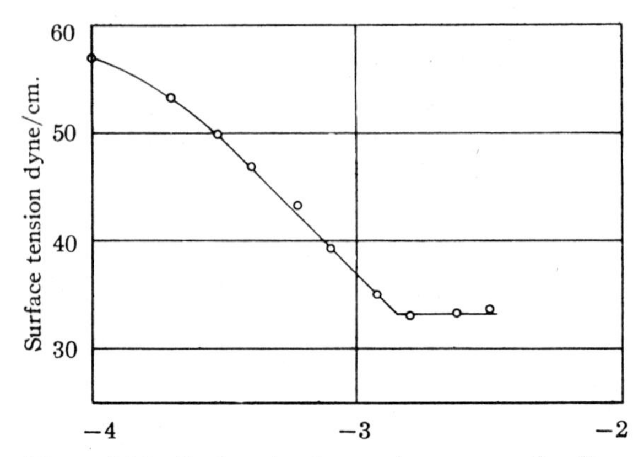

Figure 7.12. Surface tension *vs.* log concn. of sodium dodecyl sulfate solutions; *p*H 6.5, $[Na^+] = 0.1$ mole per liter (24).

Conclusions Based on Adsorption Data

The direct measurement of adsorption by radiotracer methods provides information that cannot be obtained by the indirect method based on the Gibbs adsorption isotherm. These measurements lead to the following conclusions:

(1) The usual treatment of adsorption data based on surface tension *versus* concentration measurements leads to only approximate information with regard to surface concentration. This is due to the formation of ion pairs and possibly small micelles at concentrations below the cmc. Information is not available for solving the adsorption equation containing terms for these additional species.

(2) In contrast with conclusions usually drawn from the incomplete form of the Gibbs adsorption equation, studies with radiotracers show that the monolayer is not complete at concentrations below the cmc, even in the presence of excess salt.

(3) At very low concentrations, surface hydrolysis may occur even with strong electrolytes that do not undergo hydrolysis in neutral solution.

(4) Above the cmc, multilayer adsorption has been observed. This probably results from a tendency for micelles to accumulate beneath the monolayer, with its high ionic concentration. However, there is a possibility that the observed multilayers are due to experimental error inherent in the technic.

Direct measurement of adsorption by foam fractionation did not reveal multilayer formation (25).

Selective Adsorption

Selective adsorption has been measured by radiotracer methods. It was shown that Aerosol OT — di-2-ethylhexyl sodium sulfosuccinate — will displace an adsorbed monolayer of Aerosol OTN (18). Aniansson (19) showed that at low concentrations sodium hexadecylsulfate is preferentially adsorbed from solutions containing one hundred times higher concentrations of sodium dodecylsulfate.

Selective adsorptivity was determined by a combination of foam fractionation and radiotracer methods. Sodium hexadecanoate was found to have sixty times greater selective adsorptivity than potassium dodecanate, corresponding to a selective adsorptivity factor of about 2.75 per methylene group. Experiments on mixtures of sodium *p*-dodecyl-benzenesulfonate and sodium dodecylsulfate indicated that the contribution of the benzene ring to selective adsorptivity was equivalent to about 3.5 to 4 methylene groups (26).

Adsorption Kinetics

Numerous studies have been made with regard to the rate of adsorption of long-chain polar compounds at interfaces. Typical data illustrating relatively fast rates of attaining surface equilibrium are illustrated in Figure 7.13. Surface tension measurements were made by the oscillating jet method. Slow surface aging results obtained using C^{14} — tagged sodium stearate are shown in Figure 7.14. There is still considerable uncertainty as to whether adsorption rates on fresh surfaces depend entirely upon the rate of diffusion (28, 29, 30, 31).

Slow surface aging has been considered to be caused either by steric hindrance to the surfactant ions entering a nearly complete surface layer or to an electric repulsion as an ion approaches the surface containing ions of like charge (30).

A current view is that the long aging effects previously observed were due to instrumental errors or the presence of impurities, such as heavy-metal ions or long-chain alcohols (5). The rate of adsorption of long-chain compounds appears not to be affected by the presence of the charged head-groups.

The rate of adsorption of a surfactant from water can be expressed

$$\left(\frac{dn}{dt}\right)_{\text{adsorption}} = B_1 c_d (1-\theta) \tag{7.37}$$

where n is the surface concentration, B_1 is a constant, c_d is the surfactant concentration, and θ is the fraction of the surface actually covered by molecules. In the early stages of adsorption, before θ and the desorption rate are appreciable, dn/dt is proportional to c_d.

The desorption rate is given by (7)

$$-\left(\frac{dn}{dt}\right)_{\text{desorption}} = B_2 n\, \mathbf{e}^{(ze\psi_0 - W)/kT} \tag{7.38}$$

where B_2, like B_1, is a constant depending upon the diffusion process. W is the energy of adsorption of the hydrocarbon chain, z is the valency of the long-chain ion, and ψ_0 is the potential in the surface.

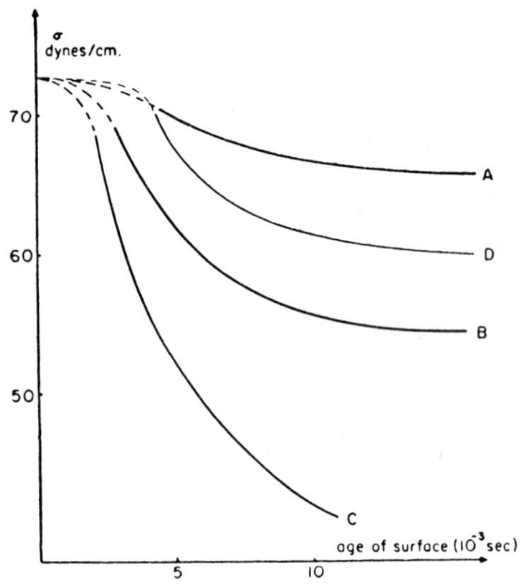

Figure 7.13. Dynamic surface tensions of some alcohol solutions (27).
A – hexyl alcohol
B – heptyl alcohol } $3.44\ 10^{-3}$ mole/liter.
C – octyl alcohol
D – heptyl alcohol $2.43\ 10^{-3}$ mole/liter.

For small intervals of time t,

$$\log\left(-\frac{1}{n}\frac{dn}{dt}\right) = \frac{ze\psi_0}{2.3\,kT} + \text{constant} \tag{7.39}$$

or

$$\log\left(-\frac{d\pi}{dt}\right) = \frac{ze\psi_0}{2.3\,kT} + \text{constant} \tag{7.40}$$

The constant includes W, and is actually constant for films at the air-water interface only when the area per molecule A is constant. At the oil-water interface, W is independent of A because inter-chain cohesion is eliminated by the oil.

In equations 7.39 and 7.40, the constant is log $B_2 - W/2.3kT$. The total energy of adsorption of the hydrocarbon chain W is 810 cal/mole per CH_2

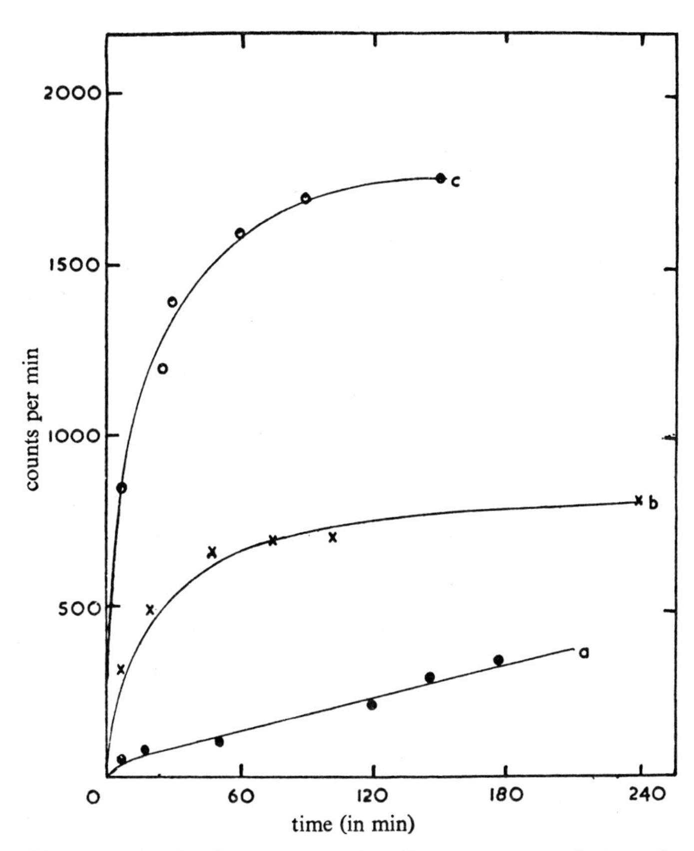

Figure 7.14. Surface ageing of sodium stearate solutions in water (28).
(*a*) 3.26×10^{-9} mole/ml; (*b*) 1.623×10^{-8} mole/ml; (*c*) 3.263×10^{-8} mole/ml.

group. For desorption in a static system at 20°C B_2 is 2×10^4 molecules per second. The value of B_1 for desorption in the same static system is 2.6×10^{18} molecules per second. The ratio $B_1/B_2 = 1.3 \times 10^{14}$ for any system.

The equilibrium surface-concentration n of a surfactant solution may be obtained by equating the adsorption and desorption rates,

$$B_1 c_d (1-\theta) = B_2 n\, \mathbf{e}^{(ze\psi_0 - W)/kT} \tag{7.41}$$

or

$$\frac{n}{1-\theta} = \left(\frac{B_1}{B_2}\right) c_d\, \mathbf{e}^{(W - ze\psi)/kT} \tag{7.42}$$

If the fraction of the surface covered θ is small ($A > 100$ Å²), the equation may be written

$$\log\left(\frac{n}{c_d}\right) = \log\frac{B_1}{B_2} + \frac{W}{2.3kT} - \frac{ze\psi_0}{2.3kT} \tag{7.43}$$

Since $\log B_1/B_2 = 14.1$,

$$\log\left(\frac{n}{c_d}\right) = 14.1 + \frac{W}{2.3kT} - \frac{ze\psi_0}{2.3kT} \tag{7.44}$$

Figure 7.15 shows that Equation 7.44 is applicable for data tabulated by Kling and Lange (32). Values for ψ_0 were calculated from Equation 7.22.

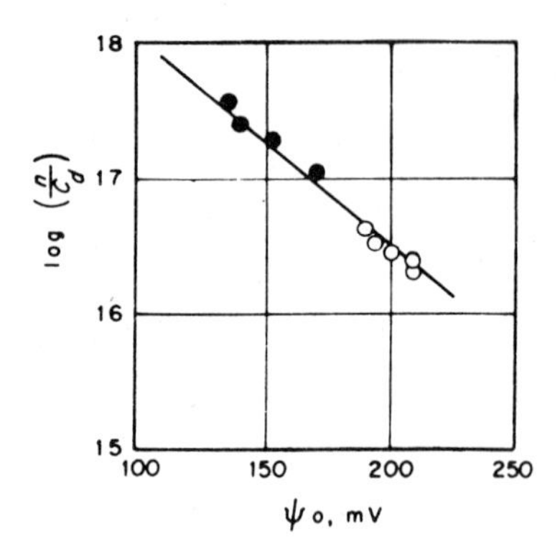

Figure 7.15. Test of isothermal adsorption Equation 7.44 for $C_{10}H_{21}SO_4^-$ monolayers adsorbed at the oil-water interface at 50°C. Points are calculated from the tabulated data of Kling and Lange, open circles referring to adsorption in the absence of salt, and closed circles to adsorption from 3.2×10^{-2} M sodium chloride solution. Units of c_d and n are moles/l and molecules/cm², respectively. The slope of the line is that required by Equation 7.44.(5)

References

1. Posner, A. M., Anderson, J. R., and Alexander, A. E., *J. Colloid Sci.* **7**, 623 (1952).
2. Hsiao, L., and Dunning, H. N., *J. Phys. Chem.* **59**, 362 (1955).
3. Matijevic, E., and Pethica, B. A., *Trans. Faraday Soc.* **54**, 1382 (1958).
4. van Voorst Vader, F., *Trans Faraday Soc.* **56**, 1067 (1960).
5. Davies, J. T., "Surface Phenomena in Chemistry and Biology", edited by J. F. Danielli, K. D. A. Parkhurst, and A. C. Riddiford, New York, Pergamon Press, 1958.
6. Schulman, J. H., and Hughes, A. H., *Proc. Roy. Soc.* (London) A, **138** 430 (1932).
7. Davies, J. T., *Trans. Faraday Soc.* **48**, 1052 (1952.
8. Davies, J. T., *Proc. Roy. Soc.* (London) **A208**, 224 (1951).
9. Crisp, D. J., in "Surface Chemistry", London, Butterworth, 1949.
10. Phillips, J. N., and Rideal, E. K., *Proc. Roy. Soc.* (London) **A232**, 159 (1955).
11. Haydon, D. A., *J. Colloid Sci.* **13**, 159 (1958).
12. Donnan, F. G., and Barker, J. T., *Proc. Roy. Soc.* (London) **A85**, 557 (1911).
13. McBaine, J. W., and Dubois, R., *J. Am. Chem. Soc.* **51**, 3534 (1929).
14. McBaine, J. W., and Humphreys, C. W., *J. Phys. Chem.* **36**, 300 (1932).
15. McBaine, J. W., and Swain, R. C., *Proc. Roy. Soc.* (London) **A154**, 608 (1936).
16. Dixon, J. K., Weith, A. J., Jr., Argyle, A. A., and Salley, D. J., *Nature* **163,** 845 (1949).
17. Hutchinson, E., *J. Colloid Sci.* **4**, 599 (1949).
18. Dixon, J. K., Judson, C. M., and Salley, D. J., "Monomolecular Layers", edited by H. Sobotka, Washington, D.C., American Association for the Advancement of Science, 1954.
19. Aniansson, G., *J. Phys. & Colloid Sci.* **55**, 1286 (1951).
20. Salley, D. J., Weith, A. J., Jr., Argyle, A. A., and Dixon, J. K., *Proc. Roy. Soc.* (London) **A203**, 42 (1950).
21. Nilsson, G., *J. Phys. Chem.* **61**, 1135 (1957).
22. Davies, J. T., *J. Colloid Sci.* **11**, 377 (1956).
23. Pethica, B. A., *Trans. Faraday Soc.* **50**, 413 (1954).
24. Roe, C. P., and Brass, P. D., *J. Am. Chem. Soc.* **76**, 4703 (1954).
25. Wilson, A., Epstein, M. B., and Ross, J., *J. Colloid Sci.* **12**, 345 (1957).
26. Shinoda, K., and Mashio, K., *J. Phys. Chem.* **64**, 54 (1960).
27. Defay, R., and Hommelen, J. R., *J. Colloid Sci.* **13**, 553 (1958).
28. Flengas, S. N., and Rideal, E. K., *Trans. Faraday Soc.* **55**, 339 (1959).
29. Hansen, R. S., and Wallace, T. C., *J. Phys. Chem.* **63**, 1085 (1959).
30. Sutherland, K. L., *Rev. Pure and Applied Chem.* (Australia) **1**, 35 (1951).
31. Rideal, E. K., and Sutherland, K. L., *Trans. Faraday Soc.* **48**, 1109 (1952).
32. Kling, W., and Lange, H., Proc. Intern. Congr. Surface Activity, 2nd., Vol. 1, p. 295, London, Butterworth, (1957).

CHAPTER 8

Surfactants

The various surface-active agents used in industry, as well as those without present commercial importance, but reported in technical literature and patents, have been thoroughly reviewed by Schwartz and his associates (1, 2). Consequently, only those surfactants that are of particular commercial importance or theoretical interest are briefly considered here.

Surface activity has been defined as the pronounced tendency of a solute to concentrate at an interface. Molecules having certain configurations exhibit surface activity. In general, these molecules are composed of two segregated portions, one of which has sufficient affinity for the solvent to bring the entire molecule into solution. The other portion is rejected by the solvent, because it has less affinity for the solvent molecules than the solvent molecules have for each other. If the forces rejecting this group are sufficiently strong, the solute molecules will tend to concentrate at an interface, so that at least part of the area of the rejected group is not in contact with solvent molecules. Conventional soaps behave as surfactants in water, in accordance with this description. The polar group has sufficient affinity for water to cause the soap to be water soluble. Rejection of the hydrocarbon chain by the water molecules results in the concentration of the soap at water-air and water-oil interfaces. Sucrose is not surface active, because the polar hydroxyl groups and the nonpolar methylene groups are not segregated.

Conventional soaps will concentrate at a metal-water interface, because the polar group has a higher affinity for the metal than it does for water. While the hydrocarbon chain may assist in the adsorption process, it is of secondary importance. Long-chain amines dissolved in hydrocarbon oils will concentrate at an oil-metal or oil-water interface, but not at an oil-air interface. The amine group does not have less affinity than the hydrocarbon chain for the hydrocarbon solvent. However, it has a stronger affinity for water and other polar molecules and atoms than for the hydrocarbon solvent. Ethanol is not considered to be a surfactant, though it will preferentially adsorb on silica gel from its solution in hexane. Water-insoluble compounds, such as octadecanol, will spread as monolayers on water, and many insoluble powders will concentrate at oil-water interfaces.

The classification of materials as surface-active agents is arbitrary. For hydrocarbon-chain compounds, it is generally considered that the molecule should contain at least about eight CH_2 groups in the chain. It is then classified as either a water- or oil-soluble surfactant. A fluorocarbon derivative with four CF_2 groups in the chain is approximately equal in surface activity to a hydrocarbon derivative with twice that number of CH_2 groups. Silicones are sometimes considered to be surfactants.

OIL-SOLUBLE SURFACTANTS

Surface-active compounds that are soluble in hydrocarbon solvents can be classified according to the following types:

(1) Long-chain polar compounds
(2) Fluorocarbon compounds
(3) Silicones

The long-chain hydrocarbons with polar groups do not lower the surface tension of hydrocarbon liquids. However, they lower the oil-water interfacial tension and they are adsorbed on polar surfaces. Typical polar groups are $-COOH$, $-OH$, $-NH_2$, $-CONH_2$, $-SH$, $-SO_3H$, and salts of long-chain carboxylic acids and sulfonates.

Short-chain fluorocarbons with polar groups are frequently sufficiently soluble in hydrocarbon oils to function as surfactants, lowering surface tension as well as interfacial tension. Longer-chain fluorocarbons attached to a hydrocarbon chain of sufficient length are soluble in hydrocarbon oils, and lower the surface tension of the oils. They do not lower the interfacial tension between the hydrocarbon solvent and polar compounds.

Silicone oils differ broadly in their chemical structure and surface-active properties. Generally, they are used as insoluble components of the system and serve as antifoaming agents. However, those of sufficiently small molecular weight to be soluble in the hydrocarbon solvent, and containing only CH_3 groups attached to silicon in the $(Si-O)_n$ skeleton can be expected to lower the surface tension of the hydrocarbon solvent. This is due to the lower surface free-energy of the CH_3 group as compared with the CH_2 group, which predominates in hydrocarbons.

If the non-aqueous solvent is not a hydrocarbon, the situation is somewhat more complicated. Various aspects of surface activity in these fluids are discussed in other chapters.

WATER-SOLUBLE SURFACTANTS

Most of the interest in surface chemistry has centered about the behavior of surfactants in aqueous systems. Thousands of compounds have been synthesized for use in wetting, foaming, antifoaming, emulsification, demulsification, detergency, corrosion-inhibition, flotation, and numerous other applications employing water. These water-soluble surfactants are generally classified according to four types:

(1) Anionic surfactants ionize in solution, with the long chain carrying a negative charge.

(2) Cationic surfactants ionize in solution, with the long chain bearing a positive charge.

(3) Nonionic surfactants do not ionize in solution.

(4) Amphoteric or ampholytic surfactants ionize in solution, with the long-chain ion carrying either a positive or negative charge, depending upon the pH of the solution.

Anionic Surfactants

Soaps The water-soluble salts of long-chain carboxylic acids have been used as surfactants since antiquity. The commercially important soaps are derived from vegetable oils and animal fats. These soaps generally contain from 12 to 18 carbon atoms per molecule. Shorter-chain fatty-acid soaps are present in soap derived from coconut oil, while spermacetti soaps contain unsaturated fatty acids with 20 and 22 carbon atoms in the chain. Only those fatty acids having an even number of carbon atoms per molecule are found in abundance in natural products.

Soaps prepared with unsaturated fatty acids are more water soluble, but somewhat poorer in detergency, than the soaps of the saturated fatty acids. Soaps of lauric and myristic acid are more water soluble and foam more abundantly than the soaps of palmitic and stearic acid.

The most commonly used soaps are sodium and potassium soaps. However ammonia is used in the preparation of certain specialty soaps, and the use of a wide variety of amines in soap preparation is expanding. These amines include morpholine, mono-, di- and tri-ethanolamine, various isopropanolamines, and water-soluble alkyl amines.

The generally good detergent action of soaps is due to the unbranched hydrocarbon chain as well as the tendency of soaps to hydrolyze and form mixed surface-films containing both soap and fatty acid. The major drawbacks of soap are incompatibility with acid, and with calcium and magnesium salts present in hard water.

Sulfonates. The volume of anionic surfactants, other than soaps, produced in the United States is about one billion pounds. Of this amount, approximately one-half the production is accounted for by sodium alkylbenzene sulfonate, sometimes referred to as sodium dodecylbenzene sulfonate or sodium propylenetetramer benzene sulfonate.

The alkylbenzene is manufactured by first polymerizing propylene to the tetramer, or dodecene–1, and then condensing with benzene. Hydrogen fluoride and less commonly aluminium chloride are used as condensing agents.

The detergent alkylate is actually a mixture of branched-chain alkylbenzenes, with 10 to 15 carbon atoms in the alkyl group. The average molecular weight is about 246, corresponding to a dodecylbenzene. The tendency has been to gradually increase the average molecular weight. Detergent alkylates corresponding to tridecylbenzene and pentadecylbenzene are also available. Sulfonates derived from the higher molecular weight alkylates are reported to have better detergency and foam stability.

The alkylate is sulfonated with oleum or sulfur trioxide, and neutralized. Most alkylbenzene sulfonate is produced as the sodium salt in combination with sodium sulfate, the ratio of active ingredient to inorganic salt varying with the method of manufacture.

Sodium dodecylbenzene sulfonate, when properly formulated, is an effective low-cost detergent. It is widely used in all types of industrial and household detergent compositions, and is the major surfactant found in heavy-duty household detergents of the high-foaming type, such as "Tide", "Fab", and "Surf". It is also used in applications requiring high foam, wetting, or emulsification properties.

Dodecylbenezene sulfonates are also sold in solution as ammonium or ethanolamine salts, where the application requires the preparation of a concentrate in which greater water solubility is required than can be obtained with the sodium salt.

The effect of surfactant concentration on surface tension and on wetting for a commercial sodium dodecylbenzene sulfonate is shown in Tables 8.1 and 8.2. In distilled water, the surface tension decreases rapidly with increasing surfactant concentration to a minimum value of 28.6 dynes/cm. With added electrolyte, the surface-tension values are considerably lower at low surfactant concentrations.

A variety of low molecular-weight sulfonates are also available. These include toluene, xylene, and isobutylnapthalene sulfonates. These materials are employed as coupling agents or hydrotropic agents to solubilize water-insoluble organic materials. They also increase the solubility of the higher-

TABLE 8.1. SURFACE TENSION VALUES OF AQUEOUS SOLUTIONS OF A COMMERCIAL MATERIAL CONSISTING OF 85 PER CENT SODIUM DODECYLBENZENE SULFONATE AND 15 PER CENT SODIUM SULFATE (3).

	Surface Tension at 25°C in Dynes cm^{-1}				
Conc., %	Distilled water	100 ppm water	300 ppm water	2% NaOH	3% H_2SO_4
0.001	66.0	48.0	43.0	42.9	42.0
0.01	38.0	32.0	28.0	29.0	28.5
0.05	29.5	26.7	27.2	27.0	27.5
0.10	28.6	26.8	27.4	27.0	27.2
0.15	28.6	26.7	27.3	27.0	27.3
0.20	28.9	26.7	27.4	27.2	27.2
0.25	29.0	26.5	27.3	—	—
0.30	28.9	26.6	27.4	—	—
0.50	29.0	26.7	27.3	—	—

TABLE 8.2. CLARKSON-DRAVES TEST. TIME REQUIRED FOR COTTON SKEINS WEIGHTED WITH A 3 G. HOOK TO SINK IN SOLUTIONS OF 85% SODIUM DODECYLBENZENE SULFONATE, 15% SODIUM SULFATE (3).

	Wetting Time at 25°C in Seconds				
Conc., %	Distilled Water	100 ppm Water	300 ppm Water	2% NaOH	3% H_2SO_4
0.05	26	33	40	50	40
0.10	8	8	8	19*	9
0.15	5	5	5	12*	5
0.20	3	3	3	9*	3

* Solution hazy.

molecular-weight surfactants in water or in solutions of electrolytes. They are frequently used in the formulation of liquid detergents.

Petroleum sulfonates are alkylaryl sulfonates produced by reacting sulfuric acid with selected petroleum distillates. Treatment of a petroleum product with sulfuric acid results in the formation of an oil layer and a water layer, containing oil-soluble and water-soluble petroleum sulfonates, respectively. The oil-soluble sulfonates are known as mahogany acids, while the water-soluble sulfonates are called green acids, because of their respective colors. These sulfonates are widely used in the preparation of wettable powders, self-emulsifiable insecticide concentrates, and cutting oils. The oil-soluble petroleum sulfonates in the form of their sodium or calcium salts are used as rust inhibitors in lubricating oils and protective oils.

Another important group of sulfonates are the Igepons, prepared by reacting a fatty acid chloride with sodium isethionate, methyltaurine, or cyclohexyltaurine. Since the latter are amides, they are more resistant to hydrolysis than Igepons prepared from sodium isethionate. These surfactants are used in the bleaching, dyeing, and scouring of textiles, fat liquoring of leather, breaking petroleum emulsions, and various detergent uses. One type is used in the manufacture of synthetic-detergent toilet bars.

TABLE 8.3. IGEPON SURFACTANTS (4).

$$RCOOH + HN(CH_3)CH_2CH_2SO_3Na \rightarrow RCON(CH_3)CH_2CH_2SO_3Na + H_2O$$

Fatty Acid — Methyltaurine

$$RCOOH + HOCH_2CH_2SO_3Na \rightarrow RCOOCH_2CH_2SO_3Na + H_2O$$

Fatty Acid — Sodium isethionate

	Fatty Acid	Organic Sulfonate
Igepon AP	Oleic acid	Sodium isethionate
Igepon AC	Coconut oil acid	Sodium isethionate
Igepon T	Oleic acid	Methyltaurine
Igepon TK	Tall oil acid	Methyltaurine
Igepon TN	Palmitic acid	Methyltaurine
Igepon CN	Palmitic acid	Cyclohexyltaurine

	Igepon Content, %	Salt Content, %
AP–75	45–48	44–47
AC–78	82–87	0.5–1.5
T–51	15–16.5	3.5–4.5
TK–42	19–21	5.5–6.5
TN–71	15–17	79.5–84.5
CN–42	25–30	7–9

Various members of the Igepon series are shown in Table 8.3. Surface tension and interfacial tension against purified mineral oil are plotted in Figures 8.1. and 8.2.

The dialkyl esters of sodium sulfosuccinic acid are among the most powerful wetting agents known. They are prepared by esterifying maleic acid with the appropriate alcohol, followed by heating with a concentrated aqueous solution of sodium bisulfite

$$\begin{matrix} HC\,COOR \\ \| \\ HC\,COOR \end{matrix} + Na\,HSO_3 \rightarrow \begin{matrix} CH_2COOR \\ | \\ CH\,COOR \\ | \\ SO_3Na \end{matrix} \qquad (8.1)$$

where R is an alkyl group.

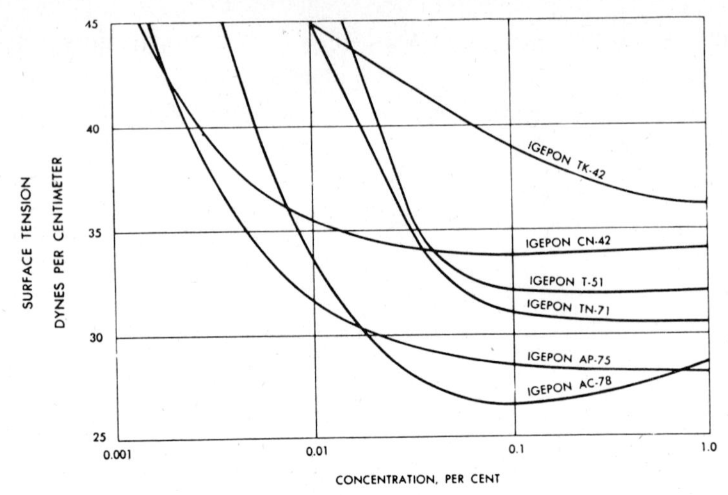

Figure 8.1. Surface tension of Igepons (4).

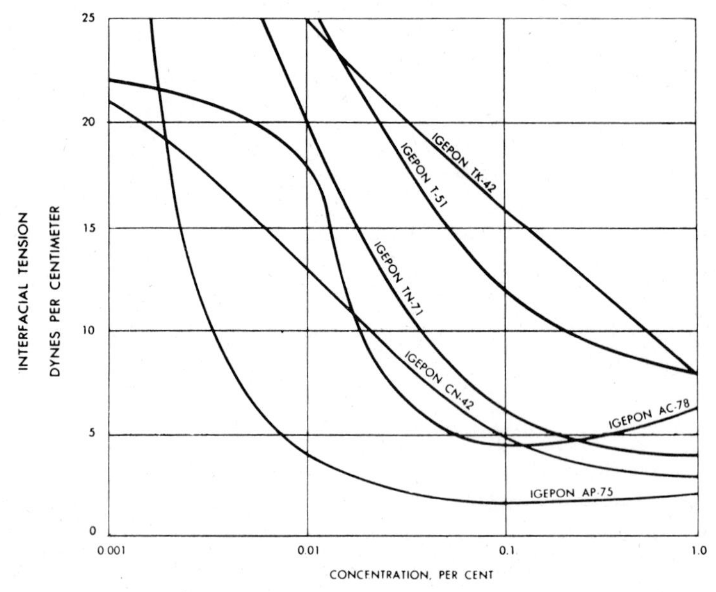

Figure 8.2. Interfacial tension of Igepons in water *versus* white mineral oil (4).

Typical members of this series are listed in Table 8.4. Aerosol OT, prepared from the 2-ethylhexanol ester of maleic acid, is best known. Aerosol MA and Aerosol AY are prepared with shorter-chain alcohols. They function as wetting agents in the presence of appreciable concentrations of electrolytes. Surface Active Agent BO, the 2-butyloctyl diester, is a better wetting agent than Aerosol OT at 54°C and higher temperatures.

TABLE 8.4. SURFACTANTS DERIVED FROM SODIUM SULFOSUCCINIC ACID (5).

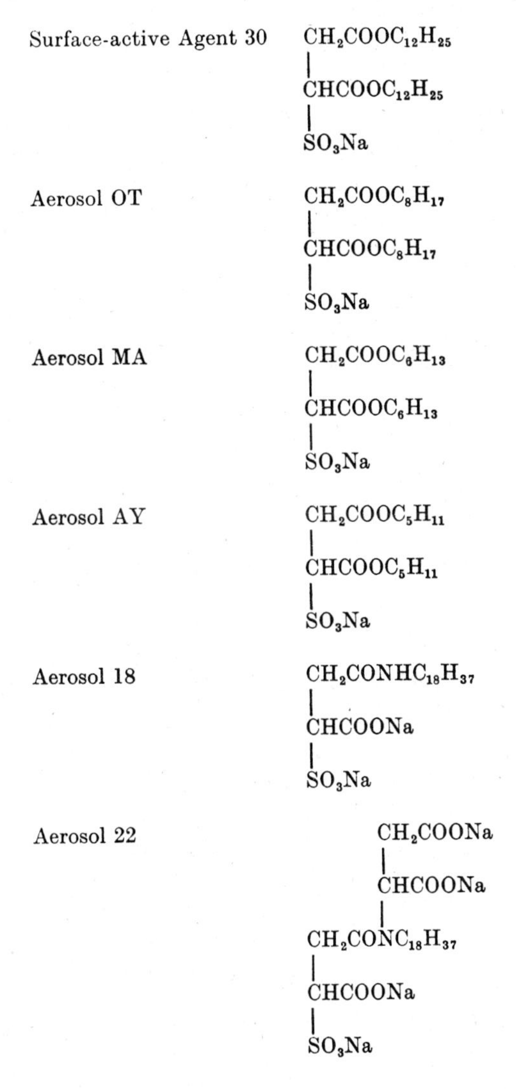

Surface-active Agent 30	$CH_2COOC_{12}H_{25}$ $CHCOOC_{12}H_{25}$ SO_3Na
Aerosol OT	$CH_2COOC_8H_{17}$ $CHCOOC_8H_{17}$ SO_3Na
Aerosol MA	$CH_2COOC_6H_{13}$ $CHCOOC_6H_{13}$ SO_3Na
Aerosol AY	$CH_2COOC_5H_{11}$ $CHCOOC_5H_{11}$ SO_3Na
Aerosol 18	$CH_2CONHC_{18}H_{37}$ $CHCOONa$ SO_3Na
Aerosol 22	CH_2COONa $CHCOONa$ $CH_2CONC_{18}H_{37}$ $CHCOONa$ SO_3Na

Aerosol 22 is unusual in that it is soluble in hot or cold water in all proportions, as well as in saturated salt solutions. It is reported to be an excellent solubilizing and antigelling agent for soaps and other surfactants. When used in emulsion polymerization, it produces emulsions of extremely fine particle size — 0.05 to 0.09 micron —, with a narrow particle-size distribution (5). Surface tension data for several of the sodium sulfosuccinate esters are illustrated in Figures 8.3 and 8.4.

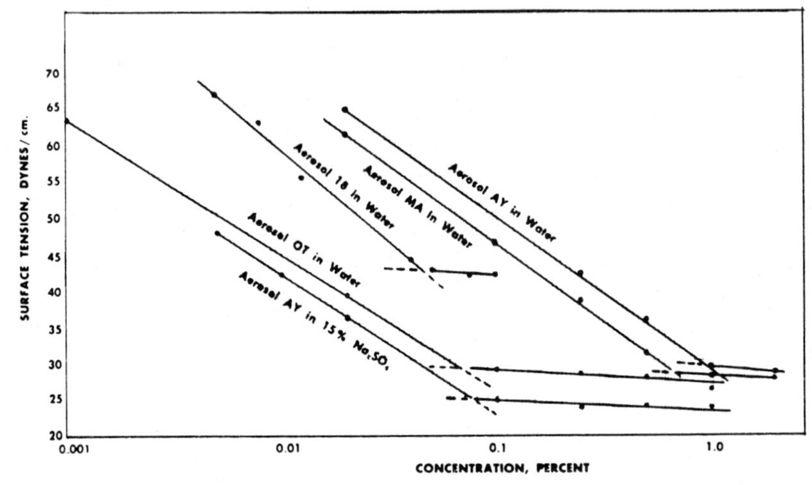

Figure 8.3. Surface tension of Aerosols at 25°C (5).

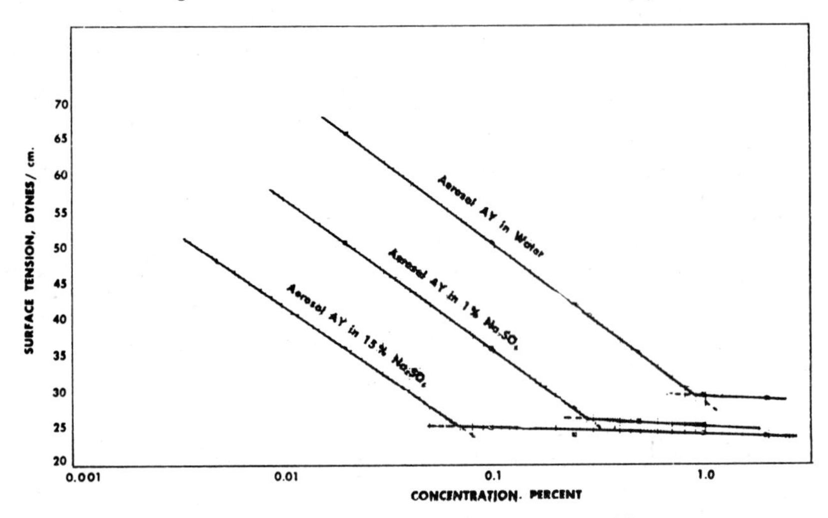

Figure 8.4. Surface tension of Aerosol AY at 25°C (5).

Another class of surfactants containing both a carboxylate and sulfonate group is exemplified by the salts of α-sulfostearic acid and α-sulfopalmitic acid. These have been recommended for the preparation of synthetic-detergent toilet bars (6).

Sulfates. One of the oldest class of synthetic surfactants is the sulfated vegetable oils. The group includes "Turkey Red Oil" or sulfated castor oil and "sulfonated red oil" which is actually a sulfated oleic acid. These oils are frequently used in fat liquoring applications of various types, for corrosion inhibition, and for the preparation of emulsion systems. At one time they were widely used in textile processing.

The alkyl sulfates are among the more important sulfates used today. They are prepared by the reaction of a fatty alcohol with sulfuric acid or chlorosulfonic acid

$$C_{12}H_{25}OH + H_2SO_4 \rightarrow C_{12}H_{25}OSO_3H \quad (8.2)$$

Alkyl sulfates derived from coconut-oil or fractionated-coconut-oil fatty alcohols are widely used in the preparation of shampoos. The sodium salt is employed in the cream-type shampoos, while triethanolamine and other amine salts are used to prepare clear liquid shampoos. At one time considerable tonnage went into the preparation of liquid and solid household detergent compositions. Alkyl sulfates derived from tallow fatty alcohols find an important use in heavy-duty household detergents. All of the alkyl sulfates find wide application in the preparation of cosmetic products. Sodium laury sulfate is used in toothpastes. The foaming action of the alkyl sulfates is highest when the alkyl group is composed of 12 carbon atoms.

TABLE 8.5. SURFACE TENSION VALUES OF TERGITOL ANIONIC WETTING AGENTS IN DISTILLED WATER (7).

	Surface Tension at 25°C in Dynes cm^{-1}		
Conc. (%)	Anionic 08	Anionic 4	Anionic 7
0.05	65.0	51.5	36.5
0.10	62.0	47.0	33.5
0.50	49.0	36.0	27.6
1.00	44.0	31.0	27.5
Anionic 08	$C_4H_9CH(C_2H_5)CH_2SO_4Na$, 38 to 40 per cent active ingredient		
Anionic 4	$C_4H_9CH(C_2H_4)C_2H_4CH(SO_4Na)CH_2CH(CH_3)_2$ 26 to 28 per cent active ingredient		
Anionic 7	$C_4H_9CH(C_2H_5)C_2H_4CH(SO_4Na)C_2H_4CH(C_2H_5)_2$ 25 to 27 per cent active ingredient		

Sulfated branched-chain alcohols are important wetting agents. Several members of this class are shown in Table 8.5. The surface tension decreases with an increase in the size of the hydrocarbon portion of the molecule.

The lower members of the series have lower surface tension values and are more effective as wetting agents in the presence of appreciable concentrations of electrolytes.

Salts of sulfated alkylphenoxyethanol and sulfated alkylphenoxyethoxyethanol are used in liquid detergent compositions. The formulas for these compounds may be represented as

$$R\text{-}C_6H_4\text{-}OC_2H_4OSO_3Na \qquad R\text{-}C_6H_4\text{-}OC_2H_4OC_2H_4OSO_3Na \qquad (8.3)$$

sodium alkylphenoxyethylsulfate — sodium alkylphenoxyethoxyethylsulfate

When the alkyl group R consists of 8 or 9 carbon atoms, these compounds have high water solubility, good foam stability, and detergent action.

Cationic Surfactants

There are two general categories of cationic surfactants. The first of these consists of long-chain primary, secondary, and tertiary amines. These are water soluble only in acidic solution, where they ionize to form a long-chain cation and a simple-salt anion. The ethoxylated amines are a variation of this type. These compounds are soluble in water over the entire pH range. However, the surfactant is ionized and carries a positive charge only in acid solution. The cationic character decreases with an increase in the molar ratio of ethylene oxide to long-chain amine. These compounds are illustrated in Table 8.6.

TABLE 8.6. COMPOUNDS WITH CATIONIC CHARACTER IN ACID MEDIA. R IS A LONG-CHAIN ALKYL GROUP. R′ AND R″ ARE GENERALLY SHORT-CHAIN ALKYL GROUPS

Compound	Type
$R\text{—}NH_2$	Primary amine
$RR'NH$	Secondary amine
$RR'R''N$	Tertiary amine
$R\text{—}NHCH_2CH_2CH_2NH_2$	Diamine
$R\text{—}N[(CH_2CH_2O)_xH][(CH_2CH_2O)_yH]$	Polyethoxylated amine
$R\text{—}N[(CH_2CH_2O)_2H]CH_2CH_2CH_2N[(CH_2CH_2O)_xH][(CH_2CH_2O)_yH]$	Polyethoxylated diamine

The second important category of cationic surfactants is the quaternary ammonium compounds. These ionize to form long-chain cations at all pH levels. They are prepared by quaternizing a tertiary amine with an alkyl halide. Several members of this class are shown in Table 8.7.

TABLE 8.7. QUARTENARY AMMONIUM COMPOUNDS.

Structure	Name
$\left[C_{18}H_{37}N(CH_3)_3 \right]^+ Cl^-$	Aquad 18–50
$\left[C_{17}H_{35}CONH(CH_2)_3N(CH_3)_2(C_2H_4OH) \right]^+ Cl^-$	Aerosol SE
$\left[C_8H_{17}C_6H_4OC_2H_4OC_2H_4{-}N(CH_3)_2(CH_2C_6H_5) \right]^+ Cl^-$	Hyamine 1622
$\left[C_{17}H_{35}CONHCH_2{-}NC_5H_5 \right]^+ Cl^-$	Zelan
$[C_{11}H_{23}COOC_2H_4NHCOCH_2N(CH_3)_3]^+ Cl^-$	Emulsept
$\left[C_{18}H_{37}N(CH_3)_2(CH_2C_6H_5) \right]^+ Cl^-$	Triton K–60

Anionic and cationic compounds are incompatible, forming precipitates in aqueous solution. The precipitated salt of the two long-chain ions can be solubilized in an excess of either component. The cationic surfactants, like the anionic compounds, are compatible with the nonionic surfactants.

The cationic agents are strongly adsorbed on negatively-charged surfaces. Since most surfaces in contact with water carry a negative charge, the cationic surfactants are more readily exhausted from solution by adsorption than other types of surfactants.

All of the long-chain cationic agents are germicidal. Quaternary ammonium compounds are among the most potent bactericides known, and are widely used for this purpose. Zelan is used as a water-proofing agent. Triton K–60 is an effective softening agent for textiles and paper. Other cationic agents are used as corrosion inhibitors, as well as for ore flotation.

Nonionic Surfactants

Fatty Alkanolamides. The older or Kritchevsky (8) type of fatty alkanolamide is prepared by heating a fatty acid with a 50 to 150 per cent molar excess of diethanolamine to form the amide,

$$RCOOH + HN\begin{matrix} \diagup CH_2CH_2OH \\ \diagdown CH_2CH_2OH \end{matrix} \rightarrow R\overset{\overset{O}{\|}}{C}N\begin{matrix} \diagup CH_2CH_2OH \\ \diagdown CH_2CH_2OH \end{matrix} \quad (8.4)$$

The product is a complex mixture containing, in addition to the amide, the alkanolamine soap of the fatty acid, amine ester, amide ester, amine diester, amide diester, substituted piperazines, and other cyclic derivatives. The more recent procedure is to react the methyl ester of the fatty acid with a slight excess of diethanolamine. This product contains a higher amide content, and is free of cyclic derivatives (9).

In addition to the diethanolamides, the fatty alkanolamides derived from monoethanolamine and monoisopropanolamine are commercially important. The fatty acid is usually lauric acid or a stripped coconut fatty acid.

Production of fatty alkanolamides was approximately 100 million pounds in 1959 (10). Their principal use is to stabilize the foaming action of anionic surfactants in the presence of soil. They are also known to enhance detergency. The dialkanolamides are used in liquid detergent compositions, while the monoalkanolamides are used in solid compositions. The monoalkanolamides are water-insoluble, but are solubilized by anionic surfactants.

Ethylene-oxide-derived Nonionic Surfactants (11). Ethylene oxide will react with any material containing an active hydrogen. Since the product of this reaction is an alcohol, which also contains an active hydrogen, it will react further to form a water-soluble ethoxy chain. If the initial material is a hydrophobic compound of suitable molecular weight, it will become a surface-active agent after a sufficient number of ethoxy groups have been added.

$$ROH + \underset{\diagdown O \diagup}{CH_2CH_2} \rightarrow ROCH_2CH_2OH \quad (8.5)$$

$$ROCH_2CH_2OH + X\underset{\diagdown O \diagup}{CH_2CH_2} \rightarrow RO(CH_2CH_2O)_xCH_2CH_2OH \quad (8.6)$$

Table 8.8 shows the structure of various commercial surfactants derived from ethylene oxide. The list is not complete. The alkylphenol-derived polyoxyethylene surfactants are widely used for industrial and household cleaning. The alkyl group is generally octyl or nonyl, with a chain of approximately

10 ethoxy groups attached to the phenyl group. Products with longer and shorter ethoxy chains are also used. Those with shorter chains have antifoaming characteristics and serve as coemulsifiers and cosolvents. They are sulfated to obtain high-foaming detergents. Alkylphenol derivatives with

TABLE 8.8. NONIONIC SURFACTANTS DERIVED FROM ETHYLENE OXIDE. R IS A LONG-CHAIN ALKYL GROUP.

Alkylphenol	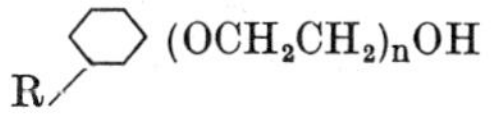 $R-C_6H_4-(OCH_2CH_2)_nOH$
Fatty alcohol	$R(OCH_2CH_2)_nOH$
Fatty acid	$RC(=O)(OCH_2CH_2)_nOH$
Fatty mercaptan	$RSCH_2CH_2(OCH_2CH_2)_nOH$
Fatty amines	$RN[CH_2CH_2(OCH_2CH_2)_nOH]_2$
	$R(R')NCH_2CH_2(OCH_2CH_2)_nOH$
Polyoxypropylene glycol	$HO(CH_2CH_2O)_a(CH(CH_3)CH_2O)_b(CH_2CH_2O)_cH$
Fatty sorbitan ester	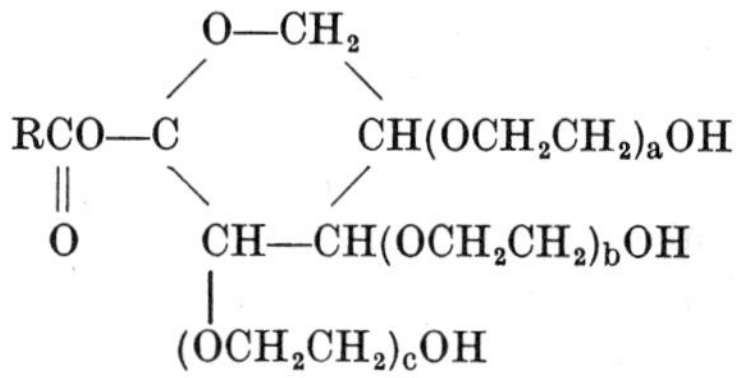 RCO—C (C=O) in ring O—CH$_2$; ring substituents $CH(OCH_2CH_2)_aOH$, $CH—CH(OCH_2CH_2)_bOH$, $(OCH_2CH_2)_cOH$

high ratios of ethylene oxide are used as latex stabilizers and dispersants for solids. The effect of increasing the ethylene oxide to hydrophobe ration on many of the properties of the surfactants is shown in Figures 8.5 and 8.6.

The tall oil-polyoxyethylene derivative is used in low foaming household detergents. Tall oil is a by-product of the Kraft process for making paper, and consists primarily of fatty and rosin acids related to oleic acid and abietic acid. The ethylene-oxide derivatives of tall oil are relatively low-cost, mediocre detergents. Ethylene oxide derivatives of fatty acids and fatty sorbitan esters are widely used in the preparation of cosmetic and pharmaceutical emulsions.

Ethers with surfactant properties are produced by reacting ethylene oxide with fatty alcohols, the branched-chain tridecyl alcohol from the Oxo process, lanolin and lecethin alcohols, among others. These ethers, like the alkylphenol derivatives, are resistant to hydrolysis. Those derived from fatty alcohols and tridecyl alcohol are used in numerous textile processes and detergent applications.

The ethoxylated amines are unusual in that they have cationic activity,

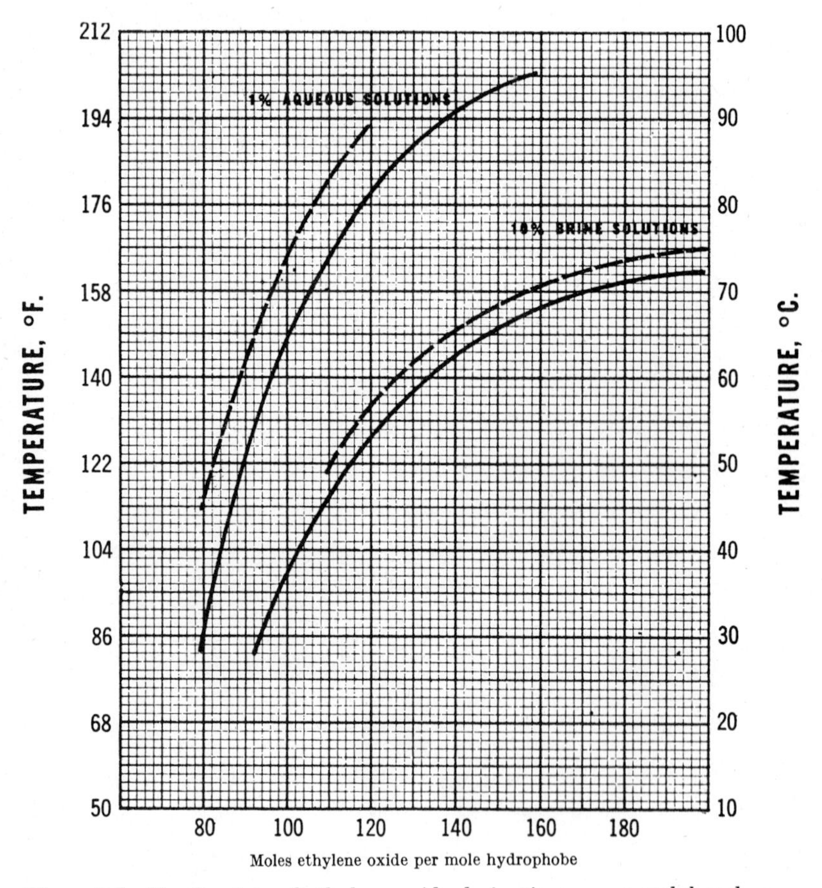

Figure 8.5. Cloud points of ethylene-oxide derivatives; —, nonylphenol; – – –, tridecyl alcohol (12).

which results in strong adsorption on surfaces. Consequently, these ethoxylated amines are used in such applications as corrosion inhibition, dispersion, flocculation, and textile softening. They are also employed as germicides, antistatic agents, and for the preparation of substantive emulsions. In other respects, they behave as nonionic surfactants.

"Block polymer" surfactants are prepared by condensing propylene oxide onto propylene glycol to obtain a series of hydrophobes ranging in molecular weight from 800 to 2500. The molecular weight of 800 is the minimum for significant surface activity. Ethylene oxide is then condensed onto both ends of the "block polymer", to give products ranging in molecular weight from about 1,000 to 27,000. In a variation of these products, the starting material is ethylene diamine, which is condensed with propylene oxide to give four

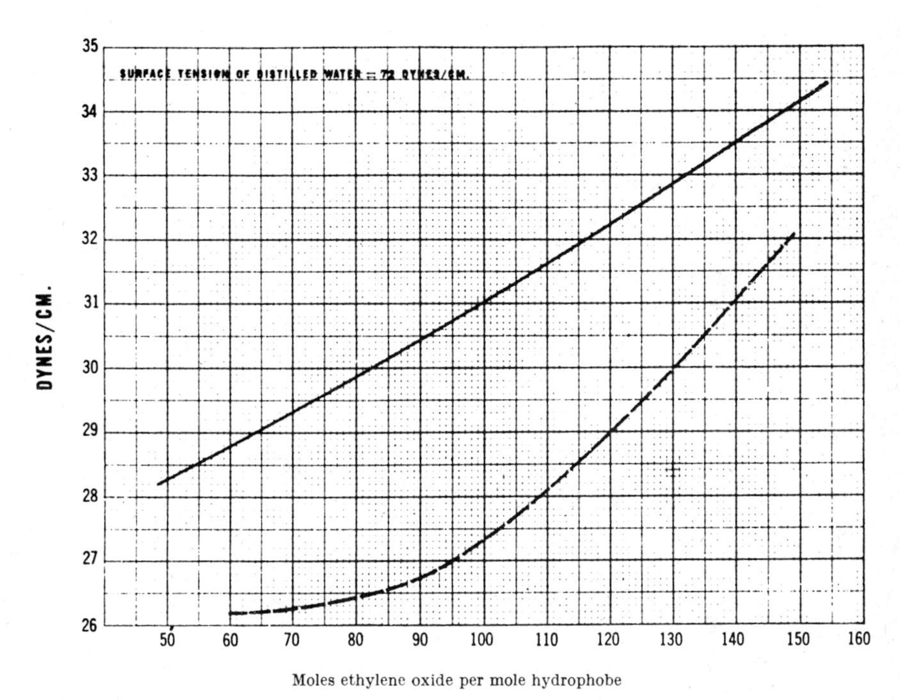

Figure 8.6. Surface tension of 0.1 weight per cent solutions of ethylene-oxide derivatives; —, nonylphenol; – – –, tridecyl alcohol (12).

polyoxypropylene branches. Polyoxyethylene adds on all four branches. These polymeric surfactants produce moderate to almost no foam. Many are good dispersing or wetting agents.

Sugar Esters (13). A new class of nonionic surfactants has been developed with a sugar group as the hydrophile. These sugar esters are prepared by an alcoholysis reaction between the methyl ester of a fatty acid and sucrose or raffinose. While diesters and higher can be obtained by this reaction, only the monoester of the C–12 to C–18 fatty acids are water soluble. There are several important differences between the sugar esters and the ethylene oxide derivatives. The latter would appear to have an unlimited degree of hydrophilic character, depending upon the number of ethylene oxide groups introduced per molecule. In fact, the hydrophilic character does not increase significantly after the addition of twenty ethoxy groups in a chain. The sucrose moiety confers more hydrophilic character to the molecule, approaching that of ionic groups.

Amphoteric Surfactants

Amphoteric surfactants contain both an anionic and a cationic group. This may be illustrated by the long-chain amino acids. These are cationic in acid solution and anionic in alkaline media.

$$RNHCH_2\,COOH \begin{array}{l} \xrightarrow{HCl} (RNH_2\,CH_2\,COOH)^+\ Cl^- \\ \xrightarrow{NaOH} (RNHCH_2\,COO)^-\ Na^+ \end{array} \qquad (8.7)$$

Amphoteric surfactants may also be prepared with a sulfate, sulfonate or phosphate as the anionic group. Lecithin is an example of the latter.

TABLE 8.9. AMPHOTERIC SURFACTANTS. R IS A LONG-CHAIN ALKYL GROUP.

$RNHCH_2CH_2COONa$

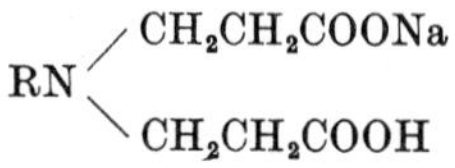

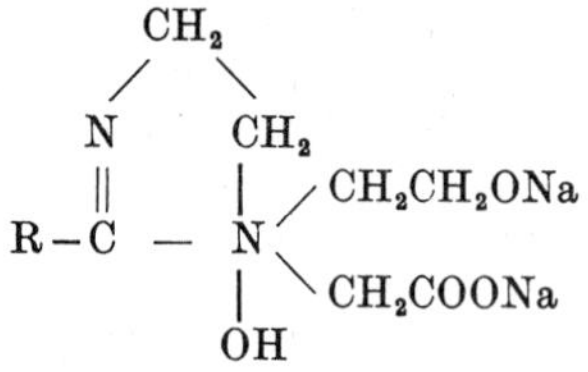

$$RN\begin{cases} CH_2CH_2COONa \\ CH_2CH_2COOH \end{cases}$$

```
      CH2
     /   \
    N     CH2
    ||    |  / CH2CH2ONa
R - C  -  N
          |  \ CH2COONa
          OH
```

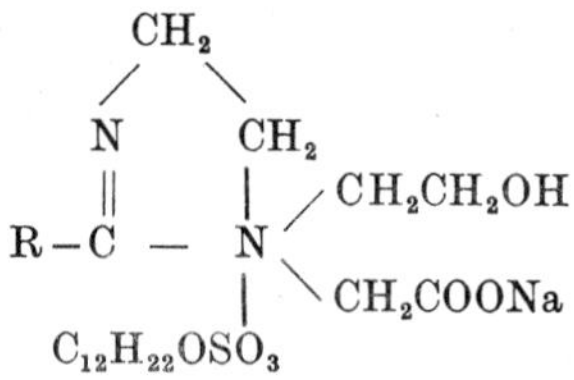

```
      CH2
     /   \
    N     CH2
    ||    |  / CH2CH2OH
R - C  -  N
          |  \ CH2COONa
      C12H22OSO3
```

Imidazoline and amino acid amphoteric surfactants are illustrated in Table 8.9.

A characteristic feature of amphoteric surfactants is the pH dependence of their surface-active properties. Solubility, foam, wetting, and depression of surface tension are at their minimum at pH values in the neighborhood of the isoelectric point. In neutral or slightly alkaline solution they exhibit the high foaming action characteristic of anionic surfactants, while showing the substantivity usually associated with cationic agents. As a consequence of their substantivity, they act as both detergents and conditioning agents in shampoos. Further, they are relatively noncorrosive to metals, unlike most anionic surfactants in dilute solution.

Miscellaneous Types

Polymeric Surfactants. Water-soluble polymers exhibit varying degrees of surface activity, depending upon their size and the distribution of hydrophobic and hydrophilic groups. Many of these gums are effective dispersing and suspending agents, as well as exhibiting some foaming action and lowering the surface tension of their solutions. Anionic polymers include sodium carboxymethylcellulose, which is used in heavy-duty detergent compositions to prevent soil redeposition onto cotton fabrics. Polyacrylates are used as auxilliary emulsifiers, and were at one time in prominence as soil-conditioning agents. Salts of the hydrolyzed copolymer of styrene and maleic anhydride are used as emulsifying and dispersing agents.

Nonionic polymer surfactants include polyvinyl alcohol, methyl cellulose, and ethoxylated phenol formaldehyde resins. The latter has been suggested for the demulsification of petroleum crude (14). Polyvinylpyridine is an example of a cationic gum. Surface activity can be enhanced by quaternizing, as shown by the studies of Strauss and coworkers (15, 16, 17) on dodecyl- and dodecylbutyl- polyvinylpyridinium salts. These materials with molecular weights in the range of 300,000 are capable of solubilizing benzene and other hydrocarbons at extreme dilutions. The single polymeric molecule acts as a micelle.

Fluorocarbon Surfactants. There are two processes available by which fluorocarbon surfactants can be produced. In the Simons process, the material to be fluorinated is electrolyzed in anhydrous HF. Perfluoroacids are produced in this manner. The other process, involving the telomerization of tetrafluoroethylene, results in the preparation of omega-hydrogen fluoroalcohols. The hydroxyl group can subsequently be converted to another polar group.

Fluorocarbons form surfaces of much lower free-energy than hydrocarbon surfaces. Consequently, fluorocarbon surfactants lower the surface tension of solutions to a greater extent than hydrocarbon surfactants. Fluorocarbons containing as few as five carbon atoms are surface active. Unlike hydrocarbon fatty acids, the perfluorinated acids are strong acids. Their salts are resistant to water hardness and acid. They also have greater thermal stability and resistance to oxidation than the hydrocarbon derivatives. Thus, the expensive perfluorinated acids are used in chromic acid electroplating baths. The conventional surfactants are rapidly oxidized in these solutions.

References

1. Schwartz, A. M., and Perry, J. W., "Surface Active Agents", New York Interscience Publishers, 1949.
2. Schwartz, A. M., Perry, J. W., and Berch, J., "Surface Active Agents and Detergents", Vol. 2, New York, Interscience Publishers, 1958.
3. The Atlantic Refining Company, "The Ultrawets", 1949.
4. Antara Chemicals, Division of General Dyestuff Corporation, "Igepals"
5. American Cyanamid Company, "Aerosol Surface Active Agents; Aerosol 22".
6. Weil, J. K., Stirton, A. J., Maurer, E. W., Ault, W. C., and Palm, W. E., *J. Am. Oil Chem. Soc.* **35**, 461 (1958).
7. Union Carbide Chemicals Company, "Tergitol Surface Active Agents", 1957.
8. Kritchevsky, W., U.S. Patent 2,089,212 (1937).
9. Levengood, S. M., and Johnson, C. A., "Symposium on Analytical Methods for Surfactants", Proceedings Chemical Specialties Manufacturers Association, 1958.
10. Goerner, J. K., *Chem. Eng. News*, p. 52, April 18, 1960.
11. "Symposium Ethylene Oxide Based Surface Active Agents", Proceedings Chemical Specialties Manufacturers Association, 1957.
12. Jefferson Chemical Company, Inc., "Surfonic Surface-Active Agents", 1958.
13. Osipow, L. I., Snell, F. D., Marra, D., and York, W. C., *Ind. Eng. Chem.* **48**, 1459, 1462 (1956); Hass, H. B., et. al., U.S. Patent 2,893, 990 (1959).
14. De Groote, M., and Keiser, B., U.S. Patents 2,499,365-6-7-8; 2,598,234.
15. Strauss, U. P., and Jackson, E. G., *J. Polymer Sci.* **6**, 649 (1951).
16. Layton, L. H., and Strauss, U.P., *J. Colloid Sci.* **9**, 149 (1954).
17. Strauss, U. P., and Gershfeld, N. L., *J. Phys. Chem.* **58**, 747 (1954).

CHAPTER 9

Properties of Solutions Containing Surfactants

Surface-active molecules are characterized by the presence of a polar and a non-polar group. The polar or active portion of the molecule is surrounded by a strong electromagnatic field and exhibits a high affinity for other polar groups and molecules, including water. The non-polar portion of the molecule has a low affinity for water and other polar molecules. The surface energy of a liquid or a solution depends upon the potential energy of the electromagnetic stray field which extends outward from the surface layer of atoms. For the stray field, and consequently the surface energy, to have a minimum value, it is necessary that the molecules present in the solution arrange themselves so that the least active portions of the various species present in solution be exposed at the surface (1). Thus, for solutions of pure surfactants in water, the surfactant molecules will tend to concentrate at the surface, with the non-polar portion directed outward.

Since there are strong forces of attraction between polar groups and molecules, it is evident that the introduction of a hydrocarbon chain into water requires that work be done to separate water molecules. The energy of the system is increased by an amount equal to the cohesional energy of the water molecules less the weaker adhesional energy between the water molecules and the hydrocarbon chain. A substance comprising molecules with polar and non-polar portions will not dissolve in water unless the energy of interaction of the polar group with water molecules is sufficient to counterbalance the tendency for the water molecules to expel the non-polar group.

The condition that the system exhibit a minimum free energy requires the smallest possible area of contact between the water molecules and the non-polar hydrocarbon chain of the surfactant molecule. There is such minimum contact when surfactant molecules concentrate at the surface and orient with their polar group directed toward the water and their hydrocarbon chain directed outward. In the bulk of the solution minimum contact between hydrocarbon chains and water molecules is accomplished by the clustering of the surfactant molecules in such a manner that the hydrocarbon chains of many molecules are in contact and are surrounded and hidden by polar groups. This aggregation of surfactant molecules is called a micelle.

When the system is in equilibrium, the distribution of molecules between two regions in which the surfactant molecules have different potential energies is given by the Boltzmann equation

$$\frac{n_1}{n_2} = \frac{p_1}{p_2} \mathbf{e}^{\frac{\Delta E}{kT}} \qquad (9.1)$$

where n_1 and n_2 are the number of molecules per unit volume in regions 1 and 2, p_1 and p_2 are the *a priori* probabilities of the molecules in the two regions and ΔE is the potential energy that must be expended to move a molecule from region 1 to region 2, while k is the Boltzmann constant and T is the absolute temperature. Region 1 refers to the unassociated surfactant molecules in the bulk of the solution. For the adsorption process, region 2 is the surface, while for micellization it is the micelle.

The relative tendency for adsorption or micelle formation can be estimated using Langmuir's principle of independent surface action (1). According to this theory, the field of force around any particular group is characteristic of the group and, as a first approximation, is independent of the nature of the rest of the molecule. Assume that molecules cluster as micelles or as an adsorbed layer so that the area per hydrocarbon chain that is exposed to water molecules or vapor is the same. Since the surface energy of hydrocarbons is less than the hydrocarbon-water interfacial energy, the potential energy of the system is less with surfactant molecules oriented in an adsorbed layer than that in micelles. Consequently, at low surfactant concentrations where micelles are absent, the surface tensions of the solutions are depressed, showing that adsorption has occurred.

Critical Micelle Concentration

Aqueous solutions of surfactants exhibit a more or less abrupt change in their physical properties over a narrow concentration range. This rapid change in properties is generally accepted to be due to the formation of oriented aggregates or micelles. The concentration of surfactant at which the concentration of micelles suddenly becomes appreciable is referred to as the critical concentration for micelle formation or cmc. That this critical micelle concentration is of fundamental importance in the selection of surfactants for specific applications is shown in subsequent chapters. In general, surface activity is due to non-micellar surfactant and the micelles act as a reservoir for the unassociated surfactant molecules and ions. At concentrations greater than the cmc value, the surface tension of the solution does not decrease further with an increase in surfactant concentration. Often detergency and foaming are at their highest at the cmc. The ability of surfactant solutions to dissolve or solubilize water-insoluble materials starts at the cmc and increases with the concentration of micelles.

For any homologous series of normal hydrocarbon-chain surfactants, the cmc value is doubled for each decrease by one in the number of carbon atoms in the long chain. Thus, as determined by the dye method, the cmc for potassium tetradecylate is 0.006 *M*. For the decylate it should be $0.0060 \times 2^4 = 0.096$, as compared with the experimental value of 0.095 (2).

For various cationic and anionic surfactants, the cmc can be expressed by the equation (3)

$$\log \text{cmc} = A - BN_c \tag{9.2}$$

where N_c is the number of carbon atoms in the long hydrocarbon chain, B is a constant and is approximately equal to 0.29 and A is a constant for a particular temperature and homologous series. Values of A and B for various ionic surfactants are shown in Table 9.1 (4).

TABLE 9.1. VALUES OF A AND B FOR VARIOUS IONIC SURFACTANTS (4).

Surfactant	Temp. (°C)	A	B
K fatty acid soaps	25	1.92	0.290
K fatty acid soaps	45	2.03	0.292
Alkane sulfonates	40	1.59	0.294
Alkane sulfonates	50	1.63	0.294
Alkyl sulfates	45	1.42	0.295
Alkyl ammonium chlorides	45	1.79	0.296
Alkyl trimethyl ammonium bromides	60	1.77	0.292

In accordance with this logarithmic relation between cmc and the number of carbon atoms or CH_2 groups, the energy required to transfer a molecule from a hydrocarbon environment to water increases by a definite amount for each additional CH_2 group in the hydrocarbon chain. This is an example of Langmuir's principle of independent surface action, since each CH_2 group produces its effect independently of the others (1).

Klevens has shown that by comparing the length of the various hydrocarbon-chain salts, a relationship can be obtained between various surfactants

(4). Chain lengths were computed from bond distances and bond angles collected by Pauling (5), using an extended zig-zag form for this comparison. The chain lengths were measured from the hydrogen of the ultimate carbon to the charged atom at the hydrophilic end of the paraffin chain. Thus, a C_{11} sulfate, a C_{12} ammonium chloride, a C_{12} sulfonate and a C_{13} fatty acid contain the same number of atoms in the linear chain. The length of the hydrophobic portion of these molecules is approximately the same. The data in

TABLE 9.2 CHANGE OF CMC WITH LENGTH OF THE LIPOPHILIC PORTION OF THE MOLECULE (4).

Surfactant Type	Lipophilic Group	Length (Å) of Lipophilic Group	CMC (moles per liter)	Temp. (°C)
Fatty acid soap	$n-C_{11}O^-$	15.3	0.050	25
Sulfonates	$n-C_{10}SO^-$	15.8	0.041	25
Sulfates	$n-C_9OSO^-$	15.6	0.052	25
Amine HCl	$n-C_{10}NH^+$	15.1	0.048	25
Fatty acid soap	$n-C_{13}O^-$	17.8	0.012	25
Sulfonates	$n-C_{12}SO^-$	18.3	0.010	30
Sulfates	$n-C_{11}OSO^-$	18.1	0.013	30
Amine HCl	$n-C_{12}NH^+$	17.6	0.014	30
Fatty acid soap	$n-C_{15}O^-$	20.4	0.0034	45
Sulfonates	$n-C_{14}SO^-$	20.7	0.0030	45
Sulfates	$n-C_{13}OSO^-$	20.7	0.0033	45
Amine HCl	$n-C_{14}NH^+$	20.2	0.0031	40
Fatty acid soap	$n-C_{17}O^-$	22.9	0.0009	55
Sulfonates	$n-C_{16}SO^-$	23.4	0.0009	55
Sulfates	$n-C_{15}OSO^-$	23.3	0.0008	55
Amine HCl	$n-C_{16}NH^+$	22.7	0.0008	55

Table 9.2 show that they have very similar cmc values. The cmc relation for these various classes of ionic surfactants can be represented by the equation

$$\log \text{cmc} = 2.26 - 0.231\ L \tag{9.3}$$

where L is the over-all length of the surface-active ion in Å.

Perhaps a more convenient equation is

$$\log \text{cmc} = 2.26 - 0.274\ N_L \tag{9.4}$$

where N_L is the number of atoms in the chain, counting from the terminal hydrogen to the charged atom at the hydrophilic end of the paraffin chain.

Effect of Salts

Salts lower the cmc value of ionic surfactants. However the salt effect is not governed by the principle of ionic strength or the Debye-Hückel relationships. The depression of the cmc depends only on the concentration of ions bearing a charge opposite to that of the surface-active ions. The nature and concentration of ions of the same charge are without effect. The fact that only ions

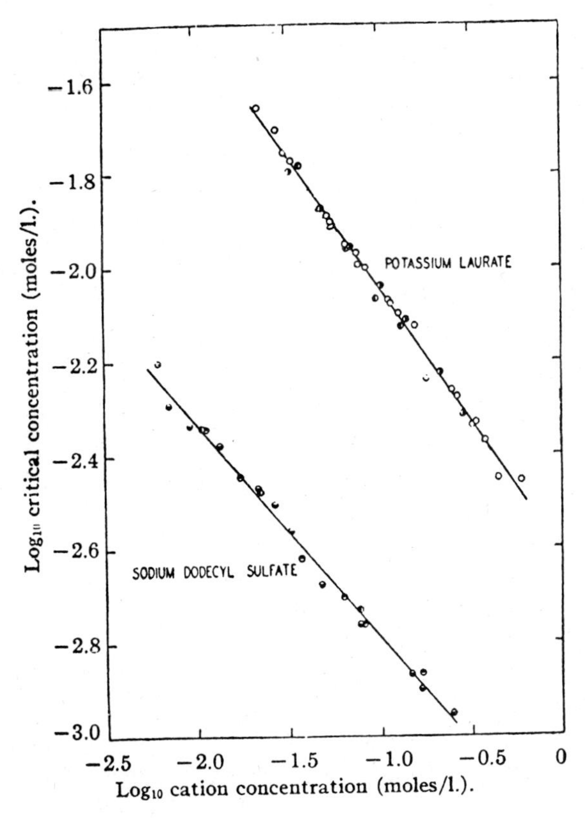

Figure 9.1. Log-log plot of the salt effect with anionic long-chain electrolytes: ○ KCl, ◒ NaCl, ◑ K_2SO_4, ◓ Na_2SO_4, ◐ $Na_4P_2O_7$ (2).

of opposite charge produce an effect on the cmc has been explained by Harkins (6) as due to the fact that micelles exhibit a substantial surface-charge density as compared with single ions in solution. Consequently, the repulsion between ions of the same charge as the micelles is sufficiently great to produce a much greater separation than that between simple ions of the same charge. By the same token, ions of opposite charge to the micelles are more strongly attracted.

The substantial effect of electrolytes on the cmc values of ionic surfactants is shown by the logarithm plot of cmc against counter-ion concentration in Figure 9.1. The counter-ion concentration is the sum of the concentration of such ions contributed by the detergent and the salt. Sodium and potassium salts have exactly the same effect on the cmc of anionic surfactants. The equations corresponding to the two curves shown in Figure 9.1. are:

$$\log \text{cmc} = -0.570 \log m^+ - 2.617 \text{ (potassium laurate)} \quad (9.5)$$

$$\log \text{cmc} = -0.458 \log m^+ - 3.248 \text{ (sodium dodecyl sulfate)} \quad (9.6)$$

where m^+ is the molar concentration of cations. According to Hobbs' theory (7) the slope should be around -0.5.

As might be expected, the cmc depends only on the length of the surface-active ion and is quite independent of the nature of the counter-ion for uni-univalent systems. Thus sodium oleate and potassium oleate have almost identical cmc values, even though the potassium soap is much more water soluble than the sodium soap. Smaller cmc values have been reported for the divalent zinc and copper salts than for the monovalent sodium, potassium, and hydrogen salts of the alkylsulfates (8). Octadecyltrimethyl ammonium oxalate has a cmc value that is only about one-third that of the bromide and chloride (4).

Effect of Structure on CMC

It has been shown that the length of the hydrophobic chain is the major factor determining the cmc value of ionic surfactants. The effect of other structural variations will now be considered.

(a) Substitution of methyl groups for hydrogen atoms attached to the nitrogen atom of cationic surfactants has only a slight effect on the cmc value. Thus *n*-tetradecyl ammonium chloride has a cmc of 0.0031 as compared with a value of 0.0040 for *n*-tetradecyl trimethyl ammonium chloride. Similarly, the introduction of a pyridine group or two hydroxyethyl groups changes the cmc only slightly. Klevens (4) has suggested that the increased bulkiness of the charged head, caused by the introduction of these groups, has a disrupting effect on the micelle which negates the influence of the branched chain.

(b) A series of alkyl sulfates was studied in which the position of the sulfate group along the chain was varied (9). With the sulfate group in the center of the chain, an increase in length by the addition of one carbon atom to each end resulted in a 2.5-fold decrease in the cmc, as compared with the four-fold decrease that would be expected, if both carbons were added at one end of the alkyl chain. As the sulfate group is moved from the terminal position of the alkyl chain toward the center of the chain, the cmc increases. However, the increase in the cmc is considerably less than would be the case if the shorter portion of the branched sulfate were removed from the molecule. In Table 9.3 comparison is made between the cmc of the branched-chain sulfates and *n*-alkyl sulfates of the same chain length as the longer portion of the branched chain.

TABLE 9.3. EFFECT OF POSITION OF $-SO_4Na$ GROUP OF *n*-TETRADECYL SODIUM SULFATES ON CMC (4).

Position of $—SO_4Na$	Tetradecyl Sulfates, cmc (moles per liter x10^{-3})	Straight Chain Sulfates	
		cmc (moles per liter x10^{-3})	No. *c* atoms
−1	1.65	1.65	14
−2	3.26	3.30	13
−3	4.52	5.8	12
−4	5.76	12.0	11
−5	7.95	25.0	10
−6	12.3	52.0	9
−7	15.8	100.0	8

(c) In contract to the 2.5-fold effect obtained by adding one carbon atom to each end of a chain, with the sulfate group in the center, the addition of a carbon to each end of the alkyl chain of dialkyl sodium sulfo-succinates results in a four-fold reduction in cmc. This is the effect that would be expected if both carbon atoms were added to one alkyl chain. Data are shown in Table 9.4.

TABLE 9.4. VALUES OF CMC FOR DIALKYL SODIUM SULFOSUCCINATES (10).

Alkyl	CMC (moles/liter)
n-Butyl	0.20
n-Amyl	0.053
n-Hexyl	0.0124
n-Octyl	0.00068
2-Ethylhexyl	0.0025

(d) The effect of one double bond is to increase the cmc by two- or three-fold. Thus potassium stearate at 60°C has a cmc value of 0.0005*M* as compared with values of 0.0012 to 0.0014*M* for potassium oleate and 0.0015*M* for potassium elaidate, both at 50°C (5). Potassium abietate has a cmc value of 0.012M at 25°C compared with 0.027M for potassium dehydroabietate at the same temperature (9). This increase in cmc is small as compared with the marked difference in solubility between saturated and unsaturated soaps.

(e) For alkylbenzene sulfonate detergents, the benzene ring is equivalent to 3.5 carbon atoms in its effect on the cmc (11). Since the length of the benzene ring is about 2.65Å as compared with about 4.35Å for 3.5 carbon atoms, the effect of the benzene ring on micelle formation is greater than predicted by Klevens' equation which relates the length of the surface active ion to the cmc.

(f) The effect of polar substitution in the hydrocarbon chain is to increase both the solubility and the cmc of the detergent. Thus, potassium 9, 10-dihydroxystearate has been reported to have a cmc value 15 to 20 times greater than potassium stearate (12, 13).

Mixtures of Surfactants

The effect of mixtures of ionic surfactants on the cmc has been examined by Harkins (14). If the two surfactants have considerably different cmc values,

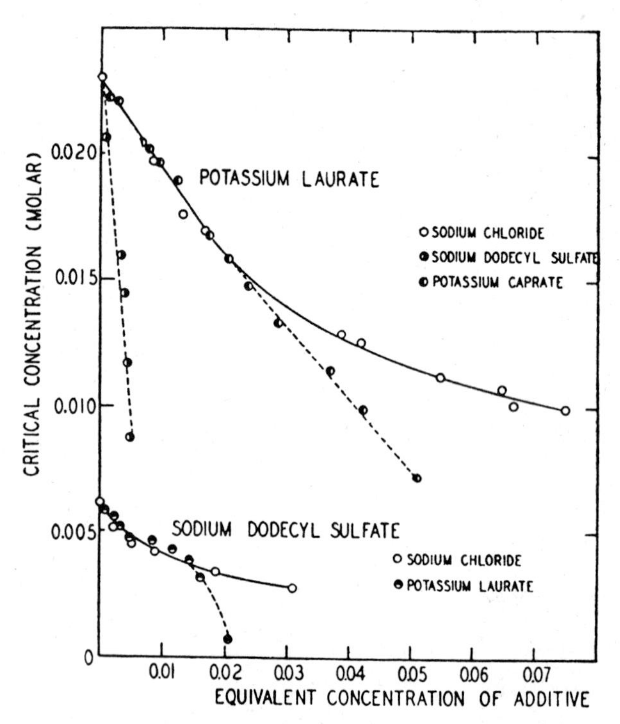

Figure 9.2. Effect of salts and other surfactants on the cmc of potassium laurate and sodium dodecyl sulfate (2).

the surfactant that has the lesser tendency to form micelles acts as though it were a salt in lowering the cmc of the other surfactant. This is shown in Figure 9.2. Potassium caprate acts as a salt in mixtures with potassium laurate at concentrations below about 0.02 equivalents. Similarly, at low concentrations, potassium laurate exerts the same effect as sodium chloride in lowering the cmc of sodium dodecyl sulfate. These detergents act as salts only at low concentrations, when they are present as unassociated electrolytes in solution. At higher concentrations, they lower the cmc to a greater extent than simple salts. When two anionic surfactants have the same cmc, as in the case of potassium myristate and sodium dodecyl sulfate, all combinations of the two have the same cmc value.

Effect of Polar Compounds on the CMC, Effect of Temperature

Long-chain alcohols, amines, and similar compounds containing a polar group produce a considerable lowering of the cmc. The effect is increased both by the concentration of the polar additive and by increasing the chain length. This is shown in Figures 9.3 and 9.4, where the cmc of potassium tetradecano-

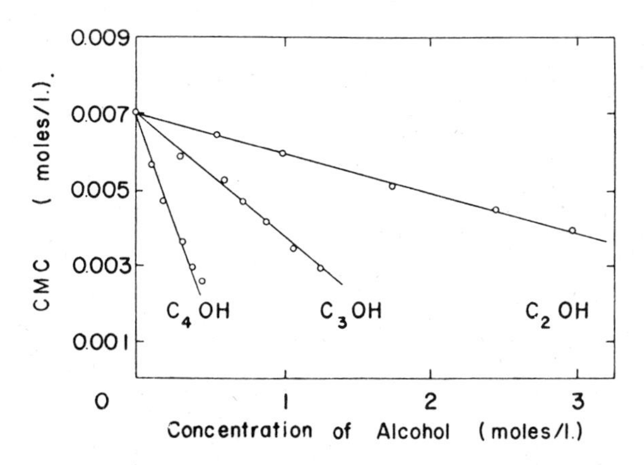

Figure 9.3. Effect of ethanol, propanol-1, and butanol-1 on the cmc of potassium myristate at 18° C (15).

ate at 18°C is plotted against the concentration of alcohols of different chain length. Figures 9.5 and 9.6 show that the logarithm of the rate of change of cmc with alcohol concentration, dC/dC_a, is a linear function of both the number of carbon atoms in the alcohol and the number of carbon atoms in the soap chain.

If the effect of temperature is considered, the situation becomes much more complicated. In the absence of additives, the cmc of ionic surfactants generally increases with an increase in temperature (16, 17, 18, 19). However, a minimum in the curve of cmc *versus* temperature for sodium dodecyl sulfate in water has been observed (20). This is shown in Figures 9.7 and 9.8 along

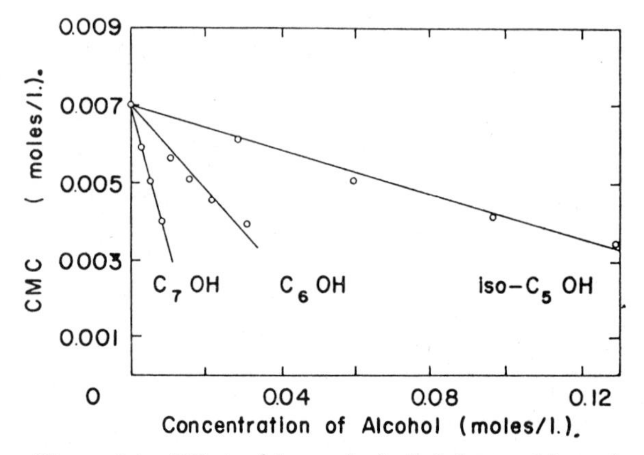

Figure 9.4. Effect of isoamyl alcohol, hexanol-1, and heptanol-1 on the cmc of potassium myristate at 18° C (15).

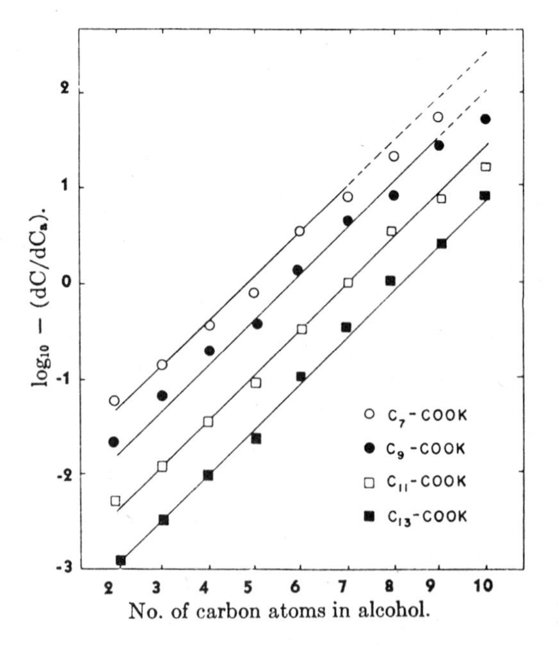

Figure 9.5. Effect of the number of carbon atoms in the alcohol on the logarithm of the rate of change of cmc with alcohol concentration (15).

with plots of sodium dodecyl sulfate in 9- per cent and 24.98- per cent ethanol solvent (21). No minimum is observed with alcohol present. However, the short-chain alcohol may increase or decrease the cmc, depending on the temperature. This may be seen more clearly in Figure 9.9. At 10°C, the cmc

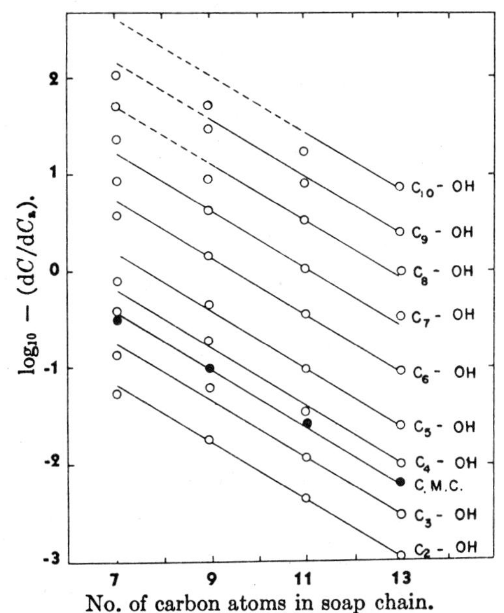

Figure 9.6. Effect of the number of carbon atoms in the soap chain on the logarithm of the rate of change of cmc with alcohol concentration (15).

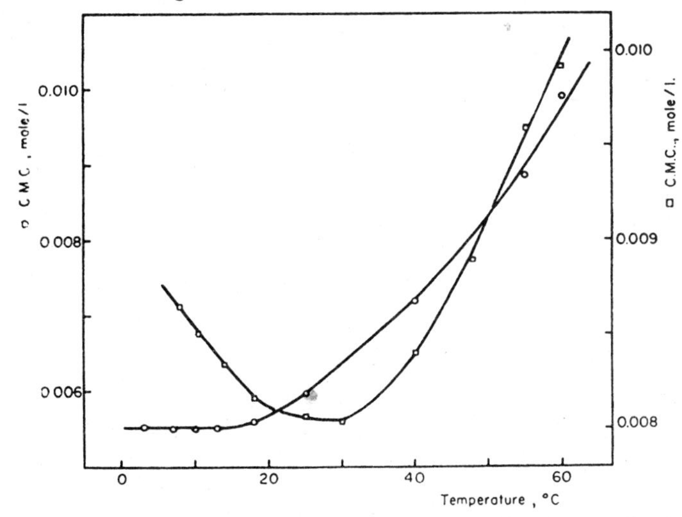

Figure 9.7. Effect of temperature on the cmc of sodium dodecyl sulfate in water (□) and in 9.27% ethanol (○) (21).

is progressively lowered with an increase in ethanol concentration, while at 55°C the cmc first decreases then increases as the ethanol content is increased.

Shape of Micelles

While the evidence that amphipathic molecules associate to form aggregates or micelles is overwhelming, the size and shape of these micelles is still in dis-

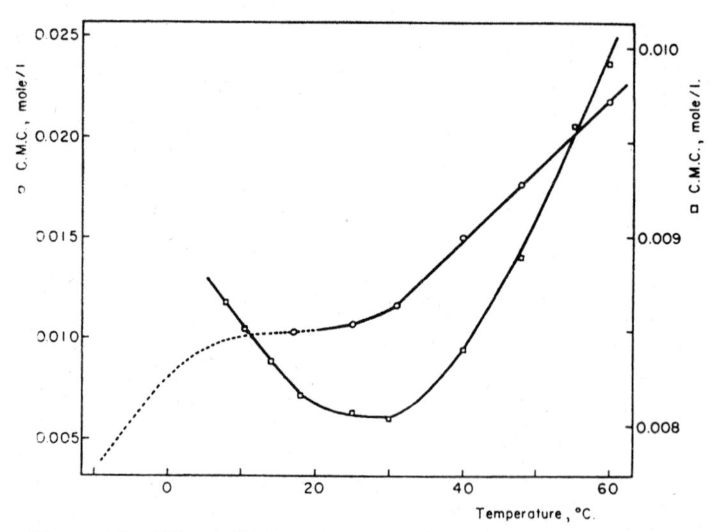

Figure 9.8. Effect of temperature on the cmc. of sodium dodecyl sulfate in water (□), and 24.98% ethanol (○) (21).

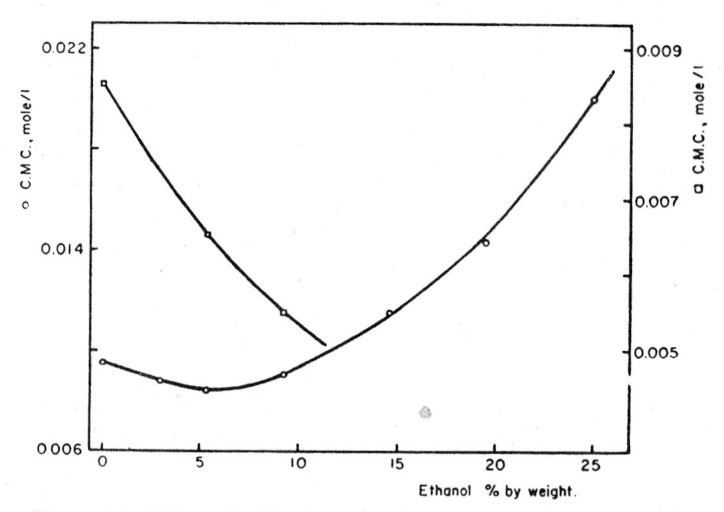

Figure 9.9. Effect of ethanol on the cmc of sodium dodecyl sulfate at 55.0° C (○) and 10.0° C (□) (21).

pute. It is reasonable to assume with McBain (22) that any combination of ions or molecules that causes a reduction in the free energy of the system can exist to some extent. The micellar size and shape that results in the greatest reduction in the free energy of the system at a given condition of concentration, temperature, and the presence of other salts and components will be most prevalent. The change in entropy on formation of micelles is very small, about 0.70 cal/deg/mole (23). This is much less than that corresponding to a phase change. Consequently, it can be expected that micelles are rather loose structures and can readily change their shape with a change in conditions.

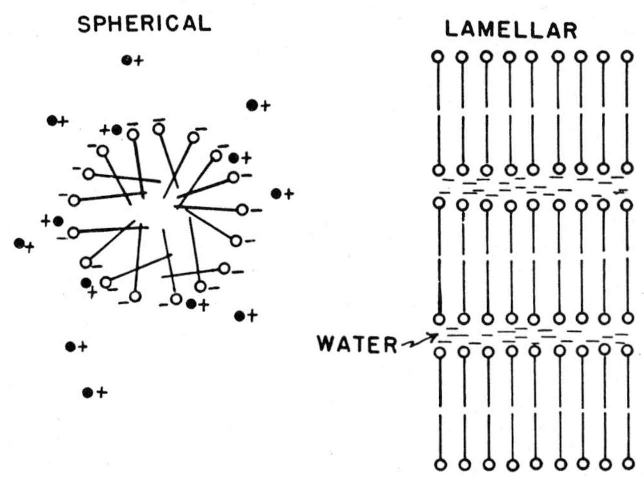

Figure 9.10. Two suggested structures of surfactant micelles.

Free-energy considerations require an arrangement of the agglomerate in aqueous solution in a manner that will expose the polar heads of the surface-active ions and molecules to water, with minimum contact between the paraffin chains and water molecules. In dilute solutions of the surfactant, this condition is fulfilled if the paraffin chains constitute a liquid pool completely surrounded by the polar heads. The simplest arrangement is the spherical micelle of Hartley (24). Geometry limits the spherical micelle to the number of units that can be packed into a sphere, and is reasonable for the 70 and 80 units per micelle found for sodium decyl sulfate and sodium dodecyl sulfate respectively. However, the 170 units per micelle found by light scattering for hexadecyl trimethyl ammonium bromide in salt solution (25, 26) appears to be too large for the micelle to be a sphere. Debye (25) has proposed a cylindrical micelle with curved ends which are approximately Hartley spheres. The body of the sphere consists of radiating discs, with the polar heads constituting the surface of the cylinder. McBain has postulated both ionic spherical micelles and neutral lamellar micelles (27). An ellipsoid of rotation was proposed by Hughes (28) and Halsey has suggested that the micelle is rod-like (29). Two forms are portrayed in Figure 9.10.

At surfactant concentrations below about 5 per cent viscosity (30, 31) and streaming birefringence (32) measurements indicate that asymmetry is absent, suggesting more or less spherical micelles. However, when the detergent concentration is increased to the range of about 5 to 20 per cent, these measurements indicate that the micelles are asymmetric or rod-like, and at higher concentrations, in the anisotropic regions, the micelles become microcrystallites of the lamellar kind (23). The tendency for asymmetry increases with carbon-chain length and with added electrolyte.

When monochromatic x-rays are passed through detergent solutions various diffraction patterns are obtained. At concentrations of about 5 per cent, the *S*-band is observed. This short band has the value of the similar spacing of a liquid hydrocarbon and thus indicates a liquid arrangement in the micelle. It does not provide important evidence concerning the geometrical structure of micelles. The *M*-band is also found at this soap concentration. This band indicates a spacing that is independent of concentration and that is closely equivalent to twice the length of the soap molecule (33). According to Harkins, who discovered the *M*-band, it is not inconsistent with a spherical micelle (34).

The x-ray long spacing, or I-band, is found at concentrations above about 10 per cent of soap. The long spacing increases with a decrease in the soap concentration as well as with solubilization of hydrocarbons. McBain (35) and initially Harkins (36) interpreted the I-band on the basis of a lamellar micelle. Subsequently, Harkins (37) considered the spacing as related to the intermicellar distance.

At present there is considerable uncertainty as to the significance of the x-ray patterns. Corrin (38), Riley and Oster (39), Fournet (40), and others (23) question that the vague patterns that are obtained are true Bragg spacings and suggest that the concept of Bragg spacings in these solutions may be meaningless.

Micellar Size and Charge

A number of methods are used to estimate the size of micelles. These include light scattering (41, 42, 43), x-ray (35, 44), diffusion (45, 46, 47, 48), and ultracentrifuge measurements (49). The method of light scattering is most widely used. Light scattering by particles that are small as compared to the wave length of light is caused by small local inhomogeneities of refractive index. These variations in refractive index arise from fluctuations in the local concentration of particles due to random thermal motion. While all species present in the colloidal solution contribute to light scattering, only the fluctuation in concentration of particles of colloidal dimensions is significant; and the contribution made by the fluctuation in concentration of these particles to light scattering is proportional to the square of the resulting fluctuation in refractive index. Light-scattering measurements have been used to calculate both micellar size and charge.

Micellar size is estimated by plotting $H\dfrac{C-C_0}{\tau-\tau_0}$ versus $C-C_0$ where H is the light scattering constant, C is the total detergent concentration in g/ml, C_0 is the cmc in g/ml, τ is the turbidity of the solution, and τ_0 is the turbidity of the solution at the cmc. The constant for light scattering is

$$H = \frac{32\pi^3\rho_0^2}{3\lambda^4 N}\left(\frac{\partial\rho}{\partial(C-C_0)}\right)_P \tag{9.7}$$

where ρ_0 is the refractive index of the solution below the cmc, λ is the wave-length in vacuum of light, N is Avogadro's number and ρ is the refractive index of the colloidal solution.

The intercept of the $H\dfrac{C-C_0}{\tau-\tau_0}$ versus $C-C_0$ plot is related to micellar size by

$$A = \frac{q}{mM} \tag{9.8}$$

where A is the intercept, m is the number of aggregates in the micelle and M is the formula weight of the detergent. The assignment of an appropriate value for q has been the subject of considerable confusion.

Debye (50) has shown that for uncharged micelles $q=1$. However, he subsequently applied the same value to q for colloidal electrolytes (51). McBain and Hutchinson (23) have suggested that $q=v$, the number of ions into which the colloidal electrolyte dissociates. The best value for q is that given by Princen and Mysels (52), based on a modification of the derivation of Prins and Hermans (53). With all salts considered to be univalent,

$$q = (n_1 + n_3)^2/n_1{}^2d_1 + 2\,n_1 n_3 d_2 + n_3{}^2d_3) \tag{9.9}$$

with

$$d_1 = 1 - p/m + p^2/4m^2 + p/4m^2 \tag{9.10}$$

$$d_2 = 1 - p/2m - fp/2m + fp^2/4m^2 + fp/4m^2 \tag{9.11}$$

$$d_3 = 1 - fp/m + f^2p^2/4m^2 + f^2p/4m^2 \tag{9.12}$$

where p is the micellar charge, n_1 and n_3 are respectively the cmc and extraneous salt concentration in moles/ml, and f is the ratio of the molar refractive index increment of the added salt to that of the detergent.

The charge on the micelle p can also be obtained from the Princen and Mysels' derivation. The slope of the line is

$$B = A\,(p^2 + p - Am\,Mp)\,/\,(2m\,M)\,(n_1 + n_3) \tag{9.13}$$

Elimination of m between Equations 9.8 and 9.13 gives

$$p = \frac{BM\,(n_1 + fn_3) + \sqrt{2B\,(n_1 + n_3)}}{A\,(1 - AME/2)} \tag{9.14}$$

where

$$E = (n_1 + fn_3)\,/\,(n_1 + n_3) \tag{9.15}$$

Once the charge has been computed, the aggregation number follows,

$$m = 1/2\,(pE + 1/AM) + 1/2\sqrt{(pE + 1/AM)^2 - (p^2 + P)E^2} \tag{9.16}$$

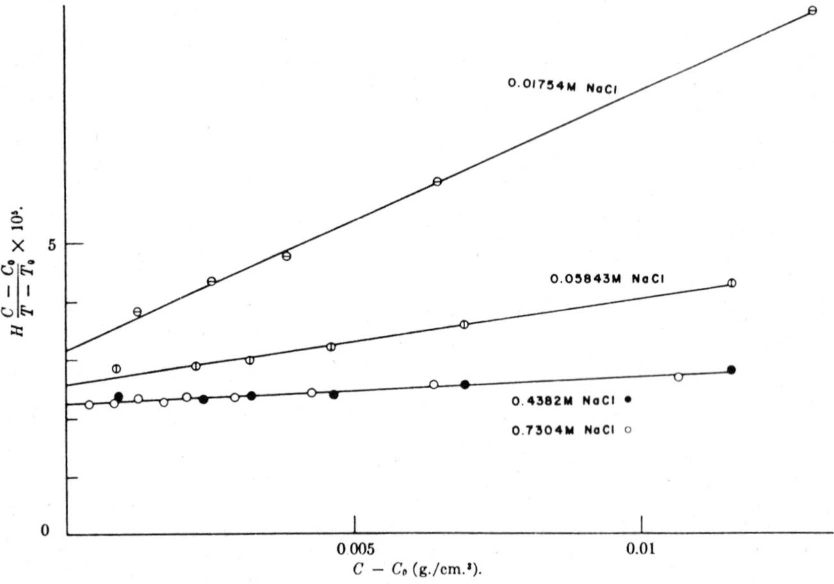

Figure 9.11. Light scattering by aqueous solutions of cetylpyridinium chloride (43).

These equations have been employed by Anacker (43) with solutions of cetylpyridinium chloride. The light scattering plot using several concentrations of sodium chloride is shown in Figure 9.11. Aggregation numbers and micellar charge data are reproduced in Table 9.5. It is interesting to

TABLE 9.5. CHARGES AND AGGREGATION NUMBERS OF CETYLPYRIDINIUM CHLORIDE MICELLES (43).

NaCl (moles/liter)	Micellar Charge	Aggregation Number
0.01754	13	95
0.05843	16	117
0.4382	28 ± 3	135 ± 1
0.7304	35 ± 3	137 ± 1

note that aggregation numbers are only 2 to 6 per cent higher than values calculated on the assumption that the micelle charge is zero, in which case the intercept $A = 1/mM$. By either method of calculation, the aggregation number increases and reaches a limiting value as the concentration of sodium chloride is increased.

The effect of chain length for a series of *n*-alkyl trimethyl ammonium bromides in water is shown in Figure 9.12, where the turbidity function is plotted versus C instead of $C - C_0$ (54). The break in each of the curves corresponds to C_0, the cmc of the detergent. The value of $H\,(C - C_0)\,/\,\tau$ at C_0 is equal to one divided by the micellar molecular weight. Thus the

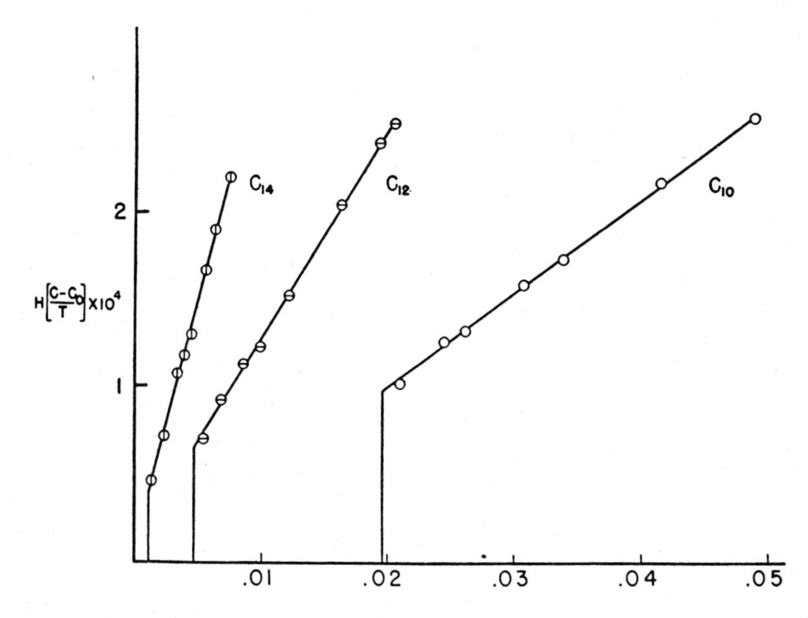

Figure 9.12. Reciprocal specific turbidities of aqueous solutions of n-alkyl trimethylammonium bromides (54).

molecular weight of the micelle increases with an increase in the hydrocarbon-chain length.

Anacker compared his results with the diffusion data of Hartley and Runnicles (47), who employed the Stokes-Einstein equation to estimate micelle radii. With sodium chloride present, the calculated radii obtained by diffusion ranged between 25.5 Å and 28.4 Å, averaging 27.2 Å. For aggregation numbers of 95 and 136, the micelle radii by light scattering correspond to 24.2 Å and 27.3 Å respectively, if it is assumed that the micelles are spherical and have a density of 0.9 g cm^{-3}. Micelle radii determined by diffusion include a hydration layer that would cause them to be about 1.5 Å greater than radii based on light scattering. This is a hydration layer of one-half molecule of water (55). Thus, agreement between the two methods of determining micellar size is reasonable.

TABLE 9.6. EFFECT OF SODIUM CHLORIDE ON THE SIZE AND CHARGE OF MICELLES (56).

Surfactant	Solution	CMC (moles per liter)	Aggregation Number (N)	Ratio of Micellar Charge to Aggregation Number (p/N)	Free Energy Change per Long-chain ion for Micelle Formation ($-\Delta G°/kT$)
Sodium dodecylsulfate	Water	0.0081	80	0.18	
	0.02*M* NaCl	0.00382	94	0.14	
	0.03*M* NaCl	0.00309	100	0.13	+15.9
	0.10*M* NaCl	0.00139	112	0.12	
	0.20*M* NaCl	0.00083	118	0.14	
	0.40*M* NaCl	0.00052	126	0.13	
Dodecylamine HCl	Water	0.0131	56	0.14	
	0.0157*M* NaCl	0.0104	93	0.13	+15.1
	0.0237*M* NaCl	0.00925	101	0.12	
	0.0460*M* NaCl	0.00723	142	0.09	
Decyl trimethyl ammonium bromide	Water	0.0680	36	0.25	+11.3
	0.013*M* NaCl	0.0634	38	0.26	
Dodecyl trimethyl ammonium bromide	Water	0.0153	50	0.21	+14.5
	0.013*M* NaCl	0.0107	56	0.17	
Tetradecyl trimethyl ammonium bromide	Water	0.00302	75	0.14	+17.6
	0.013*M* NaCl	0.00180	96	0.13	

Additional data on the effect of sodium chloride on the size and charge of micelles shown in Table 9.6 have been collected by Phillips (56). The effect of different electrolytes on the micellar molecular weights of sodium dodecylbenzene sulfonate and the nonionic iso-octylphenyl nonaethylene glycol ether is reproduced in Tables 9.7 and 9.8 from Mankowich (57). The low aggregation number of 5 found for the branched sodium dodecylbenzene sulfonate is contrasted with aggregation numbers of 57 and 24 found for sodium 3-dodecylbenzene sulfonate and sodium 4-dodecylbenzene sulfonate respectively (58).

TABLE 9.7. EFFECT OF ELECTROLYTES ON THE AGGREGATION NUMBER OF SODIUM DODECYLBENZENE SULFONATE AT 25–26°C (57).

Electrolyte	Concentration (%)	Aggregation No. of Sodium Dodecylbenzene Sulfonate
None	—	5
$Na_3PO_4 \cdot H_2O$	0.6	60
	0.9	74
	1.5	148
	2.0	248
Na_2SO_4	0.3	53
	0.6	64
	0.9	59
	1.5	210
	2.0	371
$Na_5P_3O_{10}$	0.3	51
	0.6	67
	0.9	68
	1.5	74
	2.0	83

TABLE 9.8. EFFECT OF ELECTROLYTES ON THE AGGREGATION NUMBER OF ISO-OCTYL PHENYL NONAETHYLENE GLYCOL ETHER AT 25–26°C (57).

Electrolyte	Concentration (%)	Aggregation No. of Iso-octyl Phenyl Nonaethylene Glycol Ether
None	—	135
$Na_3PO_4 \cdot H_2O$	0.3	128
	0.6	142
	0.9	154
	1.5	169
	2.0	204
Na_2SO_4	0.3	121
	0.6	128
	0.9	138
	1.5	174
	2.0	194
$Na_5P_3O_{10}$	0.3	135
	0.6	139
	0.9	145
	1.5	156
	2.0	168

McBaine and Hutchinson (23) have observed that if the size of the micelle is known, the charge can be deduced from electrical conductance data. The specific conductance of a colloidal electrolyte is given by (59)

$$\kappa = \frac{C_0}{1000}(l_c + l_a) + \frac{C-C_0}{1000\,m}\,pl_c + \frac{C-C_0}{1000m}\,4/3\,p^2 l_a \tag{9.17}$$

where κ is the specific conductance, l_c and l_a are the ionic conductances of the cation and anion respectively, p is the micellar charge and m is the aggregation number of the micelles.

The conductance κ_0 at the cmc is

$$\kappa_0 = \frac{C_0}{1000}\,(l_c + l_a) \tag{9.18}$$

Then

$$\frac{\kappa - \kappa_0}{C - C_0} = \frac{1}{1000m}\left[pl_c + \frac{p^2 l_a}{m^{1/3}}\right] \tag{9.19}$$

Energetics of Micelle Formation

A number of papers have appeared in recent years on the energetics of micelle formation (54, 56, 60, 61, 62, 63, 64, 65). Debye (54) showed that micelles form as the result of the gain in energy occasioned by removing hydrocarbon chains from the surrounding water and bringing them in contact with each other in micelles. However, energy is required to bring together the charged ends of the surfactant ions. The interplay of van der Waals and electrostatic forces determines the size and shape of micelles.

Ooshika (64) and Reich (61) criticized Debye's treatment because it was based on minimizing the free energy of the micelle rather than of the entire system. Reich pointed out that Debye's and Ooshika's treatments do not account for the formation of micelles by nonionic detergents, which would grow to infinite size if the only limitation to growth is electrical.

While the energy of hydrocarbon-chain adhesion favors micelle formation, the growth of micelles involves a decrease in the number of independent particles in the system, and hence a decrease in total entropy. If surfactant molecules are assumed to aggregate into a spherical micelle, with only the polar portions of the molecules exposed to water molecules, the introduction of an additional molecule into the micelle will result in a further decrease in total entropy, without a corresponding decrease in energy. Consequently, the free energy of the system will increase. For ionic surfactants there is the additional factor of electric repulsion (61).

The treatment by Phillips (56) is given below, because it provides results that are in reasonable accordance with experience. It is assumed that the micelles are monodispersed, with all of the micelles in a colloidal solution having essentially the same size and charge. Light scattering (54) and diffusion measurements (45, 48) have not indicated that micelles are polydispersed. However, Hoeve and Benson (62) have criticized this simplifying assumption.

The equilibrium between single ions and micelles can be represented as follows, in the case of anionic surfactants

$$mA^- + (m - p)\,\mathrm{Na}^+ \rightleftarrows M^z \tag{9.20}$$

where A^- is the surfactant anion, Na^+ is the counter ion, m is the aggregation number and z is the charge of the micelle M.

The corresponding equilibrium constant is

$$\mathrm{K}_M = \frac{[a_M]}{[a_{\mathrm{A}^-}]^m [a_{\mathrm{Na}^+}]^{m-z}} \tag{9.21}$$

where K_M is the equilibrium constant and a_M, a_{A^-} and a_{Na^+} are the activities of the micelles, surfactant anions, and counter-ions, respectively.

The true charge on the micelle z can be replaced by the effective charge p, determined by light scattering measurements. This effective charge takes into account both the number of counter-ions physically adsorbed on the surface, $m-z$, and the number bound in the double layer. According to Phillips (56) it can be interpreted as an activity coefficient. By substituting p for z, activities can be replaced by concentrations,

$$\mathrm{K}_m = \frac{[C_m]}{[C_{a^-}]^m\,[C_{\mathrm{Na}^+}]^{m-p}} \tag{9.22}$$

Phillips has shown that the equilibrium constant can be expressed in terms of micellar size and charge,

$$\mathrm{K}_m^{-1} = 3m^2 C_0^{(m-1)}\,(X + C_0)^{m-p} \tag{9.23}$$

where X is the concentration of added electrolyte and C_0 is the cmc.

The standard free energy change ΔG° per molecule associated with micelle formation is

$$-\Delta G^\circ = (kT/m)\,\ln \mathrm{K}_m \tag{9.24}$$

and

$$\frac{\Delta G^\circ}{kT} = \frac{\ln 3 + 2\ln m}{m} + \frac{m-1}{m}\ln C_0 + \frac{m-p}{m}\ln(X + C_0) \tag{9.25}$$

with concentrations expressed as mole fractions, so that ΔG° is in reference to a standard state of mole fraction unity. The free energy gain per C_{12} ion entering a micelle is found to be 14–16 kT, or about 1.25 kT per CH_2 group entering the micelle. Calculated free energy values are shown in Table 9.6 (56). Stigter and Overbeek (65) applied a correction for the charging of ions in the micelle and obtained a value of 1.02 kT per CH_2 group. Shinoda obtained 1.07 kT per CH_2 group (66).

The work of adsorption of a CH_2 group at an oil-water interface is 1.37 kT, or 810 cal/mole. The relatively close agreement in energy values indicates that the interior of the micelle has a liquid-like character. The lower value for micelle formation suggests that in the micelle exposure of the hydrocarbon chain to the aqueous phase is not completely eliminated as it is at the oil-water interface (56).

For paraffin-chain surfactants, the free energy change for micelle formation may be written

$$-\Delta G^\circ = (1.25\,kT)\,n + B \tag{9.26}$$

using Phillips' value of 1.25 kT per CH_2 group, where n is the number of CH_2 groups per detergent chain and B is a constant for a given end group.

From Equation 9.25, in the absence of electrolyte and m large

$$\Delta G^\circ = 2.3\,(2 - p/m)\,kT \log C_0 \tag{9.27}$$

Combining Equations 9.26 and 9.27, and accepting a value of 0.2 for p/m from Table 9.6,

$$-\log C_0 = 0.302\,n + B' \tag{9.28}$$

where $B' = B\,/\,(4.14\,kT)$. This predicts a linear relation between log cmc and the number of CH_2 groups. Further, the two-fold increase in the cmc with the shortening of the chain by one CH_2 group is in agreement with experimental data.

The heat of micellization can be calculated from the change in cmc with temperature, in accordance with the equation of Stainsby and Alexander (60)

$$\Delta H_M = -RT^2 \frac{d \ln C_0}{dT} \tag{9.29}$$

where ΔH_M is the change in heat content associated with micelle formation, R is the gas constant, T the absolute temperature, and C_0 the cmc.

The heat change is small and becomes negative with increase in temperature. Goddard and Benson (67) found values of about 1.0 kcal/mole at 10°C, decreasing to about −1.0 kcal/mole at 40°C for C_8, C_{10} and C_{12} sodium alkyl sulfates. Similar values—0.8 kcal/mole at 5°C and −1.9 kcal/mole at 45°C—were found for sodium dodecyl sulfate by Flockhart and Ubbelohde (20). Hutchinson, Manchester and Winslow (68) found a heat effect of about

−0.3 kcal/mole and an entropy change of about 1 cal per degree per mole for cetylpyridinium chloride. They deduced a loose structure for the micelle from these small heat and entropy effects. The standard free energy change for micelle formation can be written.

$$\Delta G^\circ = \Delta H^\circ - T\Delta S^\circ \tag{9.30}$$

where ΔH° and ΔS° are heat and enthropy changes relative to a given standard state. From the results of Hutchinson and associates, it is evident that the free energy change for micelle formation is very small. This shows the very delicate balance between the tendency for association of the paraffin chain, about 10 kcal/mole for C_{12}, and the opposition of end groups to aggregation.

Determination of Critical Micelle Concentration

Dilute solutions of both ionic and nonionic surfactants in water deviate only slightly from ideal behavior. However, over a narrow concentration range, characteristic of each material, deviation from ideal behavior becomes marked. The range of concentration, over which this dramatic change in properties occurs, is narrow and is referred to as the critical micelle concentration or cmc. At concentrations above the cmc, the solution is non-ideal. The changes that occur at the cmc can be explained through the assumption that single surfactant-ions or molecules aggregate to form micelles.

Any method that measures the deviation from ideal behavior can be used to determine the cmc of a surfactant. The change in the boiling point or the freezing point of the solution as compared with that of the pure solvent can be used to measure non-ideal behavior in accordance with the equation

$$g_1 \log N_1 = \frac{L\theta}{RT_0^2} \tag{9.31}$$

where g_1 is the osmotic coefficient and N_1 is the mole fraction of the solvent. If the freezing point depression is measured, T_0 is the freezing point of the pure solvent at a given pressure, L is the latent heat of fusion of the solvent, and $\theta = T - T_0$ is the change in the freezing point. For boiling point elevation measurements, T_0 is the boiling point of the solvent at a given pressure, L is the latent heat of evaporation of the pure solvent, and θ is the elevation of the boiling point.

For an ideal solution $g_1 = 1$, and hence for a non-ideal solution

$$g_1 = \frac{\theta_{\text{non-ideal}}}{\theta_{\text{ideal}}} \tag{9.32}$$

provided that N_1 is close to unity. The plot of freezing point against molality shown in Figure 9.13 (23) clearly shows the break in the curve corresponding to the cmc.

The osmotic pressure can be employed to determine the cmc, from the equation

$$\pi V_1^\circ = -g_1 RT \log N_1 \tag{9.33}$$

where

$$g_1 = \frac{\pi_{\text{non-ideal}}}{\pi_{\text{ideal}}} \tag{9.34}$$

for a given value of N_1 close to unity. π is the osmotic pressure and V_1° is the molal volume of the pure solvent at a given temperature and pressure.

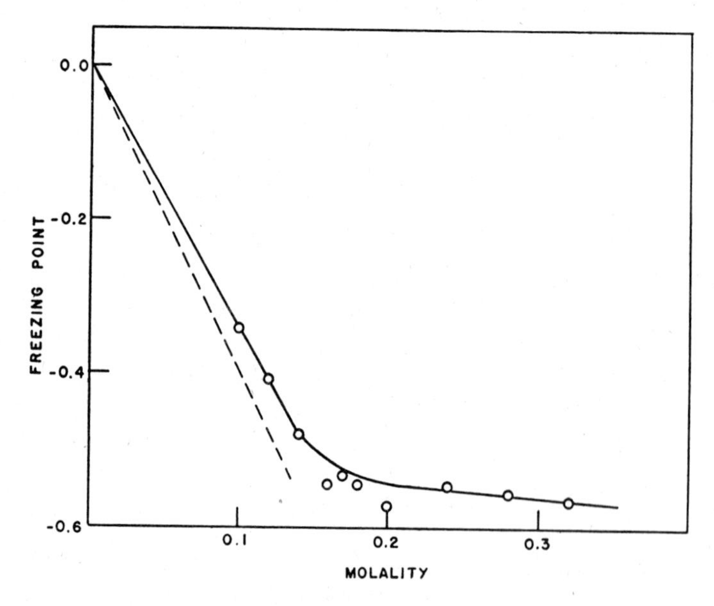

Figure 9.13. Freezing point of aqueous solutions of sodium octyl sulfate as a function of concentration. The dashed line represents the behavior of an ideal solute (23).

In addition to these osmotic properties, solubility, partial molal volume, refractivity, light scattering, solubilization, electrical conductance, color change of dyes, and surface tension have been used to determine the cmc of surfactants. Typical curves are shown in Figure 9.13, 9.14 and 9.15.

In terms of accuracy and convenience electrical conductance is the preferred method for determining the cmc, and far more accurate results have been reported using this method than any other. It is limited to ionic surfactants. Light scattering and surface tension measurements also provide accurate results, in agreement with those obtained by electrical conductance. These latter methods can be used with nonionic surfactants.

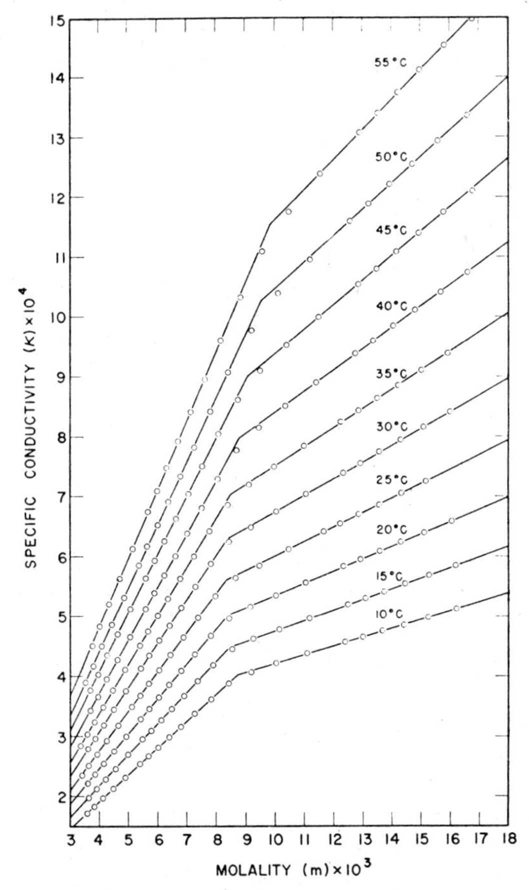

Figure 9.14. The specific conductivity (κ in mho cm.$^{-1}$) of sodium dodecyl sulphate in aqueous solution plotted as a function of the molality for temperatures from 10° to 55° C (67).

Figure 9.14 shows the specific conductance of aqueous solutions of sodium dodecyl sulfate plotted as a function of molality for various temperatures. It will be observed that the cmc first decreases and then increases with an increase in temperature. A plot of surface tension versus the logarithm of the concentration of potassium laurate is shown in Figure 9.15.

Graphs showing some measured property of the solution as a function of the concentration generally show two linear portions connected by a curve in the region where the measured property is changing rapidly. Selection of a specific concentration in the region of sudden change as the cmc value is necessarily arbitrary. The cmc is most commonly defined as the intersection of two lines extrapolating the measured property from below and above the region where the sudden change occurs.

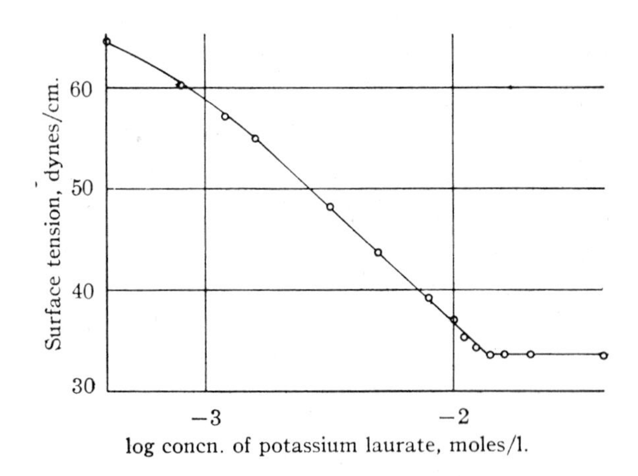

Figure 9.15. Surface tension *versus* the logarithm of the concentration of potassium laurate. Solutions at pH = 10.0, $[K^+]$ = 0.1 mole per liter (69).

Because of its simplicity, the spectral-change method involving the color change of dyes has been used more widely than any other method for estimating the cmc of surfactant solutions. While a number of dyes change color in the region of the cmc, pinacyanol chloride is most widely used as the cmc indicator. In its simplest form, the method consists of preparing a solution of the surfactant, at a concentration slightly higher than the anticipated cmc value, and about 10^{-4} or 10^{-5} molar in dye concentration. This solution is titrated with dye solution, at the same concentration as present in the surfactant solution, to a color change. With pinacyanol chloride, the visual end point is taken as the change from the initial blue color to a definite blue-purple. A refinement in the method is the use of a spectrophotometer to determine spectral changes (70). The method has been critically studied by Mukerjee and Mysels (71). They conclude that at best it can be used to determine a vague cmc range. Depending upon conditions, the value may be considerably above or below the true cmc.

Effect of Temperature on Solubility

The solubility of long-chain electrolytes varies with temperature in an unusual manner. At low temperatures, the solubility is small and it increases slowly and regularly as the temperature is increased. Then within a narrow temperature range, the solubility begins to increase very rapidly (72). Data by Adam and Pankhurst (73) showed this effect. Similar findings were made by Tartar and Wright (74, 75) and Gershman (76) for other ionic surfactants. Typical results are shown in Figures 9.16 and 9.17.

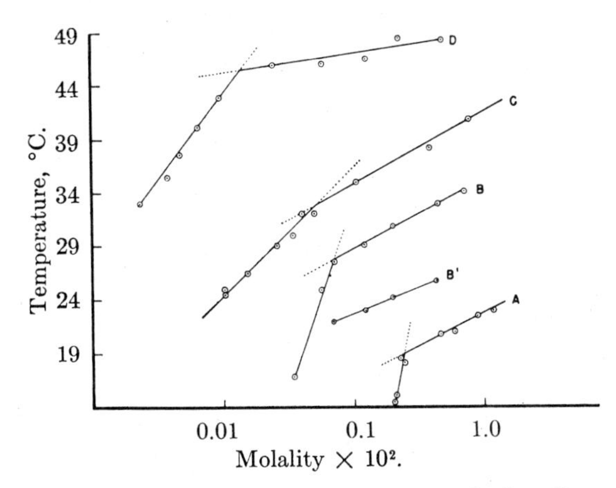

Figure 9.16. Solubilities of the branched sodium alkylbenzene-sulfonates; A, methyldecyl; B, methyldodecyl (stable phase); B′, methyldodecyl (unstable phase); C, methyltetradecyl; D, methylhexadecyl (76).

The explanation for this behavior is that the ionic surfactants in unassociated form have quite limited solubility in water, while the micelles are highly soluble. At low temperatures, the solubility is less than the cmc, but rises with an increase in temperature to the cmc. At higher temperatures, the dissolved surfactant is present largely as micelles. The temperature at which the abrupt change in solubility occurs is known as the Krafft point, and the concentration at which it occurs is the cmc at that temperature.

Raison (77) showed that the spectral-change method could be used to determine the Krafft point. Surfactant solutions containing pinacyanol are blue above the cmc and pink at lower concentrations. Since micelles can only be present above the Krafft point, the change from pink to blue which occurs on warming can be used to determine the Krafft point. The more usual procedure is to determine the temperature corresponding to the appearance of cloudiness on cooling and the disappearance of cloudiness on warming.

The effect of the length of the paraffin chain on the Krafft point is shown in Figure 9.18 from Raison (77). The Krafft point increases linearly with chain length. Values for the C_6 and C_8 sodium alkyl sulfates were obtained by extrapolation, since the methods cannot be used below 0°C.

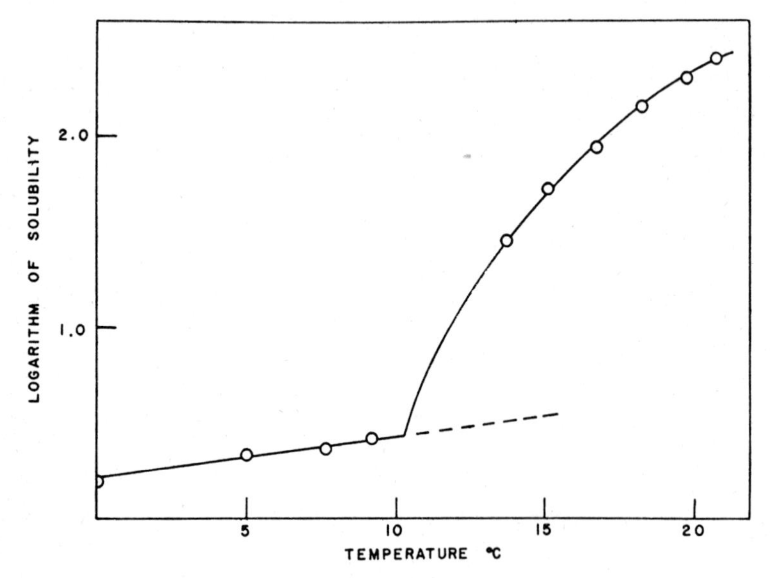

Figure 9.17. Logarithm of the solubility of sodium dodecyl sulfate in water as a function of temperature. The dashed line represents the expected behaviour in the absence of micelle formation (23).

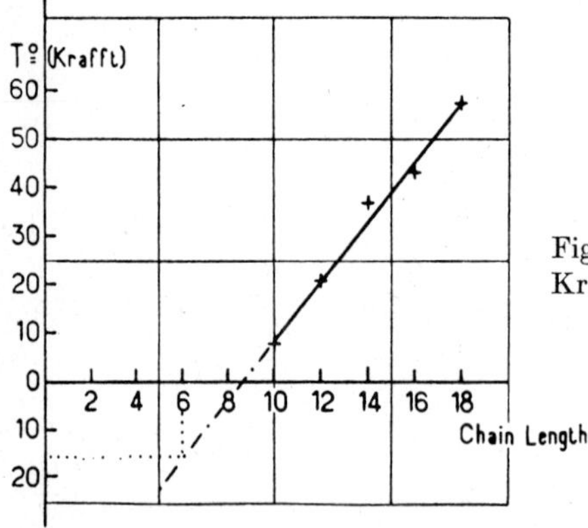

Figure 9.18. Effect of chain length on the Krafft points of sodium alkyl sulfates (77).

The Krafft point of binary mixtures of $C_{12}+C_n$ sodium alkyl sulfates plotted against the mole fraction of C_n in the mixture is reproduced from Raison in Figure 9.19. Compounds with chain lengths greater than 12 carbon atoms increase the Krafft point, while those of smaller chain-lengths decrease it. The C_2 and C_4 sodium alkyl sulfates do not form micelles, but act as simple electrolytes and increase the Krafft point slightly. While the C_6 sodium alkyl sulfate does not form micelles, it does depress the Krafft point, suggesting that it may form mixed micelles with sodium alkyl sulfates of longer chain-

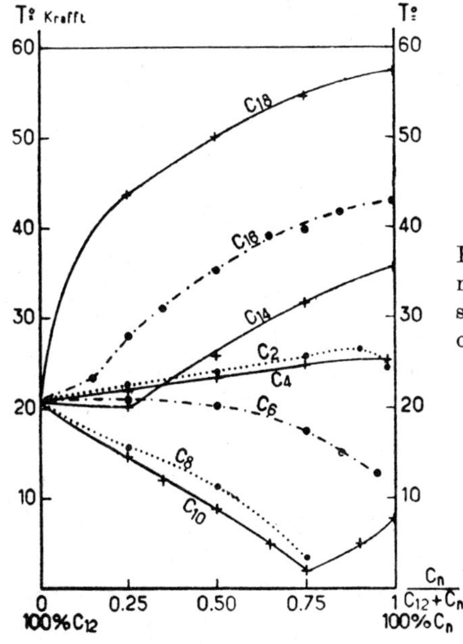

Figure 9.19. Krafft points of binary mixtures of sodium alkyl sulfates, showing the effect of the mole faction of the constituents (77).

length. The C_8 and C_{10} compounds decrease the Krafft point considerably at a mole fraction of 0.75, with the curve for the C_{10} showing a decided eutectic.

SOLUBILIZATION

Many substances that are not very soluble in water dissolve to a considerably greater extent in micellar solutions of surfactants. The increased solubility of the compound, due to the presence of micelles, is termed solubilization. Direct evidence that solubilized materials are present within the micelles is provided by x-ray measurements. When nonpolar oils are dissolved in surfactant solutions, the M-spacing increases with the amount of oil solubilized. Other evidence leaves little doubt that solubilization is due to the presence of micelles. McBain and McBain (78) established that for solubilized systems, the components are in thermodynamically stable equilibrium.

Solubilization has been broadly classified into three types, as follows:

(1) Nonpolar solubilization: The solute molecules are not oriented with respect to the water, but are dissolved in the interior of the micelles, away from the polar groups. (2) Polar-nonpolar solubilization: The solute molecules are oriented like the surfactant molecules with their polar groups directed toward the water. These solubilized molecules lie with their polar groups between the polar groups of the surfactant molecules, and their hydrocarbon chains between the hydrocarbon chains of the surfactant molecules. (3) Adsorption solubilization: The solubilized molecules are adsorbed at the polar surfaces of the micelles, but without penetration of the micelles. It is questionable whether the evidence in favor of this type of solubilization is valid.

The methods used for measuring solubilization are similar to those used for measuring solubility. It is essential that sufficient time be allowed for the system to reach equilibrium. Gas adsorption techniques are used to determine the solubilization of gases. The saturation limit of solubilization of liquids is readily determined by the appearance of opalescence. This can be determined visually, by light scattering (79), or by transmission (80) measurements. Studies on the solubilization of solids have been restricted primarily to dyestuffs. After addition of excess dye and agitation in a thermostat for periods up to several weeks, the system is allowed to settle undisturbed for a day or so and a sample is withdrawn for analysis. It is extracted with a suitable solvent, and the color is determined spectrophotometrically. Many commercial dyes contain suspending agents, and these must be removed in advance, generally by recrystallization of the dyestuffs.

The tendency for surfactant solutions to emulsify liquids and suspend solids affects the ease with which reliable solubilization data are obtained. It is helpful in determining the solubilization of liquids, since the turbidity of the system is greatly increased by the presence of emulsified liquid droplets. On the other hand, the presence of suspended solids in the sample withdrawn for analysis will result in high values.

In their monograph on "Solubilization", McBain and Hutchinson (23) treat solubilization as a distribution phenomenon between two phases: an aqueous phase and a micellar phase. As a first approximation the solubility of organic compounds in the aqueous phase is the same as in pure water. While the micelles are not a true separate phase, it is a convenient fiction.

Effect of Structure of Solubilizer

The first pertinent observation concerning the effect of the molecular structure of the solubilizer on the extent of solubilization is concerned with chain length. In all of the systems examined, an increase in the chain length in any homologous series of surfactants results in an increase in the solubilizing ability of the surfactant. Regardless of whether the solubilized molecule is polar or nonpolar, aliphatic or aromatic, it is solubilized to a greater extent by a C_{18} soap than a C_{12} soap, as shown in Table 9.9, from McBain and Richards (79).

TABLE 9.9. SOLUBILIZATION OF ORGANIC LIQUIDS BY C_{18} AND C_{12} SOAPS (79).

Organic Liquid	Molecular Volume	Moles Solubilized Per Mole Soap	
		Na Stearate	K Laurate
Cyclohexane	108.5	0.56	0.23
Toluene	106.7	0.51	0.13
Methyl *tert*-butyl ether	118	2.20	1.66
Ethylbenzene	122.5	0.40	0.20
Methyl isobutyl ketone	125.0	1.82	1.20
n-Hexane	131.3	0.46	0.18
Amyl acetate	134.1	1.71	0.89
p-Cymene	156.5	0.26	0.08
n-Octyl alcohol	157.8	0.59	0.29
Lauryl alcohol	191.7	0.13	0.03
n-Cetane	293.5	0.00	0.00
Oleic acid	339.6	0.05	0.018
Tributyrin	295.7	0.37	0.11

It was shown earlier that for straight-chain ionic surfactants with different polar groups, the cmc related more closely to the length, or to the number of atoms in the straight chain of the surfactant ion, than to the number of carbon atoms in the chain. The same relation holds approximately in the case of solubilization. Sodium dodecylsulfate, with 16 atoms in the chain, is equivalent in solubilizing capacity to myristate soap, with the same number of atoms in the chain, as shown in Table 9.10. However, Table 9.11 shows that long-chain amines are more effective solubilizers than soaps with the same chain length. Changing the counter ion has little effect on solubilization.

The effect of unsaturation in the hydrocarbon chain appears to be specific. Oleate soaps are less effective than the corresponding stearate soaps in solubilizing straight-chain or cyclic paraffins, but more effective in solubilizing aromatic or polar compounds (82). In some instances, nonionic surfactants are more effective solubilizers than the ionic surfactants. Thus Triton X100, iso-octylphenol ethoxylated with 9–10 moles of ethylene oxide per mole of hydrophobe, is more effective in solubilizing Orange OT than is potassium oleate (81). However, the nonionics studied are complex mixtures, and impurities have a considerable effect on the solubilizing capacity of surfactants.

TABLE 9.10. EFFECT OF THE NUMBER OF ATOMS IN THE LIPOPHILIC GROUP OF THE SURFACTANT ON ITS SOLUBILIZING CAPACITY (81).

		Moles Solubilized Per Mole Surfactant	
Surfactant Type	Lipophilic Group	n-Heptane	Heptyl alcohol
Fatty acid soap	n-$C_{12}O^-$	0.14	0.67
Alkyl sulfate	n-$C_{10}OSO^-$	0.17	0.74
Fatty acid soap	n-$C_{13}O^-$	0.23	—
Alkyl sulfate	n-$C_{11}OSO^-$	0.25	—
Fatty acid soap	n-$C_{14}O^-$	0.31	0.75
Alkyl sulfate	n-$C_{12}OSO^-$	0.34	0.87

TABLE 9.11. COMPARISON OF THE SOLUBILIZING POWER OF LONG-CHAIN AMINES AND SOAPS (79).

	Molecular	Moles Solubilized Per Mole Surfactant		
Organic Liquid	Volume	Dodecylamine HCl	Na Stearate	K Laurate
n-Hexane	131.3	0.75	0.46	0.18
2, 2-Dimethylbutane	133.7	0.75	0.45	0.13
n-Octane	163.3	0.29	0.18	0.08
2, 2, 3-Trimethylpentane	160.4	0.30	0.18	0.09

Effect of the Nature of the Solubilizate

The amount of material solubilized by a given surfactant solution depends upon the molecular geometry, polarity and polarizability of the solubilizate. It appears to be quite general that for any group of compounds the amount solubilized decreases with increasing molecular weight. This is shown in Table 9.11 for branched and straight-chain paraffins. Branching is without effect. The extent of solubilization of alcohols in soap solutions decreases with an increase in the chain length of the alcohols.

The introduction of a polar group into the molecule greatly increases the extent of solubilization. As shown in Table 9.9, ethers, esters, ketones, and alcohols are about twice as strongly solubilized as hydrocarbons of similar molecular weight and volume. The presence of a double bond increases the extent of solubilization, as compared with solubilizates with the same number of carbon atoms (80, 81). On the same basis of comparison, the presence of one aromatic ring increases solubilization, while two rings reduce it. Thus, the order of decreasing solubilization in potassium laurate solution is *n*-butylbenzene > *n* decane > naphthalene (81).

Effect of Added Electrolyte

Ionic surfactants with relatively short paraffin-chains require a fairly high concentration for micelle formation. Since electrolytes lower the cmc, it can be expected that they would also increase the extent of solubilization by increasing the ratio of micellar to unassociated surfactant ions. This has been shown to be the case (80). However, electrolytes generally increase the solubilizing power of homologs with small cmc values. Thus, Stearns, *et al* (80) found that for the solubilization of ethylbenzene, added salt increased solubilization by potassium myristate to a greater extent than by potassium laurate. Kolthoff and Graydon (84) found that sodium chloride increased the solubilization of Orange OT, DMAB, trans-azobenzene, and naphthalene by dodecylamine hydrochloride. Kolthoff and Stricks (85) reported that electrolytes have a variable effect, depending both on the particular solubilizer and the solubilizate. Data by Klevens (86) as well as Richards and McBain (82) suggest that the addition of an electrolyte increases solubilizing power toward nonpolar materials but decreases solubilizing power toward polar materials. McBain and Hutchinson (23) caution that the data are too scanty to conclude that the solubilization of polar compounds is actually reduced by an electrolyte.

The available evidence is to the effect that the addition of an electrolyte increases the size and alters the charge of the micelles. The assumption can be made that the micelles are in the form of spheres, and the addition of electrolyte increases the diameter of the spheres, through uncoiling of the paraffin chains. Since the ratio of surface to volume is less for larger spheres, then the effect of the polar group on the solvent action of the micelles would be expected to be reduced as the size of the micelle was increased. This would explain the increased solubilization of non-polar materials. The solubilization of polar materials will be modified due to the change in the charge on the micelles, and may be increased or decreased depending on the particular solubilizate.

Effect of Added Non-Electrolyte

In 1892 Engler and Dieckhoff (87) found that the addition of fatty acid or phenol to soap solution increased the solubilization of aromatic hydrocarbons. Green (88) found that benzene, toluene, and hexane increased the solubilization of Orange OT by potassium laurate and oleate solutions, while ethyl alcohol slightly decreased the solublizing power of the soaps. Klevens (89, 90) found that long-chain alcohols, amines and mercaptans increase the solubilization of *n*-heptane by potassium myristate solutions, as shown in Figure 9.20.

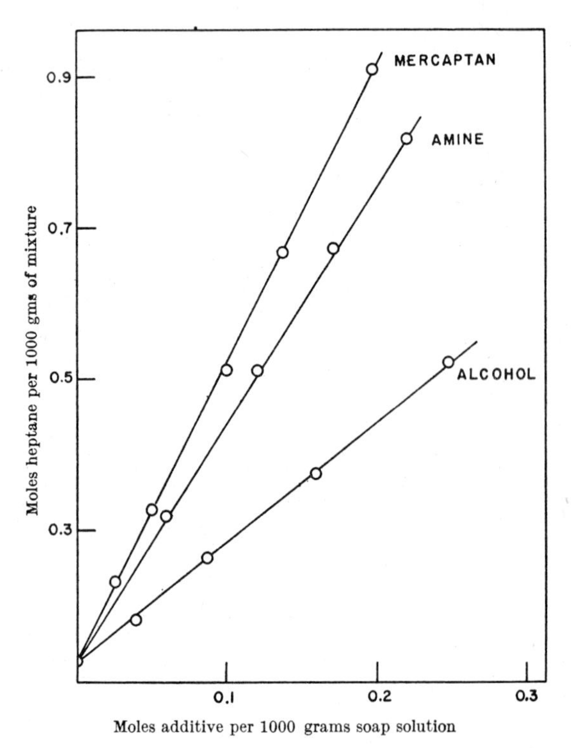

Figure 9.20. Effect of added octyl alcohol, octylamine, and octyl mercaptan on the solubilization of n-heptane in 0.35 M potassium myristate solution (23).

Solubilized organic material increases the micelle volume. Other things being equal, the increased volume of this pseudo-phase would be expected to have enhanced solvent action on other organic materials.

TERNARY SYSTEMS

Emphasis thus far has been placed on systems that are composed primarily of water, a small amount of a surface-active agent, and in some instances one or two additional components, such as a simple electrolyte or an organic compound or both. Systems containing lesser amounts of water are of considerable industrial importance, and provide insight concerning the behavior of surfactants in detergency, emulsions, foams, and other applications.

Surfactant-Water-Electrolyte Systems

Phase changes which occur upon the addition of electrolytes to soap are important in the manufacture of soap, and have been studied by McBain (91, 92. 93) and others. If potassium chloride is added to an aqueous solution of oleate soap, at a particular concentration of salt the viscosity will be observed to increase rapidly, and a viscous semisolid may form. Upon further addition of potassium chloride, the system will begin to separate into two layers, with the upper layer containing practically all the soap. If more potassium chloride is added, the upper layer will decrease in size and become more concentrated in soap.

The initial soap solution is termed an isotropic solution, since the aggregates composing the solution are structurally symmetrical. The rapid increase in viscosity observed upon addition of electrolyte corresponds to the formation of an anisotropic or liquid crystalline solution. Evidence for anisotropy can be observed using crossed polaroid discs. The phase that separates from the lower isotropic solution is also liquid crystalline. In soap-making terminology, the viscous anisotropic phase is a middle phase, from which the isotropic nigre and the liquid crystalline neat phases separate upon further addition of electrolyte.

A soap solution may form a variety of phases, even in the absence of electrolyte. A typical commercial sodium soap contains about 75 to 80 per cent tallow soap and 20 to 25 per cent coconut-oil soap. At room temperature, an aqueous solution will contain a few per cent of the soap in the form of a semisolid jelly with the remainder of the soap present as a white opaque solid, called curd. If the mixture containing less than about 27 per cent of soap is heated the soap will dissolve completely at the Krafft temperature to an isotropic liquid, nigre. This is shown as area F in Figure 9.21 for a commercial soap at 90°C (94).

At a slightly higher soap concentration, along the zero per cent electrolyte axis in the figure, a liquid-crystalline middle phase in equilibrium with nigre is found in area E. This is followed at about 37 to 52 per cent soap by area C, which is wholly middle soap. Area B contains neat soap in equilibrium with middle soap, while area A is composed entirely of neat soap.

The other areas shown in Figure 9.21 are: *D* - middle soap, neat soap, nigre; *G* - neat soap, nigre; *H* - neat soap, nigre, "lye"; *I* - nigre, "lye"; *J* - neat soap, "lye"; *K* - neat soap, curd soap, "lye". The effect of electrolyte is to lower the concentration of soap required for the formation of a liquid

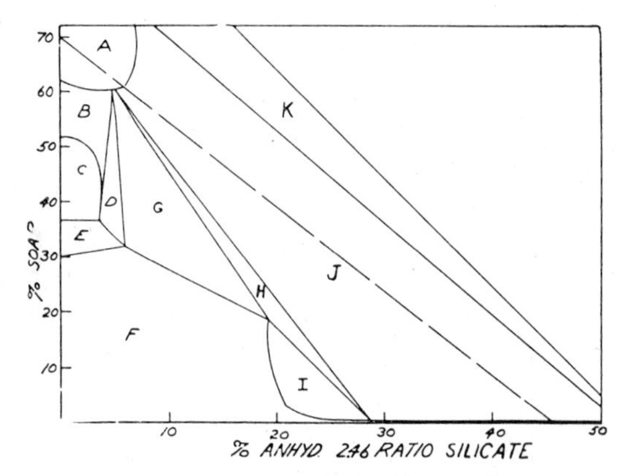

Figure 9.21. Tentative phase diagram for a commercial soap—2.46 ratio sodium silicate-water system (94).

crystalline phase. Electrolytes also raise the Krafft temperature of the soap, so that soap curd or "crystallized soap" is found at temperatures above the Krafft temperature of the salt-free soap.

Surfactant-Water-Insoluble Compounds

Solubilization data discussed earlier in this chapter were based on observations in dilute aqueous solutions. The solubilization limits for liquid were determined by the increase of turbidity upon further addition of solubilizate. This turbidity has been considered due to the emulsification of insoluble droplets of the liquid, and this assumption is correct for nonpolar oils. However, with insoluble long-chain polar compounds, the emulsion-like material formed is not simply droplets of the insoluble compound, but spherulites of a liquid-crystalline phase containing surfactant and water as well as the insoluble polar compound.

The observations which follow are applicable to all ionic surfactants in combination with long-chain polar compounds containing six or more carbon atoms in a straight chain. Any variation in the nature of the ionic group of the surfactant or the polar group of the additive has an extremely small effect.

These systems follow a similar phase equilibrium pattern, as shown in

Figure 9.22 for surfactant concentrations between about 10 and 30 per cent. Observations should be made above the melting point of the highest melting component, to avoid the separation of this material as a normal crystalline solid.

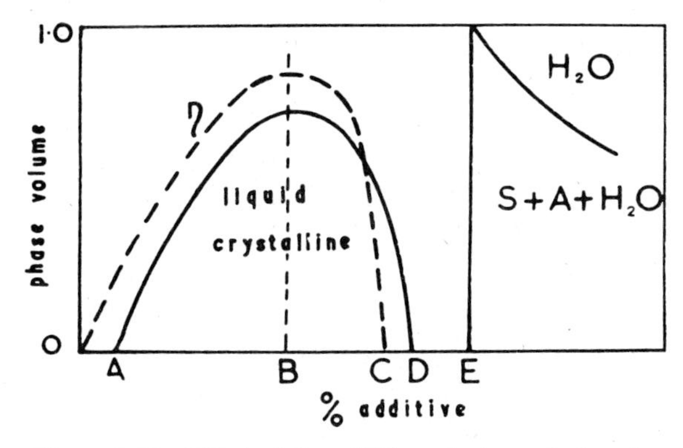

Figure 9.22. Effect of the addition of a long-chain polar compound to a surfactant solution above the cmc. $S + A + H_2O$ = surfactant + long-chain polar compound + water (95).

Point *A* in Figure 9.22 corresponds to the appearance of a liquid crystalline phase. This point also coincides with the onset of turbidity and has been considered to be the limit of solubilization of the polar additive. The additive also causes a large increase in viscosity, which reaches a maximum at *B*. At this point, the system is predominantly liquid crystalline and it no longer foams. With further addition of the polar compound, both the viscosity and the volume of the liquid-crystalline phase decrease. At *D* the system is a single isotropic phase showing a bright Tyndall cone. It is nearly as fluid as the original surfactant solution. It remains a homogeneous fluid with further additions of polar compound until *E*, where it separates into two liquid phases. The behavior at this point depends upon the nature of the surfactant. With Teepol, a branched-chain sodium alkylsulfate, the behavior at room temperature is as shown in the figure. With cetyl trimethylammonium bromide, the system is an emulsion of excess additive. It is necessary to heat about 100°C in a sealed tube to obtain water separation beyond *E*, as shown in the figure. With carboxylate soaps, heating above 100°C is also necessary, but both separated phases contain soap.

The first additions of polar compound to point *A* in Figure 9.22 results in a decrease in surface tension and in electrical conductivity, while the viscosity and the refractive index increase. The first appearance of the liquid-crystalline phase, referred to previously as the limit of solubilization, occurs at about 0.2 to 0.25 mole additive per mole of surfactant. The viscosity and refractive index reach a maximum, the surface tension is a minimum and the conductivity shows a small change in slope.

At the second break point B, the viscosity reaches a maximum that is several hundred times higher than the first maximum. The conductivity

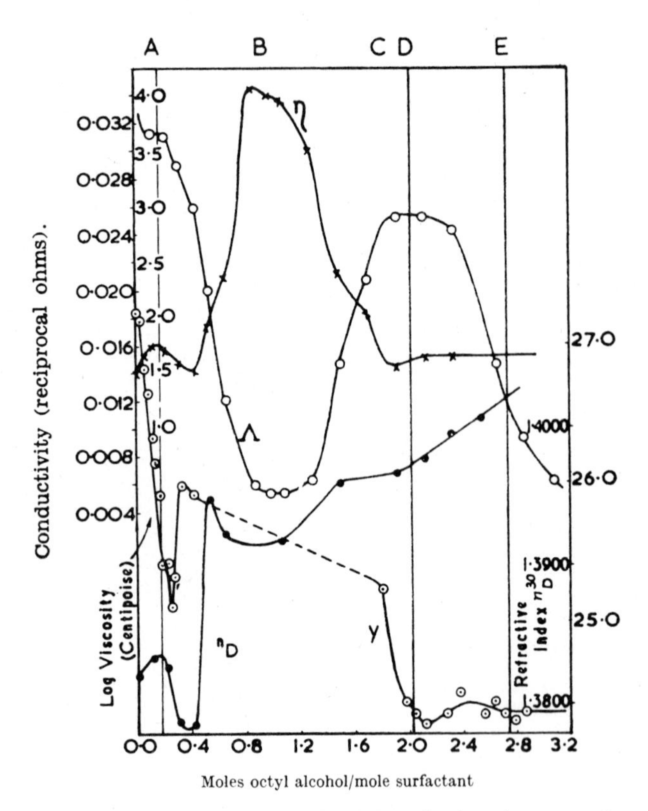

Figure 9.23. Changes in conductivity Λ, viscosity η, surface tension γ and refractive index n_0 that occur when octanol is added to an aqueous solution of Teepol, a *sec.*-alkyl sodium sulfate, above its cmc (95).

reaches a minimum and the refractive index shows a small dip. The high viscosity interfers with the measurement of surface tension, but there are indications of a maximum at B. This second break occurs at approximately an equimolar ratio of ionic surfactant and long-chain polar additive.

The reduction in conductivity and increase in surface tension between points A and B are due to the removal of solute from solution. At D, all of the solute is back in solution. Consequently, surface tension and conductivity fall between *B* and *D*. The break at *D* occurs with approximately 2 to 2.2 moles of additive per mole of surfactant.

Point *E* is the true saturation solubility of the long-chain polar compound in the surfactant solution. No abrupt change in any of the four physical properties occurs at this point. It may be noted that the volume of water is greater than that of the polar compound at *E*. These effects are illustrated in Figure 9.23.

Increasing the chain length of either additive or surfactant increases the viscosity maximum *B*, and decreases the conductivity minimum. The concentration of additive at which *B* is observed is reduced with increasing chain length. The presence of salts generally produces the expected salting-out effect. Glycols and sugars tend to increase solubility and retard phase separation.

The interaction between surfactant ions and long-chain polar molecules in these mixed micelles is much like that in mixed monolayers. Van der Waals forces acting between the long hydrocarbon chains and interaction between polar groups contribute to closer packing of the molecules. As a consequence of close packing, spherical micelles are converted to a lamellar form. The expression "complex" is generally not used to describe this interaction, since it implies chemical combination in stoichiometric proportions.

Intermicellar Equilibrium

In a series of articles, Winsor (96, 97, 98, 99) developed a theory of intermicellar equilibrium which explains many of the changes in physical properties of surfactant solutions upon the addition of polar or nonpolar components. This is shown in Figure 9.24 as an equilibrium between molecular and different micellar species. Following the diagram from left to right, at low concentrations of surfactant in water, the agent is present in molecular dispersion. An increase in the concentration of the surfactant will result in the formation of more or less spherical micelles, in equilibrium with the unassociated molecules. In some systems, a further increase in surfactant concentration will result in the formation of lamellar micelles, corresponding to a liquid-crystalline solution. Lamellar micelles can also be produced by the introduction of water-insoluble liquids in certain series of solubilized systems. Further addition of the organic liquid will result in a system that is predominantly nonaqueous, and a shift in the structure of the micelles with the polar groups now directed toward the center of the micelles. Both types of more or less spherical micelles are also shown to be in equilibrium. Solutions containing predominantly spherical micelles are isotropic.

Electrical conductivity measurements were made on a 15 per cent solution of sodium di-(2-ethylhexyl)-sulfosuccinate in hexane containing various additions of a 26.8 per cent aqueous solution of undecane-3 sodium sulfate. Results shown in Figure 9.25 can be explained in terms of intermicellar equilibria. The oil-soluble S_0 micelles are not conducting and the solution shows a high specific resistance. The shift from the S_0 micelles to the G micelles produces an initial decrease then an increase in specific resistance,

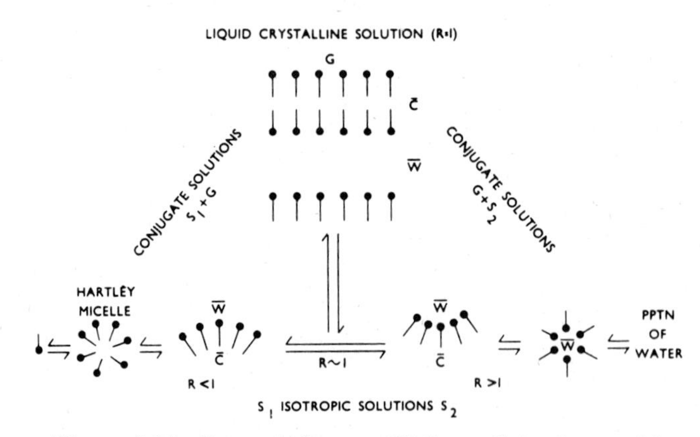

Figure 9.24. Intermicellar equilibrium. C is the amphiphilic monolayer bridging the water environment W with the organic environment O. The Hartley micelle is shown to be convex toward water in isotropic solution S_1, while the spherical micelle is convex to the organic environment in isotropic solution S_2. These spherical micelles are in equilibrium with the lamellar micelles predominating in the isotropic solution G (100).

as the lamellar micelles grow larger. The subsequent shift to the water-soluble S_w micelles results in a drastic decrease in specific resistance (98).

Comparison can be made between Winsor's systems described in Figure 9.24 and those of Lawrence in Figure 9.22. The conjugate solutions S_1+G correspond to the liquid-crystalline phase in equilibrium with an isotropic phase between points A and B. The conjugate solutions $G+S_2$ appear to correspond to liquid crystals beyond B in equilibrium with sol. The clear sol S_2 is the region between D and E. The G solution is not a gel, as suggested by Winsor, since gels are isotropic.

R-Theory of Solubilization

Winsor's (100) R-theory of solubilization derives from the observation that phase diagrams of aqueous surfactant solutions and solubilized systems

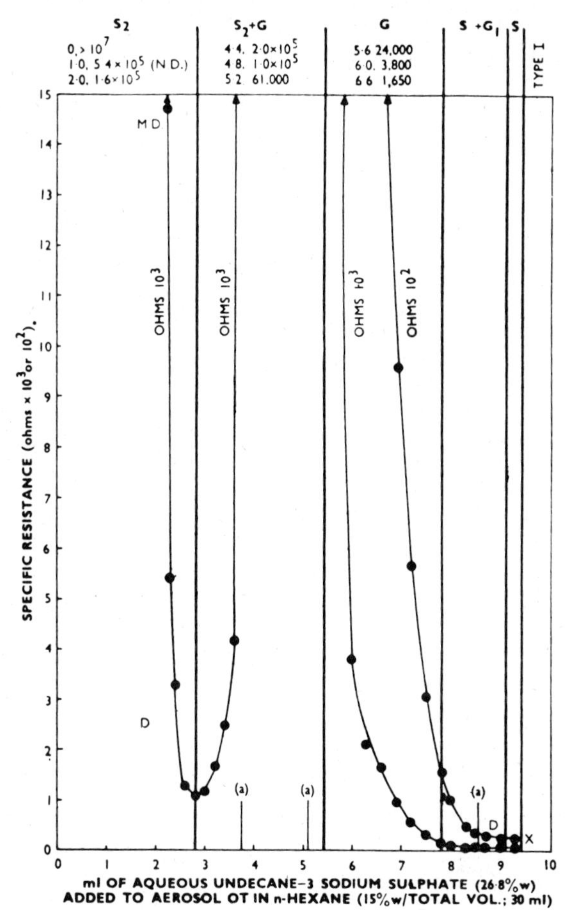

Figure 9.25. Changes in specific resistance that occur when an aqueous solution of undecane-3 sodium sulfate is added to a solution of Aerosol OT in hexane. D = opalescent dispersion in water; MD = milky dispersion in water; N.D. = No dispersion. The suggested micellar forms and the specific resistances with various additions of the aqueous solution are shown above the figure (98).

invariably show a region corresponding to the formation of a liquid crystalline solution. In accordance with Winsor's view, the liquid crystalline region results from the presence of lamellar micelles. In isotropic solutions, the micelles are more or less spherical.

Pure liquids as well as solutions containing polar molecules do not show a random arrangement of the molecules, due to molecular interaction. The arrangement becomes more disordered with an increase in temperature. For ternary systems composed of surfactant, water, and solubilizate, it is necessary to consider the organizing influence of interaction between both like and unlike molecules. The two main types of interaction operative in these systems are electrostatic and electrokinetic. Electrostatic interaction is between ions and dipoles and contributes to hydrophilic character. It is denoted by A_H in the discussion which follows. Electrokinetic interaction results from the movement of the electrons within the molecule and is the familiar van der Waals interaction responsible for the attraction between paraffin molecules. It is designated A_L.

For a binary solution, molecular interactions may be denoted:

$$A_{AA} = A_{H_{AA}} + A_{L_{AA}} \tag{9.35}$$

$$A_{BB} = A_{H_{BB}} + A_{L_{BB}} \tag{9.36}$$

$$A_{AB} = A_{H_{AB}} + A_{L_{AB}} \tag{9.37}$$

Interactions A_{AA} or A_{BB} will promote clustering of A or B molecules, respectively, and ultimately phase separation. Interaction A_{AB} will promote mixing of A and B molecules. All of these interactions are concentration and temperature dependent.

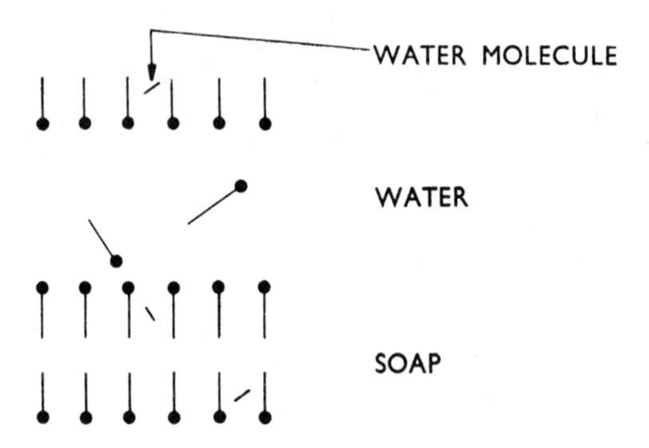

Figure 9.26. A representation of the structure of a liquid crystalline solution of potassium laurate (100).

In applying these ideas on molecular interaction to solubilized systems, Winsor considered a liquid-crystalline solution to be organized into alternate layers of liquid water and bimolecular soap leaflets as shown in Figure 9.26. The thickness of the bimolecular leaflets is approximately equal to twice the extended length of the surfactant molecule, while the thickness of the water layer depends on its concentration.

A hydrocarbon solubilized in the liquid-crystalline solution is said to be incorporated in the liquid hydrocarbon region of the surfactant leaflet, extending the thickness of the leaflet. Polar molecules are incorporated parallel to the surfactant molecules, with the polar groups among the surfactant polar groups. A mixture of surfactants produces similar bimolecular leaflets, with the thickness of the leaflet approximately twice the average extended length of the molecules composing it.

The liquid-crystalline solutions are characterized by an amphiphilic monolayer C with an aqueous environment W on one side and an oil environment O on the other, as depicted in Figure 9.27. The juxtaposition of a second monolayer, with the possible inclusion of hydrocarbon molecules, provides the oil environment. An essential condition for the stability of the lamellar micelle, and the liquid-crystalline solution, is that the monolayer C shall show no tendency to become convex or concave towards its O or W environments.

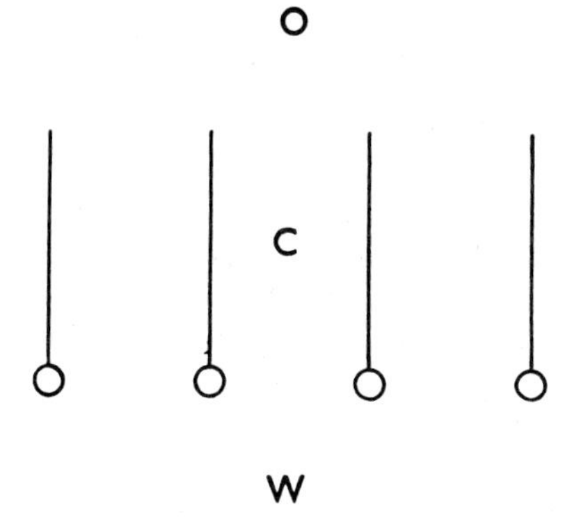

Figure 9.27. In micelle formation, the surfactant C bridges the lipophilic O and hydrophilic W regions (100).

Winsor assumes a ratio R of the relative tendencies of the C region to become convex towards O and W. In the lamellar phase $R=1$. The tendency of the C region to become convex toward the O area is assisted by the interaction between C and O molecules A_{co} and resisted by the interaction between O molecules A_{oo}. Similarly, the tendency to become convex towards W is assisted by A_{cw} and resisted by A_{ww}. The variation of R with composition is then given by

$$R = f \frac{A_{co} - A_{oo}}{A_{cw} - A_{ww}} \tag{9.38}$$

with f a constant.

When considering the effect of variations in C on the solubilization of an O or W environment, R can be considered to correspond approximately to

$$R = f\frac{A_{co}}{A_{cw}} = f\frac{A_{H_{co}} + A_{L_{co}}}{A_{H_{cw}} + A_{L_{cw}}} \tag{9.39}$$

Accordingly, if interaction between the C monolayer and W is stronger than that between C and O, R tends to be < 1 and C tends to become convex towards water. A somewhat spherical micelle forms with the water exterior to the micelle. If $A_{co} > A_{cw}$, then R tends to be > 1 and the water environment tends to be within the spherical micelle. This is shown in Figure 9.24.

Potassium laurate at a concentration in excess of 35 per cent in water forms a liquid crystalline solution, as illustrated by Figure 9.26. Since the chemical activities of water and soap are necessarily uniform throughout the phase, there must be some soap molecules present in the W region and some water molecules in the C region. On dilution of the liquid crystalline solution with water, the activity of the water is increased and the ratio $R = (fA_{co}/A_{cw})$ tends to decrease due to a mass action effect. Since the liquid crystalline solution is stable over a broad concentration range, dilution is probably accompanied initially by a change in packing of the soap molecules, with additional water molecules entering the C region, to maintain R equal to unity. On further dilution below about 35 per cent of soap, the liquid crystal breaks down and an isotropic phase is formed. According to the R-theory, this is because R is < 1 in the isotropic phase and the C region is predominantly convex toward water. This is consistent with x-ray data. The l band indicating lamellar regions diminishes in intensity and finally disappears as the isotropic solution is diluted. The M band, which has been interpreted to correspond to the diameter of spherical micelles, remains. At still lower concentrations, the micelles dissociate and the M band disappears. These phase relationships for potassium laurate solutions are shown in Table 9.12.

TABLE 9.12. PHASE—CONCENTRATION RELATIONSHIPS FOR POTASSIUM LAURATE SOLUTIONS AT ROOM TEMPERATURE (100).

Surfactant Concentration (%)	Solutions
0 –0.25	Ions and small aggregates
0.25– 35	Spherical Hartley micelles S_1 transforming to lamellar micelles at the highest concentrations
about 35	Conjugate isotropic S_1 and liquid-crystalline G
35 and above	Smectic liquid crystalline. Indefinitely extended lamellar micelles

A similar relationship is shown in Table 9.13 for hexanolamine oleate solutions. With potassium laurate solutions, concentrations of soap considerably above 35 per cent result in the crystallization of solid soap. However, with hexanolamine oleate as the soap concentration is further increased R becomes >1. At 92 per cent of soap a second isotropic solution S_2 forms with the W regions included within the C regions.

TABLE 9.13. PHASE—CONCENTRATION RELATIONSHIPS FOR HEXANOLAMINE OLEATE SOLUTIONS AT ROOM TEMPERATURE (100).

Surfactant Concentration (%)	Solutions
0–30	Isotropic S_1
30–40	Conjugate isotropic S_1 and liquid crystalline G
40–85	Liquid crystalline
85–92	Conjugate liquid crystalline G and isotropic S_2
>92	Isotropic S_2

In a series of homologous ionic surfactants, A_{Lco} and consequently R increases with increasing molecular weight of the surfactant. Liquid crystalline solutions occur at lower concentrations, the longer the hydrocarbon chain length of the surfactant.

Winsor also compared the formation of liquid crystalline solutions of tetradecane-1 and -7 sodium sulfate. In such solutions, the thickness of the hydrocarbon portions of the bimolecular layers will be approximately in the ratio of 14 to 7 and the polar groups attached per unit area of the C region will be about one-half for the -7 sulfate as compared with the -1 sulfate. The interaction between the water region and the polar groups of the C region A_{Hcw} should be less for the -7 sulfate and R greater, according to Winsor. In agreement with this, at room temperature the -7 sulfate forms a liquid-crystalline solution at about 20 per cent concentration, as compared with the greater than 40 per cent concentration required for the 1-sulfate.

Solubilization is readily considered from the point of view of the R-theory. Figure 9.28 represents the system sodium *n*-butyrate, octanol-1, water at 50°C. For compositions along the water-octanol-1 axis, with sodium *n*-butyrate absent, almost pure water and octanol-1 containing a few per cent of water separate as conjugate phases. As sodium *n*-butyrate is added miscibility increases until the concentration of sodium *n*-butyrate represented by the curve, above which they are completely miscible. Within the solubilized region above the curve, there is a range of compositions where the liquid-crystalline phase G is present. This system can be interpreted according to the R-theory.

A 55 per cent solution of sodium n-butyrate in water at 50°C is an isotropic solution with $R<1$. When octanol-1 is added to this solution, it enters the micelles with the hydroxyl groups directed toward the W region. The interaction A_{cw} is due to both sodium n-butyrate and octanol-1 in the C region. Because the ionic carboxyl group is more hydrophilic than the hydroxyl

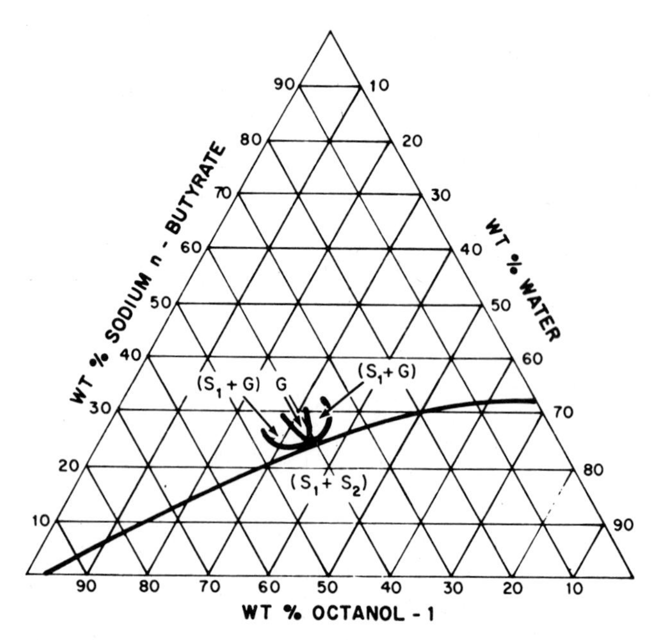

Figure 9.28. The system sodium n-butyrate-octanol-1-water at 50°C (100).

group, A_{cw} decreases as the concentration of octanol-1 is increased. Consequently, R increases to unity, resulting in the separation of a liquid crystalline phase, first as a conjugate phase in equilibrium with an isotropic phase and then as the sole phase composing the system. On further addition of octanol-1, the liquid crystals break down and an isotropic solution is formed with octanol-1 external to the micelles. In the initial isotropic solution S_1, the C region is predominantly convex towards W; in the liquid crystalline solution it is planar, and in the final isotropic solution S_2 it is predominantly concave to W.

It is also possible to pass from S_1 to S_2 without passing through a liquid crystalline phase. This can be done by raising the temperature or by adding the octanol-1 to the soap solution at a concentration selected to bypass the anisotropic phase, the so-called "channel region". This direct transition from S_1 to S_2 is shown in Figure 9.28. Although a planar C region must arise in the solution, it does not become extensive enough to lead to the formation of a liquid crystalline phase.

The system Triton X-100-benzene-water is analogous to that described. On adding benzene to a 50-per cent aqueous solution of the surfactant, the phase sequence is $S_1 \rightarrow (S_1+G) \rightarrow$ liquid crystalline solution $G \rightarrow (G+S_2) \rightarrow S_2 \rightarrow (S_2 +$ excess water phase). By selecting a different route, the liquid crystalline solution is not observed. Winsor regards benzene molecules to be distributed partly in the C region alongside the surfactant molecules, where they are polarized by induction, and partly in the O region among the hydrocarbon chains of the surfactant molecules.

In considering multicomponent systems, a number of generalizations are possible for predicting behavior. As defined by Winsor, a Type I system consists of an isotropic solution S_1 miscible with water containing a separated oil phase. A Type II system consists of an isotropic solution S_2 miscible with oil containing a separated aqueous phase.

Reduction in the water content of the system will result in an increased R, due to diminished A_{cw} by mass action effect. The observed effect is increased oil solubilization for a Type I system, or a reduction in the excess water phase for a Type II system.

Addition of an oil-soluble polar compound will result in a reduction of $A_{H_{cw}}$ and consequently, an increase in R. For a Type I system this will increase solubilization of nonpolar oils, while it will result in increased precipitation of an aqueous layer from a Type II system.

The lipophilic character of the C region may be increased by increasing the length of the hydrocarbon chain of the surfactant or by substituting an organic for an inorganic counter-ion. This will have the effect of increasing $A_{L_{co}}$ or diminishing $A_{H_{cw}}$, and causing an increase in R. This will result in increased oil solubilization for a Type I system or increased aqueous phase precipitation for a Type II system.

The effect of adding an inorganic salt is to diminish $A_{H_{cw}}$ by reducing ionic dissociation of the C layer, thus increasing R. The observed effect is increased oil solubilization or increased aqueous phase precipitation, depending upon the system type.

The reverse effects are self-evident. Thus, if the hydrophilic character of C is increased, R decreases due to either an increase in $A_{H_{cw}}$ or a decrease in

$A_{L_{co}}$. This will reduce the amount of oil solubilized in a Type I system and increase the amount of water solubilized in a Type II system.

With the more hydrophilic surfactants, $R \ll 1$ and solubilization is limited by the formation of Type I systems, with separation of an organic liquid. With the more lipophilic surfactants, $R \gg 1$ and solubilization is limited by the formation of Type II systems, with separation of water. With intermediate surfactants, or mixtures of the more hydrophilic and the more lipophilic surfactants, solubilization rises to a maximum as R equals 1. One example of this is shown in Table 9.14.

TABLE 9.14. SOLUBILIZATION OF WATER IN XYLENE BY DODECYLAMINE SALTS OF FATTY ACIDS AT 31°C. CONCENTRATION OF SURFACTANT = 1G/5 MOL XYLENE (100).

Surfactant	g H_2O/g surfactant
Dodecylamine formate	0.07
acetate	0.15
proprionate	0.57
n-butyrate	1.84
iso-butyrate	1.94
n-valerate	0.62
caproate	0.25
laurate	<0.25
palmitate	<0.25
oleate	0.12

NON-AQUEOUS SYSTEMS

Micelle Formation in Non-Polar Solvents

Oil-soluble surfactants, such as the petroleum mahogany sulfonates and the petroleum naphthanates, are used extensively as detergent additives and corrosion inhibitors in lubricating oils. These surfactants have been shown to be present in micellar form in hydrocarbon solvents (101–108). Singleterry and associates (103, 105, 106, 108, 109, 110, 111) showed that in solutions of the aryl stearates and the dinonylnaphthalene sulfonates micelles predominate at concentrations as low as 10^{-5} or 10^{-6} molar, and the size of the micelles remains fairly constant over a hundredfold or larger concentration range. Oil-soluble nonionic surfactants and amine-fatty acid soaps, on the other hand, form micelles at concentrations of about 10^{-2} molar and above, and the size of the micelles varies substantially with concentration (112, 113, 114, 115). Soaps of 2-ethylhexylsebacate form micelles in the region of 10^{-3} equivalents per liter (115).

Micelles formed in non-polar solvents can be expected to consist of an agglomeration of molecules with their polar groups directed toward the interior of the micelles. These micelles are capable of solubilizing water, as would be expected. Kitahara (107, 113, 114) determined solubilizing power and used the method to determine the cmc of dodecylammonium butyrate and caprylate soaps. Mattoon and Mathew (104) found that pure sodium bis-(2-ethylhexyl) sulfosuccinate formed spherical micelles in *n*-dodecane which persisted down to a weight concentration of 5 per cent. Saturation with water at least doubled the radii of the micelles. Using viscometry and ultracentrifuge measurements with the same system, Mathews and Hirschhorn (116) also concluded that sodium bis-(2-ethylhexyl) sulfosuccinate formed spherical micelles in the presence of moisture, and the size of the micelles increased with increasing water content. However, in the anhydrous system, the micelles appeared to be platelike.

Kaufman and Singleterry (106, 108) investigated micelle formation by a number of salts of dinonylnaphthalene sulfonate. The nature of the cation had little effect and almost all of the salts formed micelles with aggregation numbers falling within the range of 10 to 14 acid residues per micelle. The micellar size was generally insensitive to water content. Viscosity data resulted in calculated axial ratios falling between 2.45 and 1.85 for the various soaps of dinonylnaphthalene sulfonate, indicating a departure from sphericity. However, as in the case of aqueous systems, this viscosity behavior does not provide conclusive evidence of asymmetry. Solvation or surface roughness could have the same effect.

Unlike the oil-soluble naphthalene sulfonate soaps, which give low viscosity solutions, the carboxylate soaps produce high viscosities in anhydrous solution. Further these solutions in benzene are extraordinarily sensitive to traces of water (109, 110, 111). This is shown in Figure 9.29 for alkali metal phenylstearates in benzene. The s.r.u. addition—solute ratio units—refers to the moles of water added per equivalent of dissolved soap. Sodium, potassium and lithium soaps exhibit very high viscosities in anhydrous benzene, which decrease upon the addition of water. A lesser effect is obtained using the cesium soap. The lithium soap will tolerate considerably more water than the others before the viscosity drops to its minimum value (110).

Honig and Singleterry ascribe the high viscosities of the anhydrous solutions to organization of the soap molecules into extensive linear structures. A proposed structure is depicted in Figure 9.30. The basic structure is considered to be the coordinately-bonded single-chain of alternating cations and anions. The effectiveness of water, as well as other polar compounds, in

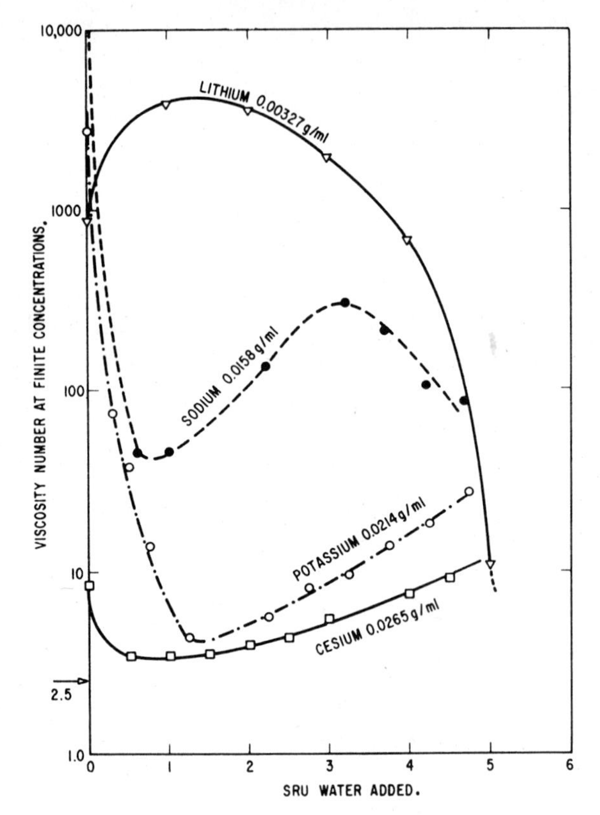

Figure 9.29. The effect of added water on the viscosity of alkali metal phenylstearate solutions at similar shearing stresses (110).

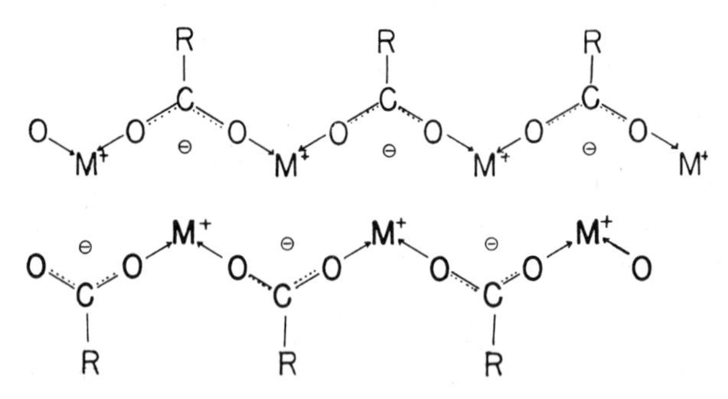

Figure 9.30. Proposed configuration for alkali metal soap micelles (111).

reducing the viscosity of the system may be due to competition between the oxygen of the added polar substance and the soap carboxylate oxygens for coordination with the metal ion.

Determination of Micellar Size by Fluorescence

An interesting method for determining the size of micelles as well as the cmc in non-aqueous systems was developed by Singleterry and associates (102, 103). The method consists of measuring the fluorescence and spectral absorption of a suitable dye. Rhodamine B is particularly suited to the detection of soap micelles in such solvents as benzene, cyclohexane, cetane, and di-(2-ethylhexyl)-sebacate. It is slightly soluble but almost non-absorbing and non-fluorescent in benzene, but fluoresces strongly in the presence of oil-dispersible carboxylate and sulfonate soaps. The flourescence appears to be due to adsorption of the dye on the soap micelles. Thus, the light emitted from a dilute solution of dye and soap illuminated with plane-polarized green light is 28 per cent polarized, as compared with a similar concentration of dye in methanol, which is only 2.3 per cent polarized. This difference in the extent of depolarization is due to the size of the fluorescing unit.

According to Perrin's (117) equation for the depolarization resulting from Brownian rotation during the excited period:

$$V = \frac{p\,(3 - p_0)\,\Gamma\,RT}{3\,(p_0 - p)\,\eta} \tag{9.40}$$

where V is the volume of the mobile unit from which the fluorescence is emitted; V is the hydrodynamic gram-molecular volume of the dye molecules, or in the case of micellar solutions, the hydrodynamic gram-micellar volume of the micelles to which the dye is adsorbed; p is the polarization of the fluorescence; p_0 is the polarization of the fluorescence of completely immobilized dye molecules; Γ is the average excited lifetime of the dye; R is the gas constant; T is the absolute temperature; and η is the viscosity of the solution. Micellar sizes determined by the fluorescence-depolarization method have been confirmed by cryoscopic (108) and osmotic pressure (103) measurements.

Figure 9.31 is a plot of apparent aggregation number versus stoichiometric concentration for sodium and barium dinonylnaphthalene sulfonates in benzene, where the apparent aggregation number was calculated from V in Perrin's equation. In the dilute range, there appears to be a decrease in micellar size. This apparent decrease results from the fact that the polarization value is a composite one, and includes the polarization of fluorescent bodies of all sizes present in the system. Rhodamine B molecules do not fluoresce in benzene. However, it is conceivable that the sulfonate molecules interact with the dye to form a complex that fluoresces. At infinite dilution of dinonylnaphthalene sulfonates the apparent aggregation number extrapolates to a value between 1.5 and 2. This suggests that a 1:1 dye-sulfonate molecular complex is the nonmicellar fluorescent unit (106).

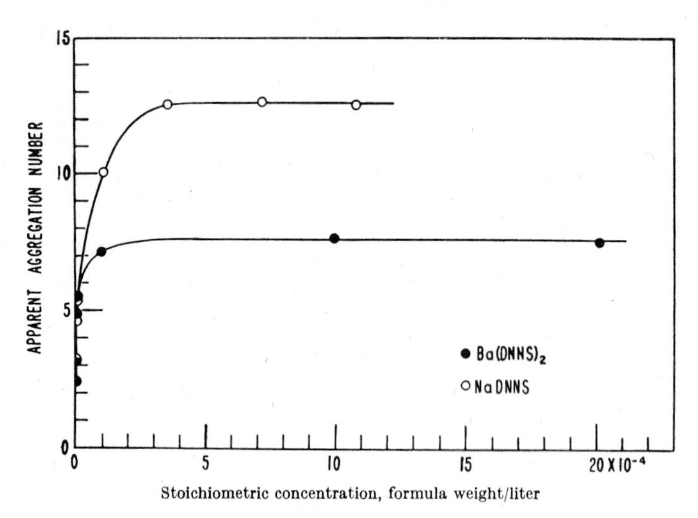

Figure 9.31. Influence of concentration on apparent aggregation numbers of dinonylnaphthalene sulfonates (106).

If fluorescence were due entirely to the dye-micelle complex, the cmc of the oil-dispersible surfactant could be determined by plotting the relative intensity of fluorescence against the square root of the surfactant concentration, with the cmc corresponding to zero intensity. However, if the monomer-dye complex also contributes to fluorescence, it is necessary to include this contribution. Kaufman and Singleterry (105) showed that

$$R\frac{Q_m}{Q_s} = \frac{p_c - p_s}{p_m - p_c} \cdot \frac{3 - p_m}{3 - p_s} \tag{9.41}$$

where R is the ratio of dye in micelles to that in the monomer-dye complex, and is effectively proportional to the concentration of micellar soap in the system. In this, Q is the quantum fluorescence efficiency, p is the fluorescence polarization, and the subscripts c, m and s refer to the composite, micellar and monomer-dye complex, respectively. A plot of R versus the stoichiometric concentration of barium dinonylnaphthalene sulfonate is shown in Figure 9.32. Extrapolation to R equals zero locates the cmc at 4×10^{-7} moles per liter (106).

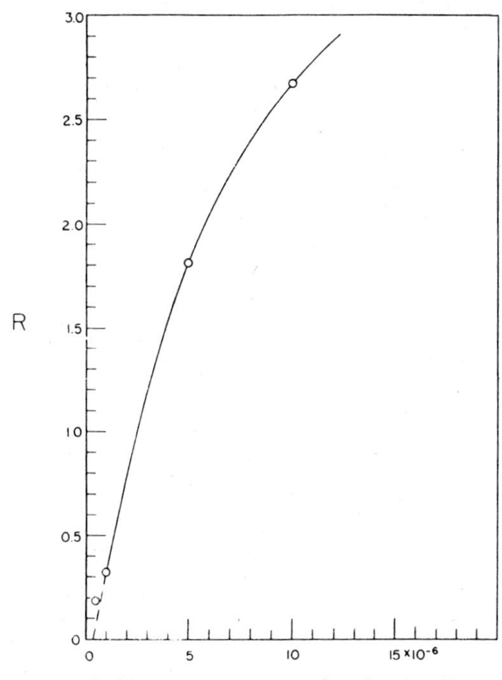

Figure 9.32. Extrapolation to the critical concentration for micelle formation of barium dinonylnaphthalene sulfonate (106).

NONIONIC SURFACTANTS

The condensation products of ethylene oxide with alkyl phenols, acids, alcohols, amines and propylene oxide condensation products are prepared as mixtures which follow a Poisson distribution law (118). The higher the degree of ethylene oxide polymerization, the wider the distribution (119). Condensation products based on the same hydrophobic group and with the same average molecular weight may have a different distribution of low- and high-molecular-weight condensation-products, depending upon reaction conditions. Consequently, they may differ in physicochemical properties.

Several investigators obtained narrow fractions of polyether surfactants by various fractionation techniques. Mayhew and Hyatt (118) used a centrifugal molecular still. They found differences between the narrow fractions and the normal mixture in some applications. Gallo (120) employed paper chromatography. Kelly and Greenwald (121) obtained fairly pure species by chromatographic adsorption on silicic acid and elution with mixed chloroform-acetone eluents. However, practically all available information on the properties of the polyoxyethylene nonionics have been obtained with normal mixtures.

Critical Micelle Concentrations

Hsiso, Dunning, and Lorenz (122) used the surface tension method to determine the cmc of a series of alkylphenols with various ethylene-oxide chain-lengths. Results shown in Figure 9.33 are for nonylphenols containing averages of 10.5, 15, 20 and 30 units of ethylene oxide per molecule. The cmc value increases with an increase in the average size of the hydrophile. Similarly, at the same molar concentration, the surface tensions increase with increasing length of the ethylene oxide chain.

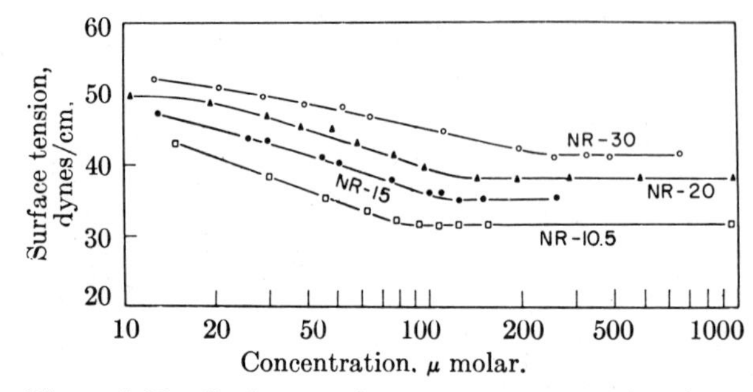

Figure 9.33. Surface tension *versus* concentration for a series of ethylene-oxide derivatives of nonylphenol. The NR number is the average number of moles of ethylene oxide per mole of alkylphenol (122).

The surface excess was calculated from Gibbs' adsorption equation and the area per molecule was calculated by assuming that the constant slope, at concentrations greater than the cmc, represents a monolayer. The molecular areas shown in Table 9.15 are considerably larger than the cross-sectional area of the benzene ring (25Å²), and increase with increasing ethylene oxide content. Thus, the ethylene oxide chain determines the areas occupied by the nonionic molecules. It appears that the molecules at the surface are close-packed, with the alkylphenol portion directed away from the solution (122).

The cmc values of these polyether nonionics are of the order of 10^{-4}

molar (123, 124), as compared with cmc values of about 10^{-2} molar for ionic surfactants with the same size hydrophobic group. Fowkes (125) has pointed out that the tendency to form micelles is proportional to the square of the concentration of ionic surfactant and only the first power of the concentration

Table 9.15. Surface excess and molecular areas of nonionic surfactants (122).

Phenol	Mole ratio ethylene oxide	C m C. μ molar	Surface excess moles/cm.^{2}	Area per molecule, Å^2.
Octyl	8.5	180–230	3.15	53
Nonyl	9.5	78–92	3.05	55
Nonyl	10.5	75–90	2.75	60
Nonyl	15	110–130	2.30	72
Nonyl	20	135–175	2.00	82
Nonyl	30	250–300	1.65	101
Nonyl	100	1000	0.95	173

of nonionic material. This can be seen by expressing the equilibrium equations for micelle formation as follows:

$$K_1 = \frac{[A^-]^n\ [C^+]^n}{[M_1]} = \frac{[A^-]^{2n}}{[M_1]} \tag{9.42}$$

$$K_2 = \frac{[N]^m}{[M_2]} \tag{9.43}$$

where K_1 and K_2 are the equilibrium constants of micelle formation for ionic and nonionic surfactants, respectively. $[A^-]$ and $[C^+]$ are the concentrations of the unassociated ionic species, $[N]$ is the concentration of unassociated nonionic molecules, and $[M_1]$ and $[M_2]$ refer to the concentrations of micelles of aggregation number n and m.

For micelles of equal aggregation number, if the equilibrium constants were the same for both micelles, a cmc of 10^{-2} molar for ionic substances would correspond to a cmc of 10^{-4} molar for nonionic surfactants.

Ross and Olivier (126) have reported on a method for the determination of the cmc of nonionic surfactants in both aqueous and nonaqueous solutions. While sufficient data are not available to assess the validity of the method, it does appear to give results that are in reasonable agreement with surface-tension measurements in aqueous solution. The cmc determinations in nonaqueous systems by this method gave results that were in fair agreement with relative differential measurements of refractive index.

TABLE 9.16. THE ABSORPTION MAXIMA OF IODINE IN DIFFERENT ENVIRONMENTS (126).

	mμ	Color Transmitted
I_2 vapor	512	Violet
I_2 in carbon tetrachloride	510	Violet
I_2 in benzene	300, 490	Red
I_2 in water	450	Brown
I_2–micelle complex	360	Yellow
I_2–KI complex in water	288, 353	Yellow

The method is based on the formation of a molecular complex between iodine and the nonionic micelle which shows an absorption maximum at 360 mμ. The absorption maxima of iodine in different environments are given in Table 9.16.

According to this method, a stock solution of iodine selected to transmit

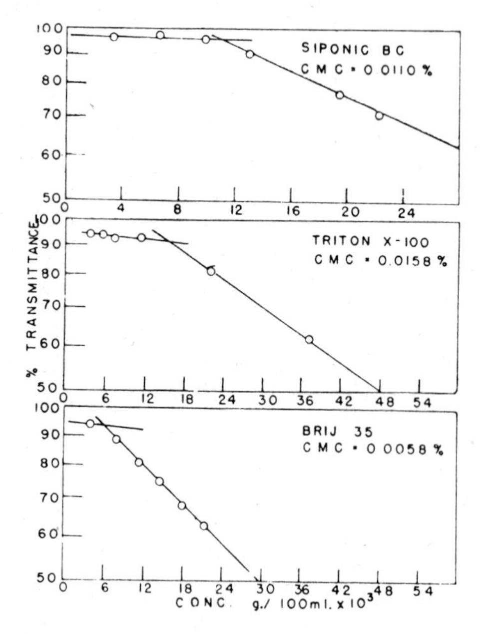

Figure 9.34. Effect of surfactant concentration on the logarithm of the per cent transmittance at 360 mμ of aqueous solutions with iodine. Siponic BC is an ethoxylated branched-chain alcohol. Triton X100 is a 9-10 mole ratio ethoxylated octylphenol, and Brij 35 is a 23 mole ratio ethoxylated lauryl alcohol (126).

80 per cent of the light transmitted by the pure solvent is used to dilute a solution of the surfactant in the same solvent, which also contains the same concentration of iodine as the stock solution. Thus a series of solutions are prepared, each containing the same concentration of iodine but different levels of the nonionic. These solutions are read using a spectrophotometer within one hour after preparation, using the iodine stock solution as a standard for 100 per cent transmittance. Typical results are shown in Figure 9.34. The break in the curve corresponds to the cmc value. The concentrations of iodine in stock solutions of different solvents used by Ross and Olivier are: for water, 30 mg per liter; for benzene 25 mg per liter; for carbon tetrachloride, 21 mg per liter.

This method was used by Becher (127) to determine the cmc values of a number of commercial nonionic surfactants. He also reported excellent agreement between the iodine method and light-scattering cmc values. Becher's results are reproduced in Figure 9.35.

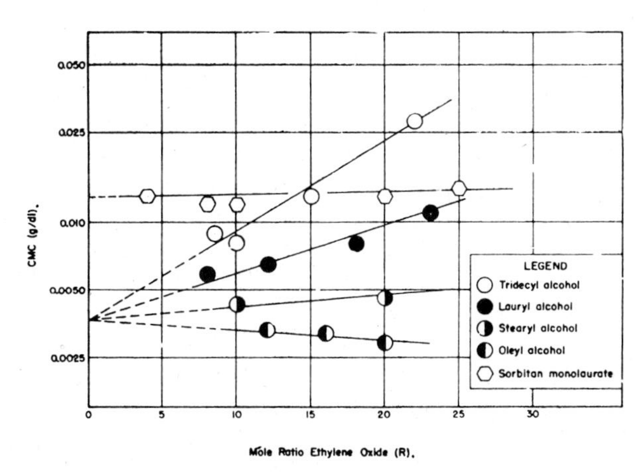

Figure 9.35. Relation between the cmc of ether-alcohols and the ethylene oxide mole ratio as determined by the iodine method (127).

Hsiao, Dunning and Lorenz (122) found that for ethylene oxide derivatives of nonylphenol, their results could be expressed by the relation

$$\ln C_0 = A + BR$$

Where C_0 is the cmc, R is the number of ethylene oxide units per mole of hydrophobe, and A and B are constants, with the value of the intercept A depending on the electrolyte content.

The equations for the lines shown in Figure 9.36 are, with C_0 expressed in units of grams deciliter^{-1} $\times$ 10^{-4}

Tridecyl alcohol:

$$\ln C_0 = 3.59 + 0.091R \tag{9.45}$$

Lauryl alcohol:

$$\ln C_0 = 3.72 + 0.038R \tag{9.46}$$

Stearyl alcohol:

$$\ln C_0 = 3.69 + 0.0068R \tag{9.47}$$

Oleyl alcohol:

$$\ln C_0 = 3.67 - 0.015R \tag{9.48}$$

Sorbitan monolaurate:

$$\ln C_0 = 4.87 \tag{9.49}$$

Within experimental error, the ethylene-oxide derivatives share a common intercept, which is 3.67 $\pm$ 0.05, while an intercept value of 4.87 was obtained for the polyolsorbitan monolaurate. The reason for the common intercept is not clear.

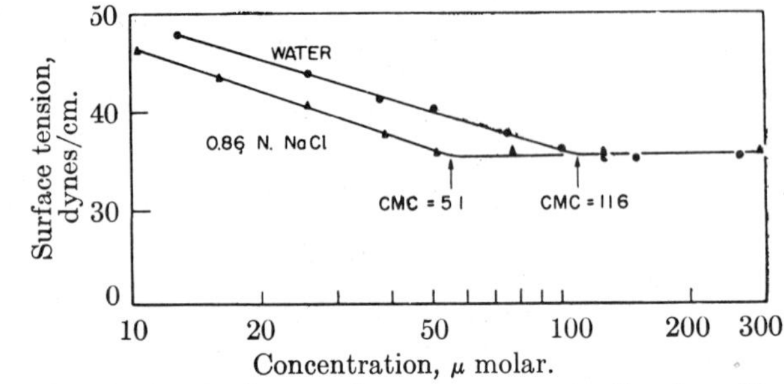

Figure 9.36. Surface tension *versus* concentration for a 15 mole ethylene oxide to nonylphenol adduct in distilled water and in 0.86 N NaCl (122).

The tendency for the cmc to increase with an increase in the ethylene-oxide chain appears not to be general, but to depend upon the size and character of the hydrophobe. With stearyl alcohol, the cmc increases only slightly with increasing ethylene oxide chain length, while it actually decreases in the case of oleylalcohol derivatives.

The effect of added electrolyte on the surface tension-concentration curve of a nonylphenol containing an average of 15 ethoxy groups in the chain is shown in Figure 9.36. These results are typical of those obtained with other molar ratios of ethylene oxide. The electrolytes lower the cmc and the surface tensions at concentrations less than the cmc. At higher concentrations of surfactant, the surface tension remains unchanged.

Kushner and Hubbard (128) failed to find evidence for a critical micelle concentration for Triton X100 by light scattering measurements. They ascribed this to the likelihood that the sample of Triton X100 had a broad distribution of molecular weights. If this were the case, the low molecular weight molecules could begin forming micelles at very low concentrations, with the higher molecular weight molecules forming micelles at somewhat higher concentrations. From their data, they concluded that the sample of Triton X100 had a monomer saturation concentration of 0.3 g per deciliter. Above this concentration essentially all detergent added to the solution becomes micellar. The light scattering data also indicated a molecular weight of close to 90,000, corresponding to approximately 140 detergent molecules to a micelle. Mankowich (57) obtained a micellar molecular weight of 81,300 by light scattering using a sample of isooctylphenyl nonaethylene glycol ether. Quite likely this material was also Triton X100.

While the cmc of ionic detergents generally increases with an increase in temperature, the cmc of polyoxyethlene-derived detergents decreases as the temperature is raised (136). This is ascribed to dehydration of the polar groups as the temperature is increased.

Solubility

The solubility behavior of these nonionics depends upon the length of the polyoxyethylene chain attached to the hydrophobe. Products containing a small ratio of ethylene oxide units to hydrophobe are soluble in hydrocarbon solvents but insoluble in water. As the length of the polar chain is increased, the surfactant becomes more water soluble and less hydrocarbon soluble, until it is miscible with water in all proportions and insoluble in hydrocarbons.

The water-soluble polyoxyethylene nonionics are less soluble in hot water than in cold water. These surfactants are thought to dissolve by the association of water molecules with the ether linkages. At higher temperatures this association is partially destroyed, causing the surfactant to become less soluble. The temperature at which the solubility of the surfactant is greatly reduced is marked by a change from a clear to a turbid solution, and is known as the cloud point. The larger the ratio of ethylene oxide units to hydrophobe, the higher the temperature at which the cloud point occurs. This is generally the case for all hydrophobic groups as shown in Figure 9.37.

The cloud points of a series of polyoxyethylene nonylphenols in the presence of electrolytes are shown in Table 9.17. These nonionics are less soluble in the presence of sodium salts and bases than in distilled water. Presumably, hydration of the ether linkages is reduced with these electrolytes present in solution. Dilute acids may raise the cloud point possibly by formation of oxonium compounds with the ether oxygens (130). While sodium chloride reduces the solubility of these nonionics in water, calcium chloride solutions appear to be better solvents than pure water. There is evidence that hydrated calcium ions are complexed by the ethereal oxygens of the polyoxyethylene chain (131).

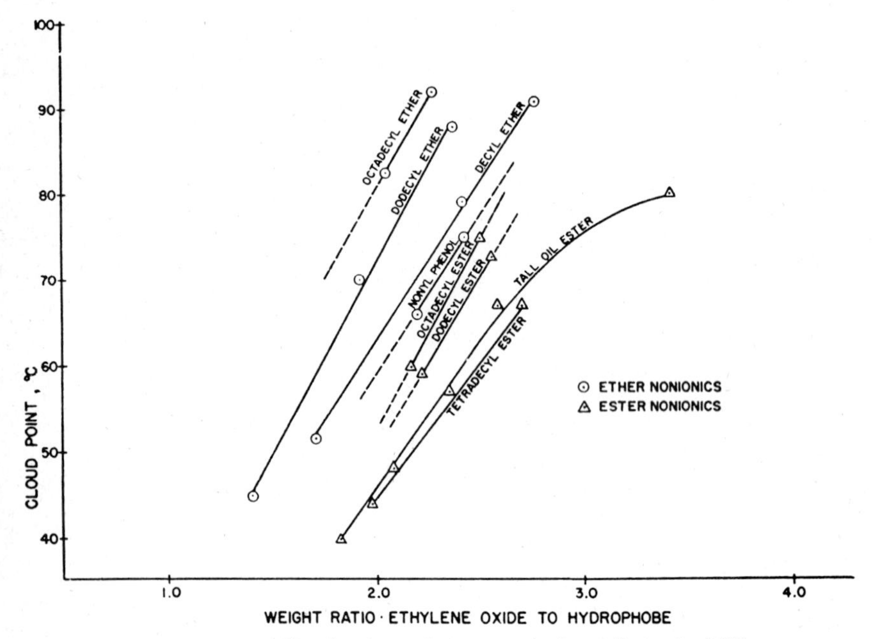

Figure 9.37. Cloud points of some nonionic surfactants (129).

The nonionic surfactants are capable of solubilizing organic compounds in the same manner as ionic surfactants. However, Weiden and Norton (132) have reported that many aromatic compounds solubilized by the nonionic have the effect of lowering the cloud point. With Triton X100 and an ethoxylated thio ether, both of which separate from aqueous solutions to form a heavier, detergent-rich phase above 60°C, they found that aromatic liquids lowered the temperature for phase separation by 5–50°C. Solubilizates that were effective in producing a detergent-rich phase at concentrations of 1 to 10 grams per liter of 2% Triton X100 included benzene, benzaldehyde, dibenzyl ether, and dibutyl phthalate.

Using benzene as the solubilizate, the detergent-rich phase appeared at a lower temperature when the concentration of detergent was decreased. This suggested to Weiden and Norton that solubilization of the benzene in the detergent micelles has the effect of decreasing the hydrophilic character of the micelles. With larger molecules, such as dibenzyl ether and dibutyl phthalate, three-component systems were obtained that were transparent at room temperature. Phase separation occurred both on raising and on lowering the temperature. At the lower temperature, solubilizate separated due to saturation of the solubilizer micelles. At the higher temperature, the detergent separated.

TABLE 9.17. CLOUD POINT OF POLYOXYETHYLATED NONYLPHENOLS IN AQUEOUS SOLUTIONS (130).

	Cloud Point °C Mole Ratio Ethylene Oxide to Nonylphenol			
Solvent	9	10.5	15	20
Water	55	72	98	>100
3% NaCl	45	61	84.5	95
3% Na_2CO_3	32	48	70	78
3% NaOH	31	45.5	67	73
3% HCl	60.5	78	>100	>100
3% H_2SO_4	51	69	96	>100
3% Na_3PO_4	43	60	83	92

When aqueous solutions of catechol and similar water-soluble phenols are combined with aqueous solutions of polyoxyethylene nonionics, separation into two layers may be observed. Livingston (133) found that this effect was produced by phenols, but not by alkali phenolates. Further, at any given pH, the strongly acidic phenols have a poorer coagulating effect on the nonionic detergents than the more weakly acidic phenols, which have a greater concentration of free phenol at the same pH. DeNavarre (134) and others have observed that the nonionic surfactants interfere with the preservative action of p-hydroxybenzoate esters.

Viscosity

The relation between viscosity and the molar ratio of ethylene oxide to hydrophobe is shown in Figure 9.38 for fatty alcohol condensates. Essentially the same results were obtained with polyoxyethylene condensates of oleyl, cetyl, and mixed sperm alcohols (135). The products that contain barely sufficient ethylene oxide groups for water solubility have very high viscosities. The viscosity decreases with an increase in the number of moles of ethylene oxide, and reaches a minimum value with 10 moles of ethylene oxide per mole of fatty alcohol. Beyond this level of ethylene oxide, viscosity increases with increasing molecular weight.

Figure 9.38 also shows that the viscosity increases with an increase in concentration of the nonionic. A further increase in concentration to about 50 to 70 per cent nonionic frequently results in gel formation. At higher concentrations of nonionic the viscosity is greatly reduced. The abnormal viscosity maximum becomes less pronounced as the temperature is increased. Salts lower the viscosity of the solution in the gel region, preventing gel formation, and raise the viscosity of the more dilute solutions.

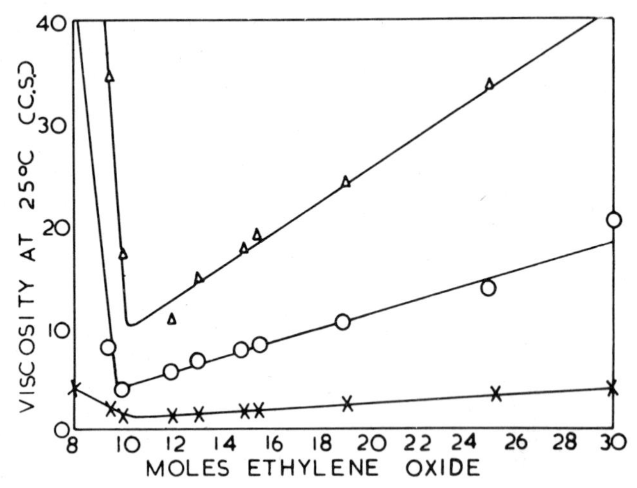

Figure 9.38. Viscosities of aqueous solutions of ethoxylated fatty alcohols. Concentration in g/liter: △-250; ○-200; X-100 (135).

Greenwald and Brown (136) explained these abnormal viscosities on the basis of x-ray and optical studies previously reported by Schulman, Matalon and Cohen (137), who had found evidence of lamellar, cylindrical and spherical aggregates of nonionic detergent molecules in aqueous solution. In these aggregates there appeared to be one water molecule bound to each ether linkage. If the nonionic solutions are likened to an emulsion, then the micelle aggregates will be in close-packed configuration at 74 per cent by volume. If the micelles are considered to be hydrated with one water molecule per ether linkage, the micelles will occupy 74 volume per cent at detergent concentrations ranging from 59 to 57 weight per cent for the three octylphenyl polyethylene oxides. This is not greatly different from the concentrations corresponding to maximum viscosity. At concentrations below the viscosity maximum, the system consists of hydrated nonionic micelles in a continuum of water saturated with nonionic molecules. Gel formation occurs at or below 74 volume per cent of hydrated micelles, depending upon the shape of the micelles, interaction between micelles and the amount of water bound. Considering the high end of the composition range, viscosity generally increases when water is added to the pure nonionic. If water links the nonionic molecules together, the viscosity should increase with increasing water content, at a rate depending on the chain length and the degree of order introduced. Beyond the amount of water required to hydrate the ether linkages, pockets of water may form to further add to the viscosity (136).

In contrast to these considerations Kushner and Hubbard (128) deduced from intrinsic viscosity data obtained with dilute solutions that Triton X100 micelles contain approximately four kinetically bound water molecules per ether linkage.

Gross Properties

Knowles and Krupin (130) observed the following generalizations with regard to the wetting of skeins of cotton yarn by ethylene-oxide derivatives of nonylphenol:

(1) Those derivatives with cloud points above 100°C in distilled water are poor wetting agents in water and in aqueous solutions containing electrolytes.

(2) Nonionic surfactants having cloud points in distilled water between 55°C and 100°C increase in wetting capacity with increase in temperature until the cloud point is reached. At higher temperatures, the wetting action is considerably reduced.

(3) The wetting capacity will vary with the composition of the solution. Electrolytes generally lower wetting efficiency.

Concerning the foaming power of these nonionics, it was found that none foamed appreciably at temperatures above their cloud point. The highest initial foams were obtained with products having the highest cloud points. As the temperature was increased, foaming power decreased. Many electrolytes tend to lower the height of foam produced (130).

FLUOROCARBON SURFACTANTS

As a class the fluorocarbons have the lowest free surface energy of any known compounds. This is illustrated by the data in Table 9–18. The surface tensions are, however, not unique from the standpoint of the Eötvös equation

$$\gamma (M/\rho)^{2/3} = K (T_c - T - 6) \tag{9.50}$$

which predicts their surface tensions fairly well from molecular weight M, density ρ, and critical temperature T_c. The values for K are in fair agreement with the accepted value for K of 2.2 (138).

TABLE 9.18. SURFACE TENSION AND EÖTVÖS CONSTANT VALUES FOR FLUOROCARBONS AND THEIR DERIVATIVES (138).

Fluorocarbons and Derivatives	Surface Tension, dynes per cm, 20°C	Eötvös Constant, K, 20°C
n-C_5F_{12}	9.87	2.0
cyclo-C_5F_{10}	11.09	2.2
$(C_4F_9)_2O$	12.2	2.4
C_8F_{18}	13.6	2.0
$(C_2F_5)_3N$	13.6	2.3
$C_2F_5(C_3F_7)_2N$	14.0	2.1
$(C_4F_9)_3N$	16.8	2.2

As one would expect, amphipathic fluorocarbon derivatives show pronounced surface activity. From Figure 9.39 it will be observed that considerably shorter chain-lengths are required for surface activity with the fluorocarbons as compared with the hydrocarbons, and surface-tension values are lower above the concentration for micelle formation.

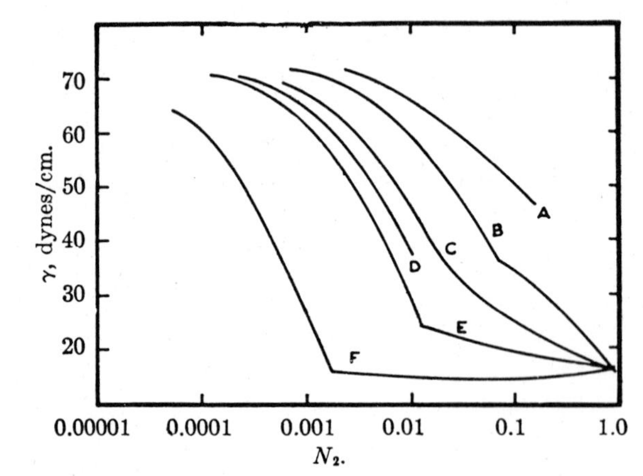

Figure 9.39. Surface tensions of aqueous solutions at 25°; A, perfluorosuccinic acid; B, perfluoroacetic acid; C, perfluoropropionic acid; D, perfluoroadipic acid; E, perfluorobutyric acid; F, perfluorocaproic acid (138).

Unlike the hydrocarbon acids, the perfluorocarbon acids behave as strong acids both above and below the cmc. The fact that perfluorocarbon acids with relatively short chains are surface active suggests a very high energy of adsorption as compared with paraffinic chains. For a monomolecular film spread on water at a surface pressure sufficient to give an area of 90Å^2 per molecule, the energy of adsorption of a CH_2 group may be taken as about 725 cal per mole (140). In comparison, the energy of adsorption of a CF_2 group at the same area per molecule is estimated at 1490 cal per mole (141). As a ratio, these values are similar to the association energies for micelle formation of 750 cal per mole per CH_2 and of 1300 cal per mole per CF_2 (56).

TABLE 9.19. VALUES OF CMC FOR A SERIES OF PERFLUOROACIDS AT 0°C TO 45°C (145).

	CMC (moles/liter)				
	0°C	18°C	25°C	35°C	45°C
CF_3COOH				~2.4	
C_2F_5COOH				1.15	
C_3F_7COOH	0.412	0.418		0.452	
$C_5F_{11}COOH$	0.048	0.0475	0.051		
$C_7F_{15}COOH$	0.0050	0.0052		0.0058	0.0065

The high association and adsorption energies of fluorocarbons are due to the very low cohesion between the fluorocarbon chains and water molecules, as compared to that between hydrocarbon chains and water molecules. Further, there is a much lower order of attraction between fluorocarbon chains than between hydrocarbon chains. This is shown by the boiling point (142) and heat and entropy of vaporization data (143) for perfluoro-carbons. As a consequence, monolayers of the perfluoroacids act as two-dimensional gases even at fairly-high surface-pressures (141).

While the cmc of hydrocarbon-chain surfactants decreases by a factor of 2 for each additional CH_2 group in the chain, for perfluorocarbon surfactants the corresponding factor is 3.1 (144). The cmc increases with an increase in temperature, as shown in Table 9.19.

Fluorinated alkanoic acids containing a terminal hydrogen atom, $H(CF_2)_n$ COOH, are similar in behavior to the perfluorinated acids. The surface tension of the pure fluoroacids is about 13 dynes/cm, and solubility decreases rapidly with chain length. The ammonium salt of $H(CF_2)_{12}$ COOH is practically insoluble in water (120). However, the cmc and its dependence on chain length is intermediate between that of the perfluoroacids and the hydrocarbon-chain soaps. Thus the 12-carbon potassium laurate, the 9-carbon ammonium fluorocarboxylate and the 6-carbon perfluoroacid have approximately the same cmc. Data are presented in Table 9.20 (120).

TABLE 9.20. VALUES OF CMC FOR A SERIES OF HIGHLY FLUORINATED ACIDS (146).

	CMC (moles/liter)
$H(CF_2)_6COONH_4$	0.25
$H(CF_2)_8COONH_4$	0.038
$H(CF_2)_{10}COONH_4$	0.009
$H(CF_2)_6COOH$	0.15
$H(CF_2)_8COOH$	0.03
$C_{11}H_{23}COOK$	0.0255
$C_{13}H_{27}COOK$	0.0066

References

1. Langmuir, I., "Phenomena, Atoms and Molecules", New York, Philosphical Library, Inc., 1950.
2. Harkins, W. D., "The Physical Chemistry of Surface Films", New York, Reinhold Publishing Corp., 1952.
3. Klevens, H. B., *J. Phys. and Colloid Chem.* **52**, 130 (1948).
4. Klevens, H. B., *J. Am. Oil Chemists' Soc.* **30**, 76 (1953).
5. Pauling, L., "The Nature of the Chemical Bond", Ithaca, New York, Cornell Univ. Press, 1940.
6. Harkins, W. D., *J. Am. Chem. Soc.* **69**, 682 (1947).
7. Hobbs, M. E., *J. Phys. Chem.* **55**, 675 (1951).
8. Lottermoser, A., and Püschell, E., *Kolloid Z.* **63**, 175 (1936).
9. Dreger, E. E., Keim, G. L., Miles, G. D., Shedlovsky, L., and Ross, J., *Ind. Eng. Chem.* **36**, 611 (1944).
10. Miller, M. L., and Dixon, J. K., *J. Colloid Sci.* **13**, 411 (1958).
11. Paquette, R. G., Lingafelter, E. C., and Tartar, H. V., *J. Am. Chem. Soc.* **65**, 686 (1943).
12. Ekwall, P., *Kolloid Z.* **101**, 135 (1942).
13. Gregory, N. W., and Tartar, H. V., *J. Am. Chem. Soc.* **70**. 1992 (1948).
14. Harkins, W. D., *J. Colloid Sci.* **1**, 469 (1946).
15. Shinoda, K., *J. Phys. Chem.* **58**, 1136 (1954).
16. Brady, A. P., and Huff, H., *J. Colloid Sci.* **3**, 511 (1948).
17. Klevens, H. B., *J. Phys. and Colloid Chem.* **51**, 130 (1947).
18. Wright, K. A., Abbott, A. D., Sivertz, V., and Tartar, H. V., *J. Am. Chem. Soc.* **61**, 549 (1939).
19. Ginn, M. E., Kinney, F. B., and Harris, J. C., *J. Am. Oil Chemists' Soc.* **37** 183 (1960).
20. Flockhart, B. D., and Ubbelohde, A. R., *J. Colloid Sci.* **8**, 428 (1953).
21. Flockhart, B. D., *J. Colloid Sci.* **12**, 557 (1957).
22. McBain, J. W., "Frontiers in Colloid Chemistry", p. 144, New York, Interscience Publishers, 1949.
23. McBain, M. E. L., and Hutchinson, E., "Solubilization", New York, Academic Press, 1955.
24. Hartley, G. S., "Aqueous Solutions of Paraffin-Chain Salts", Paris, Hermann et Cie, 1936.

25. Debye, P., and Anacker, E. W., *J. Phys. Chem.* **51**, 18 (1947).
26. Anacker, E. W., *J. Colloid Sci.* **8**, 402 (1953).
27. McBain, J. W., in "Advances in Colloid Science," Vol. 1, New York, Interscience Publishers, 1942.
28. Hughes, E. W., Am. Phys. Soc. Meeting, New York, N.Y., Sept. 1945.
29. Halsey, G. D., Jr., *J. Phys. Chem.* **57**, 87 (1953).
30. Philippoff, W., *J. Colloid Sci.* **5**, 169 (1950).
31. Klevens, H. B., 22nd National Colloid Symposium (1948).
32. Backus, J. K., and Scheraga, H. A., *J. Colloid Sci.* **6**, 508 (1951).
33. Harkins, W. D., Mattoon, R. W., and Stearns, R. S., *J. Chem. Phys.* **15**, 209 (1947).
34. Harkins, W. D., and Mittleman, R., *J. Colloid Sci.* **4**, 369 (1949).
35. McBain, J. W., and Hoffman, O. A., *J. Phys. and Colloid Chem.* **53**, 39 (1949).
36. Harkins, W. D., Mattoon, R. W., and Corrin, M. L., *J. Am. Chem. Soc.* **68**, 220 (1946).
37. Harkins, W. D., and Mittleman, R., *J. Colloid Sci.* **4**, 367 (1949).
38. Corrin, M. L., *J. Chem. Phys.* **16**, 844 (1948).
39. Riley, D. P., and Oster, G., *Disc. Faraday Soc.* **11**, 107 (1951).
40. Fournet, G., *Disc. Faraday Soc.* **11**, 121 (1951).
41. Debye, P., *J. Phys. Chem.* **53**, 1 (1949).
42. Stamm, Mariner, and Dixon, *J. Chem. Phys.* **16**, 423 (1948).
43. Anacker, E. W., *J. Phys. Chem.* **62**, 41 (1958).
44. Mattoon, R. W., Stearns, R. S., and Harkins, W. D., *J. Chem. Phys.* **16**, 644 (1948).
45. Lamm, O., *Kolloid Z.* **98**, 45 (1942).
46. Miller, G. L., and Anderson, K. J., *J. Biol. Chem.* **144**, 475 (1942).
47. Hartley, G. S., and Runnicles, D. F., *Proc. Roy Soc.* (London) **168A**, 420 (1938).
48. Vetter, R. J., *J. Phys. Chem.* **51**, 262 (1947).
49. Granath, C., *Acta Chemica Scand.* **4**, 103 (1954).
50. Debye, P., *J. Appl. Phys.* **15**, 338 (1944).
51. Debye, P., *J. Phys. and Colloid Chem.* **53**, 18 (1949).
52. Princen, L. H., and Mysels, K. J., Ninth Technical Report, Project ONR–356–254, Office of Naval Research.
53. Prins, W., and Hermans, J. J., *Proc. Kon. Ned. Akad. Wetensch* **B59**, 162 (1956).
54. Debye, P., *Ann. N. Y. Acad. Sci.* **51**, 575 (1949).
55. Mysels, K. J., *J. Colloid Sci.* **10**, 507 (1955).
56. Phillips, J. N., *Trans. Faraday Soc.* **51**, 561 (1955).
57. Mankowich, A. M., *Ind. Eng. Chem.* **47**, 2175 (1955).
58. Ludlum, D. B., *J. Phys. Chem.* **60**, 1240 (1956).
59. van Rysselberghe, P., *J. Phys. Chem.* **43**, 1049 (1939).
60. Stainsby, G., and Alexander, A. E., *Trans. Faraday Soc.* **46**, 587 (1950).
61. Reich, I., *J. Phys. Chem.* **60**, 257 (1956).
62. Hoeve, C. A. J., and Benson, G. C., *J. Phys. Chem.* **61**, 1149 (1957).
63. Nakagaki, M., *J. Chemi Soc.* (Japan) **72**, 113 (1951).
64. Ooshika, Y., *J. Colloid Sci.* **9**, 254 (1954).
65. Stigter, D., and Overbeek, J. T. G., Proc. Intern. Congr. Surface Activity, 2nd, London, Butterworth, 1957.
66. Shinoda, K., *Bull. Chem. Soc. Japan* **26**, 101 (1953).
67. Goddard, E. D., and Benson, G. C., *Can. J. Chem.* **35**, 986 (1957).
68. Hutchinson, E., Manchester, K. E., and Winslow, L., *J. Phys. Chem.* **58**, 1124 (1954).

69. Roe, C. P., and Brass, P. D., *J. Am. Oil Chemists' Soc.* **76**, 4703 (1954).
70. Herzfeld, S. H., *J. Phys. Chem.* **56**, 953 (1952).
71. Mukerjee, P. and Mysels, K. J., *J. Am. Chem. Soc.* **77**, 2937 (1955).
72. Murray, R. C., and Hartley, G. S., *Trans. Faraday Soc.* **31**, 183 (1935).
73. Adam, N. K., and Pankhurst, K. G. A., *Trans. Faraday Soc.* **42**, 523 (1946).
74. Tartar, H. V., and Wright, K. A., *J. Am. Chem. Soc.* **61**, 539 (1939).
75. Wright, K. A., and Tartar, H. V., *J. Am. Chem. Soc.* **61**, 544 (1939).
76. Gershman, J. W., *J. Phys. Chem.* **61**, 581 (1957).
77. Raison, M., Proc. Intern. Congr. Surface Activity, 2nd, London, Butterworth, 1957.
78. McBain, J. W., and McBain, M. E. L., *J. Am. Chem. Soc.* **58**, 2610 (1936).
79. McBain, J. W., and Richards, P. H., *Ind. Eng. Chem.* **38**, 642 (1946).
80. Stearns, R. S., Oppenheimer, H., Simon, E., and Harkins, W. D., *J. Chem. Phys.* **15**, 496 (1947).
81. Klevens, H. B., *Chem. Revs.* **47**, 1 (1950).
82. Richards, P. H., and McBain, J. W., *J. Am. Chem. Soc.* **70**, 1338 (1948).
83. Harkins, W. D., and Oppenheimer, H., *J. Am. Chem. Soc.* **71**, 808 (1949).
84. Kolthoff, I. M., and Graydon, W. F., *J. Phys. and Colloid Chem.* **55**, 699 (1951).
85. Kolthoff, I. M., and Stricks, W., *J. Phys. and Colloid Chem.* **53**, 424 (1949).
86. Klevens, H. B., *J. Am. Chem. Soc.* **72**, 3780 (1950).
87. Engler, C., and Dieckhoff, E., *Arch. Pharm.* **230**, 561 (1892).
88. Green, A. A., and McBain, J. W., *J. Phys. and Colloid Chem.* **51**, 286 (1947).
89. Klevens, H. B., *J. Chem. Phys.* **17**, 1004 (1949).
90. Klevens, H. B., *J. Am. Chem. Soc.* **72**, 3581 (1950).
91. McBain, J. W., in Alexander's "Colloid Chemistry", Vol. 1, p. 132, New York, Chemical Catalog Co., Inc., 1926.
92. McBain, J. W., Elford, W. J., and Vold, R. D., *J. Soc. Chem. Ind.* **59**, 243 (1940).
93. McBain, J. W., Vold, M. J., and Porter, J. L., *Ind. Eng. Chem.* 33, 1049 (1941).
94. Merritt, R. C., *J. Am. Oil Chemists' Soc.* **25**, 84 (1948).
95. Lawrence, A. S. C., "First World Congress on Surface Active Agents", Vol. 1, p. 31, Paris, Chambre Syndicate Tramagras; Hyde, A. J., Langbridge, D. M., and Lawrence, A. S. C., *Disc. Faraday Soc.* **18**, 239 (1954).
96. Winsor, P. A., *Trans. Faraday Soc.* **44**, 376 (1948).
97. Winsor, P. A., *Trans. Faraday Soc.* **46**, 762 (1950).
98. Bromilow, J., and Winsor, P. A., *J. Phys. Chem.* **57**, 889 (1953).
99. Winsor, P. A., *J. Colloid Sci.* **10**, 88 (1955).
100. Winsor, P. A., *Chemistry & Industry*, June 4, 1960, pp. 645.
101. McBain, J. W., Merrill, R. C., Jr., and Vinograd, J. R., *J. Am. Chem. Soc.* **62**, 2880 (1940).
102. Arkin, L., and Singleterry, C. R., *J. Am. Chem. Soc.* **70**, 3965 (1948).
103. Singleterry, C. R., and Weinberger, L. A., *J. Am. Chem. Soc.* **73**, 4574 (1951).
104. Mattoon, R. W., and Mathews, M. B., *J. Chem. Phys.* **17**, 496 (1949).
105. Kaufman, S., and Singleterry, C. R., Naval Research Laboratory Report 3966 (1952).
106. Kaufman, S., and Singleterry, C. R., *J. Colloid Sci.* **10**, 139 (1955).
107. Kitahara, A., *J. Colloid Sci.* **12**, 342 (1957).
108. Kaufman, S., and Singleterry, C. R., *J. Colloid Sci.* **12**, 465 (1957).
109. Honig, J. G., and Singleterry, C. R., *J. Phys. Chem.* **58**, 201 (1954).
110. Honig, J. G., and Singleterry, C. R., *J. Phys. Chem.* **60**, 1108 (1956).
111. Honig, J. G., and Singleterry, C. R., *J. Phys. Chem.* **60**, 1114 (1956).
112. Singleterry, C. R., *J. Am. Oil Chemist's Soc.* **32**, 446 (1953).

113. Kitahara, A., *Bull. Chem. Soc. Japan* **28**, 234 (1955).
114. Kitahara, A., *Bull. Chem. Soc. Japan* **29**, 15 (1956).
115. Kaufman, S., and Singleterry, C. R., *J. Phys. Chem.* **62**, 1257 (1958).
116. Mathews, M. B., and Hirschhorn, E., *J. Colloid Sci.* **8**, 86 (1953).
117. Perrin, F., *J. phys. radium* [VI] **7**, 390 (1926).
118. Mayhew, R. L., and Hyatt, R. C., *J. Am. Oil Chemists' Soc.* **29**, 357 (1952).
119. Flory, P. J., *J. Am. Chem. Soc.* **62**, 1561 (1940).
120. Gallo, V., *Boll. Chim. farm.* **92**, 332 (1953).
121. Kelly, J., and Greenwald, H. L., *J. Phys. Chem.* **62**, 1096 (1958).
122. Hsiao, L., Dunning, H. N., and Lorenz, P. B., *J. Phys. Chem.* **60**, 657 (1956).
123. Gonick, E., and McBain, J. W., *J. Am. Chem. Soc.* **69**, 334 (1947).
124. Goto, R., Sugano, T., and Koizuma, N., *J. Chem. Soc.* (Japan) **75**, 73 (1954).
125. Fowkes, F. M., *J. Phys. Chem.* **63**, 1674 (1959).
126. Ross, S., and Olivier, J. P., *J. Phys. Chem.* **63**, 1671 (1959).
127. Becher, P., *J. Phys. Chem.* **63**, 1675 (1959).
128. Kushner, L. N., and Hubbard, W. D., *J. Phys. Chem.* **58**, 1163 (1954).
129. Karabinos, J. V., Hazdra, J. J., and Kapella, G. E., *Soap and Chemical Specialties* **7**, April (1955).
130. Knowles, C. M., and Krupin, F., presented at Chemical Specialties Manufacturers Association Meeting, Washington, D.C., December 6–9, 1953.
131. Doscher, T. M., Myers, G. E., and Atkins, D. C. Jr., *J. Colloid Sci.* **6**, 223 (1951).
132. Weiden, M. H. J., and Norton, L. B., *J. Colloid Sci.* **8**, 606 (1953).
133. Livingston, H. K., *J. Colloid Sci.* **9**, 365 (1954).
134. de Navarre, M. G., "First World Congress on Surface Active Agents", Vol. II, p. 741, Paris, Chambre Syndicate Tramagras.
135. Raphael, L., "First World Congress on Surface Active Agents", Vol. I, p. 52, Paris, Chambre Syndicate Tramagras.
136. Greenwald, H. L., and Brown, G. L., *J. Phys. Chem.* **58**, 825 (1954).
137. Schulman, J. H., Matalon, R., and Cohen, M., *Disc. Faraday Soc.* **11**, 117 (1951).
138. Scholberg, H. M., Guenthner, R. A., and Coon, R. I., *J. Phys. Chem.* **57**, 923 (1953).
139. Rohrback, G. H., and Cady, G. H., *J. Am. Chem. Soc.* **71**, 1938 (1949).
140. Davies, J. T., *Trans. Faraday Soc.* **48**, 1052 (1952).
141. Klevens, H. B., and Davies, J. T., Proc. Intern. Congr. Surface Activity, 2nd, London, Butterworth, 1957.
142. Grosse, A. V., and Cady, G. B., *Ind. Eng. Chem.* (*Ind.*) **39**, 376 (1947).
143. Fowler, R. D., *et al.*, *Ind. Eng. Chem.* (*Ind.*) **39**, 376 (1947).
144. Klevens, H. B., and Raison, M., *J. Chem. Phys.* **51**, 1 (1954).
145. Klevens, H. B., and Vergnoble, J. Proc. Intern. Congr. Surface Activity, London, Butterworth, 1957.
146. Arrington, C. H. Jr., and Patterson, G. D., *J. Phys. Chem.* **57**, 247 (1953).

CHAPTER 10

Wetting

Industrial examples of wetting phenomena include the transfer of ink in printing, the removal of soil by washing, and the "drying out" of petroleum wells due to failure of the oil to penetrate sand impregnated with water. The process of mineral flotation depends upon preferential wetting to separate ore from gangue. Water-repellent treatments, corrosion inhibition, and lubrication depend upon changing the wetting properties of the solid substrate.

The Contact Angle

When a drop of liquid is placed on the surface of a solid, it may spread to cover the solid surface. Alternatively, it may remain as a stable drop on the solid. In the absence of a gravitational field, a drop which does not spread will lie on the solid surface in the form of a segment of a circle as shown in Figures 10.1a and 10.1b. In a gravitational field, if the drop is very small surface forces will predominate over the force of gravity and the shape of the droplet will not differ greatly from a true segment of a sphere. If the drop is quite large, the gravitational force will be the more important, and the sessile drop is distorted so that it is horizontal at the top as shown in Figure 10.1c. A drop of liquid of intermediate size will be partially flattened as in Figure 10.1d.

When a liquid is in contact with a clean solid surface, there is a solid-liquid interface between the two phases, while the bare surface of the solid adsorbs the vapor of the liquid until the fugacity of the adsorbed material is equal to that of the vapor and the liquid. The angle θ, measured in the liquid as shown in Figure 10.1, is known as the contact angle.

A number of methods are employed for measuring the contact angle. When the solid surface is in the form of a reasonably flat plate, one of the more suitable methods is that of the tilting plate as used by Adam and Jessop (1) and modified by Harkins and Fowkes (2). The solid is held in an adjustable holder capable of tilting the solid to any angle, with the axis of rotation at the solid-liquid interface. The plate is tilted until a position is found at which the water surface remains undistorted up to the line of contact with the solid. This is shown in Figure 10.2. The apparatus also contains provision for raising or lowering the plate in the liquid, so that the contact angle can be

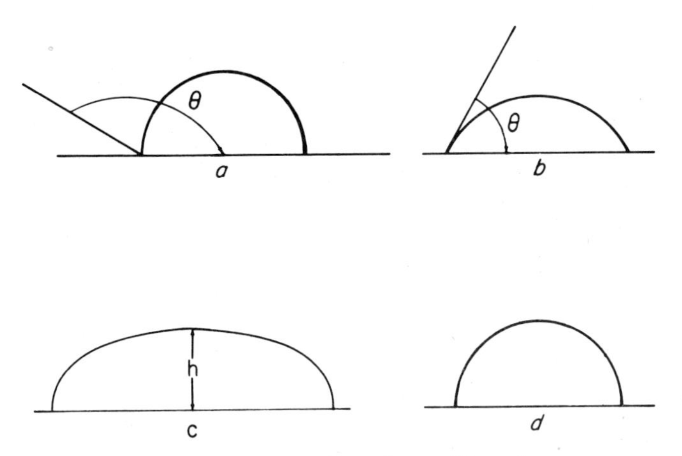

Figure 10.1. Drops of liquid on solid surface.

measured either on a portion of the plate that has been immersed in the liquid or on an unexposed portion of the plate. When the tilting plate is lowered to expose a fresh portion of the plate to the liquid, the angle measured is known as the advancing contact angle. When the plate is raised partially out of the liquid, the receding contact angle is measured. The setting of the plate to the required angle is done before, not after, raising or lowering of the plate. This requires several trials.

The more important additions made by Harkins (3) to the tilting plate apparatus were the use of glass barriers to clean the surface of the liquid, the addition of a film balance to detect the presence of impurities on the surface of the liquid, and the use of a tight cover to achieve equilibrium between liquid and vapor. Both Adam and Harkins emphasize the necessity for clean liquid and solid surfaces.

Another important method for measuring the contact angle of a liquid on a solid surface is the sessile drop method of Poynting and Thomson (4). If a small quantity of a liquid is placed on a level solid surface, provided that the liquid does not spread spontaneously, a sessile drop will be formed. If additional liquid is added to the drop, the height increases until it reaches a maximum value. Further additions of liquid increase the diameter of the drop, but not its height above the solid surface. A cross-section of the sessile drop is shown in Figure 10.1c. The relationship between the maximum height h of the sessile drop and the contact angle θ is

$$1 - \cos\theta = \rho g h^2 / 2\gamma_L \tag{10.1}$$

where ρ is the density of the liquid, g is the acceleration of gravity and γ_L is the surface tension of the liquid.

Padday (5) showed that this equation is valid provided the radius of the drop is large compared with h, so that edge effects can be neglected, and that the system has reached equilibrium.

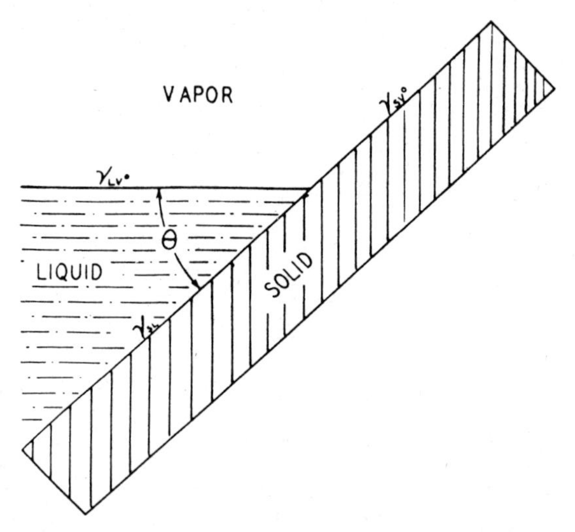

Figure 10.2. Contact angle between a solid and a liquid (3).

Bartell (6, 7) has employed a controlled-drop-volume method which makes use of a pipet with a very fine tip—approx. 0.02 mm outside diameter—from which the liquid can be forced out to increase the volume of the small drop when forming an advancing angle. Alternatively, it can be sucked back in when forming a receding angle. A greatly magnified silhouette of the drop is projected onto a ground-glass screen and a tangent is erected upon the image.

Equations of Wetting

The attraction exerted by two liquids across an interface requires that work must be done to separate them. If one considers two immiscible liquids in a column one square centimeter in diameter, the tension at the interface is γ_{ab}. When they are separated by a direct pull, work is done in forming one square centimeter of each of the two liquid surfaces, with surface tensions γ_a and γ_b. The work required to effect the separation, the work of adhesion W_A, is the difference between these forces,

$$W_A = \gamma_a + \gamma_b - \gamma_{ab} \tag{10.2}$$

The equation is due to Dupré (8).

Harkins (3) has extended the reasoning of Dupré to cover the spreading of liquids. Thus, when a drop of liquid *b* is placed on the surface of another liquid *a*, the first liquid may spread. If it does, the surface of liquid *a* disappears, while its place is taken by an equal area of interface *ab* and surface *b*, provided that liquid *b* spreads to a film of sufficient thickness that the interface *ab* and the surface *b* do not lose their identity. This change in tensions is defined as the spreading coefficient

$$S = \gamma_a - (\gamma_b + \gamma_{ab}) \tag{10.3}$$

Since liquid *b* will not spread on liquid *a* unless $\gamma_b + \gamma_{ab}$ is less than γ_a, a positive spreading coefficient is required for spreading to occur.

The work of cohesion W_c is that necessary to separate liquid in a column one square centimeter in area to give two surfaces, each one square centimeter in area

$$W_c = 2\gamma_b \tag{10.4}$$

Combining equations 10.2, 10.3 and 10.4 gives

$$S = W_A - W_c \tag{10.5}$$

The work of adhesion for a solid and a liquid has been defined by Harkins as

$$W_{A(SL)} = \gamma_S + \gamma_L - \gamma_{SL} \tag{10.6}$$

where γ_S is the surface tension of the solid, free from any adsorbed surface film. Since γ_S and γ_{SL} cannot be determined directly, recourse is made to an equation which is generally ascribed to Young;

$$\gamma_{Se} = \gamma_{SL} + \gamma_L \cos\theta \tag{10.7}$$

where γ_{Se} is the surface tension of the solid covered with an adsorbed film from the liquid.

The Young equation is most readily visualized as an equilibrium between force vectors, as shown in Figure 10.3. The surface tension of the solid will favor spreading of the liquid. This is opposed by the solid-liquid interfacial tension and the vector of the surface tension of the liquid in the plane of the solid surface. Since the surface tension of the liquid acts in a direction tangent to the liquid drop at the solid-liquid-air point of contact, the appropriate vector is equal to $\gamma_L \cos\theta$.

Following the treatment of Harkins, Equation 10.7 can be written

$$\gamma_S - \gamma_{Se} = \gamma_S - \gamma_{SL} - \gamma_L \cos\theta \tag{10.8}$$

The quantity $\gamma_S - \gamma_{Se}$, which is π_e, is the decrease in the surface tension of the solid which occurs when it becomes covered with an adsorbed film in equilibrium with its vapor. This quantity can be determined from the adsorption isotherm.

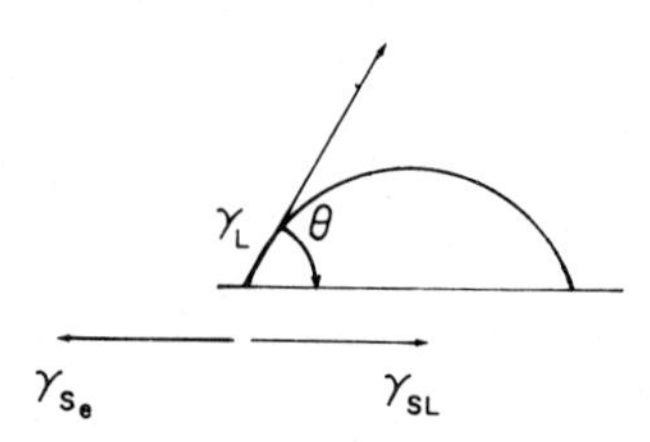

Figure 10.3. Vector forces.

Then

$$\gamma_S - \gamma_{SL} = \pi_e + \gamma_L \cos\theta \tag{10.9}$$

or

$$W_{A(SL)} = \pi_e + \gamma_L (1 + \cos\theta) \tag{10.10}$$

Johnson (9) has recently commented on the controversy concerning the Young equation. Bikerman (10) denied the validity of the equation on both theoretical and experimental grounds. Pethica and Pethica (11) questioned the validity of the equation in a gravitational field. Johnson has shown the basic reasons for this confusion and has provided thermodynamic proof as to the validity of the equation. While Johnson's derivation will not be repeated here, certain portions of his discussion should be emphasized.

It is frequently assumed that the surface tension γ of an interface and the specific surface free-energy F_S are numerically equal. In fact, the two are defined as follows:

$$\gamma = (\partial F/\partial A)_{T,V,ni} \tag{10.11}$$

and

$$F_S = (F - F^{\alpha} - F^{\beta}) / A \tag{10.12}$$

where F is the free energy of the system,

A is the surface area of the interface,

T is the temperature,

V is the volume,

n_i is the number of moles of component i,

F^{α} is the free energy of a unit of volume in the homogeneous part of α multiplied by the volume of α, and

F^{β} is the free energy of β defined analogously to F_{α}.

The surface tension and surface free-energy per unit area are related by the equation

$$F_S = \gamma + \sum_{i=1}^{m} \Gamma_i \mu_i \qquad (10.13)$$

where Γ_i is the surface excess of component i per unit area,

$$\mu_i = (\partial F/\partial n_i)_{T,V,SL,n_j}$$

the chemical potential of component i_j and m is the number of components in the system.

For a pure liquid, the surface excess of the solvent can be considered to be zero and the surface tension is numerically equal to the specific free surface energy. For those systems in which adsorption is important, γ and F_S are definitely not equal. Young's equation is valid in terms of γ but not F_S.

In dealing with systems containing surfaces it is frequently assumed that the necessary and sufficient condition for equilibrium is that the free surface-energy of the system be at a minimum. Actually, the total free energy of the system, at constant T, V, and mass must be at a minimum. This misconception frequently leads to erroneous equations.

While we have in Figure 10.3 shown the Young equation as a static equilibrium of a group of vector tensions acting at a point, it should be understood that this is an oversimplification. Other vectors are undoubtedly involved in establishing the condition for equilibrium. The thermodynamic treatment is more meaningful.

Spreading Coefficient

Zisman (12) found that for pure liquids that do not spread on low-energy solid surfaces, the value π_e could be neglected in the Dupré Equation 10.10, so that to a good approximation

$$W_A = \gamma_L (1 + \cos\theta) \qquad (10.14)$$

It must be noted that this equation is only valid when the contact angle of a liquid on a solid is the same in air as in air saturated with the vapor of the liquid. Otherwise π_e must be added to the right side of the equation.

Boyd and Livingston (13) define the initial and final spreading coefficients of a liquid on a solid surface as

$$S_{LS} = \gamma_S - \gamma_{SL} - \gamma_L \qquad (10.15)$$

and

$$S_{LSe} = \gamma_{Se} - \gamma_{SL} - \gamma_L \qquad (10.16)$$

where S_{LS} is the initial spreading coefficient,

S_{LSe} is the final spreading coefficient,

γ_S is the surface tension of the solid free from an adsorbed film, and

γ_{Se} is the surface tension of the solid covered with an adsorbed film of material in equilibrium with the liquid and its vapor.

Then the final spreading coefficient corresponds to an equilibrium value. Applying the Young Equation 10.7 to 10.16 results in

$$S_{LSe} = \gamma_L (\cos\theta - 1) \tag{10.17}$$

It also follows that

$$S_{LS} - S_{LSe} = \gamma_S - \gamma_{Se} = \pi_e \tag{10.18}$$

TABLE 10.1. SURFACE TENSION VALUES OF VARIOUS LIQUIDS AND THE CONTACT ANGLES FORMED BY THESE LIQUIDS ON FLUORINATED POLYMER SURFACES: 80–20 COPOLYMER OF TETRAFLUOROETHYLENE AND CHLOROTRIFLUOROETHYLENE, 60–40 COPOLYMER OF TETRAFLUOROETHYLENE AND CHLOROTRIFLUOROETHYLENE, KEL-F A POLYCHLOROTRIFLUOROETHYLENE, AND TFE-E A 50:50 INTERPOLYMER OF TETRAFLUOROETHYLENE AND ETHYLENE. VALUES DETERMINED AT 20°C (12).

Liquid	Surface Tension, γ_L dynes/cm	Contact Angle θ on Fluorinated Polymers			
		80–20 Copolymer	60–40 Copolymer	Kel–F	TFE–E
n-Alkanes					
Hexadecane	27.6	37	24	Spr	12
Tetradecane	26.7	35	23	Spr	9
Dodecane	25.4	35	19	Spr	Small angle
Decane	23.9	27	5	Spr	Spr
Nonane	22.9	26	Spr	Spr	Spr.
Heptane	20.3	8	Spr	Spr	Spr
Hexane	18.4	Spr	Spr	Spr	Spr
di(*n*-alkyl) Ethers					
Octyl	27.7	43	29	Spr	16
Heptyl	27.0	41	22	Spr	9
Amyl	24.9	33	9	Spr	Spr
Butyl	22.8	25	Spr	Spr	Spr
Propyl	20.5	8	Spr	Spr	Spr
Ethyl	17.0	Spr	Spr	Spr	Spr
Miscellaneous					
Water	72.8	100	94	90	93
Glyceryl	63.4	96	87	82	85
Formamide	58.2	91	75	82	79
Methylene iodide	50.8	84	76	64	69
α-Brom naphthalene	44.6	67	63	48	60
Tricresyl phosphate	40.9	67	56	44	55
Benzyl phenylundecanoate	37.7	64	54	37	48
tert-Butyl naphthalene	33.7	55	45	18	39
di(2-Ethylhexyl) phthalate	31.2	54	39	6	32
Benzene	28.8	45	25	Spr	20

Spreading of Pure Liquids on Low-energy Solids

In a classic series of experiments starting about 1946, Zisman and his associates determined the relationships between chemical constitution and the conditions required for spreading. In these studies specularly smooth surfaces were employed. The equilibrium advancing contact-angles of the various liquids on the clean solid surfaces were obtained by gently placing a drop on the surface and adding small increments of liquid to the drop until the advancing contact angle reached a maximum and reproducible value. When working with liquids which attacked the plastic surfaces, only the initial contact angle was reported. Contact angle measurements were made with a goniometer eyepiece.

In Table 10.1 the data of Fox and Zisman (12) are reproduced, showing the surface tension and the contact angle for a number of pure liquids on several low-energy surfaces. The liquids are arranged in the order of decreasing surface tension for each series. The low-energy surfaces shown are TFE-E a 50:50 interpolymer of tetrafluoroethylene and ethylene, Kel-F a polychlorotrifluoroethylene, and 80–20 and 60–40 copolymers of tetrafluoroethylene and chlorotrifluoroethylene, respectively. It should be observed that as the surface tension of the liquid γ_L decreases, the contact angle θ on a given surface decreases and the spreading coefficient S_{LSe} becomes less negative.

Figure 10.4 is a plot of cosine θ against the surface tension for many of

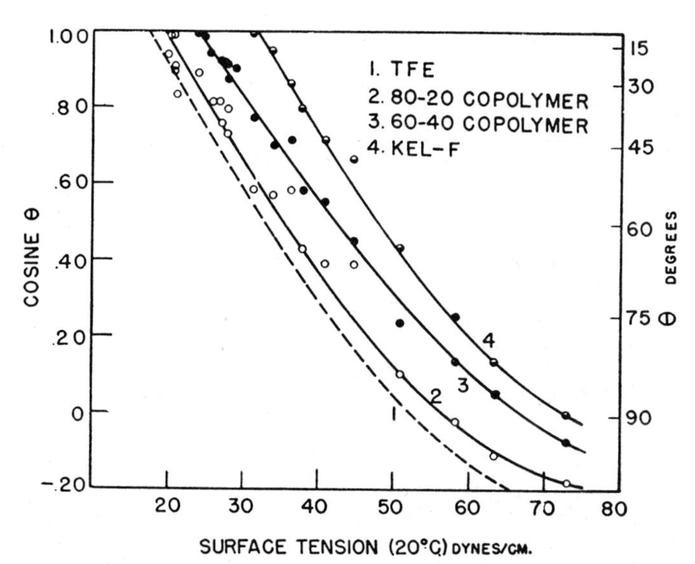

Figure 10.4. Surface tension *versus* the contact angle of various liquids on fluorinated polymers (12).

these liquids on the smooth solid surfaces. Polytetrafluoroethylene, TFE, is also included in the figure. The intersection of each curve of Figure 10.4 with the ordinate cos $\theta=1$ is equivalent to the critical surface tension, γ_c, which has been defined as that value of the liquid surface tension below which liquids spread on a given polymer. Thus for TFE, γ_c is 18 dynes/cm.

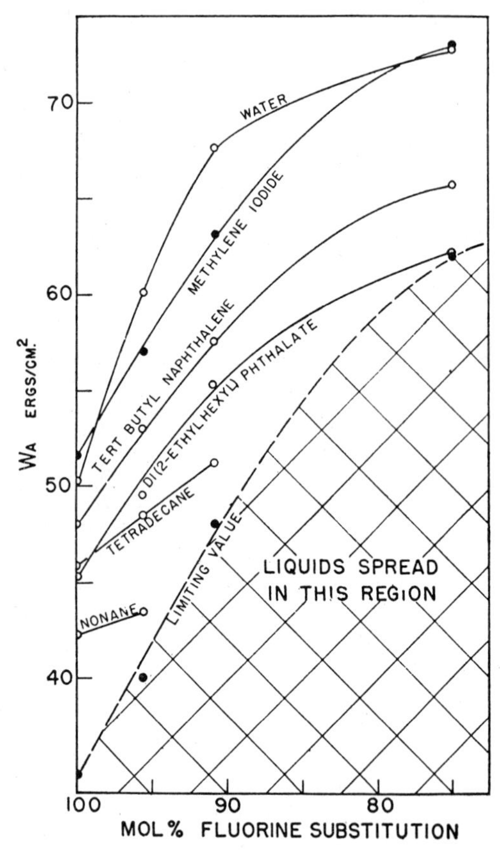

Figure 10.5. Work of adhesion of various liquids on fluorinated polymers *versus* mole per cent fluorine substitution in the polymers (12).

Very few liquids have surface tension values low enough for the liquid to spread on this polymer. Kel-F has the largest critical surface tension of the polymers of this series, approximately 31 dynes/cm. It will be observed that as the chlorine content in the polymer increases, the contact angle for a given liquid decreases. Similarly, γ_c of the polymer increases with decreasing fluorine content.

This is seen further in Figure 10.5 where the work of adhesion of some of the liquids on fluorinated polymers is plotted against mole per cent of fluorine substitution. With increasing fluorine content, there is decreased adhesion of liquids for the polymers. Also the region for spreading is quite small at

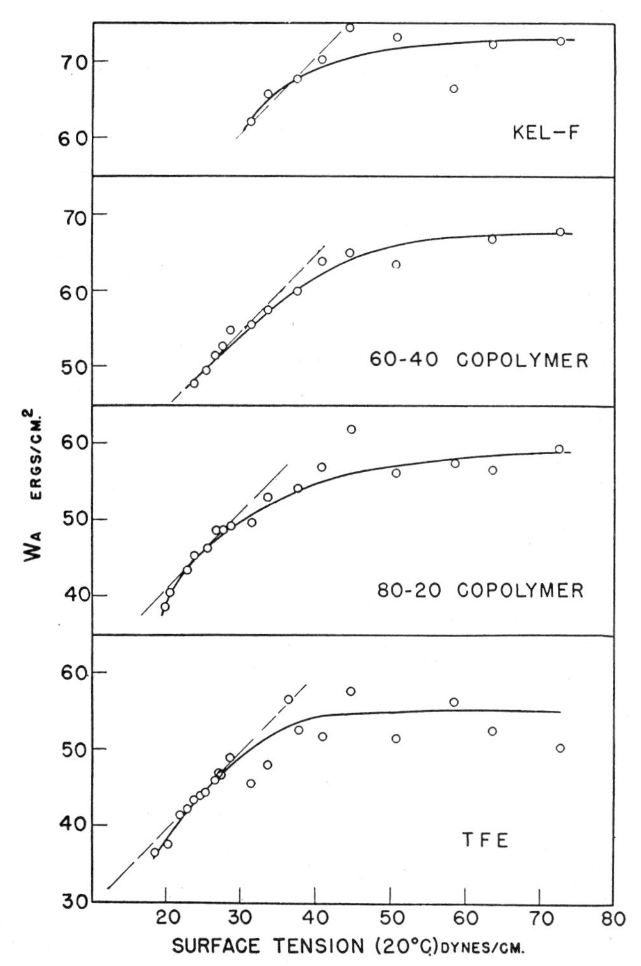

Figure 10.6. Surface tension *versus* work of adhesion of various liquids on fluorinated polymers (12).

100 mole per cent of fluorine substitution, but increases with decreasing fluorine content.

The work of adhesion is expressed as in 10.6

$$W_A = \gamma_L + \gamma_S - \gamma_{SL} \tag{10.19}$$

TABLE 10.2. SURFACE ENERGY RELATIONS OF VARIOUS LIQUIDS ON HYDROCARBON SURFACES AT 20°C (14).

Liquid	Surface Tension (dynes/cm)	*n*-Hexatriacontane Contact Angle (deg.)	*n*-Hexatriacontane Final Spreading Coefficient (ergs/cm²)	Paraffin Contact Angle (deg.)	Paraffin Final Spreading Coefficient (ergs/cm²)	Polyethylene Contact Angle (deg.)	Polyethylene Final Spreading Coefficient (ergs/cm²)
n-Alkanes							
Hexadecane	27.6	46	−8.4	27	−3.0	Spr.	
Tetradecane	26.7	41	−6.6	23	−2.1	Spr.	
Dodecane	25.4	38	−5.4	17	−1.1	Spr.	
Decane	23.9	28	−2.8	7	−0.2	Spr.	
Nonane	22.9	25	−1.8	Spr.		Spr.	
Di(*n*-alkyl) ethers							
Decyl	28.4	54	−11.7	—		—	
Octyl	27.7	50	−9.9	23	−2.2	Spr.	
Heptyl	27.0	45	−7.9	20	−1.6	Spr.	
Amyl	24.9	41	−5.0	11	−0.5	Spr.	
n-Alkyl benzenes							
Hexylbenzene	30.0	47	−9.6	—		—	
Butylbenzene	29.2	45	−8.6	—		—	
Propylbenzene	29.0	45	−8.5	—		—	
Ethylbenzene	29.0	45	−8.5	—		—	
Methylbenzene	28.5	41	−7.0	—		—	
Benzene	28.9	42	−9.4	24	−2.5	Small angle	(*ca.*) 0
Esters							
Tricresyl phosphate	40.9	72	−28.3	62	−21.7	34	−7.0
Benzyl phenylundecanoate	37.7	62	−20.0	52	−14.5	28	−4.4
Di(2-ethylhexyl) phthalate	31.2	52	−12.0	36	−9.5	5	−0.1
Pentaerythritol tetracaproate	30.4	56	−13.4	—		—	
Tri(2-ethylhexyl) tricarballylate	29.6	56	−13.0	—		—	

Halogenated liquids							
Methylene iodide	50.8	77	−39.4	66	−30.1	52	−19.5
sym-Tetrabromoethane	49.7	74	−36.0	—		—	
Aroclor 1242	45.3	73	−32.1	—		—	
α-Bromonaphthalene	44.6	67	−27.2	47	−14.2	35	−8.1
sym-Tetrachloroethane	36.3	60	−18.1	36	−7.0	10	−0.6
Std. fluorolube	25.1	43	−6.7	38	−5.3	Spr.	
FCD-330 (fluorinated hydrocarbon)	20.2	45	−5.9	33	−3.2	Spr.	
FCD-329 (fluorinated hydrocarbon)	16.0	—		Small angle	(*ca.*) 0		
Miscellaneous							
Water	72.8	111	−98.9	108	−95.3	94	−77.9
Glycerol	63.4	97	−71.1	96	−70.1	79	−51.3
Formamide	58.2	92	−60.2	91	−59.2	77	−45.1
tert-Butyl naphthalene	33.7	55	−14.4	38	−7.1	7	−0.2
Carbon disulfide	31.4	53	−12.5	—		—	
n-Heptylic acid	28.3	49	−9.5	—		—	
Methylphenylsiloxane (102 cstokes)	26.1	49	−8.7	—		—	
Polymethylsiloxane (35 cstokes)	19.9	(*ca.*) 20	−0.2	Spr.		Spr.	

Since the free energy decrease ΔF_s on immersion of a solid in a pure liquid is

$$\Delta F_S = \gamma_S - \gamma_{SL} \tag{10.20}$$

then

$$W_A = \gamma_L + \Delta F_S \tag{10.21}$$

Figure 10.6 is a plot of liquid surface tension against the work of adhesion of these liquids on the fluorinated polymers. If the free energy decrease ΔF_S on immersion of a given solid in the liquid were the same for all liquids, the plot would be a straight line with a 45° slope and the intercept of the line with the W_A ordinate would equal ΔF_S. It will be seen from the curves that at low values of γ_L the majority of points do lie on a straight line with a positive 45° slope. At higher γ_L values each curve rises to a fairly constant level characteristic of the polymer. This maximum value of W_A appears to be identical with the value given by the liquid which has a 90° contact angle on the given polymer.

Similar studies by Fox and Zisman (14) were conducted on smooth surfaces of polyethylene, paraffin, and single crystals of pure n-hexatriacontane platelets. The members of this interesting series differ in the ratio of methyl to methylene groups in the surface. The polyethylene surface may be considered as comprising entirely methylene groups if the molecular weight is sufficiently high. The polyethylene used in the study was reported to have a molecular weight of about 240,000. Paraffin surfaces are composed of small crystals oriented at random and contain both methyl and methylene groups. Single crystals of pure n-alkanes are thin platelets whose surfaces comprise methyl groups in the closest possible packing.

Measurements of the contact angle θ and calculated values of the final spreading coefficients for a large number of liquids on the three solids are shown in Table 10.2. It will be observed that for the same liquid, the contact angle decreases as the proportion of methylene groups in the solid surface increases. A surface that is composed predominently of methyl groups has a lower energy and resists wetting more than one composed largely of methylene groups.

The contact angle for a given solid decreases with decreasing surface tension of the liquid, even between nonhomologous groups. Similarly, the final spreading coefficient increases with decreasing surface tension in every homologous group, as well as between groups.

Plots of cos θ versus surface tension are shown in Figures 10.7 and 10.8. While all of the points fall on one line for polyethylene, they are grouped on two lines for paraffin and four lines for n-hexatriacontane. Thus, the critical surface tension for spreading, $\cos\theta=1$, is not independent of the nature of the liquid for these hydrocarbon surfaces. With the paraffin surface, the upper line of Figure 10.7 was obtained with liquids which dissolved the paraffin on prolonged contact. However, contact angle measurements were completed before the paraffin visibly showed the solubility effects. Similarly,

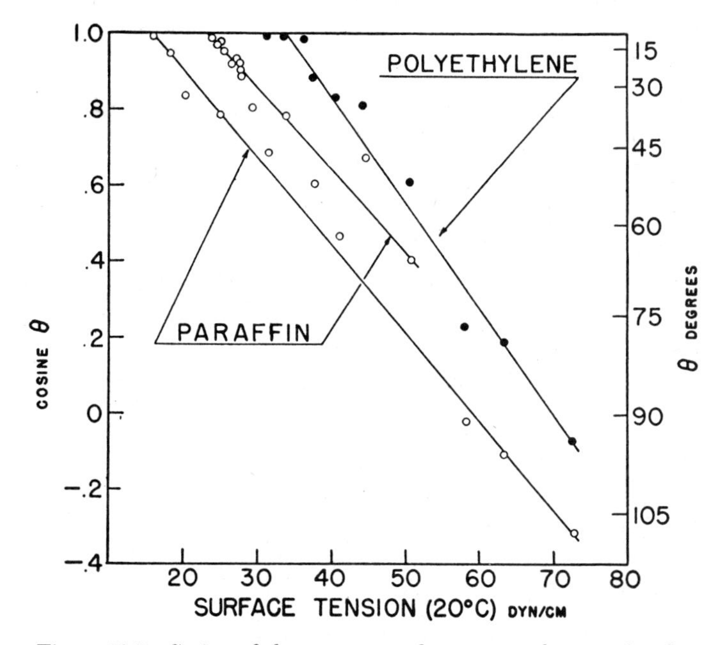

Figure 10.7. Cosine of the contact angle *versus* surface tension for polyethylene and paraffin (14).

in Figure 10.8, the upper lines were obtained with the *n*-alkanes and *n*-alkyl benzenes which have the most solvent effect on *n*-hexatriacontane.

The reason for the multiple curves obtained with hydrocarbon surfaces as compared with the single curve obtained with fluorinated surfaces may be seen by writing the Young Equation 10.7 in the form

$$\frac{\gamma_{Se}-\gamma_{SL}}{\gamma_L}=\cos\theta \tag{10.22}$$

In a plot of γ_L *versus* $\cos\theta$ all points will fall on a single straight line only when $\gamma_{Se}-\gamma_L$ is the same for all liquids on a particular solid. In order for $\cos\theta$ to equal unity, it is necessary that γ_L be the same as $\gamma_{Se}-\gamma_{SL}$. Liquids which dissolve hydrocarbons must have low values of γ_{SL}, while those liquids which have no solvent effect on the surface would be expected to produce a relatively high solid-liquid interfacial tension. Then, from the curves obtained with *n*-hexatriacontane it can be seen that the strongly polar liquids and fluorocarbons fall on the line for liquids of high γ_{SL}. The halocarbons and esters fall on a line corresponding to a somewhat smaller γ_{SL}. The line for alkanes and alkyl ethers indicate a still smaller γ_{SL}. All the alkyl benzenes appear to produce the smallest γ_{SL} of the liquids investigated.

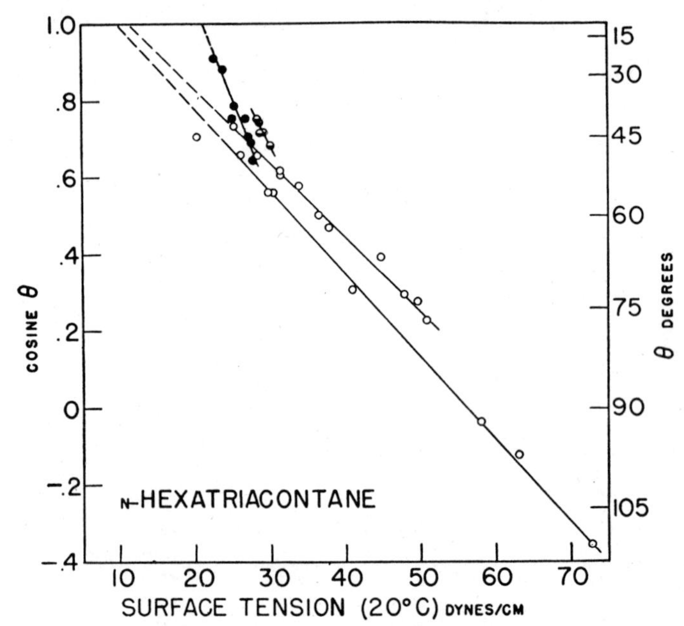

Figure 10.8. Cosine of the contact angle *versus* surface tension for hexatriacontane. ○ miscellaneous liquids; ● *n*-alkanes and di-*n*-alkyl ethers; ◒ *n*-alkyl benzenes (14).

The fact that only one curve was obtained with the fluorinated polymers and polyethylene suggests that no liquid was found which had a sufficiently low interfacial tension with these polymers and whose surface tension was sufficiently large for nonspreading.

In Table 10.3 is reported wettability data for nylon, polyethylene terephthalate and polystyrene (15). A plot of liquid surface tension versus the cosine of the contact angle for several liquids on the polymers is shown in Figure 10.9. Data obtained with polyethylene are included for comparison. For nylon the points for the hydrogen-bonding liquids and the halogenated liquids collect on two different straight lines. The same situation is found for thio-liquids on polystyrene, while the data for polyethylene terephthalate plot on a single straight line.

The data obtained with nylon show that the presence of an amide group in a surface that is otherwise aliphatic hydrocarbon in composition results in increased wetting by hydrogen-bonding liquids. That the amide group has less effect on the wetting by the halogenated liquids can be seen by comparing results for polyethylene. The shift in the lines for the hydrogen-bonding liquids is more than double the shift for halogenated liquids.

TABLE 10.3. WETTABILITY OF HIGH POLYMERS AT 20°C (15).

Liquid	Surface tension, dynes/cm	Contact Angle, degrees Nylon (6, 6)	Polyethylene terephthalate	Polystyrene
Water	72.8	70	81	91
Glycerol	63.4	60	65	80
Formamide	58.2	50	61	74
Thiodiglycerol	54.0	38	46	62
Methylene iodide	50.8	41	38	35
Aroclor 1242	45.3	19	17	21
α-Bromonaphthalene	44.6	16	15	15

The greater wettability of polystyrene as compared with polyethylene is due to the presence of benzene rings in the polymer. That this shift in wettability is small, results from the fact that the benzene ring is partially buried in the surface. The greater wettability of styrene by halogenated liquids is due to the solvent action of these liquids.

Polyethylene terephthalate is more wettable by hydrogen-bonding liquids than polyethylene or polystyrene because of the presence of the ester groups and the benzene rings. In polyethylene terephthalate, the benzene rings are in the principal chain and more exposed at the surface than in the case of polystyrene. Polyethylene terephthalate and nylon exhibit a much greater reistance to the solvent action of the halogenated liquids than polystyrene, and consequently a more normal contact angle with these liquids. Normal wetting behavior has been defined as that resulting from forces of adhesion at the sold-liquid interface of the order of van der Waals forces (16).

It has been reported (17) that in the stretched polyethylene terephthalate films used in this study, the benzene rings all lie in a plane normal to the surface of the film. Because of this planar structure, the cohesional forces between benzene rings in neighbouring chains act to create a close-packed array of polymer chains. This would explain the resistance to solvent attack and to moisture penetration. Also, the $C = O$ bond of the ester group is said to lie perpendicular to the surface of the plastic. If the direction of this bond is away from the surface, it would explain the failure of hydrogen-bonding liquids to cause increased wetting.

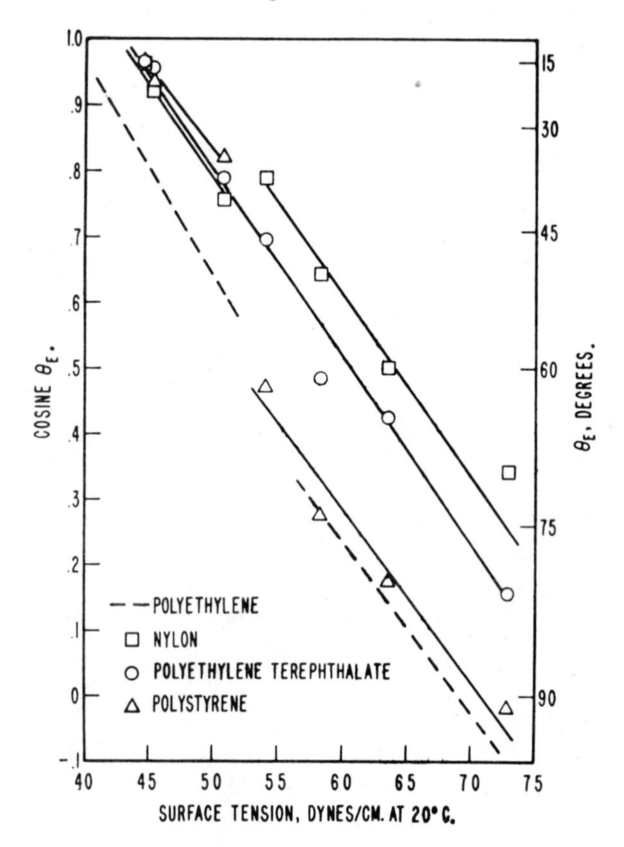

Figure 10.9. Liquid surface tension *versus* cosine of the contact angle for several liquids on the surfaces of high polymers (15).

Spreading of Pure Liquids on Monolayers

Since a number of organic liquids are nonspreading on low-energy surfaces, it was of interest to determine whether high-energy surfaces like platinum could be made nonwettable by coating with thin films of materials with lower surface energies.

Zisman and his associates (18) correctly reasoned that they could deposit a monolayer of an organic compound on a solid surface, without adhering solution, by dissolving the compound to be adsorbed in a less adsorbable solvent with a surface tension greater than γ_c of the monolayer. Thus, they found that they could deposit monolayers free of adhering solution by immersing clean, polished platinum discs in solutions of a large number of organic compounds in water. The observed contact angles of several of the liquids studied on various polar monolayers as well as on thin films of nonpolar liquids are given in Table 10.4, with the liquids listed in decreasing

TABLE 10.4. CONTACT ANGLES AT 20°C OF VARIOUS LIQUIDS ON THIN FILMS ADSORBED ON PLATINUM (18).

Liquid	Surface tension, dynes/cm	Surface coated with adsorbed monolayer of: 2-Ethyl-hexyl amine	α-Amyl-myristic acid	Sebacic acid	Aniline	Surface coated with thin film of hexadecane
Water	72.8	77	79	54	55	79
Glycerol	63.4	66			47	42
Methylene iodide	50.8	53		36		
Tetrabomoethane	49.7	49	55	35		
Tricresyl phosphate	40.9	40	44	22	20	31

order of their surface tension. Figure 10.10 shows plots of cos θ versus liquid surface tension for the data in the table. Contact angles of pure liquids on single crystals of naphthalene and anthracene are shown in Table 10.5.

A common misconception of the past was that organic substrates were readily wetted by all organic liquids. Subsequently, it was thought that a solid surface had to be covered with methyl groups to show oleophobic properties. From the above data, it can be seen that any organic surface will be nonwetting or oleophobic for liquids with a surface tension higher than the critical surface tension of the exposed surface.

TABLE 10.5. CONTACT ANGLES AT 20°C OF PURE LIQUIDS ON SINGLE CRYSTALS OF NAPHTHALENE AND ANTHRACENE (18).

Liquid	Surface tension (dynes/cm)	Contact angle on single crystals of: Cleaved anthracene	Cleaved naphthalene	Sublimed naphthalene
Water	72.8	94	95	92
Glycerol	63.4	77	79	80
Formamide	58.2	73	77	72
Ethylene glycol	47.7	62	65	60

Table 10.6 shows the critical surface tension of a variety of surface structures. The lowest known critical surface tension is obtained with a surface composed entirely of CF_3 groups, as in the case of perfluorolauric-acid mono-

layers. With monolayers formed from shorter-chain perfluorinated acids, packing is less close and some CF_2 groups are exposed, leading to an increase in γ_c. These surfaces are oleophobic to all liquids.

Just as the CF_3 group has a lower surface energy than CF_2, the methyl group CH_3 has a lower surface energy than the methylene group CH_2. Thus, polyethylene is more readily wetted than an octadecylamine monolayer. A close-packed aromatic structure such as a crystal face of naphthalene, with only the ring edges exposed has a lower γ_c value than a benzoic acid mono-

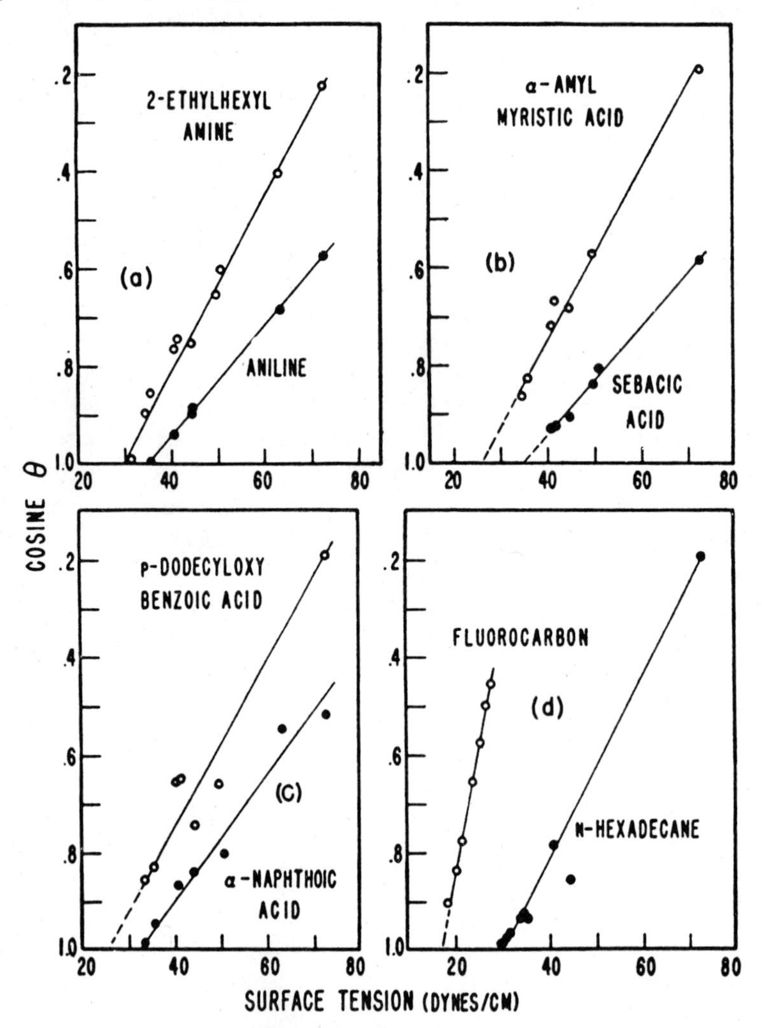

Figure 10.10. Liquid surface tension *versus* contact angle of liquids on various monolayers and thin liquid films at 20° C (18).

TABLE 10.6. THE CRITICAL SURFACE TENSION OF LOW ENERGY SOLID SURFACES.

Surface	Chemical Structure of Surface	Critical Surface Tension γ_c (dyne/cm)
Perfluorolauric acid, monolayer	CF_3, close packed	5.6
Perfluorbutyric acid, monolayer	CF_3, less closely packed	9.2
Perfluorokerosene, thin liquid film	CF_2, some CF_3	17.0
Polytetrafluoroethylene, solid	CF_2	18.2
Octadecylamine, monolayer	CH_3, close packed	22.
α-amyl myristic acid, monolayer	CH_3 and CH_2	26.
2-ethyl hexyl amine, monolayer	CH_3 and CH_2	29.
n-hexadecane, crystal	CH_2, some CH_3	29.
Polyethylene, solid	CH_2	31.
Naphthalene, crystal	⬡, edge only	25.
Benzoic acid, monolayer	⬡, edges and faces	53.
2-naphthoic acid, monolayer	⬡, edges and faces	58.
Polystyrene, solid	CH_2, some ⬡	32.8–43.3
Polyethylene terephthalate, solid	⬡, CH_2, ester	43.0
Nylon, solid	CH_2, amide	42.5–46.0

layer where both aromatic edges and faces are exposed. The order of increasing wettability of surfaces is $CF_3 < CF_2 < CH_3 <$ aromatic ring edge $< CH_2 <$ aromatic ring face.

Wetting of incomplete monomolecular layers was studied by Bartell and Ruch (19). Monomolecular films of fatty amines were deposited on chromium-plated slides by adsorption from solution in *n*-hexadecane. The films were then partially depleted by dipping in ether or benzene. Figures 10.11 and 10.12 show contact angles of several liquid alkanes on these surfaces, as a function of the per cent depletion of the monomolecular film. Resistance to wetting by normal hydrocarbons at low depletion, with a rapid decrease in the contact angle at about 50-per cent depletion is striking. The investigators hypothesize that with only partial depletion, the liquid alkanes fill the voids and thus retain the methyl surface. As the per cent depletion increases, the fatty amines are no longer capable of supporting the solvent molecules perpendicular to the plane of the surface and a sufficient proportion of methylene groups are exposed to permit at least partial wetting. With trimethylbenzene, which cannot orient to fill the voids, the contact angle decreases progressively with increasing per cent depletion. These results are

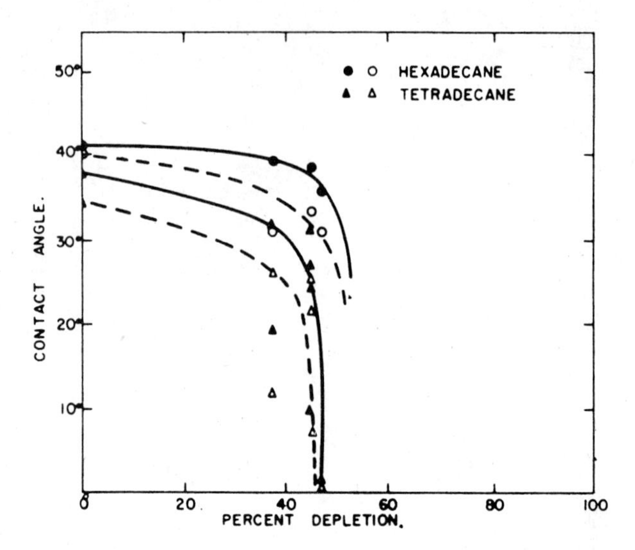

Figure 10.11. Contact angles of n-hexadecane and n-tetradecane on depleted monolayer of n-octadecylamine. Solid points represent advancing angles, open points represent receding angles (19).

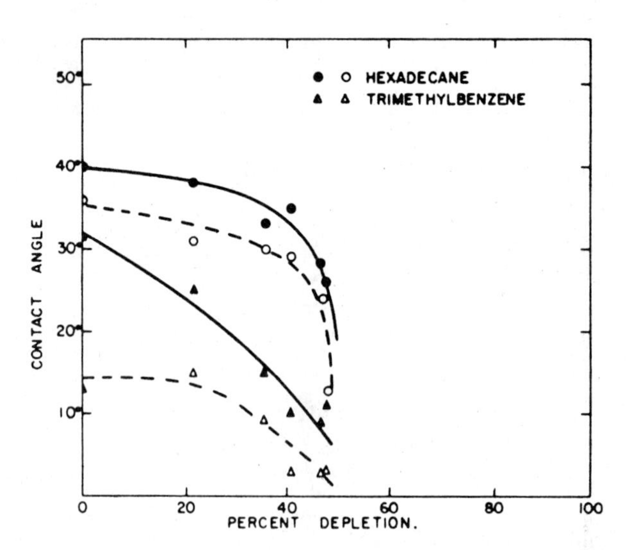

Figure 10.12. Contact angles of n-hexadecane and 1,3,5-trimethylbenzene on depleted monolayer of n-octadecylamine. Solid points represent advancing angles, open points represent receding angles. (19)

not general, but depend on a delicate balancing of surface tensions. When the surface tension of the liquid substantially exceeds the critical surface tension of the solid containing the adsorbed film, complete wetting does not occur until the adsorbed film is essentially completely removed.

Non-Spreading of Mixtures of Organic Liquids

While pure liquids will spread on high energy surfaces, such as platinum, mixtures of these same liquids frequently will not spread. Thus, when hexadecane is mixed with lower surface-tension liquids such as methylsiloxanes or fluorocarbons, the latter are adsorbed on the platinum to form low-energy surfaces on which the hexadecane will not spread. A mixture of sym-tetrachloroethane and hexadecane will not spread on clean platinum. The explanation given is that the hexadecane is preferentially adsorbed to give a polyethylene-like surface with a γ_c less than the surface tension of the mixture (18).

When a drop of *n*-decane was placed on a smooth, clean surface of nylon it spread spontaneously. A second drop of *n*-decane saturated with perfluorolauric acid—10^{-5} mole per cent—was placed on top of the thin film of *n*-decane left by the first drop. The *n*-decane drew together in less than one minute to form a sessile drop with a contact angle of about 40°. Ellison and Zisman (15) explain this behavior on the basis of adsorption of an oriented monolayer of perfluorolauric acid on nylon through hydrogen-bonding of the acid groups to the amide groups at the surface. The contact angle of 40°, as compared with that of 70° obtained with *n*-decane on a close-packed film of perfluorolauric acid adsorbed on a platinum foil, results from the spacing of the amide groups to make unlikely the formation of a close-packed film of the perfluorinated acid. Substitution of a fluorocarbon oil for the perfluorolauric acid in a similar experiment did not result in non-spreading of the *n*-decane solution. This is evidence that the adsorption of perfluorolauric acid on nylon was by the carboxylic acid group rather than by the fluorinated chain. Similarly, sessile drops were not obtained when the experiment using the saturated solution of perfluorolauric acid in *n*-decane was repeated with polyethylene, polyvinyl chloride, and polyethylene terephthalate. These polymer surfaces do not afford sites for oriented adsorption of perfluorolauric acid through the carboxylic acid group (15).

Wetting by Aqueous Solutions

It has been stated or implied that aqueous solutions of surfactants wet solids by a mechanism in which the surfactant is adsorbed with the hydro-

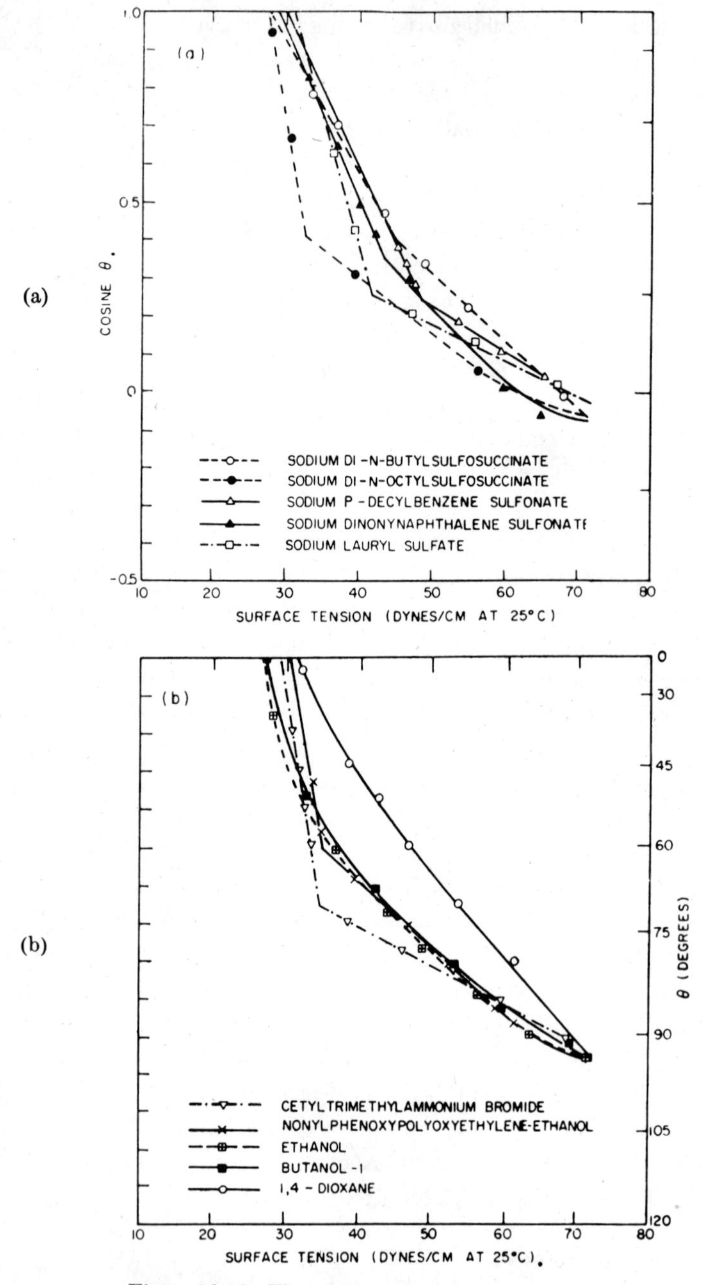

Figure 10.13. Wettability of polyethylene by aqueous solutions: (a) of anionic wetting agents; (b) of other compounds (20).

carbon chain in contact with paraffin or similar low-energy surfaces and the polar group directed to the aqueous phase. Bernett and Zisman (20) attempted to explain the wetting by aqueous solutions on the same basis as wetting by pure organic liquids. Spreading is then caused to a first approximation by the surface tension of the aqueous solution, and γ_{SL} the solid-solution interfacial tension plays only a minor role. Then, the ability of an aqueous solution to spread on a low-energy surface should depend upon the critical surface tension value of the surface and the concentration of wetting agent that must be dissolved in water to depress the surface tension of water below this value.

TABLE 10.7. COMPARISON OF THE BREAKPOINT CONCENTRATIONS FOR VARIOUS WETTING AGENTS AND SURFACES (20).

	Concentration, moles/l		
Compound	Polyethylene	Teflon	cmc
Na-di-*n*-butyl sulfosuccinate	5.0×10^{-2}	4.2×10^{-2}	2×10^{-1}
Na-di-*n*-octyl sulfosuccinate	6.8×10^{-4}	4.5×10^{-4}	6.8×10^{-4}
Na-*p*-decyl benzenesulfonate	1.2×10^{-3}	1.2×10^{-3}	3×10^{-3} at 50°
Na lauryl sulfate	3.0×10^{-3}	2.7×10^{-3}	8×10^{-3}
Cetyltrimethylammonium bromide	4.3×10^{-4}	3.8×10^{-4}	1×10^{-3} at 60°

The liquid surface tension versus cos θ curves for a number of aqueous solutions of wetting agents and several organic solvents were obtained on smooth polyethylene and Teflon surfaces, as shown in Figures 10.13 and 10.14. At the critical surface tension, cos θ =1, there is only a narrow spread for the various solutions on each of these surfaces. For Teflon γ_c is between 16.5 and 19.5 dynes/cm, while for polyethylene, γ_L is between 27.5 and 31.5 dynes/cm. These values are in reasonable agreement with those found with pure organic liquids. Only solutions of sodium di-*n*-octyl sulfosuccinate gave surface tensions lower than 30 dynes/cm and were able to spread freely on polyethylene. None of the aqueous solutions had surface-tension values low enough to spread on Teflon. Thus, observations agree with the assumption that, to a first approximation, spreading of aqueous solutions of wetting agents on low-energy solids depends upon the surface tension of the solution and the value of γ_c of the solid.

It will be observed that there is an abrupt change in the slope of the γ_L *versus* cos θ curves for the aqueous solutions. In all cases, the concentration

(a)

(b)

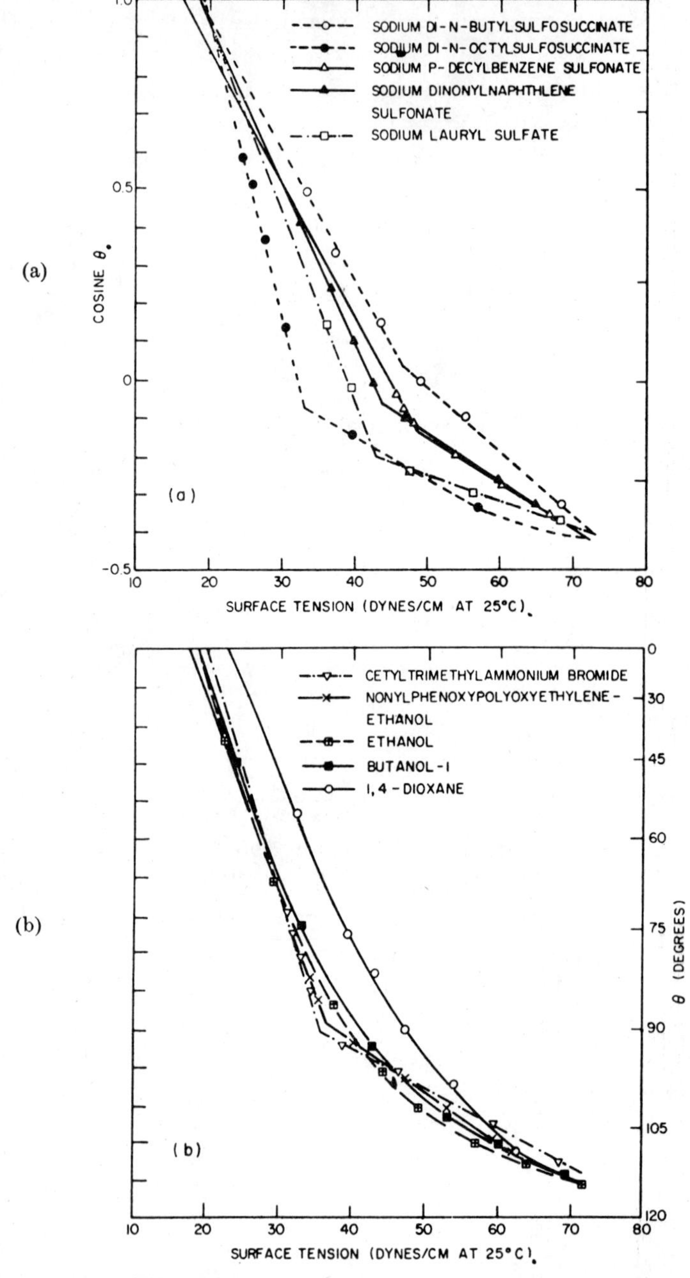

Figure 10.14. Wettability of Teflon by aqueous solutions: (a) of anionic wetting agents; (b) of other compounds (20).

of wetting agent corresponding to the surface tension at which the discontinuity occurs is approximately the same as the critical micelle concentration, cmc, of the wetting agent. From the data shown in Table 10.7, it appears that the cmc is always slightly higher.

The results suggest that at very low concentrations of wetting agent, where the liquid surface phase is essentially water, γ_{SL} cannot be neglected. However, at concentrations sufficiently high for the liquid surface to approximate close packing of adsorbed solute molecules, the wetting properties are the same as that of a pure liquid having the same outermost hydrocarbon structure, with γ_{SL} constant and small in value.

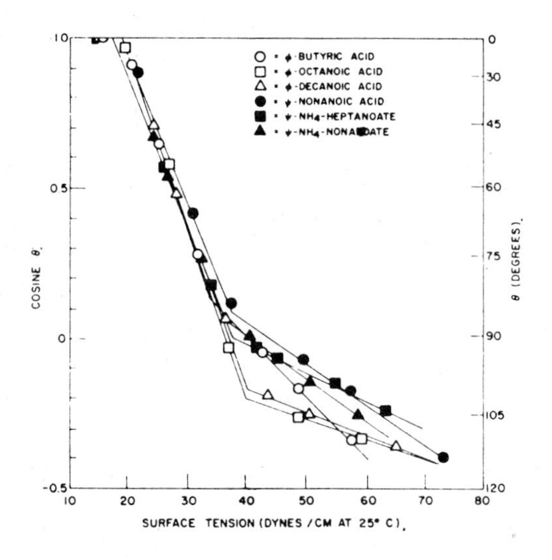

Figure 10.15. Wettability of Teflon by aqueous solutions of highly fluorinated compounds (21).

Bernett and Zisman (21) also conducted wetting studies on low-energy solids using aqueous solutions of perfluoro acids and ω-monohydroperfluoro acids or their potassium salts. Results obtained were drastically different from those that were found with the conventional surfactants. When the usual procedure was employed of placing small drops of the aqueous solutions on clean polyethylene or Teflon surfaces, erratic results were obtained. It was demonstrated that this was due to the adsorption of the solute on the plastic surfaces leaving a drop of a higher surface tension resting upon the adsorbed film. It was also found that the fluorinated solute molecules adsorbed more strongly on polyethylene than on Teflon. This was to be expected since the change in free surface energy in going from a surface consisting of CH_2 groups to one consisting of CF_3 groups is much greater than in going from a surface consisting primarily of CF_2 groups.

To circumvent the difficulty of erratic contact angle values, a modified method was employed. A large drop of the test solution was allowed to remain undisturbed for two minutes on the surface, and was then removed by blotting. Placing of a second drop of the same solution in the same area gave a stable, reproducible contact angle. The results shown in Figures 10.15 and 10.16 were obtained by this method.

The critical surface tension γ_c for Teflon, as shown in Figure 10.15, fell in

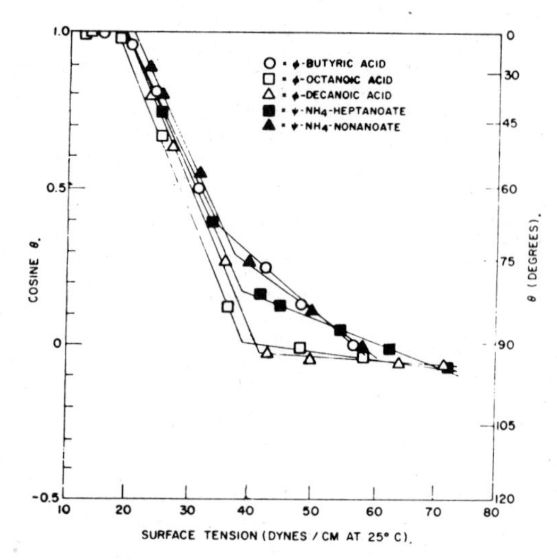

Figure 10.16. Wettability of polyethylene by aqueous solutions of highly fluorinated compounds (21).

the same region as that found with the conventional surfactants, 17–19 dynes/cm. Several of the fluorinated acids depressed the surface tension of water below this critical level and completely wetted the Teflon.

The critical surface tension for polyethylene had previously been found to be approximately 31 dynes/cm. However, Figure 9–16 shows that the graphs of cos θ *versus* γ_L intercepted the line cos $\theta = 1$ at values between 19.0 and 21.5 dynes/cm. These unusual results are caused by the adsorption of the fluorinated solutes on this low-energy surface.

In the previously reported studies with conventional wetting agents, the discontinuities in slope in the cos θ *versus* γ_L curve of each solution occurred approximately at the cmc. With the fluorinated solutes, the discontinuities occurred at concentrations considerably lower than the cmc of the solutes, and at correspondingly higher surface tensions. This is shown in Table 10.8.

TABLE 10.8. COMPARISON OF SURFACE TENSIONS AND CONCENTRATIONS AT THE BREAKPOINTS OBTAINED FROM SURFACE TENSION VERSUS CONCENTRATION AND FROM CONTACT ANGLE VERSUS SURFACE TENSION CURVES (21).

	γ_{LV} at discontinuity (dynes/cm)			Concentration at discontinuity (wt %)		
	From γ_{LV} vs. c	From cos θ vs. γ_{LV}		From γ_{LV} vs. c	From cos θ vs. γ_{LV}	
Solute	(at cmc)	Polyethylene	Teflon	(cmc)	Polyethylene	Teflon
ϕ-Butyric acid	25.5	34.5	34.5	14.8	8.0	8.0
ϕ-Octanoic acid	15.3	39.5	39.5	0.38	0.1	0.1
ϕ-Decanoic acid	20.5	42.0	40.0	0.042	0.011	0.012
ψ-Nonanoic acid	21.8	—	38.0	0.5	—	0.19
ψ-NH_4-Heptanoate	26.9	39	37.5	4.0	1.6	1.7
ψ-NH_4-Nonanoate	27.5	38	36	1.3	0.78	0.70

ϕ = perfluoro; ψ = ω-monohydroperfluoro

Thus, a graph of surface tension *versus* concentration for aqueous solutions of perfluorooctanoic acid established a cmc of 0.38 weight per cent. The surface tension at this concentration was 15.3 dynes/cm. However, the discontinuity in the cos θ *versus* γ_L curves on both Teflon and polyethylene occurred at a concentration of 0.1 weight per cent, corresponding to $\gamma_L =$ 39.5 dynes/cm. Bernett and Zisman showed that the discontinuities in the cos θ *versus* γ_L curves were due to the adsorption of the solutes at these low concentrations causing a lowering in the critical surface tension of wetting of the solid, and therefore decreasing the wetting tendencies of the solutions.

The wettability behavior of fluorinated wetting agents differs from that of the more usual wetting agents. The pronounced adsorption of fluorinated solutes on low-energy surfaces causes a maximum in the adsorption *versus* concentration curve to occur at a much lower concentration at the solid-solution interface than at the solution-air interface. With conventional wetting agents, the adsorption maximum occurs at about the same concentration at both interfaces.

Wetting of High-energy Surfaces by Organic Liquids

As shown by Equation 10.3, the necessary condition for spreading is that $\gamma_S > \gamma_{SL} + \gamma_L$. Harkins and Feldman (22), extrapolating from measurements of the spreading coefficients of liquids on water and mercury, concluded that practically all liquids should spread on clean metals and high melting solids. This appears to be reasonable, since solids such as diamond, silica and most metals are believed to have free surface-energies at ordinary temperatures ranging from several hundred to several thousand ergs/cm² (23). In contrast, most liquids and low melting solids such as organic polymers and waxes have free surface-energies ranging from about 25 to 1000 ergs/cm².

Zisman and his associates (24) have shown that there are conditions where liquids will not spread on a high energy solid. Thus, when the liquid is a polar-nonpolar molecule of certain types, or it contains a solute of one of these types, it will deposit through adsorption at the solid-liquid interface a low-

TABLE 10.9. CONTACT ANGLES OF ALIPHATIC CARBOXYLIC ESTERS AND ETHERS AT 20°C AND 50 PER CENT RELATIVE HUMIDITY (24).

Compound	Surface tension, dynes/cm. at 20°	On metals, degree		On non-metals, degree	
		18/8 Steel	Brass	Silica	Sapphire
Aliphatic diesters					
Dibutyl citraconate	30.4	0	0	10	0
Dibutyl pyrotartrate	29.3	0	0	11	0
Bis-(2-ethylhexyl) adipate	30.2	0	0	23	15
Dibutyl β-methyladipate	30.4	0	0	11	0
Bis-(2-ethylhexyl) β-methyladipate	30.1	0	0	22	8
Bis-(2-ethylhexyl) sebacate	31.1	0	0	16	11
Bis-(1-methylheptyl) sebacate	31.0	0	0	17	6
Dioctyl sebacate	32.2	0	0	13	11
Bis-(3,5,5-trimethylhexyl) sebacate	29.9	0	0	12	10
Bis-(2-ethylhexyl) glutarate	29.4	0	0	21	13
Bis-(3,5,5-trimethylhexyl) glutarate	28.4	0	0	26	22
Bis-(2-ethylhexyl) β,β′-thiodipropionate	31.3	0	0	33	24
Bis-(2-(2-ethylbutoxy)-ethyl) azelate	34.3	0	0	0	0
Diethylene glycol dicaproate	30.8	0	0	29	20
Dipropylene glycol dicaproate	29.5	0	0	15	5
Triethylene glycol bis-(2-ethylhexanoate)	30.3	0	0	23	15
Polyethylene glycol bis-(2-ethylhexanoate)	28.2	0	0	23	21
1,6-Hexanediol bis-(2-ethylhexanoate)	30.2	0	0	5	0
1,10-Decanediol bis-(2-ethylhexanoate)	31.4	0	0	0	0
Aliphatic monoesters					
Decyl acetate	28.3	0	0	0	0
Amyl caproate	27.0	0	0	0	0
Decyl caproate	28.8	0	0	11	0
Undecyl 2-ethylhexanoate	27.5	0	0	0	0
Methyl laurate	28.3	0	0	19	8
Amyl laurate	28.2	0	0	5	0
Decyl laurate	—	0	0	17	13
Aliphatic ethers					
Ucon (DLB-50-B)	28.3	0	0	0	0
Ucon (DLB-100-B)	29.9	0	0	0	0

energy surface on which it will not spread. When the adsorbed film is close-packed with terminal CH_3, CF_2H, or CF_3 groups, only liquids with low surface-tensions can spread on the resulting surfaces. When the adsorbed molecules are branched or cyclic structures, all liquids will spread except those with high surface-tensions. Many classes of pure liquids, including the branched and straight-chain aliphatic acids and alcohols are autophobic (25). The liquid is unable to spread on its own adsorbed film. When a solid is coated with a non-polar adsorbed film, liquids which have higher surface

TABLE 10.10. CONTACT ANGLES OF CYCLIC CARBOXYLIC ESTERS AND ETHERS AT 20°C AND 50 PER CENT RELATIVE HUMIDITY (24).

Compound	Surface tension, dynes/cm. at 20°	On metals, degree		On non-metals, degree	
		18/8 Steel	Brass	Silica	Sapphire
Cyclic diesters					
2-Methyl-1,3-pentanediol bis-(phenylacetate)	39.4	8	10	11	4
Bis-(2-phenylethyl) β-methyladipate	41.3	9	15	8	7
Bis-(cyclohexane-ethyl) β-methyladipate	35.8	3	9	17	8
Dipropylene glycol bis-(hydrocinnamate)	40.1	14	17	9	4
Dibutyl phthalate	33.6	6	0	11	11
Bis-(2-ethylhexyl) phthalate	31.3	8	2	28	27
Bis-(2-ethylhexyl) tetrahydrophthalate	30.7	0	0	19	17
Bis-(2-ethylhexyl) isophthalate	30.8	0	0	19	11
Bis-(2-ethylhexyl) terephthalate	32.0	0	0	10	0
Dibutyl pinate	30.5	0	0	7	0
Diheptyl pinate	31.0	0	0	23	8
Cyclic monoesters					
Butyl phenylundecanoate	32.9	0	0	0	0
Benzyl phenylundecanoate	38.0	9	11	18	6
1-Methyl-4-ethyloctyl hydrocinnamate	31.7	0	0	21	19
2-Ethylhexyl β-(3-phenylpropylmercapto)-propionate	34.7	0	0	16	11
Cyclic ethers					
Benzyl phenylundecyl ether	36.5	3	12	14	3
1,9-Bis-(phenylmethoxy)-nonane	39.2	18	23	0	0
1,5-Bis-(3-phenylpropoxy)-pentane	38.7	9	20	0	0
1,4-Bis-(3-phenylpropoxy)-3-methylbutane	37.5	5	11	0	0
1,6-Bis-(2-phenylethoxy)-3-methylhexane	38.0	17	10	5	0
1,11-Diphenyl-6,10-dimethyl-1,5,8,11-butoxyundecane	38.5	7	7	0	0
1,5-Bis-(3-phenylpropoxymercapto)-pentane	43.2	12	12	6	5
Phenoxyphenylcetane	33.3	2	0	10	4
α-Naphthyl ethyl ether	39.3	10	7	5	4

tensions than the critical surface tension γ_c of the adsorbed film will not spread.

Data obtained by Fox, Hare and Zisman (24) for the contact angles of a large number of pure organic liquids on specularly polished, clean high-energy surfaces are presented in Tables 10.9 through 10.13. While the liquid surface tension alone did not determine whether spreading would occur on these surfaces, in all but one case, the liquids that were non-spreading on all four types of surfaces had surface tensions in excess of 33.5 dynes/cm.

Liquid aliphatic hydrocarbons spread on all high-energy surfaces because they give surface tensions less than 30 dynes/cm while the critical surface tensions of their own adsorbed films would not be less than that of polyethylene, 31 dynes/cm. Similarly, liquid polymethylsiloxanes give an adsorbed monolayer with an outermost surface of methyl groups which are not

TABLE 10.11. CONTACT ANGLES OF SATURATED HYDROCARBONS AT 20°C AND 50 PER CENT RELATIVE HUMIDITY (24).

Compound	Surface tension, dynes/cm. at 20°	On metals, degree		On non-metals, degree	
		18/8 Steel	Brass	Silica	Sapphire
Open-chain compounds					
Hexadecane	27.6	0	0	0	0
V-120 polyethylene	27.8	0	0	0	0
SS-903 polyethylene	30.4	0	0	0	0
SS-906 polyethylene	30.7	0	0	4	0
Cyclic compounds					
9-(α-(*cis*-0.3.3-Bi-cyclooctyl)-methyl)-heptadecane	31.2	0	0	0	0
1,7-Dicyclopentyl-4-(3-cyclopentylpropyl)-heptane	34.6	2	0	8	6
3,3′-Dicyclopentyldicyclopentane		0	0	0	0
1,3-Dicyclopentylcyclopentane	34.6	0	0	0	0
1-Cyclohexyl-2-(cyclohexylmethyl)-pentadecane	32.7	0	0	0	0
4-Cyclohexyleicosane	31.6	0	3	0	0
9-*n*-Dodecylperhydrophenanthrene	34.2	0	0	6	2
1-α-Decalylhendecane	32.7	0	0	0	0
1,1-Di-(α-decalyl)-hendecane	35.1	3	3	7	1
1-α-Decalyl-2-cyclohexylethane	32.0	0	0	32	8
1,1-Dicyclohexylethane	33.0	0	0	5	0
2,6-Dimethyl-4-(α-decalylmethyl)-heptane	35.6	0	0	0	0
2,6-Dimethyl-4-cyclohexylmethylheptane	27.0	0	0	0	0
Isoamyl-3,3,5-trimethylcyclohexane	25.3	0	0	0	0
Decalin	30.5	0	0	0	0
2-Ethyldecalin	30.5	0	0	0	0

TABLE 10.12. CONTACT ANGLES OF AROMATIC HYDROCARBONS AT 20°C AND 50 PER CENT RELATIVE HUMIDITY (24).

Compound obsd.	Surface tension, dynes/cm. at 20°	On metals, degree 18/8 Steel	Brass	On non-metals, degree Silica	Sapphire
8-*p*-Tolylnonadecane	30.7	0	0	0	0
p-Di-*sec*-amylbenzene	28.1	0	0	0	0
p-Octadecyltoluene	31.5	0	0	8	0
p-Dodecyltoluene	29.9	0	0	0	0
1,1-Diphenylethane	38.0	0	0	0	0
Amyldiphenyl	34.6	0	0	0	0
α-Methylnaphthalene	36.4	0[a]	3[a]	0[a]	3[a]
t-Butylnaphthalene	33.7	3[a]	4	3[a]	4[a]
Monoamylnaphthalene	34.3	3[a]	4	3[a]	4[a]
Nonylnaphthalene	32.5	6	0	9	10
Dinonylnaphthalene	31.5	2	0	18	12

[a] Evaporated

TABLE 10.13. CONTACT ANGLES OF AROMATIC CHLORINE AND PHOSPHATE COMPOUNDS AT 20°C AND 50 PER CENT RELATIVE HUMIDITY (24).

Compound	B.p. at 0.6 mm.	$n^{20}D$	Surface tension, dynes/cm. at 20°	Platinum, degrees	Silica, degrees	Sapphire, degrees
Phosphate esters						
Tricresyl phosphate			40.9	*ca.* 2	*ca.* 2	*ca.* 2
Tri-*o*-cresyl phosphate			—	7	14	18
Tri-*o*-chlorophenyl phosphate			45.8	7	19	21
Di-(*o*-chlorophenyl)-monophenyl phosphate			43.6	9	11	24
Biphenyl mono-(*o*-xenyl) phosphate			44.3	9	16	17
Chlorinated hydrocarbons						
x-Chlorobiphenyl ether			40.6	0	0	0
x-Dichlorobiphenyl ether			42.8	12	0	*ca.* 2
Chlorinated biphenyl (Aroclor 1242)			41.9	12	10	6
Chlorinated biphenyl (Aroclor 1248)			42.9	11	14	7
Chlorinated biphenyl (Aroclor 1254)			45.8	14	29	12
Aroclor 1248—Fraction I	130–143°	1.6259	—	5	19	5
Aroclor 1248—Fraction II	143–147°	1.6302	—	11	11	7
Aroclor 1248—Fraction III	147–152°	1.6340	—	13	14	6
Aroclor 1248—Fraction IV	152–156°	1.6385	—	15	13	22

closely packed. Since γ_c of hexatriacontane is about 21 dynes/cm, γ_c for the silicone monolayer must exceed 21. Surface tensions of 19 to 20 dynes/cm are obtained with polymethylsiloxane liquids. Consequently, it follows that these liquids must spread on all high-energy surfaces.

Autophobic liquids have been defined as those compounds which are non-spreading by virtue of the fact that they adsorb unaltered to form low-energy films on which the bulk of the liquid will not spread. Autophobic liquids shown in Tables 10.9 through 10.13 include the chlorinated diphenyls, derivatives of triphenyl phosphate and certain aromatic hydrocarbons, esters, and ethers. Non-spreading properties can be induced in all pure liquids by the addition of a minor concentration of a polar compound which preferentially adsorbs to form a suitable low-energy surface.

The fact that a large number of esters are non-wetting on hydrated silica and sapphire surfaces has been shown to be due to hydrolysis of the esters *in situ* to form low-energy adsorbed films of organic acids or alcohols.

Effect of Temperature on Wettability (26)

Solutions of *n*-alkyl amines, acids, and alcohols in organic solvents deposit films that are not wetted by, or are oleophobic to, all but the lowest-boiling hydrocarbons and other nonpolar liquids. For any given solute-solvent combination at a given concentration there is a critical temperature τ_w above which the film is no longer unwetted by the solution. A plot of the critical temperature *versus* concentration in cetane for 18-carbon acid, alcohol, amine, and amide is shown in Figure 10.17.

Oleophobic behavior requires close packing and a high degree of orientation in the monolayer. Wetting occurs when a sufficient number of molecules develop kinetic energy in excess of a potential barrier. The increased probability of desorption with an increase in temperature can be compensated by an increase in the concentration of solute C. From the Boltzmann distribution law

$$C = A \exp(-U/R\tau_w) \tag{10.23}$$

where U is the energy of adsorption. For any given chain length, the nature of the polar group has a noticeable effect on U. In a homologous series of compounds, the energy of adsorption increases with an increase in chain length.

When the oleophobic film is prepared from a melt, a critical temperature of wetting τ_w was found to show a linear relation with the number of carbon atoms N for each homologous series. The data are graphed in Figure 10.18.

The energy of adsorption can then be considered to be made up of two terms, U_0 due to the attraction of polar groups for the adsorbing surface, and

U_c due to the cohesive forces between adjacent methylene groups of the hydrocarbon chains. If the total number of carbon atoms per molecule is sufficiently large and the size of the polar groups is compatible with the close packing of most of the methylene groups, then as a first approximation, $U_c = uN$, where u is the average energy of cohesion per methylene group. Then

$$U = U_0 + uN \qquad (10.24)$$

The condition for wetting of the film from the melted, oleophobic polar compound is

$$\frac{U_0 + uN}{R} = b^2 \tau_w \qquad (10.25)$$

where b^2 is a constant.

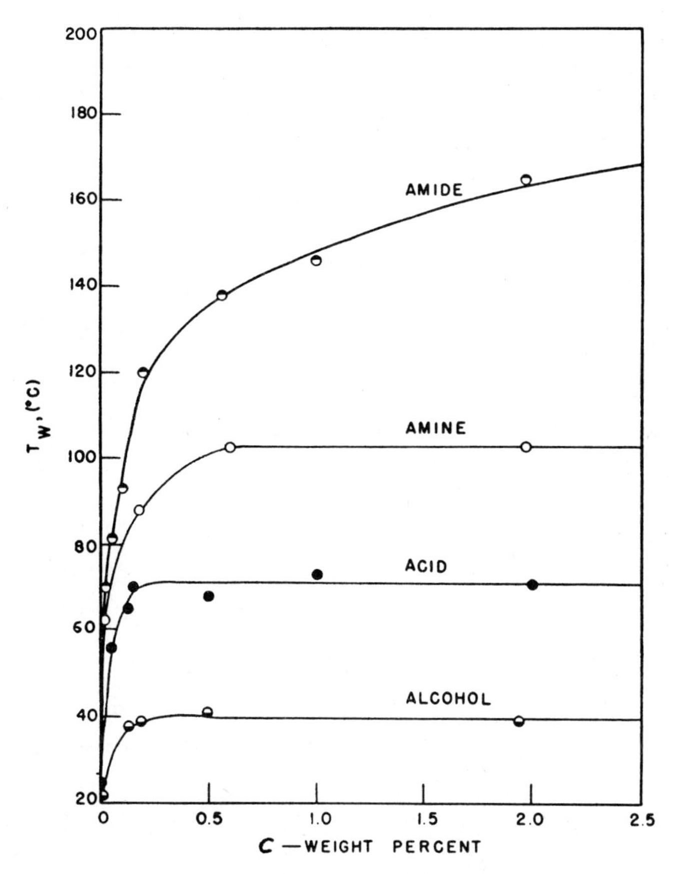

Figure 10.17. Effect of concentration and polar group on 18-carbon compounds dissolved in cetane (26).

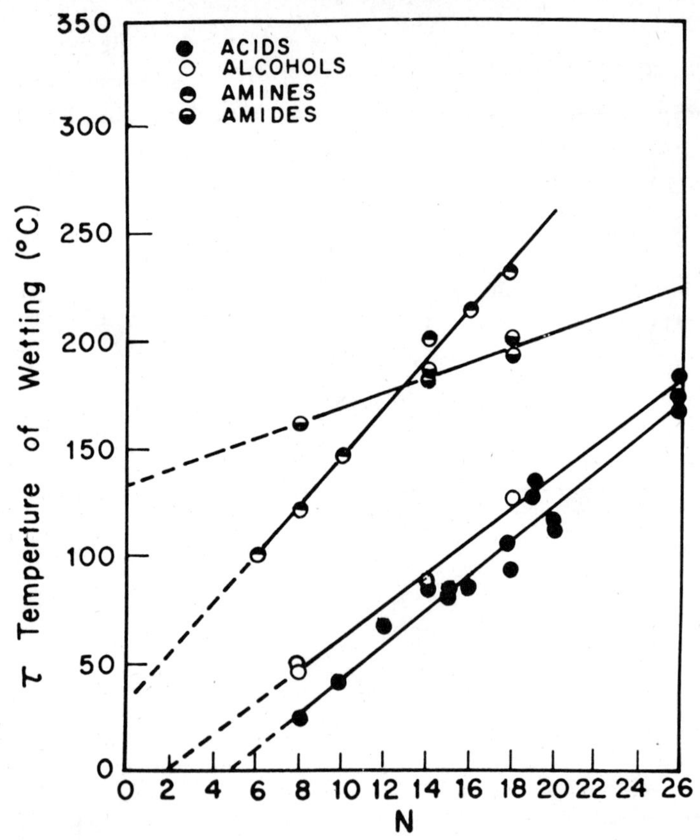

Figure 10.18. Effect of the number of carbon atoms in the hydrocarbon chain on the temperature of wetting of oleophobic films prepared from molten pure compounds (26).

Spreading on Organic Liquids

Jarvis and Zisman (27) investigated the conditions for spreading of fluorine-containing compounds on organic liquids. It was found that the compounds which exhibited large spreading coefficients on various of the organic liquids always contained organophilic as well as organophobic groups in the molecule. The organophilic groups were generally hydrocarbon chains, while the organophobic groups were fluorocarbon chains. The fluorinated compounds are designated as follows:

$F(CF_2)_n COOH$	ϕ - acid
$F(CF_2)_n CH_2OH$	ϕ' - alcohol
$H(CF_2)_n COOH$	ψ - acid
$H(CF_2)_n CH_2OH$	ψ' - alcohol

The compounds investigated included esters of carboxylic acids and fluorinated alcohols, esters of fluorinated acids and non-fluorinated alcohols, fluorinated organosilanes, perfluoroalkanes and their monochloro derivatives, and fluorocarbon chains attached to various polar groups, such as alcohol, sulfonate, mercaptan, ether, phosphate and sulfide. The liquids upon which the fluoro-compounds were spread included a double end-blocked polypropylene oxide (Ucon Fluid DLB 44E), *n*-hexadecane, white mineral oil, nitromethane, and tricresyl phosphate, among others.

As noted, the various fluorocarbon chains constitute the organophobic part of the molecule, while the hydrocarbon groups are organophilic. Increased substitution of fluorine in the organophobic group in a molecule decreased the surface tension and generally increased the spreading coefficient on an organic liquid substrate. In general, the greater the organophobic/organophilic ratio in the molecule, the greater the spreading coefficient. Spreadability and surface activity increased, the lower the surface tension of the fluorochemical and the greater the surface tension of the organic substrate or solvent.

In the design of an agent which is to be surface active in organic solvents, Jarvis and Zisman emphasize the necessity for selecting constituents that contribute least to surface tension and give a low value of W_c and a high value of W_A. Compounds based on ϕ-acids and ϕ'-alcohols give lower surface-tension values than corresponding compounds containing ψ-acids or ψ'- alcohols. Increased branching of either the organophobic or organophilic portion of the molecule will increase both surface tension and solubility in organic substrates, and it will also tend to increase the reversible work of cohesion, W_c. An aromatic nucleus will have similar effects. Polar groups generally increase the reversible work of cohesion and the surface tension of the molecule.

In continuing their studies, Jarvis and Zisman (28) investigated the surface activity of partially-fluorinated carboxylic esters dissolved in organic solvents. It was found that as little as 1 per cent by weight or 0.01 mole/liter of suitable fluorine-containing solutes were able to reduce the surface tension of some organic liquids by as much as 50 per cent. Larger surface tension depressions were obtained, the higher the surface tension of the solvent. The fluoroesters investigated adsorbed primarily as monomolecular films at the surface of each organic solvent. The adsorbed films were in a gaseous or otherwise expanded state.

In Figure 10.19 are shown a number of surface tension *versus* concentration curves for partially fluorinated carboxylic esters in various organic solvents.

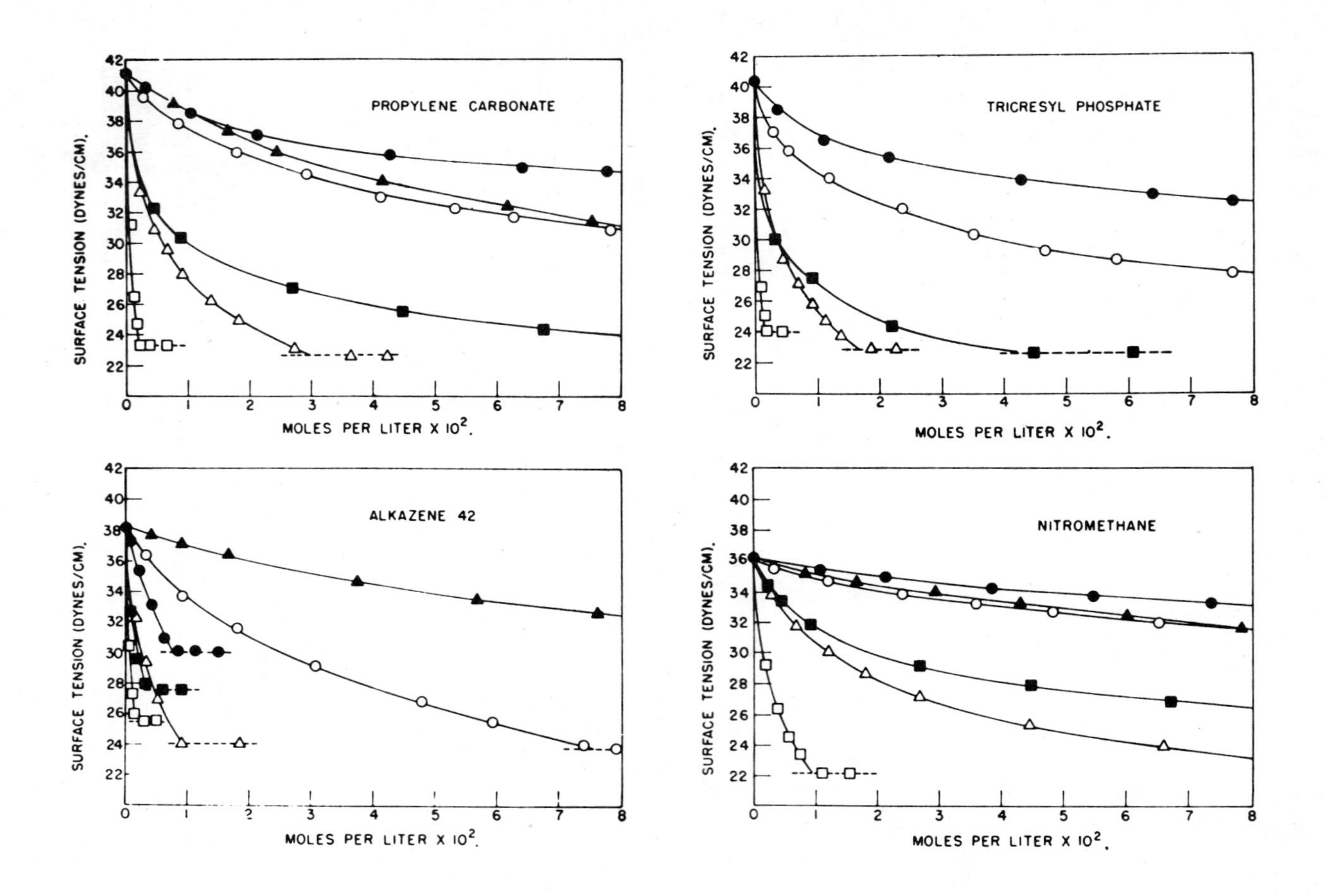

PROPYLENE CARBONATE
TRICRESYL PHOSPHATE
ALKAZENE 42
NITROMETHANE
SURFACE TENSION (DYNES/CM).
MOLES PER LITER X 10^2.

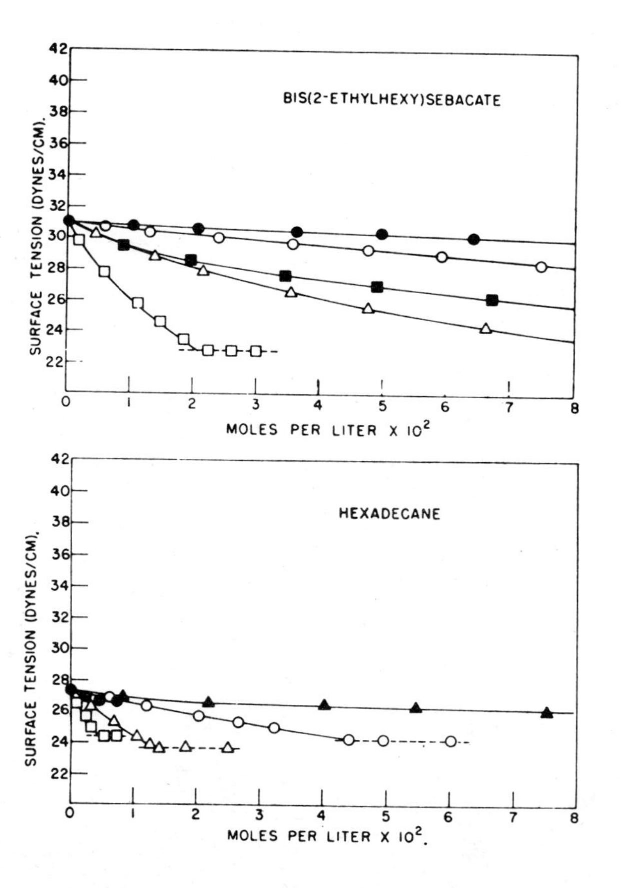

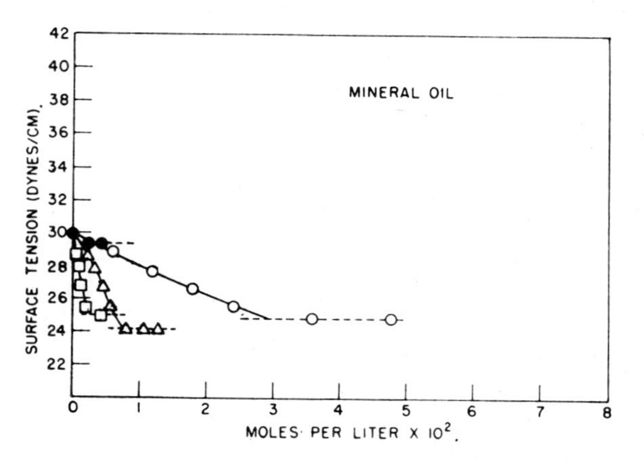

Figure 10.19. Surface tension-concentration isotherms of partially fluorinated carboxylic esters in organic liquids; ○, bis-(ϕ'-butyl) 3-methylglutarate; △, bis-(ϕ'-hexyl) 3-methylglutarate; □, bis-(ϕ-octyl) 3-methylglutarate; ●, bis-(ψ'-heptyl) 3-methylglutarate; ▲, hexyl ϕ-butyrate; ■, 1,2,3-trimethylolpropane tris-(ϕ-butyrate) (28).

It will be observed that as the chain-length of the fluorocarbon group is increased, the initial surface tension depression is greater. This effect is also related to the solubility of the surfactant in the organic solvent. The lower the solubility the greater the initial slope $(\partial\gamma/\partial_c)_{\text{initial}}$. However, a large initial slope does not imply a large maximum surface tension depression, since the surfactant may not be sufficiently soluble.

Derivatives of ϕ'-alcohols are generally more surface active than the corresponding derivatives of the ψ'-alcohols in polar solvents, probably because these solvents are capable of associating with the terminal hydrogen of the ψ'-alcohols. In non-associating organic solvents, the differences between ϕ'-alcohols and ψ'-alcohols tend to disappear. The data show that no single surfactant is most effective in all organic solvents. Instead, there is the necessity for a proper solubility balance; if the solubility is either too high or too low the maximum surface-tension lowering is decreased.

Jarvis and Zisman (29) have calculated a work of adsorption per CF_2 group spread on water at areas per molecule greater than 150Å^2 of 850 cal per g-mole. This is compared with a value of 625 cal per g-mole for the work of adsorption per CH_2 group at a water-air interface reported by Langmuir. The reversible work of adsorption per CF_2 group at organic liquid-air interfaces is shown in Table 10.14 (10).

TABLE 10.14. REVERSIBLE WORK OF ADSORPTION PER —CF_2— GROUP FOR PARTIALLY FLUORINATED (ϕ-ALKYL) 3-METHYLGLUTARATE DIESTERS ADSORBED AT ORGANIC LIQUID-AIR INTERFACE (29).

Organic Solvent	Solvent surface tension, dynes/cm	Reversible work per —CF_2— group, cal/g-mole: From butyl and hexyl derivatives	Reversible work per —CF_2— group, cal/g-mole: From hexyl and octyl derivatives
Propylene carbonate	41.1	420	420
Tricresyl phosphate	40.4	360	360
Alkazene 42	38.2	340	310
Nitromethane	36.2	330	320
Bis-(2-ethyhexyl) sebacate	31.0	210	200
Mineral oil	29.9	210	190
Hexadecane	27.4	180	200

Hysteresis of Contact Angle

It has been observed by numerous investigators (1) that the contact angle that is obtained when the liquid drop is advancing on a clean solid surface is larger than the receding angle obtained when the liquid is being withdrawn from the surface of the solid. The difference between advancing and receding angles is called the hysteresis of the contact angle. Harkins (3) has stated that if equilibrium is attained, there can only be one value of the contact angle between a given solid and liquid. He ascribed "advancing" and "receding" angles to improper preparation of the surface and poor technics in making measurements. There is little doubt that this situation holds in many instances. Table 10.15 indicates the variety of surfaces for which no hysteresis was found (30).

TABLE 10.15. THE CONTACT ANGLE MADE BY WATER ON VARIOUS SOLID SURFACES FOR WHICH NO HYSTERESIS WAS FOUND. THE CONTACT ANGLE ON PARAFFIN VARIES ACCORDING TO THE COMPOSITION OF THE PARAFFIN (30).

Solid	Contact Angle
Ceylon graphite	85.7
Talc	87.8
Stibnite	84.2
Paraffin	108–111

TABLE 10.16. CONTACT ANGLES OF WATER DROPS ON GOLD SURFACES IN VARIOUS ATMOSPHERES AT 25C (31)

Atmosphere	θa degrees	θr degrees
Water vapor	7 ± 1	0
Water vapor + pur air	6 ± 1	0
Water vapor + benzene vapor	84 ± 2	82
Water vapor + pure air + benzene vapor	86 ± 1	83 ± 1
Water vapor + lab. air	65	30
Water vapor + outdoor air	13	0

However, contact-angle hysteresis will frequently be observed under practical conditions of measurement. The two factors responsible for hysteresis appear to be surface roughness and adsorption effects (31). That the latter factor includes contamination and non-equlibrium effects is evident from such data as those shown in Table 10.16. Pure water on a pure high-energy surface such as gold should give a contact angle $\theta=0$. Advancing angles of 6 and 7, as compared with a previously reported (32) value of 40, indicate that an almost clean surface had been obtained. The effect of various degrees of contamination on hysteresis is evident from the table.

Surface Roughness

Bartell and Shepard (33, 34, 35) determined the effect of surface roughness on contact angles and contact-angle hysteresis by the direct approach of measuring contact angles on uniformly roughened surfaces. They defined roughness in terms of the average height h to which the asperities rise above the horizontal surface plane and the mean angle of inclination ϕ of the sides of

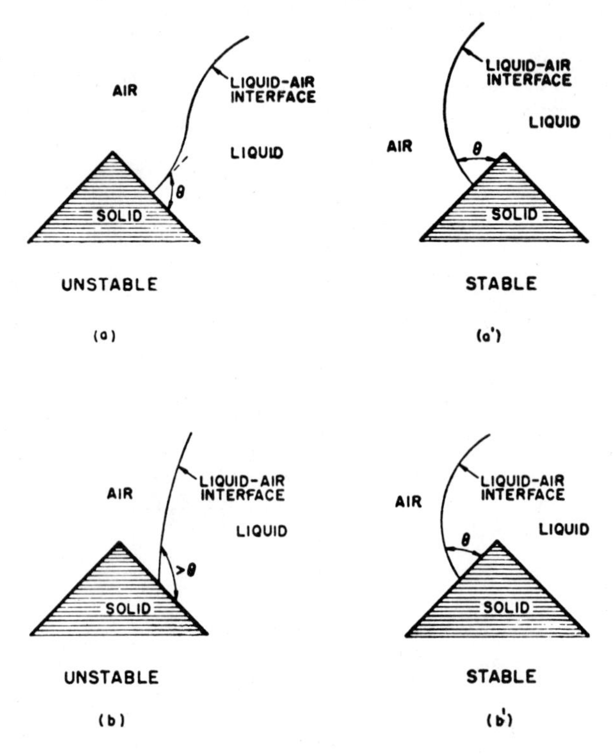

Figure 10.20. (a) If at the solid-liquid-air interface the stable contact angle, θ, with respect to the pyramid face is formed, the curvature of the liquid-air interface will be distorted tending to cause the liquid to advance (a)′. (b) If the curvature of the liquid-air interface is constant, the angle made by the liquid with the solid on this pyramid face will be greater than the stable angle, θ, and the liquid will tend to advance (b′) (33).

the asperities with respect to the horizontal plane of the solid surface. They obtained uniformly coarse surfaces by ruling smooth blocks of paraffin at selected spacings with precison-ground V-shaped tools in one direction, and again at right angles to produce a series of regular pyramids. They found it necessary to use drops that were very large, 4.5 to 6 cm, in comparison to the dimensions of the roughness in order to obtain stable and reproducible contact angles. They observed that along the drop edges on all the rough surfaces, the solid-liquid-air interface was very irregular. The advancing drop never came to rest on the pyramid faces oriented toward the center of the drop. The reason for this is shown in Figure 10.20.

For the roughened paraffin surface studied, they found that the height of the asperities had no effect on either advancing or receding contact angles. However, the angle of inclination of the asperities had a profound effect. An increase in the angle of inclination ϕ generally produced an increase in the apparent advancing contact-angle and a decrease in the apparent receding contact-angle. The exceptions to this generalization are the receding contact-angles for water and $3M$ calcium chloride and the advancing contact-angle for methanol at the paraffin-liquid-air interface. The hysteresis of the contact angle generally increases with an increase in the angle of inclination ϕ, as shown in Table 10.17. The paraffin-methanol-air system is an exception. This may be related to the fact that it had the greatest solvent action on paraffin of the liquids studied.

TABLE 10.17. COMPARISON OF AVERAGE HYSTERESIS VALUES IN DEGREES (35).

Liquid	Smooth surface $\theta a - \theta r$	$\phi = 30°$ $\theta a - \theta r$	$\phi = 45°$ $\theta a - \theta r$	$\phi = 60°$ $\theta a - \theta r$
Water	11°	30°	48°	64°
3 M $CaCl_2$	10	24	34	49
Glycerol	7	23	44	77
Ethylene glycol	7	28	54	103
Methyl cellosolve	20	64	82	93
Methanol	15	58	50	0

According to Wenzel (36, 37) within a given geometrical area a roughened surface will contain a larger true surface area than a smooth surface. Wenzel therefore contends that the Young equation should be modified by multiplying the energy change by a roughness factor r, which is the ratio of actual surface area to geometrical surface area:

$$r\,(\gamma_S - \gamma_{SL}) = \gamma_L \cos\theta' \tag{10.26}$$

where θ' is the contact angle on the roughened surface. The roughness factor r is always greater than one except on an ideally smooth surface when it is equal to one. The contact angle θ observed on a smooth surface is related to that observed on a roughened surface as follows:

$$r(\cos\theta) = \cos\theta' \tag{10.27}$$

From this relation, it follows that if a contact angle on a smooth surface is less than 90°, roughening the surface would decrease the observed contact angle. If the angle on a smooth surface is more than 90°, roughening the surface should increase the observed contact angle.

Though the Wenzel equation is widely accepted (38), Shepard and Bartell found that it did not correlate with their experimental values. Further, the contact angle depended upon whether the interface at the periphery of the drop was roughened. If the drop rested on a roughened surface, but the solid-liquid-air interface extended onto a smooth portion of the paraffin, the contact angle was the same as that obtained on a smooth surface. The Young equation is valid with respect to the angle of contact at a liquid-solid-air interface. At any point on an actual surface, the plane of the solid surface may be oriented at an angle to the horizontal direction of the surface, from which the angle is measured. The measured angle is the mean of all the angles being made, with respect to the horizontal line of reference, along the line of contact. The mean angular deviation of the actual plane of contact from the reference plane can be much greater on a roughened surface than on a smooth surface.

Contact Angles Involving Two Liquid Phases

Contact angles discussed up to this point involved a solid surface, a liquid, and a gaseous phase. These contact angles relate to wetting phenomena pertinent to lubrication, ore flotation, spreading of herbicides on plant leaves, textile wetting, and so forth. There are many industrial processes that involve the displacement of one liquid by another on a solid surface. Many corrosion inhibitors function by displacing water from metal surfaces. In detergent processes, the oil or grease film covering the substrate may be rolled back by the advancing aqueous detergent solution. In industrial processes generally, it is important to distinguish between advancing and receding contact angles, because non-equilibrium conditions are commonplace.

The Young equation is applicable to any two fluid phases on a solid surface, and thus, can be applied to the contact angle formed by water, a water-insoluble organic liquid, and a solid surface. Ray and Bartell (39) showed the interrelationship between this contact angle and the contact angles formed by each of the two liquids separately on the solid surface.

In the ideal case, where there is no hysteresis, we have for the two solid-liquid-air systems

$$\gamma_{sa} - \gamma_{sw} = \gamma_{wa} \cos \theta_{swa} = A_{swa} \tag{10.28}$$

$$\gamma_{sa} - \gamma_{so} = \gamma_{oa} \cos \theta_{soa} = A_{soa} \tag{10.29}$$

subtracting 10.29 from 10.28

$$\gamma_{so} - \gamma_{sw} = A_{swa} - A_{soa} \tag{10.30}$$

For the solid-water-organic liquid system

$$\gamma_{so} - \gamma_{sw} = \gamma_{wo} \cos \theta_{swo} = A_{swo} \tag{10.31}$$

Substituting 10.30 in 10.31

$$A_{swa} - A_{soa} = A_{swo} \tag{10.32}$$

In these equations the subscripts refer to the phases:

sa=solid-air, *so*=solid-organic liquid, *sw*=solid-water, *wa*=water-air, *oa*=organic liquid-air, *wo*=water-organic liquid, *soa*=solid-organic liquid-air, *swa*=solid-water-air, and *swo*=solid-water-organic liquid.

In the more complicated case involving hysteresis, with water advancing, the relationships follow. For the solid-liquid-air systems

$$\gamma_{sa} - \gamma_{sw} = \gamma_{wa} \cos \theta^{a}_{swa} = A^{a}_{swa} \tag{10.33}$$

and

$$\gamma_{s^1a} - \gamma_{so} = \gamma_{oa} \cos \theta^{r}_{soa} = A^{r}_{soa} \tag{10.34}$$

Subtracting 10.34 from 10.33

$$\gamma_{so} - \gamma_{sw} + (\gamma_{sa} - \gamma_{s^1a}) = A^{a}_{swa} - A^{r}_{soa} \tag{10.35}$$

For the solid-water-organic liquid system

$$\gamma_{so} - \gamma_{s^1w} = \gamma_{wo} \cos \theta^{a}_{swo} = A^{a}_{swo} \tag{10.36}$$

or, adding and subtracting γ_{sw}

$$\gamma_{so} - \gamma_{sw} + (\gamma_{sw} - \gamma_{s^1w}) = A^{a}_{swo} \tag{10.37}$$

From 10.35 and 10.37

$$A^{a}_{swa} - A^{r}_{soa} - (\gamma_{sa} - \gamma_{s^1a}) = A^{a}_{swo} - (\gamma_{sw} - \gamma_{s^1w}) \tag{10.38}$$

where the superscripts *a* and *r* distinguish between advancing and receding contact angles respectively, and the s^1 designates that the solid was previously in contact with the organic liquid. Otherwise, the subscripts have the same meanings as before. Similar interrelationships can be found for the case where water is receding. Table 10.18 shows that the equations are justified by experimental data.

TABLE 10.18. COMPARISON BETWEEN OBSERVED AND CALCULATED CONTACT ANGLES ON CELLULOSE DERIVATIVES (39).

System	Contact angles, degrees			
	Advancing, θ^a		Receding, θ^r	
	Obsd.	Calcd.	Obsd.	Calcd.
Acetate-FM2, foil, side A				
Methylene iodide	111	109.5	67	69.5
α-Bromonaphthalene	106	107	69	67
α-Chloronaphthalene	110		67.5	65
Bromobenzene	119.5		64.5	59
Acetate-FM2, foil, side B				
Methylene iodide	105	104.5	59	58.5
α-Bromonaphthalene	105	102	55	55.5
α-Chloronaphthalene	104.5		54	52.5
Bromobenzene	115		46	45
Acetate-FM2, film				
Methylene iodide	103	103	70	69
Triacetate, film				
Methylene iodide	101.5	103	81	78
α-Bromonaphthalene	106	107	82	79.5
Bromobenzene	113		81	77
Acetopropionate-H, foil, side A				
Methylene iodide	119	119	77.5	77.5
α-Bromonaphthalene	128	121	80	81
Bromobenzene	132.5		75	76
Acetopropionate-H, foil, side B				
Methylene iodide	116	118	69	65
α-Bromonaphthalene	124.5	121.5	65	66.5
Bromobenzene	123		56	59
Tripropionate, film				
Methylene iodide	120	119.5	84.5	85
Acetobutyrate-H, foil, side A				
Methylene iodide	127	127.5	83	80
α-Bromonaphthalene	131	129	86	82
Bromobenzene			79	78.5
Acetobutyrate-H, foil, side B				
Methylene iodide	128	128.5	80	76.5
α-Bromonaphthalene	128	127	74	79
Bromobenzene	141.5		72.5	70.5
Tributyrate, film				
Methylene iodide	138	138.5	100	96
Ethyl cellulose-H, foil, side A				
Methylene iodide	121	124	65.5	63
Ethyl cellulose-H, foil, side B				
Methylene iodide	118	118.5	58	58

Capillary Wetting

The rise of a liquid in a capillary tube depends upon the contact angle and the curvature of the tube. These in turn determine the curve of the liquid surface. The pressure difference across the curved surface depends upon the free energy of the surface and provides the hydrostatic pressure under which the liquid flows up the tube (1). If the radius r of a cylindrical tube is sufficiently small that the liquid meniscus is essentially spherical, then the height of rise h of the meniscus is given by

$$\frac{2\gamma\cos\theta}{r} = gh(\rho_1 - \rho_2) \tag{10.39}$$

where γ is the surface tension, θ is the contact angle, g is the gravitational constant, and $(\rho_1 - \rho_2)$ is the difference in density between the rising liquid and the surrounding fluid.

The rate at which a liquid will rise in a vertical cylindrical capillary-tube, assuming the absence of wall slippage effects is given by the Washburn (40) equation:

$$\frac{dh}{dt} = \frac{2\gamma\cos\theta - gh\rho}{r} \cdot \frac{r^2}{8\eta h} \tag{10.40}$$

where $\rho = \rho_1 - \rho_2$, the difference in density between the rising liquid and the surrounding fluid, η is the viscosity of the fluid, and t is the time.

For flow through horizontal capillary tubes, the effect of gravity can be neglected, and the equation takes the form

$$\frac{dh}{dt} = \frac{2\gamma r\cos\theta}{8\eta h} \tag{10.41}$$

Flow in capillaries is important in such diverse applications as the flow of sap in plants, recovery of petroleum from oil sands, catalysis, and the wetting of textile fabrics (41).

Wetting of Cotton Yarn by Aqueous Surfactant Solutions (42)

The wetting of hydrophobic cotton yarn by aqueous surfactant solutions is an example of the general problem of the wetting of porous hydrophobic solids. If the cotton yarn is treated as a parallel bundle of smooth-walled capillaries, the derived rate of penetration parallel to the threads would be proportional to γ cos θ, the product of the liquid surface tension and the cosine of the contact angle on the wall of the capillary. However, the direction of penetration is perpendicular to the cotton fibers immersed in the aqueous solution, and the rate of penetration is not proportional to γ cos θ.

The logarithm of the sinking time of cotton skeins immersed in the aqueous solution is known to be a linear function of the logarithm of the concentration of the wetting agent (43). Fowkes (42) showed that the logarithm of sinking time is also a linear function of the contact angle or of the surface tension.

Rates of wetting were determined by Fowkes using the Draves-Clarkson sinking test (43). This test employs a 5-gram skein of gray unboiled cotton yarn of 54-inch loops containing 120 threads. Each thread in cross section contains 100–200 cotton fibers. The skein is attached to a 3-gram hook and totally immersed in the solution of wetting agent. It is held submerged by

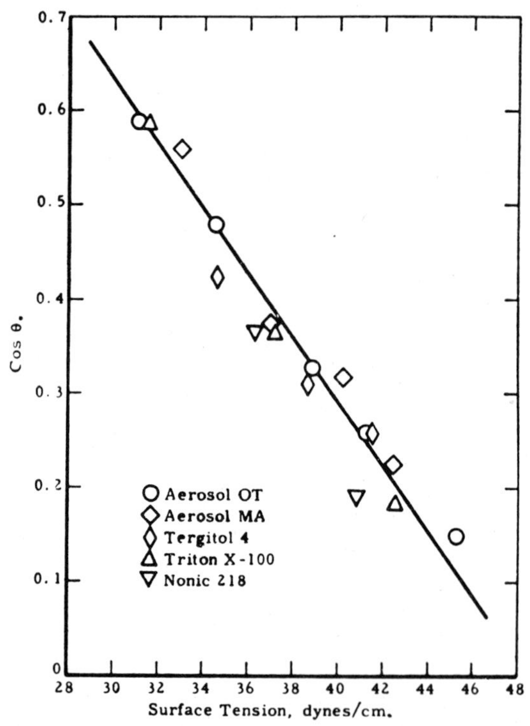

Figure 10.21. Contact angles of various anionic and nonionic surfactants on paraffin wax (42).

means of a weight tied to the hook by a thread. The penetrating solution displaces the air in the skein. The sinking time is the time required for the immersed skein with attached hook to sink in the solution. The yarn has a surface of natural oils and waxes and is hydrophobic. In this study, the oils were rinsed out with benzene to leave a waxy surface. The contact angle of water on the waxy cotton was 107°, the same as on paraffin wax.

In addition, a yarn-bundle wetting test (44) was employed, with the yarn floated on the aqueous solution. By both methods, the logarithm of the sinking time was found to be proportional to $\cos \theta$.

As shown in Figure 10.21, the surface tension of solutions of paraffin-chain surfactants is related to the contact angle on wax by

$$\cos \theta = 1.68 - 0.0358\gamma \tag{10.42}$$

The surface tension can then be related to the sinking time t_s,

$$\log t_s = A + B\gamma \tag{10.43}$$

where A and B are constants.

However, the surface tension and contact angle used in these equations represent the values at the front of the penetrating liquid. They will undergo change during the wetting experiment, due to adsorption of the surfactant on the fibers.

The surface tension of solutions of wetting agents is related to the logarithm of the concentration c, such as by the following integrated form of the Gibbs adsorption equation,

$$\gamma = C - \frac{2.303\, RTZ \log c}{N\sigma} \tag{10.44}$$

where C is the intercept of γ at $\log c=0$, Z is the number of particles—1 for nonionic, 2 for ionic surfactants—per molecule in solution, N is Avogadro's number, and σ is the surface area per adsorbed molecule. Sinking time can be related to the concentration of wetting agent by substituting this expression for γ in Equation 10.43,

$$\log t_s = A + BC - BD \log c \tag{10.45}$$

where D equals 2.303 RTZ/Nσ, and thus varies with σ/Z for different wetting agents.

Fowkes determined the extent of adsorption of surfactants on paraffin wax using the Gibbs adsorption equation. The calculated value for each surfactant corresponds to the area of hydrophobic cotton interface per adsorbed molecule σ_i as

$$\sigma_i = \frac{2.303\, RTZ}{N} \cdot \frac{d \log c}{d(\gamma \cos\theta)} \tag{10.46}$$

These values are compared in Table 10.18 with the area per molecule in the surface of the solution σ_s for five commercial wetting agents. Values for σ_s were computed from the Gibbs equation,

$$\sigma_s = \frac{-2.303\, RTZ}{N} \cdot \frac{d \log c}{d\gamma} \tag{10.47}$$

TABLE 10.19. CALCULATED AREAS PER MOLECULE OF ADSORBED SURFACTANTS.

Agent	Mol. Wt.	Z	Area per Molecule at Surface, σ_s Å²	Area per Molecule at Interface, σ_i Å²
Nonic 218	562	1	61.5	53.2
Triton X100	577	1	52.4	51.9
Aerosol OT	444	2	137	154
Aerosol MA	388	2	103	118
Tergitol 4	292	2	132	135

Aerosol OT is the sodium salt of di-2-ethylhexyl sulfosuccinate. Aerosol MA is the dihexyl derivative. Tergitol 4 is the sodium salt of a highly branched alkanol sulfate. Triton X100 is a condensate of octyl phenol with ethylene oxide with an average of about 9 ethoxy units. Nonic 218 is an ethylene oxide condensate with a dodecyl mercaptan. The polyether chain also has about 9 units (42).

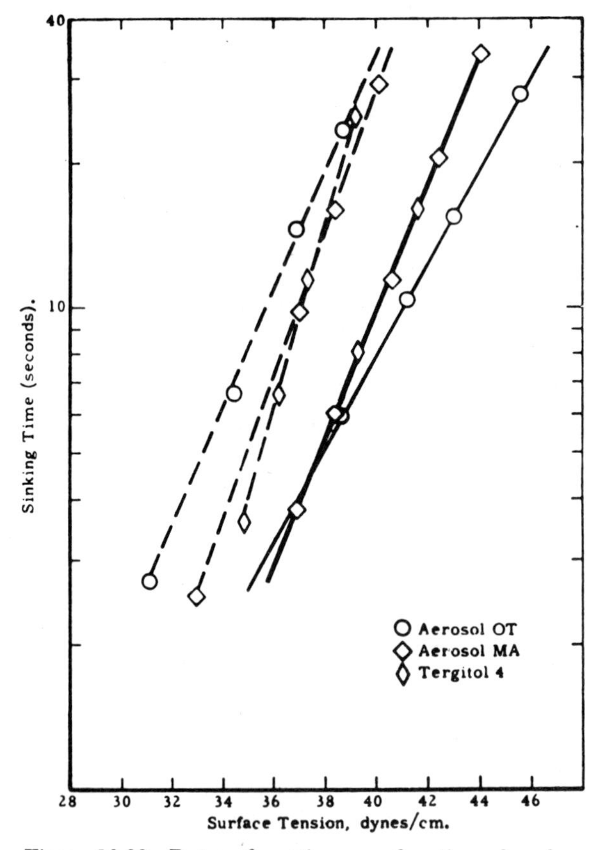

Figure 10.22. Rates of wetting as a function of surface tension: solid line, Draves test with a 3-g hook; dashed line, yarn bundle test (42).

Table 10.19 shows that the nonionic surfactants, Triton X100 and Nonic 218, are extracted from solution by adsorption to a much greater extent than the ionic surfactants tested. Consequently, they show a reduced rate of wetting. As the advancing solution front is depleted by the adsorption of surfactant, additional surfactant must be brought up by diffusion to maintain a low surface tension.

Figure 10.22 shows the rate of wetting as a function of surface tension. The surface tension required by Figure 10.22 for a given sinking time may be represented by γ_1, which is attained with a concentration c_1 of a surfactant. The concentration $c-c_1$ is required to provide fast enough diffusion to maintain γ_1 at the advancing front of the penetrating solution. Thus, from Figure 10.22 a Draves-Clarkson sinking time of 10 seconds requires a surface tension of 40–41 dynes/cm. For Nonic 218, 40.1 dynes/cm is achieved at a concentration of 0.0057 per cent. However, a 0.04 per cent concentration is required for a 10 second wetting time. Then $c-c_1$ equals 0.0343. In general, $c-c_1$ varies inversely with S_i, the area per gram of adsorbed agent. That is, the greater the extent of adsorption, the larger the excess concentration $c-c_1$ required for a given wetting time.

WETTING BY LIQUID METALS (45)

A number of technological problems require knowledge of the wetting properties of liquid metals on metal and oxide surfaces. These include soldering, brazing, and heat transfer. The surface properties of metals are also important in the metallurgical studies of grain growth and the growth of large single crystals. The principles previously discussed, concerning the spreading of liquids on surfaces, are equally applicable to liquid metals. However, there is one fundamental difference. The liquid metals have considerable chemical reactivity, either due to the inherent nature of the metal, as in the case of the alkali metals, or because of the high temperature required to maintain them liquid. In predicting wetting or non-wetting by liquid metals it is necessary to consider the likelihood of chemical reaction and a change in the composition of the substrate surface.

In order for a liquid to spread on a solid, a necessary, though not sufficient, condition for spreading is that the liquid have a lower surface tension than the solid.

Surface Tension of Solids

Several methods have been employed to estimate the surface energy of solids. One method compares the heat of solution of a fine powder of known large surface area with the heat of solution of large particles of the same material (46, 47). Another method compares the specific heat of a fine powder of known large surface area with the specific heat of the bulk solid (48). These methods suffer from the shortcoming that small particles of colloidal dimensions frequently have a highly distorted lattice. Consequently, their heat content apart from their surface energy is considerably greater than that of larger particles (49).

The surface free energy of solid metals has been determined accurately by high-temperature creep measurements on very thin wires (50, 51, 52). The "viscous" creep rates of very fine wires is determined as a function of load, and extrapolated to zero creep rate. For a single crystal cylinder

$$\gamma_s = w/\pi r \tag{10.48}$$

where γ_s is the surface tension of the solid, w is the load at zero creep rate and r is the radius of the wire. For polycrystalline wire a correction term is applied for the grain boundary tensions acting perpendicular to the wire axis

$$\gamma_s = \frac{w}{\pi r} + \frac{n}{l} r\gamma^* \tag{10.49}$$

where n is the number of grains per wire, l is the length of the wire, and γ^* is the grain boundary tension computed from the dihedral angles of the grain boundaries. The determinations are carried out in a controlled atmosphere in equilibrium with the vapor of the metal. The temperature should be within about 60°C of the melting point to obtain practical creep rates.

The surface free energies of solids are generally only a little higher than those of the corresponding liquids, and can be estimated from the liquid surface tension values. Methods that have been used include the maximum bubble pressure method (53), sessile drop method (54, 55) and pendant drop method (56). However, the measurement of liquid surface tensions at temperatures of about 1000°C involves a number of experimental difficulties, due to contamination by the atmosphere or the substrate.

Bondi deduced that the maximum increase in surface free energy upon solidification of a liquid metal should be of the order

$$\Delta\gamma = \Delta h_f \,/\, A_w \tag{10.50}$$

where Δh_f is the heat of fusion per atom, and A_w is the area per atom in the interface. In Table 10.20 calculated surface tension values have been compared with those determined by the creep method. Agreement is within the experimental error of the method. Bondi suggested that the equation appears to be applicable to metals and possibly other valence bond crystals. However, for molecular solids, $\Delta\gamma$ should be about 1/8 to 1/3 $(\Delta h_f/A_w)$, depending upon the crystal structure.

At temperatures substantially below the melting point, equilibrium in

TABLE 10.20. SURFACE TENSION OF SOLIDS DETERMINED BY THE CREEP METHOD AND CALCULATED FROM THE SURFACE TENSION OF THE LIQUID AND EQUATION 10.50 (45).

Metal	Temperature (°C)	Surface tension of metal, by creep method	Crystal Plane	Surface tension calculated by Eq. 10.50	Increase in surface tension on solidification, calculated by Eq. 10.50
Silver	900	1140 ± 90	100	1150	220
			111	1180	250
Gold	1300	1400 ± 65	100	1396	256
		1510 ± 100	111	1438	298
Copper	1050	1430	100	1509	329
		1670	111	1560	380
γ-Iron	1400		100	2070	370
			111	2127	427
Tin	150	704	100	765	128
			001	672	35

TABLE 10.21. SURFACE TENSION OF LIQUID METALS (45).

Metal	Temperature (°C)	Surface Tension of Liquid Metal (dynes/cm)
Aluminum	700	900
Antimony	635	383
Bismuth	300	376
Cadmium	370	608
Copper	1140	1120
Gallium	30	735
Gold	1120	1128
Iron	1530	1700
Steel		950–1220
Lead	350	442
Magnesium	700	542
Mercury	20	476.1
Potassium	64	119
Selenium	220	105.5
Silver	995	923
Sodium	100	206.4
Thallium	313	(446)
Tin	700	538
Zinc	700	750

the surface becomes very slow and the surface energy of the solid is not likely to have a uniform value. Chalmers, King, and Shuttleworth (57) observed that mechanical scratches on silver did not disappear until the temperature was raised to 60°C below the melting point. Electron diffraction suggests that the appearance of disorder in the surface layer, or "surface melting" may begin at 0.7 T_m°C where T_m is the melting point (58).

TABLE 10.22. SURFACE TENSION OF LIQUID METAL OXIDES (45).

Metal Oxide	Temperature (°C)	Surface Tension of Liquid Metal Oxides (dynes/cm)
Al_2O_3	2050	580
B_2O_3	900	79.5
FeO	1420	585
La_2O_3	2320	560
Na_2SiO_3 (glass)	1000	310
PbO	900	132
SiO_2	1400	(200 to 260)
ZnO	1300	(455)

TABLE 10.23. SURFACE TENSION OF FUSED SALTS (45).

Salt	Temperature (°C)	Surface Tension of Fused Salts (dynes/cm)	$d\gamma_L/dT$
LiF	840	255	−0.10
CsF	692	107	−0.08
LiCl	608	140	−0.07
NaCl	801	114	−0.07
NaI	660	88	−0.04
KCl	780	97	−0.07
KI	681	78	−0.11
CsI	620	91	−0.06
Li_2SO_4	852	224	−0.06
Na_2SO_4	884	196	−0.07
Cs_2SO_4	1015	113	−0.09
$LiNO_3$	254	118	−0.13
$CsNO_3$	414	92	−0.094
$LiBO_2$	845	265	−0.064
KBO_2	946	137	−0.027
$NaPO_3$	620	209	−0.059
KPO_3	820	161	−0.066
$BiCl_3$	271	66.2	−0.13
$BiBr_3$	250	66.5	−0.10
$SnCl_2$	307	97.0	−0.09
$CdCl_2$	602	77.2	−0.02

Because intermolecular forces are very large in metals and metal oxides, the surface energies of these systems would be expected to be high. As shown in Tables 10.21, 10.22 and 10.23, the surface tension of liquid metals exceed those of organic liquids by a factor of ten or more. Those of fused metal oxides and salts, while generally less than liquid metals, are considerably higher than for organic liquids.

As in the case of organic liquids, since surface free energy is related to

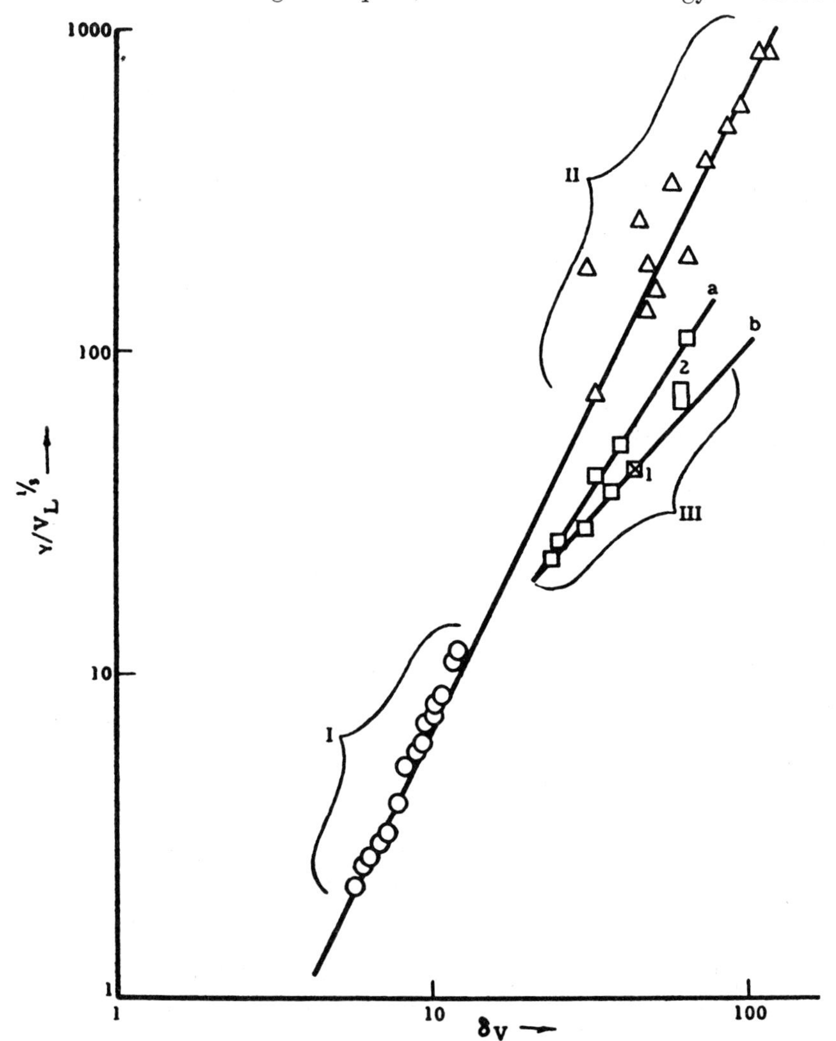

Figure 10.23. Relation between surface free energy and cohesive energy for: I, spherical organic molecules and non-metallic elements; II, metals; and III, inorganic salts and oxides (1 = lead monoxide; 2 = silicon dioxide) (45).

intermolecular forces, the free energy of surface formation per mole should be approximately proportional to the energy of vaporization ΔE_v. In Figure 10.23 is plotted $\gamma/V_L^{\frac{2}{3}}$ which is proportional to the free energy of surface formation *versus* $(\Delta E_v/V_L)^{\frac{1}{3}}$, where V_L is the molar volume of the liquid. The data for metals straddles a curve which is a straight-line extension of the curve for organic substances and non-metallic elements.

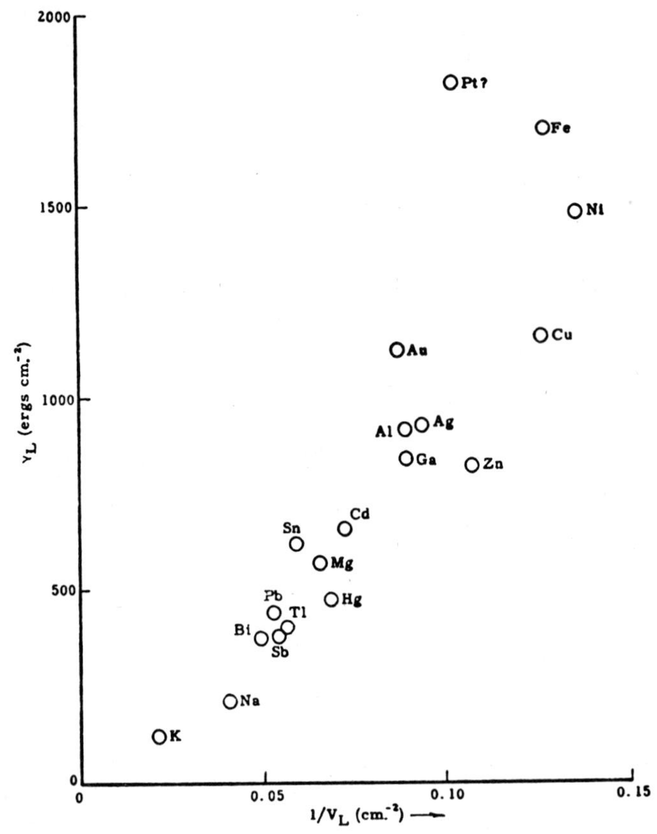

Figure 10.24. Plot of surface tension of liquid metals against their inverse atomic volume (45).

Atterson and Hoar (59) showed that there is a fairly good correlation between the surface tension of metals at their melting point and the inverse atomic volume, as represented in Figure 10.24. Within a given group of the Periodic Table, the reduction in surface tension with increasing atomic radius is quite general. The same effect is observed with corresponding groups of metal derivatives (60).

Surface Tension of Mixtures

The surface free energy of mixtures of metals is generally dominated by the component with the lower value. This is to be expected, since the metal component with the lower surface free energy will concentrate at the surface, provided that it does not form an intermetallic compound. Low surface free energy metals, such as the alkali metals, are surface-active in mercury and other metals belonging to this group. The lowering of surface tension by alkali and alkaline earth metals is shown in Figure 10.25. However, if the components form compounds that are less stable in the surface layer than in the bulk, the surface tension of the mixture may be higher than that of the pure components. With iron-nickel alloys, the surface tension varies linearly with composition (55).

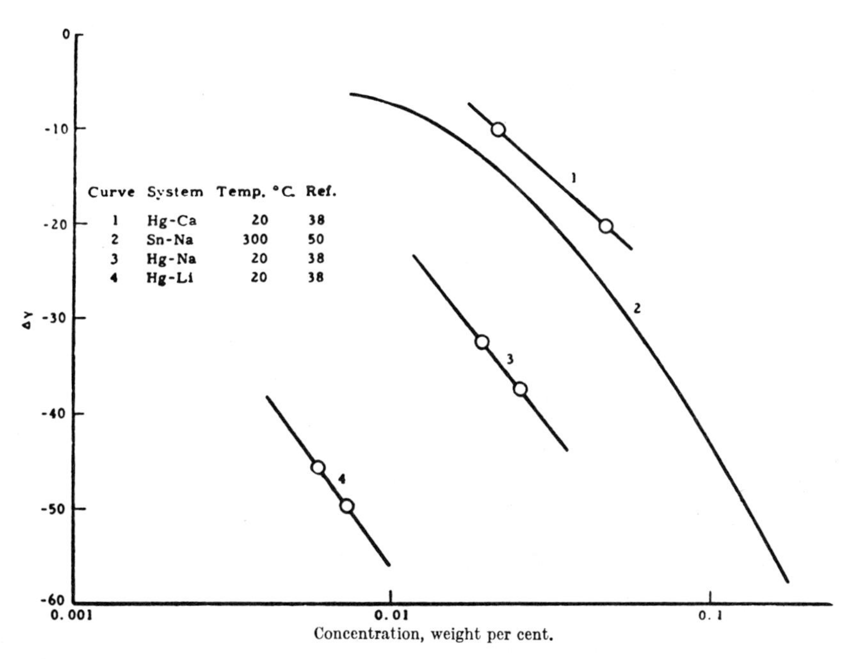

Figure 10.25. Effect of the concentration, in weight per cent, of alkali and alkaline earth metals on the surface tension of liquid metals (45).

As shown in Figure 10.26, oxygen, sulfur and nitrogen depress the surface tension of pure iron, while carbon is without effect (61). The excess surface concentrations, calculated from the slopes of the curves, gave values of 7.62 Å^2 and 14.4 Å^2 as the area per atom for oxygen and sulfur, respectively. The value for oxygen is in reasonable agreement with the value of 8.12 Å^2

per atom in the plane of maximum packing in FeO, while the calculated concentration of sulfur in the surface is somewhat higher than the corresponding value of 11.56 Å² per atom in the plane of maximum packing in FeS (61).

The temperature coefficient of surface free energy $d\gamma/dT$ is negative for pure metals as it is with organic compounds. For alloys, the temperature coefficient may be positive or negative depending on the effect of temperature on the distribution of the components at the surface.

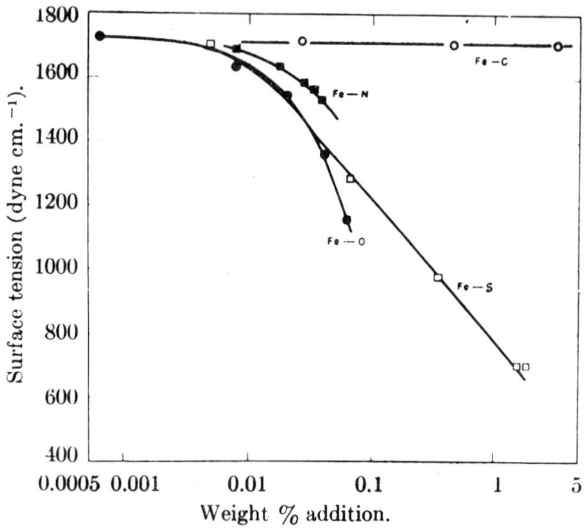

Figure 10.26. Effect of carbon, nitrogen, sulfur, and oxygen on the surface tension of liquid iron (61).

Surface Tension of Inorganic Compounds

The generalities concerning the surface properties of metals are also applicable to metal oxides and salts. There is a trend of decreasing surface free energy with increasing ionic radius of the components of the melts. As shown in Figure 10.27, several oxides depress the surface tension of slags and glasses. The effect of sodium monoxide or phosphorous pentoxide on ferrous oxide is similar to that of surfactants in water. However, other oxides may have a negligible effect or increase the surface tension of the system, probably due to compound formation.

Interfacial Tension

Interfacial tension data for a number of fused salts on liquid metals are presented in Table 10.24. In general, the interfacial tension between the fused salt and the liquid metals is sufficiently small to give large, positive

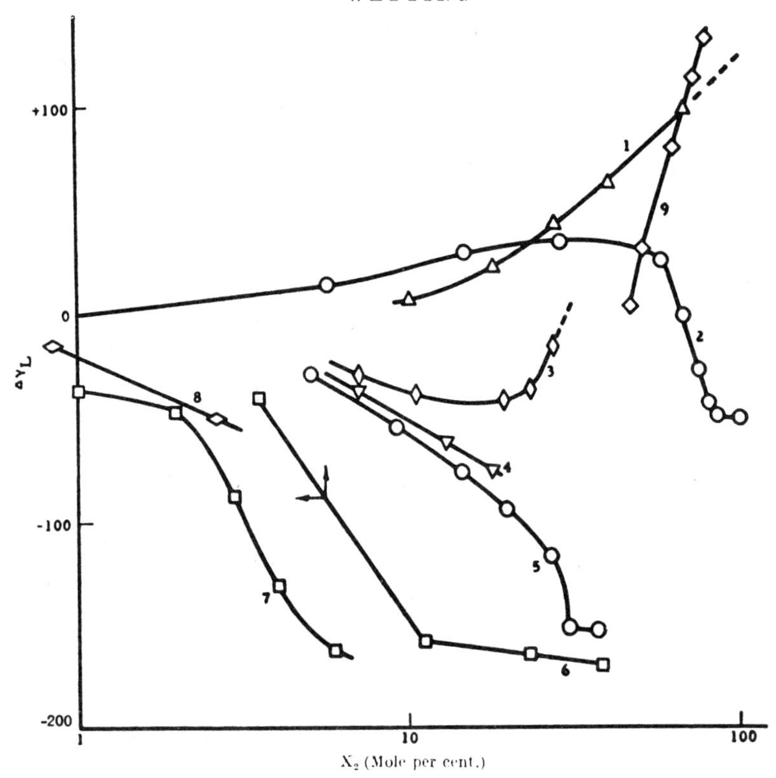

Figure 10.27. Change in surface-tension ($\Delta\gamma_L$) of fused metal oxides by addition of various other metal oxides (X_2 = mole per cent of solute) (45).

Curve	System	temperature °C
1	SiO_2 in PbO	1000
2	B_2O_3 in PbO	900
3	CaO in FeO	1412
4	TiO_2 in FeO	1410
5	SiO_2 in FeO	1420
6	P_2O_5 in FeO	1480
7	Na_2O in FeO	1410
8	MoO_3 in glass	1380
9	ZnO in B_2O_3	1300

spreading coefficients. The salts spread on the liquid metals with which they have been paired in the table. Heavy metal salts tend to give high spreading coefficients. They also promote emulsification of metal into a glass melt. Potassium chloride greatly increases the interfacial tension between metals and their chlorides.

The interfacial tensions of liquid metals against organic compounds are much larger, and the spreading coefficients smaller, then for the fused inorganic compounds in contact with the liquid metals.

Since metal derivatives generally have a lower surface tension than the

TABLE 10.24. INTERFACIAL TENSION AND INITIAL SPREADING COEFFICIENTS OF FUSED SALTS ON LIQUID METALS (45).

System	Temperature (°C)	Surface Tension of Liquid Metal, γ_a	Surface Tension of Fused Salt, γ_b	Interfacial Tension, γ_{ab}	Spreading Coefficient, $S_{b/a}$
$Cu-Cu_2S$	1131	1120		90	
$Cu-Cu_2S$	1215			76	
$Sn-SnCl_2$	275	615	100	342	173
$Sn-(ZnCl_2+NH_4Cl)$	250	616		420	
$Pb-PbCl_2$	510	430		190	
$Pb-PbCl_2$	600	423		170	
$Pb-(PbCl_2+KCl)$	460	434	105	228	101
$Pb-(PbCl_2+KCl)$	600	423	95	208	120
$Cd-CdCl_2$	605	585	77	4	504
$Cd-CdCl_2$	725	573	75	3.9	494
$Cd-(0.32\ CdCl_2+0.68\ KCl)$	605	585	81	131	373
$Cd-(0.32\ CdCl_2+0.68\ KCl)$	725	573	71	125	377

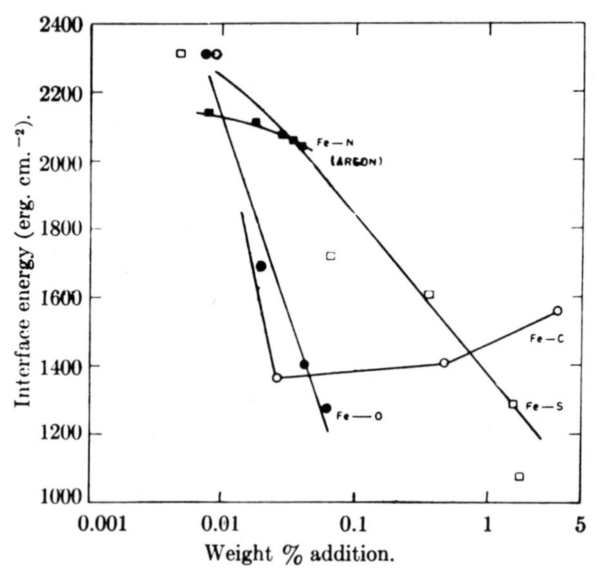

Figure 10.28. Effect of carbon, nitrogen, sulfur, and oxygen on the interfacial tension between liquid iron and aluminum oxide (61).

metal itself, it is not surprising that reactive gases reduce the surface tension of metals. Even inert gases will lower the surface tension, due to physical adsorption. Data plotted in Figure 10.28 show that iron alloys of carbon, sulfur, oxygen and nitrogen lower the interfacial tension between liquid iron and aluminium oxide (61).

Wetting of Solid Surfaces

Reliable data on the wetting of solids by liquid metals are difficult to obtain, because of contamination by oxide films at the high temperatures of measurement. While the presence of an oxide layer lowers the surface tension of the metal, it does not assure permanent non-wetting. Many liquid metals will in time lift the oxide layer off the base metal either by diffusion through it, or by migration starting from pin holes in the oxide layer.

Wherever the liquid metal and the solid substrate form intermetallic compounds or solid solutions, initial spreading invariably occurs. Sometimes

TABLE 10.25. EXAMPLES OF WETTING BY LIQUID METALS IN A HYDROGEN ATMOSPHERE. IN ALL INSTANCES CONTACT ANGLE $\theta = 0$, EXCEPT * WHERE WETTING IS NOT CLEARLY DEFINED (45).

Solid Substrate	Liquid	Temperature (°C)	Nature of Interaction
γ-Fe	Al	700	Compound
γ-Fe	Sn*	920–1000	Compound
α-Fe	Bi*	600	None
α-Fe	Zn	500	Compound
Cu	Ag	850	Solid solution
Cu	Cd*	350	Compound
Cu	Sn	400	Compound
Cu	Te	400	Compound
Cu	Zn	300	Solid solution
Ni	Ag	1000	Solid solution
Ni	Bi	310–470	Compound
Ni	Cd	400	Compound
Ni	Pb	358–700	Solid solution
Ni	Sb	700	Compound
Ni	Sn	358	Compound
Ag	Cd	400	Solid solution
Ag	Sb	550	Solid solution
Ag	Sn	300	Solid solution
Ag	Zn	500	Solid solution
Au	Ag	1000	Solid solution
Au	Cd	350	Compound
Au	Pb	400	Compound
Au	Sn	275	Compound
Au	Zn	450	Compound

only a specific intermetallic compound formed at the interface is easily wetted. Thus, tin-lead alloy gives a stable coating on copper at temperatures below 380°C, the decomposition temperature of the copper-tin-η-phase. Above that temperature, the coating is unstable and breaks up into individual droplets (62). For systems that form intermetallic compounds, wetting at one temperature does not provide assurance that wetting will also be observed at a higher temperature. Data on wetting by liquid metals are shown in Tables 10.25 and 10.26.

The addition of a second component to a liquid metal can facilitate wetting by reducing the surface tension of the liquid metal or by lowering the interfacial tension against the solid, or by doing both. The addition of lead to tin reduces its surface tension. Nickel forms solid solutions with both iron and copper, and it is wetted by lead and tin. The addition of as little as 0.1 per cent of nickel to lead enables lead to form stable coatings on copper and steel. Similarly, it assists the wetting of tin on copper (62). In general, an additive that will form an intermetallic compound with the substrate and which is wetted by the liquid metal will promote wetting of the substrate by the liquid metal. Data on intermetallic compounds have been compiled by Pauling (63).

Spreading is sometimes promoted by immersing the liquid metal and the base metal under another liquid. Where an improvement in wetting occurs, it is generally due to a chemical or electrochemical effect. Thus, many organic liquids can reduce the metal oxide at elevated temperatures. Aqueous alkaline solutions will dissolve the oxide layer on tin and promote spreading by mercury (64). Admixture of suitable heavy metal salts in a flux can result in electrolytic deposit of the heavy metal on the base metal (62).

TABLE 10.26. EXAMPLES OF NON-WETTING BY LIQUID METALS (45).

Solid Substrate	Liquid	Temperature, °C	Atmosphere	Nature of Interaction
α-Fe	Bi	400	Hydrogen	None
α-Fe	Cd	400	Hydrogen	None
α-Fe	Pb	400	Hydrogen	None
Cu	Bi	350	Hydrogen	None
Cu	Pb	350	Hydrogen	None
Ni	Cu	>1100	Vacuum	None
Ni	Pb	<358	Hydrogen	
Ni	Sn	<340	Hydrogen	
W	Ag		Vacuum	None
Mo	Bi	400	Hydrogen	
Mo	Pb	400	Hydrogen	

References

1. Adam, N. K., "The Physics and Chemistry of Surfaces", London, Oxford Univ. Press, London, 1941.
2. Fowkes, F. M., and Harkins, W. D., *J. Am. Chem. Soc.* **62**, 3377 (1940).
3. Harkins, W. D., "The Physical Chemistry of Surface Films", New York, Reinhold Publishing Corp., 1952.
4. Poynting and Thomson, 'Properties of matter', Text Book of Physics, p. 156, London, 1905.
5. Padday, J. F., Proc. Intern. Congr. Surface Activity, 2nd, Vol. 3, London, Butterworth, 1957.
6. Bartell, F. E., and Zuidema, H. H., *J. Am. Chem. Soc.* **58**, 1449 (1936).
7. Ray, B. R., and Bartell, F. E., *J. Colloid Sci.* **8**, 214 (1953).
8. Dupré, Théorie Mécanizue de la Chaleur, 369 (1869).
9. Johnson, R. E., Jr., *J. Phys. Chem.* **63**, 1655 (1959).
10. Bikerman, J. J., Proc. Intern. Congr. Surface Activity, 2nd, Vol. 3, London, Butterworth, 1957.
11. Pethica, B. A., and Pethica, T. J. P., Proc. Intern. Congr. Surface Activity, Vol. 3, London, Butterworth, 1957.
12. Fox. H. W., and Zisman, W. A., *J. Colloid Sci.* **7**, 109 (1952).
13. Boyd, G. E., and Livingston, H. K., *J. Am. Chem. Soc.*, **64**, 2383 (1942).
14. Fox, H. W., and Zisman, W. A., *J. Colloid Sci.* **7**, 428 (1952).
15. Ellison, A. H., and Zisman, W. A., *J. Phys. Chem.* **58**, 503 (1954).
16. Ellison, A. H., Fox, H. W., and Zisman, W. A., Ibid **57**, 622 (1953).
17. Miller, R., and Willis, H., *Trans. Faraday Soc.* **49**, 433 (1953).
18. Fox, H. W., Hare, E. F., and Zisman, W. A., *J. Colloid Sci.* **8**, 194 (1953).
19. Bartell, L. S., and Ruch, R. J., *J. Phys. Chem.* **63**, 1045 (1959).
20. Bernett, M. K., and Zisman, W. A., Ibid **63**, 1241 (1959).
21. Bernett, M. K., and Zisman, W. A., Ibid **63**. 1911 (1959).
22. Harkins, W. D., and Feldman, A., *J. Am. Chem. Soc.* **44**, 2665 (1922).
23. Harkins, W. D. in Alexander's "Colloid Chemistry", Vol. 6, New York, Reinhold Publishing Corp., 1946.
24. Fox, H. W., Hare, E. F., and Zisman, W. A., *J. Phys. Chem.* **59**, 1097 (1955).
25. Hare, E. F., and Zisman, W. A., Ibid **59**, 335 (1955).
26. Brophy, J. E., and Zisman, W. A., *Ann. N. Y. Acad. Sci.*, **53**, 836 (1951).
27. Jarvis, N. L., and Zisman, W. A., *J. Phys. Chem.* **63**, 727 (1959).
28. Jarvis, N. L., and Zisman, W. A., *J. Phys. Chem.* **64**, 150 (1960).
29. Jarvis, N. L., and Zisman, W. A., *J. Phys. Chem.* **64**, 157 (1960).
30. Fowkes, F. M., and Harkins, W. D., *J. Am. Chem. Soc.* **62**, 3377 (1940).
31. Bartell, F. E., and Smith, J. T., *J. Phys. Chem.* **57**, 165 (1953).
32. Bartell, F. E., and Cardwell, P. H., *J. Am. Chem. Soc.* **64**, 494 (1942).
33. Bartell, F. E., and Shepard, J. W., *J. Phys. Chem.* **57**, 211 (1953).
34. Bartell, F. E., and Shepard, J. W., Ibid **57**, 455 (1953).
35. Bartell, F. E., and Shepard, J. W., Ibid **57**, 458 (1953).
36. Wenzel, R. N., *Ind. Eng. Chem.* **28**, 988 (1936).
37. Wenzel, R. N., *J. Phys. Chem.* **53**, 1466 (1949).
38. Moilliet, J. L., and Collie, B., "Surface Activity", New York, D. Van Nostrand Co., 1951.
39. Ray, B. R., and Bartell, F. E., *J. Phys. Chem.* **57**, 49 (1953).
40. Washburn, E. W., *Phys. Rev.* **17**, 273 (1921).

41. Minor, F. W., Schwartz, A. M., Buckles, L. C., and Wulkow, E. A., *Am. Dyestuff Reptr.* **49**, 419 (1960).
42. Fowkes, F. M., *J. Phys. Chem.* **57**, 98 (1953).
43. Draves, C. Z., and Clarkson, R. G., *Am. Dyestuff Reptr.* **20**, 201 (1931); also, Year Book of Am. Assoc. Textile Chemists and Colorists.
44. Edelstein, S. M., and Draves, C. Z., *Am. Dyestuff Reptr.* **38**, 343 (1949).
45. Bondi, A., *Chem. Revs.* **52**, 417 (1953).
46. Fricke, R., *Kolloid-Z.* **96**, 213 (1941).
47. Fricke, R., In "Handbuch der Katalyse", Vol. 6, p. 108. Vienna, J. Springer, 1943.
48. Jura. G., *J. Chem. Phys.* **12**, 1335 (1949).
49. Hüttig, G. F., *Kolloid-Z.* **124**, 160 (1951).
50. Udin, H., In "Metal Interfaces", p. 114 American Society for Metals, Cleveland, Ohio (1952).
51. Udin, H., Shaler, A. J., and Wulff, J., *J. Metals* **1**, No. 2, Trans. 186 (1949).
52. Udin, H., *et al.*, *J. Metals* **3**, 1206, 1209 (1951).
53. Sauerwald, F., Schmidt, B., and Pelka, F., *Z. anorg. Chem.* **223**, 84 (1935).
54. Becker, G., Hardus, F., and Kornfeld, H., *Arch. Eisenhuttenw* **20**, 363 (1949).
55. Kingery, W. D., and Humenik, M., Jr., *J. Phys. Chem.* **57**, 359 (1953).
56. Davis, J. K., and Bartell, F. E., *Anal. Chem.* **20**, 1182 (1948).
57. Chalmers, B., King, R., and Shuttleworth, R., *Proc. Roy. Soc.* (London) **A193**, 465 (1948).
58. Hüttig, G. F., "Handbuch der Katalyse", Vol. 6, p. 390, Vienna, J. Springer, 1943.
59. Atterton, D. V., and Hoar, T. P., *Nature* **167**, 602 (1951).
60. Weyl, W. A., *Trans. Soc. Glass Technol.* **32**, 247 (1948).
61. Halden, F. A., and Kingery, W. D., *J. Phys. Chem.* **59**, 557 (1955).
62. Bailey, G. L. J., and Watkins, H. C., *J. Inst. Metals* **80** (2), 57 (1951/2).
63. Pauling, L., and Ewing, F. J., *Rev. Modern Phys.* **20**, 112 (1948).
64. Tammann, G., and Arntz, F., *Z. anorg. allgem. Chem.* **192**, 45 (1930).

CHAPTER 11

Emulsions

When two immiscible liquids are combined and shaken, one of the liquids takes the form of droplets dispersed in the other liquid. This is called an emulsion. Once agitation has ceased, the dispersed droplets begin to rise or fall, depending upon whether the density of the discontinuous or internal phase is less or more than that of the continuous phase. This change in the distribution of globules throughout the external or continuous phase is termed sedimentation or creaming. Some authors refer to sedimentation as the downward movement of the globules and creaming as the upward movement (1). The terms may be used interchangeably, since the phenomena are without basic difference. When two globules collide, they will frequently coalesce to form one larger globule. In time, coalescence will be complete and the dispersed phase will no longer exist as such. It will become a continuous phase separated from the other by a single interface.

If an emulsifying agent is added to the two immiscible liquids, which are then shaken, one liquid will again become dispersed in the other. However, there will be a number of differences. The droplets will be smaller, and, consequently, they will sediment more slowly. Collisions will result less frequently in coalescence. Given sufficient time, two continuous phases separated by a single interface will again result. Depending upon the system, sufficient time may mean anything from seconds to years.

An emulsion is a system consisting of one or more immiscible liquids dispersed in another. Generally, there are only two immiscible liquids. The presence of at least two liquid phases in an emulsion distinguishes it from a solution or a solubilized system. Practically all emulsions are thermodynamically unstable. However, this is not an essential requirement of emulsions. Sedimentation, which is one form of instability, can be prevented by so adjusting the densities of the two phases that they are identical. The interfacial free-energy always has a positive value. Since coalescence results in a decrease in interfacial area and, consequently, in a decrease in the interfacial free-energy of the system, emulsion droplets will tend to coalesce. Harkins (2) has observed interfacial tensions less than 0.01 dyne/cm with paraffin oil emulsified in sodium oleate solution, when the soap was formed *in situ.* With olive oil in place of the paraffin oil, the interfacial tension fell to 0.002 dyne/cm. With such low interfacial tensions, emulsification occurs spontaneously. Thermodynamic calculations made on the basis of free energy changes show that in some instances where the interfacial tension is very low the emulsion may be stable (2).

Practically all of the emulsions that have been studied or that are of industrial importance contain water. It is common practice to describe emulsions in accordance with the way in which the aqueous phase is distributed throughout the emulsion. If water constitutes the continuous phase, the emulsion is called an oil-in-water emulsion, or O/W. The term oil is applied to any water-immiscible liquid phase, regardless of its characteristics. Conversely, if the aqueous phase is the internal or dispersed phase, the system is called a water-in-oil emulsion, or W/O.

There are also more complex systems. For example, oil may be dispersed in an aqueous phase and each of the oil globules may contain a number of tiny water droplets. Similarly, water-free emulsions are sometimes encountered. Two immiscible organic liquids can form emulsions. While mercury has been emulsified in water, it hardly appears appropriate to refer to mercury as the "oil" phase.

Frequently, if the volume of one of the immiscible liquids is very much smaller than the other, the liquid occupying the smaller volume will become the dispersed phase. However, there are numerous exceptions. Emulsions have been reported in which the internal phase has constituted as much as 99 per cent of the entire volume (3).

A number of simple tests are available for determining whether an emulsion is of the O/W or W/O type. These are based on the fact that many of the properties of the emulsion are more like those of the continuous phase than the internal phase. A few of these tests follow:

(a) *Dye test.* Select two dyes, one soluble only in the oil phase and the other only in the water phase, and sprinkle a few grains of each on different portions of the emulsion. The dye that is soluble in the continuous phase will impart a uniform color to the emulsion. The other dye will produce discontinuous spots of color.

(b) *Dilution test.* Place small portions of the emulsion in contact with water and with a solvent for the oil phase. The emulsion will disperse more readily in the liquid that is miscible with the continuous phase.

(c) *Electrical conductivity.* If the oil phase is continuous, the emulsion will be nonconducting. Otherwise, the electrical conductivity will approximate that of the aqueous phase.

A large variety of different types of materials are used to produce stable emulsions. These include water-soluble and oil-soluble soaps and synthetic surfactants of the nonionic, anionic, and cationic types. Many naturally-occurring emulsifying agents, such as the phospholipids are used. Both natural and synthetic gums and thickening agents serve as emulsifiers. Finely divided solids, which are partially wetted by both phases, are also effective. Hydrosols, extremely fine dispersions of small quantities of oil in water, are stabilized by traces of electrolytes (4).

PHYSICAL PROPERTIES OF EMULSIONS

Only the more important of the physical properties of emulsions, that determine their characteristics and usefulness for intended applications, will be considered. These are particle size and size distribution, optical and electrical properties and viscosity. Emulsion stability will be discussed separately.

Particle Size and Size Distribution

Many authorities place the lower size limit of emulsion droplets at the smallest size that can be seen in an ordinary light microscope, about 0.1 micron. The largest size of an emulsified droplet is sometimes placed at the limit of visibility of the naked eye, about 50 microns. This arbitrary assignment of particle-size range for emulsions is one of convenience. Thus, the particle-size distribution of an emulsion can be described within the limitations of a single optical system. For most commercially important emulsions, the particle size falls predominantly in the range of 0.1 to 10 microns. However, transparent emulsions containing droplets of the order of 0.01 micron in diameter are known and are used commercially (6). There isn't any fundamental basis for fixing an upper or lower limit to the size of emulsified droplets.

Emulsions are ordinarily not homodispersed, though a method has been devised for preparing latices by emulsion polymerization where the diameters of the globules fall predominantly within the very narrow range of 0.05 to 0.09 micron (7). With most emulsions, the particle-size distribution is Gaussian. The change in the size-distribution curve with time, leading to a more diffuse distribution in the direction of larger particles, is a measure of the stability of the emulsion. Typical size-distribution curves are shown in Figure 11.1 and 11.2.

The Brownian motion of the droplets is also dependent on their size. Since the droplets must be in kinetic equilibrium with the molecules of the medium (9),

$$(\tfrac{1}{2}m\overline{v^2}) \text{ particle} = (\tfrac{1}{2}m\overline{v^2}) \text{ molec. medium} = \frac{3}{2}\frac{RT}{N} \tag{11.1}$$

where m is the mass of the particle or of a molecule of the medium and $\overline{v^2}$ is the mean of the square of its velocity. However, the particle varies in

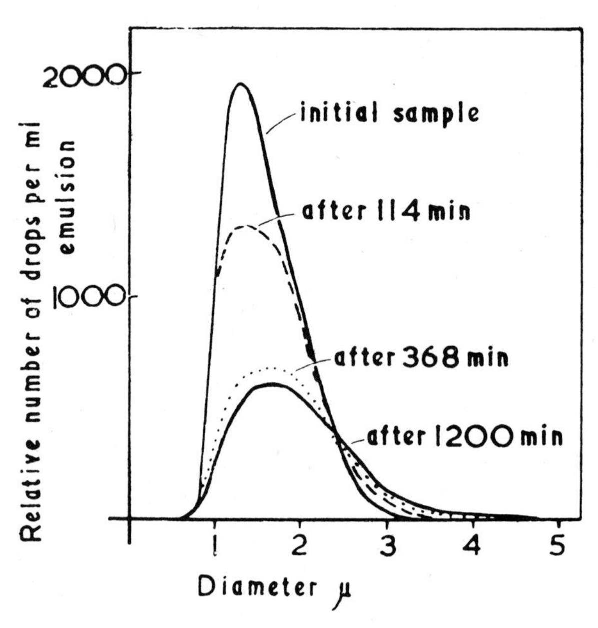

Figure 11.1. Variation of size-frequency distribution with time for unstabilized emulsions (8).

direction many million times per second, and only the displacement as the result of a long zigzag path can be seen and measured. Both Einstein (10) and von Smoluchowski (11) deduced the following equation relating velocity to displacement:

$$\overline{\Delta X^2} = 2t \cdot \frac{RT}{N} \cdot \frac{1}{6\pi\eta r} \tag{11.2}$$

where $\overline{\Delta X^2}$ is the mean of the square of the displacement during the time t projected on a chosen X direction, η is the viscosity coefficient of the system, and r is the radius of the particle, supposed spherical.

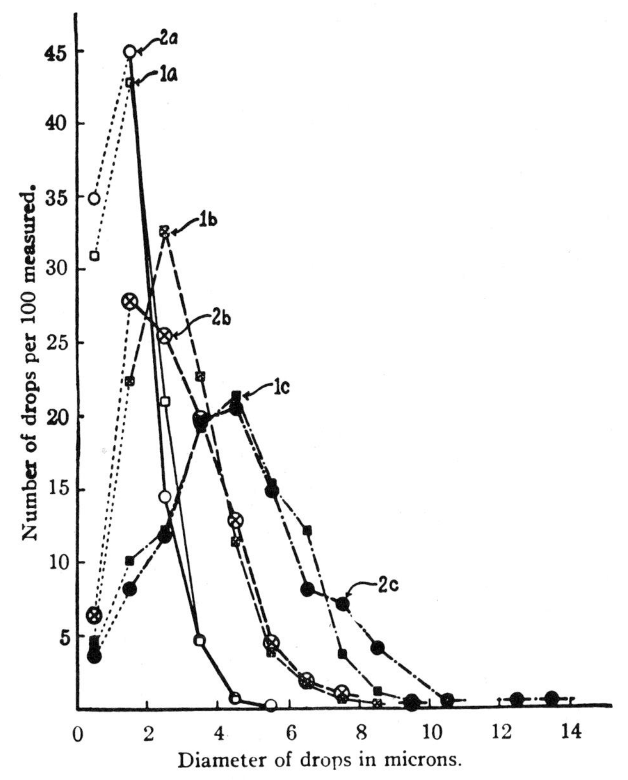

Figure 11.2. Change in size distribution with aging. 0.005*M* sodium oleate and octane: 1*a*, measured on first day; 1*b*, measured on third day; 1*c*, measured on seventh day. 0.005*M* cesium oleate: 2*a*, measured on first day; 2*b*, measured on third day; 2*c*, measured on seventh day (2).

Optical Properties

Disregarding the effect of dyes and pigments, the appearance of an emulsion is determined by the particle size of the dispersed droplets and the difference in the refractive indices of the two phases. If the refractive indices are the same, the emulsion is clear or transparent regardless of particle size. If different, the emulsion increases in opacity with a decrease in particle size, the maximum opacity occurring when the particle diameter is about one micron. A further decrease in particle size results in increasing transparency. The effect of particle size on appearance is shown in Table 11.1.

The laws of Lambert and Beer are applicable to emulsion systems (13):

$$I_x = I_0 e^{-Kcl} \quad (11.3)$$

where I_0 is the intensity of a parallel, monochromatic beam of light entering the emulsion, I_x is the intensity of the light beam at its exit, l is the thickness of the emulsion layer, K is a proportionality constant, and c is the concentration of particles. The equation is only applicable when the particle-size distribution is the same throughout the length of the emulsion, and the concentration is not so high that the results are affected by multiple-scattered light.

TABLE 11.1. EFFECT OF PARTICLE SIZE ON THE APPEARANCE OF EMULSIONS (12).

Particle size	Appearance
Macro globules	Two phases may be distinguished
Greater than 1 μ	Milky white emulsion
1 to approx. 0.1 μ	Blue-white emulsion
0.1 to 0.05 μ	Gray semitransparent
0.05 μ and smaller	Transparent

When a beam of light passes through an emulsion, light is scattered in a sideways direction at the boundary of the dispersed particles. If the beam of light is monochromatic, the scattered light is also monochromatic and of the same wave length. This is called the Tyndall effect. It frequently serves as a means of distinguishing emulsions from homogeneous fluids which do not exhibit this effect, except to an extremely small extent. The Tyndall light is polarized, even when the emulsion is illuminated with ordinary light. Further, if polarized light is used, the direction of the vibrations is rotated 90° (14).

The relationship between particle size and the intensity of scattered light is extremely complex. The equation of Rayleigh, for example, applies only for spherical particles which are not conductors and which are extremely small as compared with the wave length of the incident light. Consequently, it is not applicable to most emulsion systems. According to this theory:

$$\frac{I_s}{I_0} = 24\pi^3 \cdot \frac{V^2}{\lambda^4} c \left(\frac{n_1^2 - n_0^2}{n_1^2 + 2n_0^2}\right)^2 \tag{11.4}$$

where: I_0 is the intensity of the incident light; I_s is the total scattered light; V is the volume of one particle; λ is the wave length of the scattered light; c is the number of particles per cm^3; n_0 is the refractive index of the medium; n_1 is the refractive index of the particle material.

The intensity of scattered light increases with both particle concentration and particle size. With still larger particles, there is an increasing number of maxima and minima in scattered light intensity (14).

The theory of light scattering by large spherical particles has been discussed by Gumprecht and Sliepcevich (15). Based on the development of the Mie theory (16), the intensity of a parallel beam of light is reduced in traversing a dispersion of uniform, spherical particles as follows:

$$\frac{I_x}{I_0} = \mathbf{e}^{\frac{K\pi D^2 cl}{4}} \tag{11.5}$$

where: D is the diameter of the uniform particles; I_0 is the intensity of the incident, parallel beam; I_x is the intensity of the beam after traversing the medium; l is the distance traversed through the medium; c is the particle concentration; K is the scattering coefficient.

The scattering coefficient K is a function of m, the index of refraction of the droplet relative to that of the surrounding medium, and α which equals $\pi D/\lambda$. Values of K have been calculated and are available for several values of m and for values of α ranging from less than 1 up to 400 (17). This covers the usual emulsion range. The variation of K with α is shown in Figure 11.3. Here the refractive index of the droplets is 1.50 and that of the medium is 1.00.

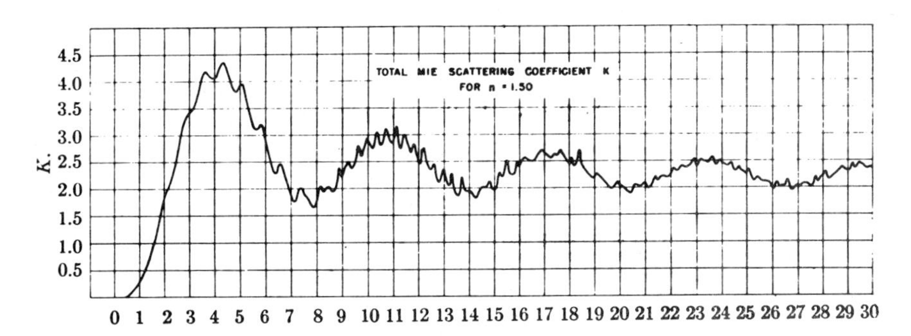

Figure 11.3. Total Mie scattering coefficient K for $n = 1.50$ as function of the size parameter a. The results are obtained using the exact Mie theory. K shows 5 major oscillations, the amplitude of which decreases with a. Superimposed on these major oscillations are ripples first visible at $a \sim 2$. A second set of ripples begins at $a \sim 7.5$ (18).

An equation has also been developed by which the particle-size distribution of emulsified droplets can be determined from the transmission equation and Stokes' law,

$$D = k / \sqrt{t} \tag{11.6}$$

where D is the diameter of the largest particle present in the light beam after the elapsed time t from the start of settling; and

$$k = \sqrt{\frac{18\,h\eta}{g\,(\rho_1 - \rho_2)}} \tag{11.7}$$

where: h is the settling height; η is the viscosity of the medium; g is the acceleration of gravity; $(\rho_1 - \rho_2)$ is the difference in density of the dispersed particles and the surrounding medium.

$$RKN = \frac{8}{\pi l k^3} \cdot \frac{1}{I_x} \cdot \frac{dI_x}{dt} \cdot t^{5/2} \tag{11.8}$$

R is a proportionality factor relating the total light scattering K to the apparent scattering. It is very close to one for small values of α and small angles of diffracted light. N is defined as the number of particles of average diameter D per unit volume of the dispersion per unit range of particle diameters.

If a curve is plotted of I_x *versus* t, the quantity RKN can be calculated from the value of I_x and the slope of the curve at a given time t. The value of D can be calculated from Stokes' law at the same given time t. Therefore, RKN can be obtained as a function of D. Since R and K are known functions of D, N can be calculated as a function of D to obtain a size-frequency

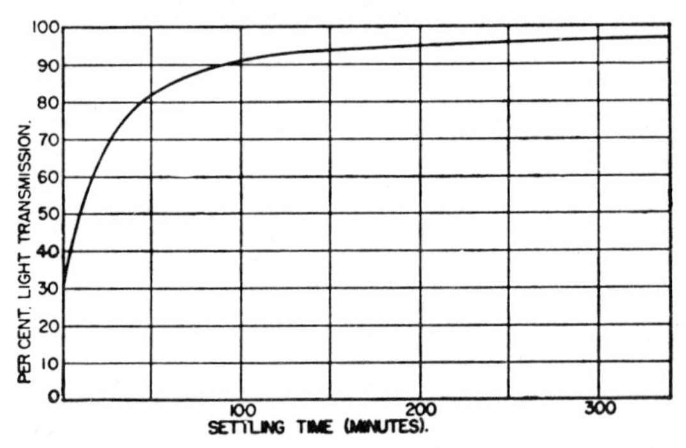

Figure 11.4. Per cent light transmission *vs.* settling time for kerosene fog (15).

distribution curve. A curve of light transmission *versus* time and the calculated size-frequency distribution curve for a kerosene fog are shown in Figure 11.4 and 11.5.

Electrical Properties

The dielectric constant of an emulsion can be expressed as follows (19):

$$\bar{D} = D_0\,[1 + 3\phi\,(D_1 - D_0)\,/\,(D_1 + 2D_0)] \tag{11.9}$$

where $\bar{D}$, D_0 and D_1 refer to the macroscopic dielectric constant of the emulsion, the continuous phase, and the globules, respectively, and ϕ is the volume fraction of the dispersed phase. This, and other expressions that have been developed, agree with experimental results to within a few per cent for values of ϕ up to about 0.30.

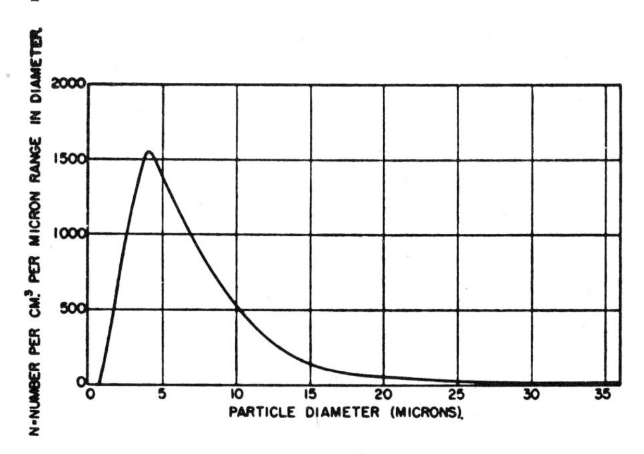

Figure 11.5. Size-frequency distribution curve for kerosene sprayed at 700 p.s.i. (15).

Emulsions in which water is the continuous phase show a higher conductivity than those in which the oil phase is continuous. The conductivity of emulsions is approximately that of the continuous phase. It has been found that the conductivity of a petroleum emulsion containing 50 per cent of water as the internal phase had a conductivity two or three times higher than the dry petroleum (20). Conductivity increased several fold with an increase in temperature.

Emulsified droplets frequently possess a definite electric charge. As a consequence, when the emulsion is placed in an electrical field, the droplets will be observed to move toward the electrode carrying a charge opposite to that of the droplets. This migration is called electrophoresis. Data are usually reported in terms of mobility, $u = v/E$, the velocity v in microns per second under a potential E of one volt per cm.

The potential of the moving particle, arising from the electric charge, is called the zeta potential ζ. The relationship between the electrophoretic mobility and the zeta potential, for a nonconducting droplet, is given by Henry's equation (21):

$$u = \frac{v}{E} = \frac{D\zeta}{6\pi\eta} f(\kappa r) \qquad (11.10)$$

where η is the coefficient of viscosity of the continuous phase, D is the

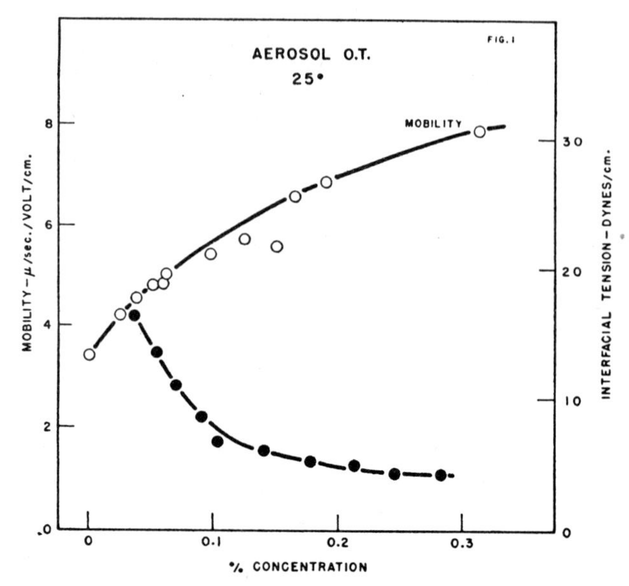

Figure 11.6. Mobility of Nujol droplets in solutions of Aerosol OT. Interfacial tension of aqueous solutions of Aerosol OT against Nujol (22).

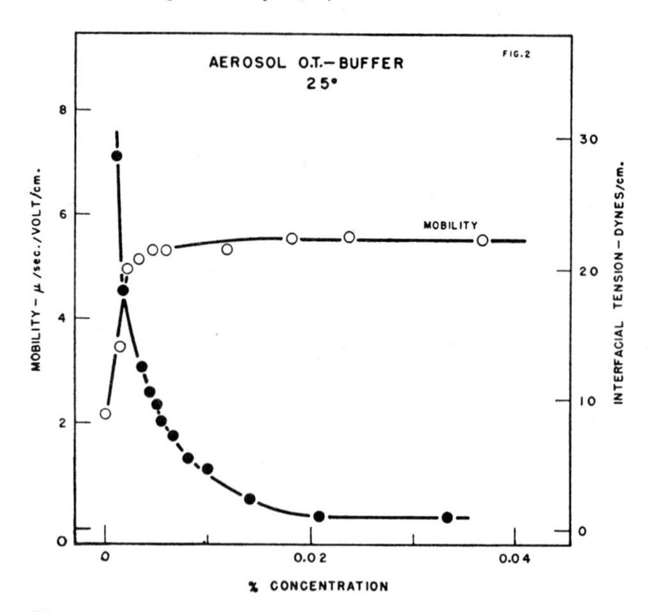

Figure 11.7. Mobility of Nujol droplets in solutions of Aerosol OT. Interfacial tension of aqueous solutions of Aerosol OT against Nujol. Solutions made up in 0.05 *M* acetate buffer, pH 6 (22).

dielectric constant of the continuous phase, r is the droplet radius, and κ is the Debye-Huckel function. The value of the Henry function $f(\kappa r)$ approaches 1.5 when $\kappa r \gg 1$, and it is 1.0 when $\kappa r \ll 1$. For most emulsions, where the electrolyte concentration is low, about $10^{-4}M$ or less, and the droplet radius is about 1 micron, $f(\kappa r) = 1.5$. Then $u = \dfrac{D\zeta}{4\pi\eta}$.

Figures 11.6 and 11.7 show electrophoretic mobility measurements of paraffin oil droplets in solutions of Aerosol OT. Interfacial-tension values are shown for comparison. The effects produced by inorganic electrolyte are clearly evident. Curves based on the electrophoretic mobility of carbon black in benzene are given in Figure 11.8. The suspending agents were the calcium salt of diisopropylsalicylic acid (Ca dips) and tetraisoamylammonium picrate (tiap).

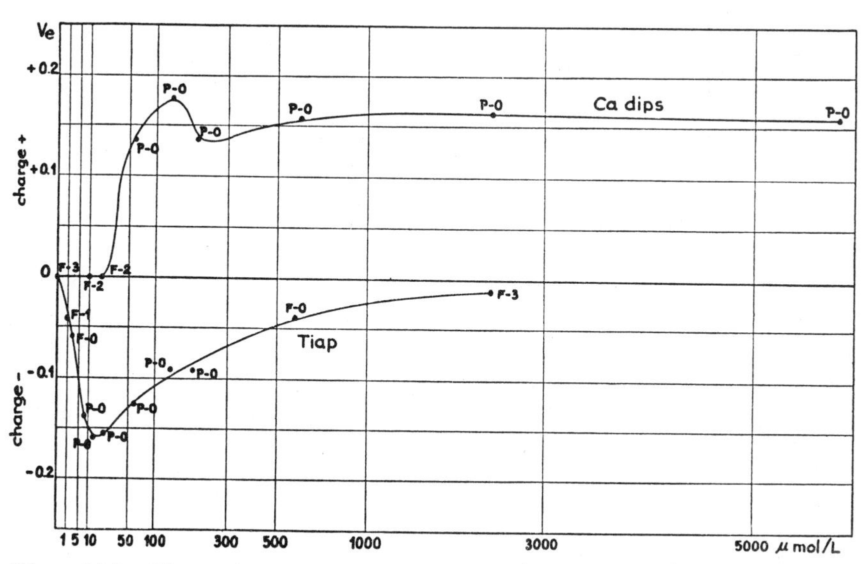

Figure 11.8. Electrophoretic velocity v as a function of the concentration of calcium diisopropylsalicylate and tetraisoamylammonium picrate (23).

Rheology

Emulsions exhibit various types of flow properties. Frequently, these viscosity characteristics determine the ultimate usefulness of the emulsion. The simplest type of viscous behavior for liquids is referred to as Newtonian flow. For such Newtonian liquids, the rate of shear is directly proportional to the shearing stress, as shown in Figure 11.9,

$$f = \eta s \tag{11.11}$$

where f is the applied force or shearing stress; s is the rate of shear; η is the coefficient of viscosity.

There are many non-Newtonian liquids and the stress-shear diagrams can exhibit various shapes. Figure 11.10 is characterized by the fact that the curve goes through the origin, but it is not a straight line. The viscosity f/s decreases as the stress is increased. In Figure 11.11, the shape of the curve

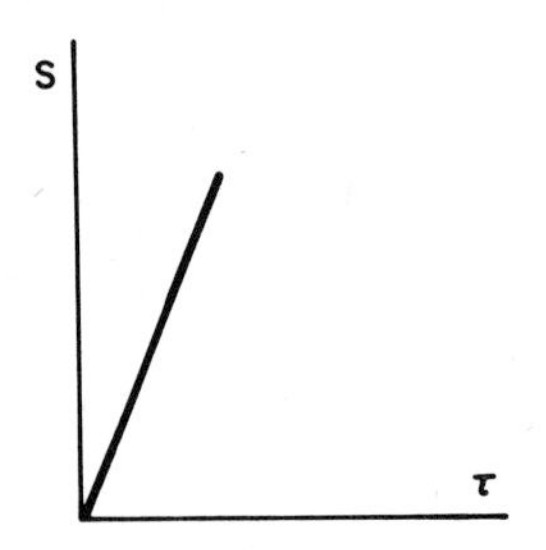

Figure 11.9. Relation between rate of shear, S, and shearing stress, for a Newtonian liquid (24).

is similar, except that it does not pass through the origin. The emulsion does not begin to flow until the shearing stress has reached the value f_1, which is called the yield value. This is called a plastic or pseudoplastic system.

Another type of flow, called dilatancy, is shown in Figure 11.12. Above the value f_d, the viscosity increases with an increase in the shearing stress. This flow behavior is sometimes observed with suspensions. Finally, in many dispersions, viscosity changes with time as well as with the shearing stress. An emulsion may be fluid and exhibit Newtonian flow. After standing undisturbed, it may become plastic, exhibiting a yield value. This variation of viscosity with time is called thixotropy.

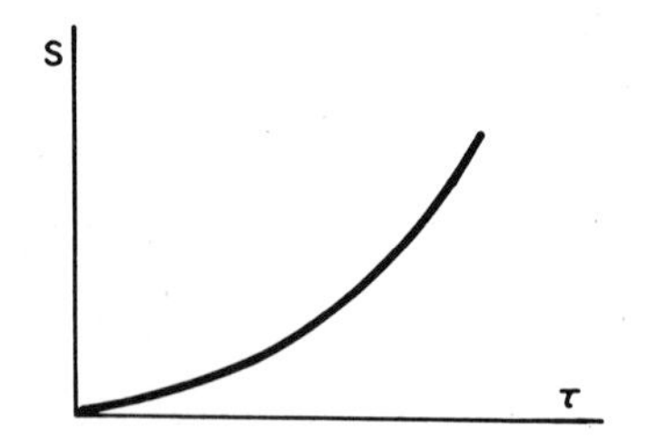

Figure 11.10. Non-Newtonian liquid. Differential viscosity decreasing at increasing shearing stress (24).

All of these flow characteristics can be exhibited by emulsions. If an aqueous solution of an emulsifying agent exhibits Newtonian flow, the dispersion of a small amount of oil in the solution will not change the flow character, though the viscosity will increase. Colloidal dispersions of many emulsifiers, such as stearate soaps in water, exhibit non-Newtonian flow. Where these fluids constitute the continuous phase of an emulsion, the flow is also non-Newtonian.

Plasticity is the result of the agglomeration of dispersed particles. A yield value is necessary to separate particles to cause flow. In the case of thixotropy, the adhesion of particles occurs slowly. Moreover, the adhesion

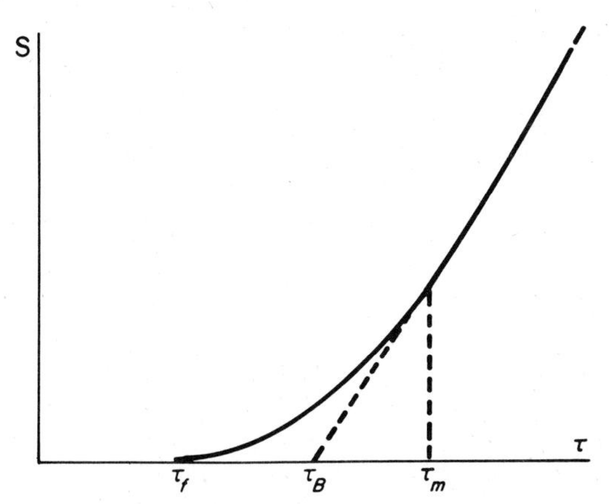

Figure 11.11. Plastic system τ_f = First yield value.
τ_B = Bingham yield value.
τ_m = maximum yield value.

is so weak that it can be completely destroyed by shaking or stirring. With dilatancy there is no adhesion, but a mechanical hindrance of the particles to flow. On the application of a large shearing stress, the particles pack together.

The fundamental equation relating the viscosity of a suspension with

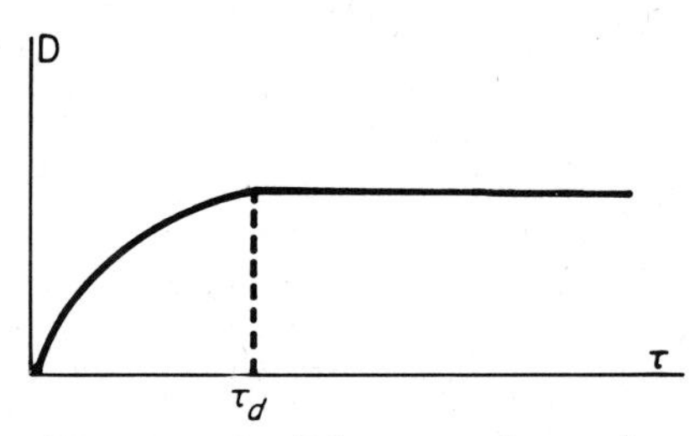

Figure 11.12. Dilatancy. Increasing shearing stress does not increase the rate of shear.

that of the suspending liquid and the volume fraction of dispersed rigid spheres is due to Einstein (25),

$$\eta = \eta_0 (1 + 2.5\phi) \tag{11.12}$$

where η is the viscosity of the suspension, η_0 is the viscosity of the external phase, and ϕ is the volume fraction of rigid spheres in the suspension. This is a limiting law, whose validity increases as ϕ approaches zero.

Innumerable modifications of the Einstein equation have been reported. Many are applicable for the systems studied. However, the following considerations will show the futility of attempting to derive a useful generalization.

(1) The addition of an emulsified phase can profoundly affect the viscosity of the external phase. The concentration of emulsifying agent will decrease due to adsorption on the globules and possibly by dissolution in the globules. Solubilization can affect the viscosity contribution of micelles. Partial solubility of the globules in the external phase can alter the solubility of the emulsifying agent in that phase.

(2) The actual volume fraction occupied by the globules will depend upon the miscibility of the two phases. Adsorbed surfactant as well as the electroviscous effect will increase the effective volume of the globules. Thus, van der Waarden (26) concluded that the dispersed-phase droplets, regardless of the actual droplet size, are surrounded by a rigid layer 30–35Å thick. This result was obtained with O/W emulsions stabilized by sodium naphthalene sulfonates. In the case of small droplets, 0.2 micron in diameter, this is a 120 per cent increase in volume. Richardson (27) found that viscosity at high rates of shear varies inversely as the mean globule size.

(3) The droplet-size distribution will influence the viscosity. Richardson (28) showed that more-uniform emulsions have a higher viscosity than those emulsions having a broad droplet-size range. This is also evidenced by the fact that the maximum volume that can be occupied by uniform spheres is 74 per cent of the total volume. Since smaller spheres can fill the voids between the larger spheres, the maximum volume occupied by the internal phase can approach 100 per cent if the droplets are nonuniform in size.

(4) If the droplets are readily distorted, the viscosity is lowered. This depends on the viscosity of the internal phase, the viscosity of the interfacial film, and the interfacial tension. If the globules readily assume an oblong shape, they offer less resistance to flow and they can more easily slide past each other (28).

(5) The degree of flocculation of globules will greatly influence viscosity. If the volume concentration of globules is low, flocculation will

produce the hydrodynamic effect of larger droplets with a larger total volume-fraction, the additional volume being contributed by a portion of the internal phase entrapped between agglomerated globules. At high concentrations, flocculation can immobilize the emulsion, resulting in plastic flow with a finite yield value. The shape of the viscosity curve will depend upon the extent of flocculation.

The information required to predict the viscosity of an emulsion is rarely if ever available. This information includes the effective volume of the internal phase, the droplet-size distribution, the effect of the composition of the dispersed phase on the viscosity of the external phase, the energy required for distorting the globules and the extent of flocculation of the globules.

In general, it can be said that the viscosity of an emulsion is increased if the volume fraction of the internal phase is increased, if the globule size is reduced, or if the globule size distribution is made more uniform. Increasing the surface potential in forming different emulsions can reduce viscosity by preventing flocculation. However, if the droplets are already dispersed, the viscosity will be increased by increase of the surface potential. Similarly, an increase in emulsifying-agent concentration can reduce viscosity by dispersing a flocculated system, or increase viscosity by increasing the effective volume of the globules and their rigidity, or by increasing the viscosity of either the internal or the external phase.

PREPARATION OF EMULSIONS

Various methods are employed for the preparation of emulsions. If the emulsifying agent is a soap, smaller particle-size globules are generally obtained if the soap is formed *in situ.* The fatty acids are dissolved in the oil phase and the alkaline materials are dissolved in the water phase before combining the two phases. With other types of emulsifiers, the lipophilic emulsifier is dissolved in the oil phase, as well as the hydrophilic emulsifier, if it is soluble in the oil. Otherwise it is dissolved in the water.

In the preparation of oil-in-water emulsions, the finest particle-size is ordinarily obtained by first forming and then inverting a water-in-oil emulsion. Thus, the aqueous phase is added slowly to the oil phase with agitation to form a water-in-oil emulsion. The viscosity of the emulsion will increase to a semi-solid consistency during the addition of the aqueous solution. On further slow addition the emulsion will suddenly thin out and become fluid. This corresponds to the point of inversion of the emulsion to the oil-in-water type. The remainder of the water phase can then be added rapidly.

If the volume of the oil phase is small as compared with that of the water

TABLE 11.2. TYPICAL SURFACTANTS USED IN THE PREPARATION OF EMULSIONS (3).

Chemical type	Hydrophile-lipophile balance[a]	Sensitive to
Anionic:		
Sodium stearate	HH	Acids, salts, cationic agents
Potassium laurate	HH	Acids, salts, cationic agents
Morpholine oleate	HH	Acids, salts, cationic agents
Sodium lauryl sulfate	HH	Cationic agents
Sodium 2-ethylhexyl sulfate	HH	Cationic agents
Sodium xylenesulfonate	HH	Cationic agents
Sodium naphthalenesulfonate	HH	Cationic agents
Sodium alkylnaphthalenesulfonate	HH	Cationic agents
Sodium sulfosuccinate	HH	Cationic agents
Sodium oleic acid sulfonate	H	Cationic agents
Sodium castor oil sulfonate	H	Cationic agents
Glycerol monostearate containing a sodium fatty alcohol sulfate	H	Cationic agents, hot alkali
Glycerol monostearate containing a soap	N–H	Acids, salts, cationic agents
Lithium stearate	L	Acids, salts, cationic agents
Magnesium oleate	LL	Acids, salts, cationic agents
Aluminum stearate	LL	Acids, salts, cationic agents
Nonionic:		
Polyoxyethylene fatty alcohol ethers	HH–H	—
Polyglycol fatty acid esters	HH–H	Hot alkali
Polyoxyethylene modified fatty acid esters	HH–L	Hot alkali
Polyoxyethylene-polyol fatty acid esters	HH–LL	Hot alkali
Polyoxypropylene fatty alcohol ethers	H–L	—
Polypropylene glycol fatty acid esters	H–L	Hot alkali
Polyoxypropylene modified fatty acid esters	H–L	Hot alkali
Polyoxypropylene-polyol fatty acid esters	H–L	Hot alkali
Polyol fatty acid monoesters	H–LL	Hot alkali
Lecithin	H–L	—
Polyhydric alcohol fatty acid di-, tri-, etc., esters	L–LL	Hot alkali
Cholesterol and fatty acid esters	L–LL	Hot alkali
Lanolin	L–LL	Hot alkali
Oxidized fatty oils	L–LL	Hot alkali
Cationic:		
Quaternary ammonium salts	HH	Alkalies, free fatty acids, anionic agents
Amine hydrochlorides	HH	Alkalies, free fatty acids, anionic agents

[a] HH = strongly hydrophilic; H = slightly hydrophilic; N = neutral; L = slightly lipophilic; LL = strongly lipophilic.

phase, the usual practice is to add the oil to the water to form a coarse emulsion, and then to homogenize or colloid mill. In the preparation of water-in-oil emulsions, the water is added slowly to the oil phase, with good agitation. Emulsions with smaller size globules are obtained by milling or homogenizing. In all instances, before combining the two phases, each is heated to dissolve components and maintain them in solution. The phase that is to be poured is ordinarily kept a few degrees higher than the other to allow for cooling during the pouring step. After the emulsion is formed, agitation is continued during cooling in many instances. A number of different types of equipment for emulsification are available. Procedures have been recommended by Brown and Myers (29) for the preparation of various emulsions. Surfactants frequently used in the preparation of emulsions are listed in Table 11.2.

HYDROPHILE-LIPOPHILE BALANCE (HLB)

One of the more useful concepts developed for the control of the emulsion type by the selection of emulsifying agents is the "HLB" system (30).

TABLE 11.3. ESTIMATED HLB VALUES FOR SEVERAL EMULSIFYING AGENTS (30).

Emulsifier	Estimated HLB
Anionic	
T. E. A. Oleate	12
Lecomene C	12.7*
Sodium Oleate	18
Potassium Oleate	20
Cationic	
Atlas G-251	25–35
Non-Ionic	
Oleic Acid	App. 1
Span 85	1.8
Arlacel C	3.7
Span 80	4.3
Span 60	4.7
Span 20	8.6
Tween 81	10.0
Tween 60	14.9
Tween 80	15.0
Tween 20	16.7
Other values listed in booklet, "Surface Active Agents," published by Atlas Powder Co.	

* Tentative value.

Surfactants are classified according to the size and strength of the hydrophilic and lipophilic portions of the molecule. The balance in the size and strength of these two opposing groups is called hydrophile-lipophile balance or HLB.

According to this system, an emulsifier that is lipophilic in character is assigned a low HLB number, while an emulsifier that is hydrophilic in character is assigned a high number. The values that have been assigned to emulsifiers range from one to forty. The midpoint in lipophilic-hydrophilic character is about ten (30). When two or more emulsifiers are blended, the HLB values are intermediate.

TABLE 11.4. REQUIRED HLB VALUES FOR THE PREPARATION OF EMULSIONS CONTAINING VARIOUS OILS (30).

Oil or Wax	Required HLB for O/W Emulsion	Required HLB for W/O Emulsion
Cottonseed Oil	7.5	—
Carbontetrachloride	9*	—
Paraffin (household)	9	4
Microcrystalline wax (Micropac Q, S-V)	9.5*	?
Mineral Oil, White, light (Marcol GX)	10	4
Mineral Oil, White, heavy (Nujol)	10.5	4
Mineral Seal Oil	10.5*	?
Petrolatum, white (White Perfecta)	10.5	4
Silicone Oil (G.E.)	10.5*	?
Kerosene	12.5*	?
Naphtha	13*	?
Cetyl Alcohol	13	?
Orthodichlorobenzene†	13*	?
Beeswax, white	10–16	5
Carnauba Wax	14.5*	?
Candelilla Wax	14.5*	?
Lanolin, U. S. P., anhyd.	15	8
Dimethyl Phthalate	15*	?
Orthophenylphenol	15.5*	?
Stearic Acid	17	—

* Tentative.
† Plus 3.5% of pine oil and isopropanol.

The HLB system as originally conceived is empirical. Each oil to be emulsified and each emulsifier is assigned an HLB number on the basis of a series of emulsification tests. According to Griffin, about 75 emulsions must be prepared in order to assign an HLB value to an emulsifier. Typical values for emulsifiers and oils are given in Tables 11.3 and 11.4. The equation that expresses the relationship between the HLB value of the emulsifier combination and the required HLB of the oil is

$$\frac{W_A \mathrm{HLB}_A + W_B \mathrm{HLB}_B}{W_A + W_B} = \mathrm{HLB}_{\mathrm{oil}} \tag{11.13}$$

where W_A and W_B refer to the amounts by weight of emulsifier A and B required to give a reasonably stable emulsion with the particular oil. HLB_A and HLB_B are the assigned HLB values for emulsifiers A and B, and $\mathrm{HLB}_{\mathrm{oil}}$ is the required HLB of the oil for the type of emulsion being studied.

If the required HLB value for the oil is not known, a series of emulsions are prepared using a predominantly lipophilic and a predominantly hydrophilic emulsifier with known HLB values, such as Span 40 and Tween 40. The total weight of the emulsifiers is maintained constant for the series. The required HLB value of the oil is then calculated from the proportions of the two emulsifiers that produced the most stable emulsion of the series.

The HLB values of certain classes of nonionic surfactants can be calculated from analytical or composition data. For fatty-acid esters of polyhydric alcohols, approximate values can be obtained from the equation,

$$\mathrm{HLB} = 20\left(1 - \frac{S}{A}\right) \tag{11.13}$$

where S is the saponification number of the ester, and A is the acid number of the acid. An equation based on composition is

$$\mathrm{HLB} = \frac{E + P}{5} \tag{11.14}$$

where E is the weight per cent of ethoxy groups and P is the weight per cent of polyhydric alcohol present in the molecule.

TABLE 11.5. SURFACTANT APPLICATIONS AS RELATED TO HLB VALUES (30).

HLB Range	Use
4–6	W/O emulsifiers
7–9	Wetting agents
8–18	O/W emulsifiers
13–15	Detergents
15–18	Solubilizing

However, the HLB system does not provide information as to the specific emulsifiers, or even the types of emulsifiers, that will produce the most stable emulsions. In order to obtain commercially acceptable emulsions, it is necessary to evaluate a large number of emulsifier combinations. The HLB system is helpful in providing information concerning the preferred proportions for various emulsifier combinations.

Griffin has observed that HLB values indicate the possible usefulness of surfactants in other applications besides emulsification, as shown in Table 11.5. However, this is only a generalization. It does not follow that every surfactant with an HLB value between 13 and 15, for example, is a useful detergent.

TABLE 11.6. COMPARISON OF HLB FROM EXPERIMENT AND FROM GROUP NUMBERS (31).

Surface-active Agent	HLB from Experiment	HLB from Group Numbers
Na Lauryl sulphate	40	(40)
K Oleate	20	(20)
Na Oleate	18	(18)
Tween 80 (sorbitan "mono-oleate" + 20(CH_2—CH_2—O) groups)	15	15.8
Sorbitan "mono-oleate" + 10(CH_2—CH_2—O) groups	about 13.5	12.5
Tween 81 (sorbitan "mono-oleate" + 5(CH_2—CH_2—O) groups)	10	10.9
$C_{18}H_{37}N(CH_2CH_2OH)(CH_2—CH_2—O—CH_2—CH_2OH)$	10	(10)
Sorbitan monolaurate	8.6	8.5
Methanol	—	8.4
Ethanol	—	7.9
n-Propanol	—	7.4
Sorbitan "mono-oleate" + 2(CH_2—CH_2—O) groups	about 7	7.0
n-Butanol	7.0	7.0
Sorbitol "mono-oleate"	about 7	7.2
Sorbitan monpalmitate	6.7	6.6
Sorbitan monostearate	5.9	5.7
Span 80 (sorbitan "mono-oleate")	4.3?	5.0
Propylene-glycol monolaurate	4.5	4.6
Sorbitan distearate	about 3.5	3.9
Glycerol monostearate	3.8	3.7
Propylene glycol monostearate	3.4? (may contain soap)	1.8
Sorbitan tristearate	2.1	2.1
Cetyl alcohol	1	1.3
Oleic acid	1	(1)
Sorbitan tetrastearate	about 0.5	0.3

Calculated values for "mono-oleates" assume these are 1.38 oleates.

While the HLB system was developed empirically, Davies has shown that it does have a theoretical basis, as discussed subsequently. Calculated HLB values are compared with experimental values in Table 11.6.

EMULSION STABILITY

Although many emulsions appear to be stable for an indefinite time, they are not stable in the strict thermodynamic sense of the word. The large interfacial area represents a considerable amount of free energy which is only at a minimum when the dispersed droplets have coalesced into one large liquid mass.

Three interrelated forms of instability may be observed with emulsions. These are sedimentation, flocculation, and coalescence. Since the emulsion globules ordinarily have a density different from that of the dispersion medium, they will tend to accumulate at the bottom or top of the emulsion. The rate of sedimentation depends upon the difference in density between the globules and the dispersion medium, the particle size, and the viscosity of the continuous phase. The sedimentation velocity is given by Stokes' equation,

$$v = \frac{4/3\, \pi r^3 g\, (d_1 - d_2)}{6\pi \eta r} \qquad (11.15)$$

where v is the rate of sedimentation, r is the droplet radius, g is the acceleration of gravity, d_1 is the density of the droplet, d_2 that of the liquid, and η is the viscosity of the dispersion medium.

A sedimentation velocity of the order of 1 mm in 24 hours is usually sufficiently counteracted by accidental thermal convection (32). Application of the Stokes' equation to an O/W emulsion at room temperature, with the density of the globules 0.8 that of the water, and with $\eta = 0.01$, leads to the conclusion that a particle radius of 0.16 micron will have a sedimentation velocity of 1 mm per day and will not cream.

Flocculation is an agglomeration or sticking together of particles in the form of a loose and irregular cluster in which the individual particles can still be recognized. Flocculation as well as coalescence increases the effective particle size and thus leads to an increased rate of sedimentation. Though flocculation has been considered to be irreversible, it has been pointed out that this is not always the case with emulsions. Agglomerated droplets can frequently be redispersed by mild shaking (33). Forces responsible for flocculation in emulsions are London-van der Waals forces of attraction, frequently counteracted by electrostatic repulsion. In that event, electrolytes influence flocculation according to the rule of Schulze and Hardy. In addition, flocculation may occur due to interaction of the flocculating agent with the emulsifier.

Since coalescence cannot take place until two or more droplets are in contact, it must be preceded by flocculation. The rate of coalescence depends upon the number of points of contact between the globules. With emulsifier present, contact points can rupture at an extremely slow rate. It is generally agreed that the barrier to the coalescence of agglomerated droplets is an interfacially adsorbed film of emulsifying agent (24). The energy required for the coalescence of hydrocarbon-oil droplets in an aqueous solution of sodium oleate has been shown to be of the order of 6.5 kcal/mole (8).

FLOCCULATION

Interaction of the flocculating agent and the emulsifier to cause agglomeration is specific and generally depends on insolubilizing the emulsifier. An emulsion stabilized by soap can be flocculated by lowering the pH or by the addition of calcium salts. An emulsion stabilized by an anionic surfactant such as sodium lauryl sulfate can be agglomerated by the addition of a cationic surfactant or a dye cation. However, these effects are not of fundamental importance. Consideration will instead be given to the distribution of electrostatic charges about an oil-water interface, the origin of these charges, and the interaction of forces of attraction and repulsion, the net effect of which is the flocculation or dispersion of particles.

Electric Double-layer at the Oil-water Interface

If two liquid phases are placed in contact, one of which contains salt ions, there will generally be an unequal distribution of anions and cations between the two phases. If the phases are oil and water, the oil phase will ordinarily develop a negative charge since anions are generally more soluble than cations in oil. This observation has given rise to the rule of Coehn to the effect that the sign and magnitude of the charge depends on the difference in magnitude of the dielectric constants of the two phases. This rule is without foundation (23).

The potential difference between the interiors of the two liquid phases depends entirely upon the equilibrium distribution of the ions between the phases. It does not depend upon the concentration of the ions. The potential drop across an oil-water interface is shown in Figure 11.13a. A diffuse double layer is present on both sides of the interface, and the potential drop occurs partly in the oil phase and partly in the water phase. At the interface itself there is a boundary potential χ, caused by the orientation of dipoles (32).

The introduction of an ionic surfactant will not change the potential difference between the interior of the two phases, except by its equilibrium distribution between the phases, or by altering that of the ions already present. The most profound effect of the ionic surfactant is the potential induced at the interface by its adsorption. This is shown in Figure 11.13b.

The adsorption of surfactant ions at an interface is followed by the adsorption of counter-ions in a Stern layer. According to Davies, the counter-ions

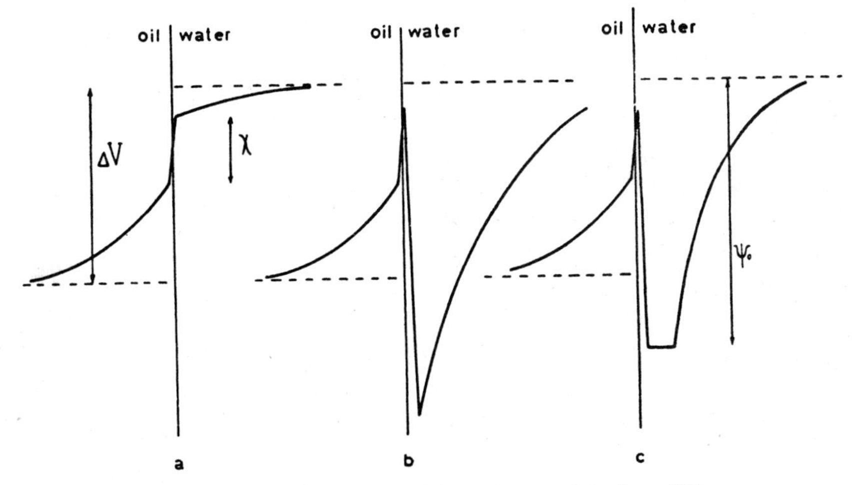

Figure 11.13. The potential at an oil-water interface (33).
a. in the absence of surface active ions.
b. after addition of soap ions, in solution of very low ionic strength.
c. in the presence of soap ions and a large amount of salt.

are situated between the ionic heads of the surfactant molecules. The extent of counter-ion adsorption, in relation to adsorbed surfactant ions, depends upon the concentration of such ions of opposite charge in solution. The flattening effect of counter-ions on the potential is shown diagramatically in Figure 11.13c.

Surface-charge Distribution

Several methods are available for estimating the distribution of surface charges about emulsified oil droplets containing adsorbed surfactant ions. The double layer at the oil-water interface can be considered to comprise an adsorbed layer of surfactant ions having a charge σ, a layer composed of counter-ions with charge σ_1, and a diffuse layer containing ions of both species with counter-ions in excess by the amount σ_2. Since the entire double layer is electrically neutral,

$$\sigma - \sigma_1 = \sigma_2 \tag{11.16}$$

The number of adsorbed ions is numerically equal to the charge density in electronic charges,

$$\frac{\sigma}{e} = \Gamma_{s^-} \tag{11.17}$$

where Γ_{s^-} is the surface excess of surfactant ions, and e is the value of an electronic charge, $4.77 + 10^{-10}$ esu.

The value of Γ_{s^-} can be obtained from the Gibbs adsorption isotherm, by radiotracer methods, or from a suitable equation of state. Powell and Alexander (22) calculated Γ_{s^-} from an equation of state. For Aerosol OT, they employed the equation

$$\pi (A - 40) = 210 \tag{11.18}$$

where π is the surface pressure in dynes per cm and A is the area per molecule, the reciprocal of which—with attention to units—gives the surface excess, and consequently σ.

The charge in the diffuse layer σ_2 was calculated from the equation for a sphere, relating charge to potential. With Debye-Huckel approximations,

$$\sigma_2 = \frac{D(1 + \kappa r)}{4\pi r} \zeta \tag{11.19}$$

where D is the dielectric constant, κ is the Debye-Huckel function, r is the radius of the sphere, and ζ is the zeta potential. For the emulsion droplets under consideration, $\kappa r \gg 1$.

Thus, by measuring the electrophoretic mobility and the particle size of the emulsified oil droplets, and by selecting a suitable equation of state for the interfacial film, the charge densities at the interface, in the Stern layer, and in the diffuse layer can be calculated. The results of such calculations have been expressed in Table 11.7 as the number of counter-ions adhering per 1000 long-chain ions on the droplets (22). The values for sodium dodecyl sulfate were calculated from the data of Powney and Wood (36) using the equation of state,

$$\pi (A - 15) = 615 \tag{11.20}$$

If the electrolyte concentration is low, oil globules in an emulsion stabilized by an ionic emulsifier generally have electrophoretic mobilities corresponding to zeta potentials of about 100 to 150 mv. Verwey (35) calculated that one ion per 1000 Å^2 of interface can produce a potential drop of this magnitude. Powney and Wood (36) found that the electrophoretic mobility of oil droplets in soap solutions reaches a maximum much below the cmc of the soap. Thus, except for the initial quantity of surfactant ions adsorbed, further adsorption has only a negligible effect on the surface potential. At the cmc, the interfacial area per adsorbed surfactant ion is generally about

TABLE 11.7. COUNTER-IONS IN THE STERN LAYER (22).

Soap	Percentage Concentration of Aqueous Solution	Number of Gegen Ions Adhering per 1000 Long Chain Ions on Droplet
Aerosol OT	0.01	994
	0.02	991
	0.04	984
Aerosol AY	0.01	986
	0.02	980
	0.04	968
Aerosol OT (in buffer solution)	0.001	933
	0.002	892
	0.005	874
Sodium dodecyl sulphate	0.01	970
	0.07	931
	0.19	900

50 Å^2, or about 1/20th of the required interfacial area calculated by Verwey for a zeta potential of 100 to 150 mv. The difference is due to the adsorption of counter-ions in an amount equal to about 95 per cent of the adsorbed surfactant-ions. This is in reasonable agreement with the results shown in Table 11.7.

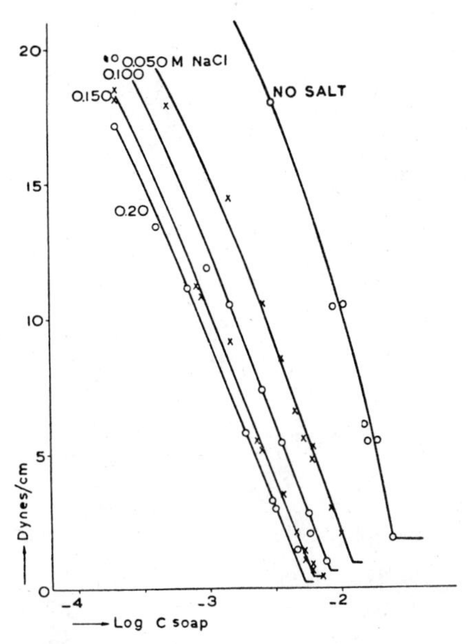

Figure 11.14. Interfacial tension of Aerosol MA solutions containing sodium chloride, against oil mixture (33).

A somewhat different method of calculation was used by van den Tempel (33), who related the three charge-density values to the surface potential in the Stern layer ψ_δ. The surface excess of surfactant ions, and consequently σ, was calculated from the usual forms of the Gibbs adsorption equation. The interfacial-tension curves against mineral oil for solutions of Aerosol MA and sodium laurate with added salt, that were used for these calculations, are shown in Figures 11.14, 11.15, and 11.16.

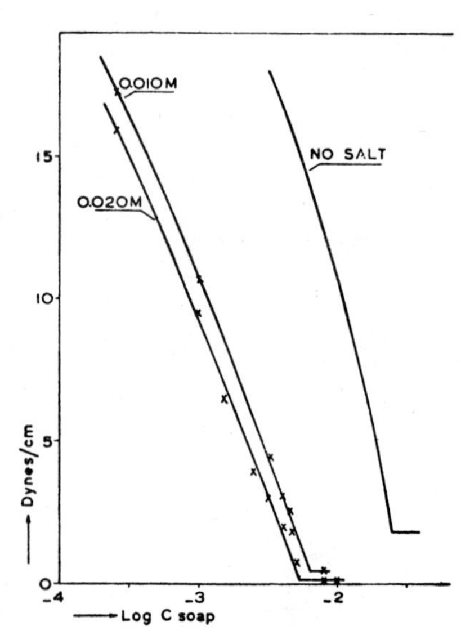

Figure 11.15. Interfacial tension of Aerosol MA solutions containing magnesium chloride, against oil mixture (33).

The values for σ_1, which correspond to the surface concentration of counter-ions in the Stern layer, were calculated using the Langmuir-Stern equation,

$$\frac{\sigma_1}{e} = \frac{N_1}{1 + \dfrac{0.6 \times 10^{24}}{18c_+} \exp\left(\dfrac{ze\psi_\delta}{kT}\right)} \tag{11.21}$$

where N_1 is the number of available positions in 1 cm² of the Stern layer, c_+ is the number of counter-ions of valency z in 1 cm³ of the homogeneous solution, e is the elementary charge, and ψ_δ is the potential in the Stern layer with respect to a point in the aqueous phase far from the interface.

The factor $\dfrac{0.6 + 10^{24}}{18}$ is the number of water molecules per cm³ and corresponds to the number of positions available to the counter-ions in 1 cm^3 of the homogeneous solution. An estimate of N_1 is 10^{15}, when the depth of this layer is taken as $3 + 10^{-8}$ cm. The charge in the diffuse part of the double layer σ_2 can be related to the potential at the boundary between

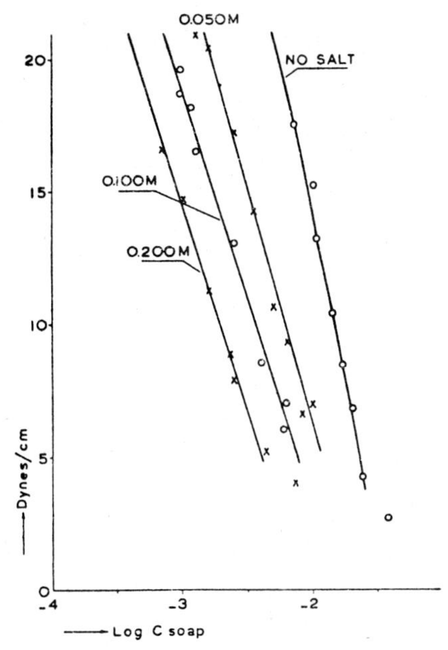

Figure 11.16. Interfacial tension of sodium laurate solutions containing sodium chloride, at pH = 11.0, against oil mixture (33).

the Stern and Gouy layers ψ_δ by the general equation, which avoids Debye-Huckel assumptions,

$$\sigma_2 = \sqrt{\frac{DkT}{\pi} \sum c_i z_i^2} \, . \, \sinh \left(\frac{ze\psi_\delta}{2kT} \right) \tag{11.22}$$

where c_i is the concentration of ionic specie i with valency z_i.

Curves can then be constructed for σ, σ_1, and σ_2 as a function of ψ_δ for any system in which the concentration of counter-ions c_+ is known. The actual value of σ can be obtained from interfacial-tension measurements, and the value of ψ_δ for the system thus determined. A set of curves relating the charge-density values to the potential in the Stern layer is shown in Figure 11.17. Values of ψ_δ calculated from these curves are plotted against salt concentration in Figure 11.18. The potential in the Stern layer decreases almost linearly with increasing salt concentration. The same decrease in potential can be obtained with one-tenth the concentration of bivalent counter-ions as monovalent counter-ions.

The zeta-potential ζ of oil droplets in the systems described above was determined from electrophoretic mobility measurements and the values are shown in Table 11.8. The relation between ζ in Table 11.8 and ψ_δ in Figure 11.18 can be compared, since $-kT/e = 25$ mv. It will be observed that the

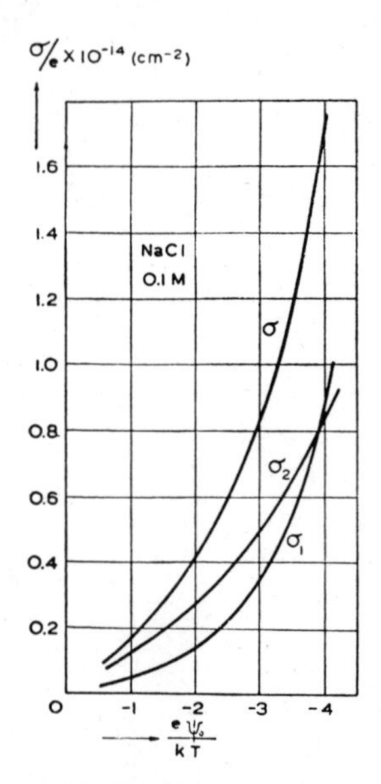

Figure 11.17. Relation between the potential in the Stern layer, ψ_δ, the amount of adsorbed soap σ/e, and the partition of the counter ions between Stern- and Gouy-layer, in a 0.100 M solution of a 1–1 valent electrolyte (33).

values for the two potentials are in reasonable agreement for monovalent cations, while the zeta-potential values are somewhat lower than the calculated potential in the Stern layer for bi- or tervalent cations.

Unfortunately, conclusions concerning the surface excess determined from surface tension—or interfacial tension—*versus* concentration curves may not be correct. They are based on the faulty assumption (37) that only unassociated surfactant ions are present in solution until a critical concentration where micelles suddenly appear. Adsorption determined by other methods shows more complex relationships (38).

TABLE 11.8. ZETA POTENTIAL OF OIL DROPS EMULSIFIED IN AQUEOUS AEROSOL MA SOLUTION (33).

Soap Conc. (mol/l)	Electrolyte Conc. (mol/l)		ζ(mV)
0.00031	—		103
,,	NaCl	0.050	131
,,	,,	0.100	124
,,	,,	0.150	114
,,	,,	0.200	117
0.00088	—		140
,,	NaCl	0.100	126
,,	,,	0.150	105
0.00088	$MgCl_2$	0.005	66
,,	,,	0.010	58
0.00088	Luteo-cobaltic	0.0002	15
,,	chloride	0.0003	12

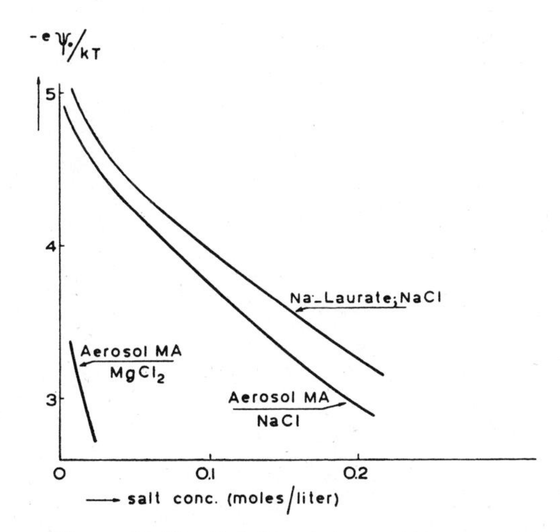

Figure 11.18. Relation between the potential in the Stern-layer and the salt concentration in the aqueous phase (33).

Potential Energy of Interaction

The stability of hydrophobic sols has been explained by taking into account the electrostatic repulsion and the London-van der Waals attraction between the particles (39). Typical potential energy curves as a function of distance between the particles are shown in Figures 11.19, 11.20 and 11.21.

Figure 11.19 shows the electrostatic energy of repulsion between two charged particles. At substantial distances, the energy of repulsion is zero

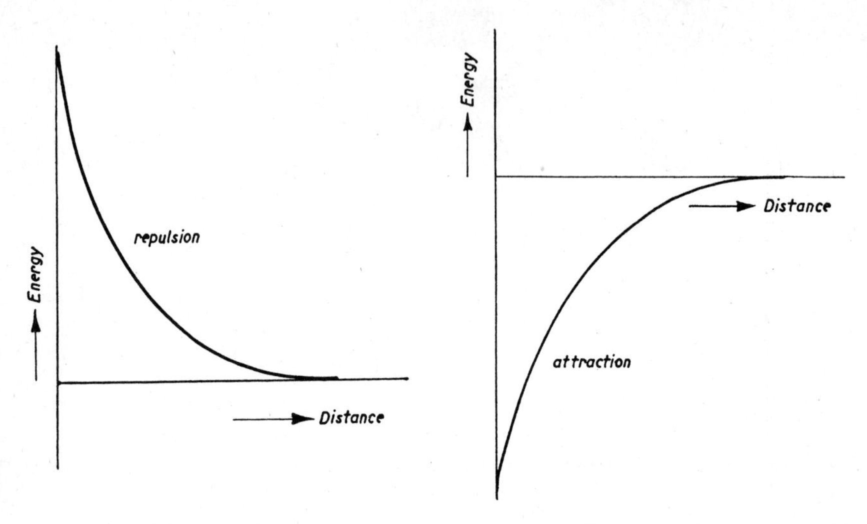

Figure 11.19.

Figure 11.20.

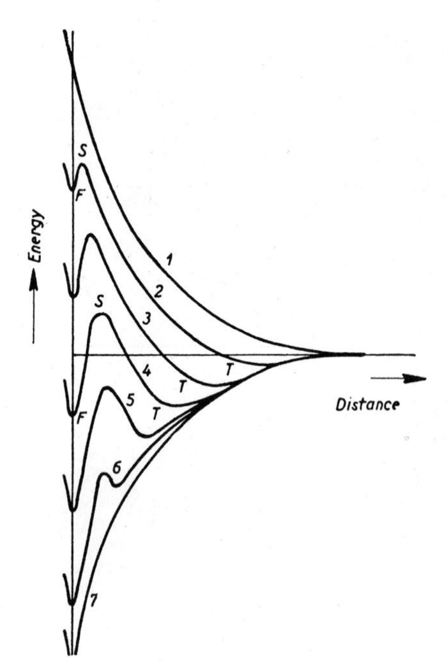

Figure 11.21

Figure 11.19–11.21. Examples of potential curves representing the energy of interaction between colloidal particles (40).

and it increases as the particles approach. Figure 11.20 shows the London-van der Waals attraction between particles. The opposing energy terms are combined in Figure 11.21 to show the interaction of forces. For stability of a sol, it is necessary that a potential-energy maximum of sufficient height—$\gg kT$—exist. This is the case with curves 1, 2, 3 and 4, while curves 5, 6 and 7 represent unstable dispersions. Secondary minima occurring at fairly large interparticle distances are exhibited by curves 2, 3, 4, 5 and 6.

The rate of flocculation is determined by the height of the maximum in the potential-energy curve, and, consequently, by the potential drop in the double layer and by the electrolyte concentration. The Schulze-Hardy rule, which expresses the influence of the valency of the coagulating ions on this rate, has been shown to be a consequence of the large potential-drop in the double layer surrounding the colloidal particles.

The flocculation of a hydrophobic sol is nonreversible, that is the particles can not be redispersed by stirring. In the case of emulsion droplets both irreversible and reversible flocculation have been observed, with the latter type more commonly experienced (33). With concentrated emulsions, reversible flocculation is evidenced by the thickening of the emulsion. Slight agitation restores the original viscosity. Experimental study of the flocculation of emulsified droplets is much more difficult than that of hydrophobic sols because of the reversibility of flocculation as noted above, the wide range of droplet sizes encountered in an emulsion, the greater stability of the smaller droplets under certain conditions, and the complicating factor of coalescence.

It has been found that the flocculation of emulsified droplets is influenced by the valency of the counter-ion in accordance with the rule of Schulze and Hardy. However, the mechanism is somewhat different from that with hydrophobic sols. With the solid sol-particles, the valency of the counter-ions is manifest through the reduced thickness of the double layer. With emulsified droplets the potential drop is sharply reduced also; a bivalent ion is as effective as a monovalent ion at one-tenth the concentration (33).

The equations that have been developed in accordance with the theory of Verwey and Overbeek for the stability of hydrophobic sols lead to very high potential energy barriers for relatively large emulsified droplets where ψ_δ is of the order of 100 mv, indicating that flocculation will not occur. Nevertheless, flocculation has been observed to occur in these systems at an appreciable rate. Van den Tempel (33) has offered strong arguments to the effect that the observed flocculation actually occurs in the secondary minimum of the potential energy curve.

In accordance with this theory, the interparticle distance H_m at which the potential energy V has a minimum value is:

$$H_m^2 \cdot \mathbf{e}^{-\kappa H_m} = 1.80 \times 10^{-9} \cdot \frac{V^2}{\kappa \gamma^2} \tag{11.23}$$

where $1/\kappa$ is the thickness of the double layer, v is the valence of the counter-ions, and

$$\gamma = \frac{\mathbf{e}^{\left(\frac{z}{2}\right)} - 1}{\mathbf{e}^{\left(\frac{z}{2}\right)} + 1} \tag{11.24}$$

where $z = ve\psi_\delta/kT$. This relationship is shown in Figure 11.22 for two values of the surface potential.

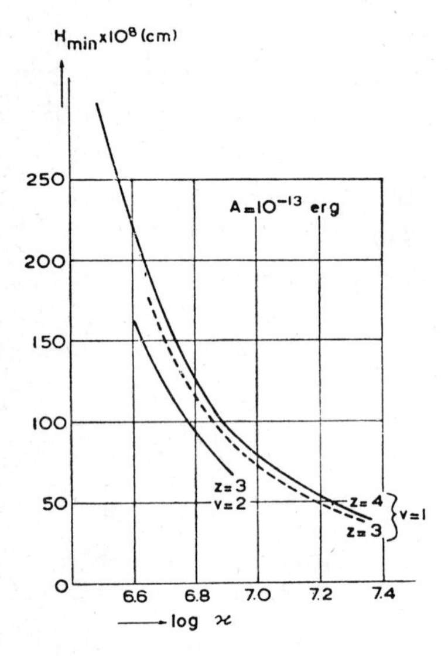

Figure 11.22. Interparticle distance at the "secondary minimum" of the potential energy, as a function of the electrolyte concentration (33).

The valency of the counter-ion greatly influences the interparticle distance of the secondary minimum because it influences both v and κ. The same value of H_m is obtained with $0.020M$ solution of a salt of a bivalent counter ion as with $0.100M$ solution of a 1–1 valent electrolyte.

Since the negative value of the potential energy in the secondary minimum is due almost entirely to the van der Waals attraction, the depth is about inversely proportional with H_m. At a given electrolyte concentration, the depth would be approximately proportional to the particle size.

If flocculation occurs in the secondary minimum, the rate of flocculation should be slightly higher than that corresponding with rapid flocculation, since there is no energy barrier to flocculation and attraction prevails at relatively large distances between the particles. This higher rate of flocculation has been observed (33).

Van den Tempel (33) noted that flocculation of relatively large oil droplets with a surface potential of 100 mv would seem improbable because of the very large potential barrier. If flocculation across this barrier did occur the distance between the surfaces of the oil globules would become less than 1Å, flocculation would be expected to be practically irreversible and coalescence would probably occur immediately after flocculation.

If flocculation occurs in the secondary minimum the effect of the valency of the counter-ions is due to a decreased distance between particles in the secondary minimum and an increased depth of the minimum. This is interpreted as an increase in the degree of flocculation, such that there is less tendency for the flocculated particles to redisperse by thermal motion. This is reasonable if the depth of the minimum is of the order of kT.

Approximate Equation for Flocculation Rate

Davies (31) found the following equation to be valid for the rate of coalescence of charged emulsion drops.

$$\text{Rate} = A_1 \, \mathbf{e}^{-0.24\psi_0^2/RT} \tag{11.25}$$

where A is the coalescence rate of an unstabilized emulsion, with $\psi_0 = 0$. The surface potential ψ_0 is expressed in millivolts, and the gas constant R is in calories.

Since the rate is a function of exponential $(-\psi_0^2)$, the equation is an expression of the rate of flocculation, rather than the rate of coalescence. For the examples in which the equation proved valid, all flocculated droplets must have coalesced rapidly.

The value of ψ_0 used in Davies equation was calculated from the theory of Gouy, rather than Stern. Accordingly,

$$\sigma = \sqrt{\frac{DkT}{\pi} \sum c_i z_i^2} \; . \; \sinh\left(\frac{ze\psi_0}{2kT}\right) \tag{11.26}$$

where c_i is the concentration of ionic specie i with valency z_i. The surface charge density σ is related to the number of adsorbed surfactant ions by the equation

$$\sigma = \frac{e}{A} \tag{11.27}$$

where e is the electronic charge and A is the area in $Å^2$ per charged group in the monolayer. Then

$$\psi_0 = \frac{2kT}{e} \sinh^{-1} \left(\frac{\sqrt{\dfrac{\pi e^2}{DkT}}}{A \sqrt{\sum c_i z_i^2}} \right) \tag{11.28}$$

Since,

$$\frac{kT}{e} = 25.2 \text{ mv at } 20°\text{C} \tag{11.29}$$

and

$$\sqrt{\frac{\pi e^2}{DkT}} = 134 \text{ mv at } 20°\text{C} \tag{11.30}$$

then,

$$\psi_0 = 50.4 \sinh^{-1} \left(\frac{134}{A\sqrt{\sum e_i z_i^2}} \right) \tag{11.31}$$

The surface potential calculated from the Gouy theory has been found empirically to be related to the zeta potential ζ calculated from the mobilities of oil droplets in surfactant solutions. The relation,

$$\zeta = 0.55\psi_0 \tag{11.32}$$

has been claimed to hold reasonably well (41). However, Anderson (42) did not find this relation valid. Instead, the zeta potential corresponded closely with the potential in the Stern plane, as discussed previously.

COALESCENCE

An emulsion is not considered broken unless coalescence follows flocculation. While flocculation can be a reversible process, coalescence is essentially irreversible. In the breaking of practically all technical emulsions, coalescence is the slow step in the process. Consequently, the rate of coalescence determines the stability of the emulsion. It can be expected that the factors responsible for coalescence are not the same as for flocculation.

There is an energy barrier that stabilizes emulsions against coalescence that is quite different from that which retards flocculation (8). It is generally recognized that this energy barrier is due to the interfacial film of adsorbed emulsifier (2). However, there is disagreement as to the mechanism responsible for this barrier. Thus, coalescence may occur as the result of bridging. For example, if two water droplets in a W/O emulsion are in contact, they are separated by a double layer of hydrocarbon chains corresponding to the lipophilic portion of the emulsifier. Water molecules can traverse this chain to form a bridge. Once the droplets are thus connected, they can coalesce with a decrease in free energy. The energy barrier is the energy required for the water molecules to pass between neighboring hydrocarbon chains.

Alternatively, localized displacement of the emulsifier at the interface can take place (41). Displacement in the plane of the interface could occur if the surface were covered by an incomplete monolayer. However, it would be resisted by the resulting local inequalities in interfacial tension. With a larger surface excess of emulsifier, the viscosity or elasticity of the adsorbed film would increase the barrier to such displacement.

Displacement of the emulsifier from the interface into one of the liquid phases is also possible. With flocculated droplets, the close juxtaposition of the interfaces hinders displacement into the continuous phase. This steric hindrance to displacement into the discontinuous phase does not exist. Since emulsion type depends upon the solubility characteristics of soluble emulsifiers and the relative wettability of insoluble emulsifiers and powders, it would appear that the main factor determining stability is the resistance to displacement into the discontinuous phase (31).

Information is rarely adequate to judge whether the energy barrier to coalescence in any specific instance is due to bridging or displacement of one type or another. Coalescence will proceed along whichever route offers the lowest potential energy barrier.

Emulsion Stability and Emulsion Type

This discussion is a substantial modification of the treatment by Davies (31). In the case of an O/W emulsion stabilized by a nonionic emulsifier, the emulsifier acts by binding water tightly to the surface of the dispersed oil globules. If coalescence is due to desorption of the emulsifier into the globules, this water of hydration must be displaced. The total energy barrier $\sum E_h$ required for this displacement will depend upon the number and type of the hydrated groups on each molecule of the emulsifying agent, and on the fraction of the interface covered, θ. Then the rate of coalescence is

$$\text{Rate}_1 = A_1 \, \mathbf{e}^{-\theta \Sigma E_h/RT} \tag{11.33}$$

where A_1 is the hydrodynamic collision factor, including the phase-volume of oil and the viscosity of water.

Alternatively, if coalescence depends upon the desorption of the nonionic emulsifier into the continuous water phase, then the energy barrier is determined by the energy required to transfer the hydrocarbon chain into the water medium. This is about 800 cal/mole per CH_2 group. With this assumption, the rate of coalescence is

$$\text{Rate}_2 = A_1 \, \mathbf{e}^{-800n\theta/RT} \tag{11.34}$$

where n is the number of CH_2 groups in the hydrocarbon chain.

It is known that hydrophilic emulsifiers stabilize O/W emulsions. As the number of hydrated groups on the molecule is increased, emulsion stability increases to a maximum value, and then decreases. Since the degree of adsorption θ will decrease, while the hydration energy will increase with the number of hydrated groups, Rate_1 will give the correct relationship between emulsion stability and the hydrophilic character of the emulsifier. Rate_2 will decrease as the number of hydrated groups on the molecule is increased, and is incorrect for this case.

For W/O emulsions stabilized by nonionic emulsifiers, the same rate equations are applicable, except that Rate_1 now refers to the desorption of the emulsifier into the continuous oil phase. Rate_2 applies to the desorption of the emulsifier into the water globules. It is known that W/O emulsions become less stable as the emulsifier is made more hydrophilic. This situation can only be described by Rate_2.

Thus, the coalescence of both O/W and W/O emulsions stabilized by nonionic emulsifiers can be described in terms of the energy required for the desorption of the emulsifier into the dispersed globules. Then, the rate of coalescence of O/W emulsions is

$$\text{Rate}_{\text{O/W}} = A_1\, \mathbf{e}^{-\theta \Sigma E_h / RT} \tag{11.35}$$

For W/O emulsions, the rate of coalescence is

$$\text{Rate}_{\text{W/O}} = A_2\, \mathbf{e}^{-800 n\theta / RT} \tag{11.36}$$

where A_2 is the appropriate hydrodynamic collision factor.

The kinetic equations can be used to determine whether an emulsion will be of the O/W or W/O type. In the process of emulsification, both types may form simultaneously. The type that is finally observed will depend upon the relative stability of the two types. Thus, if the O/W system is stable, and the W/O emulsion breaks very rapidly, the emulsion type will be O/W.

Combining the equations for $\text{Rate}_{\text{W/O}}$ and $\text{Rate}_{\text{O/W}}$ leads to

$$\frac{\text{Rate}_{\text{W/O}}}{\text{Rate}_{\text{O/W}}} = \frac{A_2}{A_1}\, \mathbf{e}^{\theta(\Sigma E_h - 800 n)/RT} \tag{11.37}$$

If $A_1 = A_2$, an O/W emulsion will form if $\sum E_h > 800n$. Otherwise, the emulsion will be of the W/O type.

Davies expressed the hydrophilic-lipophilic balance (HLB) of the emulsifier in the form

$$\text{HLB} = \sum \text{(hydrophilic group numbers)} - n\ \text{(group number per CH}_2\text{ group)} + 7 \tag{11.38}$$

The rate equation in logarithmic form is

$$RT \ln \left(\frac{\text{Rate}_{W/O}}{\text{Rate}_{O/W}} \right) = RT \ln \left(\frac{A_2}{A_1} \right) + \theta \left(\sum E_h - 800n \right) \tag{11.39}$$

If the viscosities and phase volumes of oil and water are equal, so that $A_2 = A_1$,

$$RT \ln \left(\frac{\text{Rate}_{W/O}}{\text{Rate}_{O/W}} \right) = \theta \left(\sum E_h - 800n \right) \tag{11.40}$$

Table 11.9 is a listing of various HLB group numbers. If the group number for CH_2 is inserted in the HLB equation, we have the following comparison,

$$\frac{\text{HLB}-7}{0.475} = \frac{\sum (\text{hydrophilic group number})}{0.475} - n \tag{11.41}$$

and

$$\frac{RT}{800\theta} \ln \left(\frac{\text{Rate}_{W/O}}{\text{Rate}_{O/W}} \right) = \frac{\sum E_h}{800} - n \tag{11.42}$$

This leads to the following relations, provided that $A_2 = A_1$, showing that the HLB system has a kinetic basis,

$$\ln \left(\frac{\text{Rate}_{W/O}}{\text{Rate}_{O/W}} \right) = 2.9\,\theta\,(\text{HLB} - 7) \tag{11.43}$$

TABLE 11.9. HLB GROUP NUMBERS FOR HYDROPHILIC AND LIPOPHILIC GROUPS. THERE IS AN UPPER HLB VALUE OF ABOUT 20 FOR HIGH MOLECULAR-WEIGHT POLYETHYLENE-OXIDE DERIVATIVES (31).

Hydrophilic groups	Group number
$—SO_4'Na^+$	38.7
$—COO'K^+$	21.1
$—COO'Na^+$	19.1
N (*tertiary amine*)	9.4
Ester (*sorbitan ring*)	6.8
Ester (*free*)	2.4
—COOH	2.1
Hydroxyl (*free*)	1.9
—O—	1.3
Hydroxyl (*sorbitan ring*)	0.5
Lipophilic groups	
—CH—	
$—CH_2—$	
$—CH_3—$	−0.475
=CH—	
Derived groups	
$—(CH_2—CH_2—O)O$	+0.33
$—(CH_2—CH_2—CH_2—O)—$	−0.15

and

$$\text{hydrophilic group number for single hydrated group} = \frac{E_h}{1680} \tag{11.44}$$

If the HLB of the emulsifier is close to 7, or if θ is small, neither type of emulsion is favored. For a typical O/W emulsifying agent, HLB=11, and with $\theta \sim 1$, $\text{Rate}_{O/W} = 10^{-5}\,\text{Rate}_{W/O}$.

The energy of hydration of a single OH group is about 4,000 cal/mole. The energy of transfer of an OH group from water to a hydrocarbon oil is 3200 cal/mole. This value is equal to that calculated from Equation 11.44, when the group number of the hydroxyl group is taken as 1.9. Thus, E_h is the energy of transfer of a hydrophilic group from water to the oil phase.

Davies has also related the emulsion type to the partition coefficient of the emulsifier between oil and water. The free-energy change on transferring emulsifier from the water phase to the oil phase is

$$\Delta F_{w \to o} = -RT \ln (c_o/c_w) \tag{11.45}$$

where c is the emulsifier concentration, and the subscripts O and W refer to oil and water, respectively. The work of transfer $\Delta F_{w \to o}$ is composed of terms from the hydrophilic and lipophilic parts of the molecule. Then

$$\Delta F_{w \to o} = \sum \Delta F_{w \to o}(\text{hydrophilic groups}) - 800n \tag{11.46}$$

and

$$\frac{RT}{800} \ln \left(\frac{c_w}{c_o}\right) = \frac{\sum \Delta F_{w \to o}(\text{hydrophilic groups})}{800} - n \tag{11.47}$$

Comparing this with the kinetic results, and assuming $A_1 = A_2$, we have

$$\frac{\text{Rate}_{W/O}}{\text{Rate}_{O/W}} = \left(\frac{c_w}{c_o}\right)^{\theta} \tag{11.48}$$

This relation corresponds to the qualitative Bancroft rule, to the effect that emulsifying agents stabilize O/W emulsions if they are more soluble in the water phase, and *vice versa*.

It follows from this discussion that when a mixture of nonionic emulsifiers is more effective than a single emulsifier with the appropriate HLB value in stabilizing an emulsion, it is due to the effect on θ, the fraction of the interfacial area covered by the emulsifying agents. This appears to be reasonable for O/W emulsions, where the hydrophilic portion of the water-soluble nonionic emulsifiers is large and bulky. There would be less steric hindrance to the further adsorption of molecules carrying a small number of hydrophilic groups.

A similar line of reasoning should be applicable to the coalescence of oil globules stabilized with ionic emulsifiers. The rate of coalescence should depend upon the fraction of the surface covered by emulsifier, and the energy required to transfer a single ionic group from water to oil, as well as the contribution of the electrical potential at the interface. Thus,

$$\text{Rate}_{\text{O/W}} = A_1\, \mathbf{e}^{-\theta(\Sigma E_h - \beta\psi_0)/RT} \tag{11.49}$$

where ψ_0 is the double layer potential, and β includes the contribution of neighboring globules on the potential as well as the valence of the ionic group and the electron charge e. The term $\sum E_h$ refers to the energy required to transfer surfactant ion and counter-ion together into the oil phase.

In the coalescence equation, a high potential ψ_0 has an adverse effect on stability. This is necessary, since the electrical repulsion will favor desorption of the emulsifier.

The effect of inorganic electrolytes can be quite complex. Flocculation will be favored, as previously discussed. To the extent that θ is increased and ψ_0 decreased, the added electrolyte will reduce the rate of coalescence. However, ion-pair formation can reduce $\sum E_h$ the energy required to transfer the emulsifier into the oil phase. Thus, it is not surprising that electrolytes have been observed to both increase and decrease the coalescence rate.

Assuming coalescence and not flocculation to be the slow step, it is possible to write a general equation for the rate of coalescence of stabilized emulsions. The change in free energy on transferring a species from the interface to the interior of a globule is

$$\Delta F_{i \to g} = - RT \ln (c_g/\Gamma) \tag{11.50}$$

where c_g and Γ refer respectively to the concentration of the species in the globule and at the interface. The rate of coalescence is

$$\text{Rate} = A\, \mathbf{e}^{-\theta\Delta F_{i\to g}/RT} \tag{11.51}$$

or

$$\text{Rate} = A\, \mathbf{e}^{\theta \ln(c_g/\Gamma)} \tag{11.52}$$

This reduces to the simple expression

$$\text{Rate} = A \left(\frac{c_g}{\Gamma}\right)^{\theta} \tag{11.53}$$

This equation for the rate of coalescence should be applicable to both O/W and W/O emulsions, regardless of whether the emulsifier is ionic or nonionic. Unfortunately, experimental data are not available to test its validity. Qualitatively, it is known that combinations of water-insoluble long-chain polar compounds and ionic surfactants form mixed surface films. These mixtures also produce more stable O/W emulsions than with ionic surfactants alone. It can be expected that the increased stability is due to the increase in θ, the fraction of the interfacial area covered by the adsorbed emulsifying agents.

Mixed Surface Films

There is abundant evidence that the mixed surface-films from combinations of water-insoluble long-chain polar compounds and ionic surfactants exhibit high viscosity at the water-air interface. The state of the surface film depends upon the hydrocarbon chain length of the polar compound. The nature of the polar group appears to be without importance. In general, if the polar compound contains less than about six or seven carbon atoms, the surface film is gaseous. Polar compounds containing seven to ten carbon atoms in the molecule give liquid-condensed films of low surface viscosity, when present in combination with the ionic surfactant. If the polar compound contains twelve or more carbon atoms per molecule, condensed films of high viscosity are obtained.

Combinations of ionic surfactants with short-chain polar compounds, giving gaseous surface films, produce unstable O/W emulsions. With medium chain-length polar compounds, the emulsion stability is enhanced, provided that the polar compound is present in about five to ten times the molar amount of the ionic surfactant. The long-chain polar compounds are effective emulsion stabilizers, when present at the same molar concentration as the ionic surfactant. The most stable O/W emulsions are obtained when the ionic surfactant and the water-insoluble polar compounds have the same hydrocarbon chain length (44).

Combinations of long-chain water-insoluble polar compounds and ionic surfactants that form high-viscosity surface films produce stable foams. These combinations also form relatively stable O/W emulsions. The stabilization of foams by long-chain polar compounds can be ascribed to a mechanical effect. However, this mechanism need not apply to emulsions. The presence of an oil phase reduces van der Waals interaction between the hydrocarbon chains in the surface film. Consequently, the viscosity at the water-oil interface is much less than at the water-air interface. At lower temperatures high viscosity at the water-oil interface has been observed, and this has been related to emulsion stability (45).

It is evident that the mixed films at the water-oil interface are more closely packed than films composed entirely of ionic surfactant. Ions alone repel one another and give gaseous films. As shown in Equation 11.53, increased surface coverage θ would be sufficient to explain the greater emulsion stability.

While there is a lack of direct evidence for ion-dipole interaction between the ionic group of the water-soluble surfactant and the polar group of the water-insoluble compound, the close proximity of these groups suggests such interaction. For these mixed films, it is likely that an additional energy term should be included in Equation 11.49. This corresponds to the ion-dipole interaction energy that must be overcome in transferring the ionic or polar group from the interface to the interior of the oil globule.

Mechanical Stabilization of Emulsions

In the previous sections it was tacitly assumed that the emulsifying agents are soluble in either or both phases of the emulsion. Under these conditions, the Gibbs adsorption equation is applicable. However, there are a large number of practical emulsions where at least one of the emulsifying agents is insoluble in both the oil and the water phases. This situation is more common than is generally realized.

Lecithin is commonly regarded as an oil-soluble emulsifier. However, if the system contains water, the lecithin becomes hydrated and forms an anisotropic phase insoluble in both bulk phases. Hydrated sorbitan esters of the long-chain fatty acids quite frequently swell to an anisotropic structure insoluble in both oil and water phases. Any emulsifier system that exhibits optical anisotropy at the concentration at which it is used can be considered to function as an insoluble emulsifier. Combinations of emulsifiers that give mixed interfacial films, such as a soap and a long chain alcohol, form a liquid crystalline or anisotropic phase when present at a sufficiently high concentration.

The anisotropic phase forms an interphase of considerable rigidity surrounding the emulsified droplets. Emulsions produced by means of an anisotropic phase are generally characterized as creams, ointments or plastic structures. The ratio of emulsifier components can be varied over broad limits without adversely affecting the stability of the emulsion, which is due to the mechanical barrier contributed by the interphase.

An insoluble interphase can be used to prepare both water-in-oil and oil-in-water emulsions. The emulsion type depends upon the wettability of the interphase. If it is preferentially wetted by the oil phase, the emulsion will be oil continuous. Otherwise, the emulsion will be water continuous. If the interphase is ionized, the emulsion will be water continuous.

Two other general types of emulsifiers stabilize emulsions due to a mechanical barrier. These are (1) finely-divided solids, and (2) polymeric materials.

The solid powders that stabilize emulsions, such as soot and clay, are insoluble in both phases. The emulsion type depends upon the wettability of the powder. The phase that preferentially wets the powder forms the continuous phase of the emulsion. Thus, a W/O emulsion will form with carbon black. The emulsion can be broken by the addition of a wetting agent, which makes the carbon particles wettable by water. The particles then pass into the water droplets.

Alberts and Overbeek (46) studied the stability of water-in-benzene emulsions stabilized with oil-soluble, ionizing emulsifiers. It was found that the emulsions with the highest surface-potentials were the least stable. These included emulsions stabilized by magnesium-, barium-, and zinc-petroleum sulfonates, and calcium- and aluminum-oleates. Ferric-, magnesium-, lead-, and cupric-oleates produced low surface-potentials and stable emulsions. It was found that the low surface-potentials were due to partial hydrolysis of the soaps. The products of hydrolysis are insoluble in both water and benzene, accumulate at the interface, and interfere with the formation of an electrical double layer in the oil phase. It appears that many soaps that stabilize W/O emulsions, do so as the result of hydrolysis to form particles that are insoluble on both phases.

Numerous observations have been made to the effect that proteins and other water-soluble polymers give high-viscosity films at the oil-water interface. The presence of a mechanical barrier at the interface is sufficient to explain their action in stabilizing emulsions.

Review of Factors Involved in Emulsion Stability

For emulsions of either the O/W or W/O type stabilized by nonionic emulsifiers, the HLB system provides a method by which stable emulsions can be prepared. The required HLB of the oil phase can be determined by preparing emulsions using mixtures, in varying proportions, of a water-soluble and water-insoluble nonionic emulsifier, both with known HLB values. Once the required HLB value of the oil phase is known, different types of emulsifier pairs can be tested by the preparation of emulsions, to obtain the combination that gives the most stable emulsion. As an example, the required HLB value for a mineral oil-in-water emulsion is 10. Sorbitan monooleate has an HLB value of 7, while sorbitan monooleate with 20 ethylene oxide groups has an HLB value of 15. Then $7X+15\,(1-X) = 10$, $X = 0.625$. The proper combination is 0.625 parts by weight of sorbitan monooleate plus 0.375 parts by weight of the ethoxylated sorbitan monooleate. The emulsion could be compared with one containing glyceryl monostearate and an ethoxylated stearic acid, in proper proportions, as the emulsifier pair.

The emulsion can be broken by adding an emulsifier that will throw the emulsifier combination out of balance with the required HLB for the oil. It was shown that the HLB system does have a theoretica basis. The stability of emulsions stabilized with nonionic emulsifiers can be described in terms of the energy required to desorb the emulsifier into the discontinuous phase.

Emulsions stabilized with nonionic emulsifiers are frequently in a flocculated state, because of the absence of an electrical barrier. However, these systems may remain stable, without evidence of appreciable coalescence, for many years.

With emulsions stabilized by ionic emulsifiers, flocculation is resisted by the charge on the particles. Frequently, flocculation is the slow step, followed quickly by coalescence. For such systems, the Verwey-Overbeek theory is a reasonable description. The log (rate of coalescence) $\sim 0.24\psi_0^2/RT$. Added electrolytes have an adverse effect on stability.

Resistance to coalescence is enhanced by the addition of a long-chain polar compound to the emulsion containing an ionic surfactant. Thus, the addition of cetyl alcohol will stabilize an O/W emulsion containing sodium cetyl sulfate. Soaps at their natural pH are better emulsifiers than soaps in more alkaline solution. This is due to the presence of the free fatty acid at the lower pH. In general, the most stable emulsions are obtained using an additive of the same hydrocarbon-chain length as the ionic emulsifier.

Combinations of polar additives and ionic surfactants that produce stable emulsions invariably produce films at the water-air interface with high surface viscosity, showing that the monomolecular films are condensed. Surface viscosity at the water-oil interface has been observed by some workers (45), but not by others (47). It is evident that with these mixed surface films, the interfacial area is more completely covered, than when the ionic surfactant is the only emulsifier. As shown in Equation 11.53, this greater coverage is sufficient to account for enhanced emulsion stability, without the necessity for assuming a mechanical barrier.

It has been shown that polymers that stabilize O/W emulsions also produce monomolecular films at the oil-water interface with pronounced surface viscosity (48). Emulsion stability may be due to the mechanical barrier or to the difficulty of desorbing a polymeric material. Insoluble emulsifying agents and finely divided solids also stabilize emulsions due to a mechanical barrier. The emulsion will be W/O if the insoluble material is preferentially wet by the oil phase. Otherwise it will be O/W.

MICRO EMULSIONS

The emulsions thus far discussed are macro emulsions, with the droplet diameter seldom smaller than 0.5 micron. Micro emulsions can be readily produced with the diameter of the droplets in the range of 100 to 600 Å (49). These small droplet sizes form spontaneously, without the necessity for homogenizing or colloid milling. Further, the micro emulsions are thermodynamically stable. Since the diameter of the droplets is less than 1/4 the wavelength of light, the systems are transparent (50).

Negative Surface Tension (49)

Bowcott and Schulman (50) showed that in order to form micro emulsions it is necessary that the concentration of emulsifiers must be greater than that required to reduce the oil-water interfacial tension to zero. This indicates that the interfacial tension γ_i must have a metastable negative value, which would give a negative free energy variation. This would cause droplets to break up spontaneously, stabilize the dispersed phase, and prevent phase separation.

The total interfacial tension can be defined

$$\gamma_i = \gamma_{ow} - \pi \tag{11.54}$$

where γ_{ow} is the oil-water interfacial tension in the absence of the emulsifying agents, and π is the spreading pressure of the added emulsifiers at the oil-water interface.

If $\pi > \gamma_{ow}$, there will be a negative interfacial tension and micro emulsions will form due to the free energy available to increase the interfacial area. The equilibrium condition will be reached when $\pi = \gamma_{ow}$ or at zero interfacial tension. If $\pi < \gamma_{ow}$, macro emulsions will form. Since the interfacial tension is positive, droplets will tend to coalesce.

The metastable negative interfacial tension cannot be measured directly, since the interface emulsifies spontaneously. However, π can be calculated directly if a counter tension is placed on the interfacial measuring device, such as a Wilhelmy plate, to prevent surface breakup.

$$\gamma_i = \gamma_{ow} - \pi = \gamma_{aw} - \gamma_{oa} - \pi_m \tag{11.55}$$

The right term is that obtained by the pull on a Wilhelmy plate at a duplex film interface, with π_m the measured surface pressure. γ_{aw} is the surface tension of the air-water interface, and γ_{oa} is the surface tension of the oil-air interface. From the experimentally determinable terms, π and γ_i can be calculated.

The duplex films under consideration consist of a mixed monolayer of emulsifier molecules penetrated by the oil phase molecules. These are formed by first penetrating a monolayer composed of a long-chain fatty acid or fatty alcohol by an anionic surfactant in the aqueous phase. This produces a high surface pressure, π. This mixed film is now penetrated by nonpolar oil molecules which have been mixed into the system. Since the ratio of oil molecules to the amphipathic molecules is made high, about 50:1, on compression of the mixed film on a Langmuir trough, some of the oil molecules will be incorporated in the mixed film and others will be squeezed out. The latter will spread on the mixed monolayer in the form of a very thin oil film, about 150 to 500 Å thick. Color interference patterns are observed as the mixed monolayer is compressed. The tension at the newly formed oil-air interface γ_{oa} holds the interface together and permits π_m to be measured, although π is already greater than γ_{ow}. The thickness of the duplex film is approximately equal to the diameter of oil droplets in a micro emulsion.

Mechanism of Formation of Micro Emulsions (49)

Surface pressures of about 35 dynes/cm can be achieved by spreading a long-chain alcohol or cholesterol on water. Penetration of the monolayer by an ionic surfactant, such as the salt of a long-chain sulfate and an alkyl amine can increase the surface pressure to more than 60 dynes/cm. If the surface pressure is held constant at a value between 35 and 60 dynes/cm, immediate expansion of the interfacial area takes place as the ionic surfactant penetrates the monolayer as the air-water interface.

Similarly, penetration of the mixed monolayer by the oil molecules at the oil-water interface causes an expansion of the interfacial area, and this is the basis for the formation of micro emulsions. The penetrating pressure in this case must be greater than the oil-water interfacial tension γ_{ow}. This is approximately 50 dynes/cm for alkanes and 35 dynes/cm for aromatic hydrocarbon compounds. Thus, it is easier to obtain negative surface tensions and micro emulsions with aromatic hydrocarbon oils, provied that the oil molecules can penetrate the interfacial mixed film.

Micro emulsions can be prepared using aqueous solutions containing 10 to 40 per cent of ionic surfactant. In these concentrated solutions, the surfactant micelles have a definite lamellar structure that can be swelled to only a limited extent by nonpolar oils. Above the Krafft temperature, these lamellar micelles are completely dispersible in water. When these micelles are penetrated by polar amphipathic molecules, such as medium-chain alcohols, they are able to swell almost without limit with both oil and water so long as there is a condensed interphase. This almost unlimited swelling is observed if the alcohol chain length is between 5 and 8. The effect of decanol is borderline, while the higher alcohols are without effect. The medium chain-length alcohols appear to penetrate between the anionic surfactant molecules and disorder the regular condensed two dimensional packing in the micelles to produce a liquid interphase. This effect can also be brought about by raising the temperature, or by using an ionic surfactant with a large counter ion, as in the case of ethanolamine oleate (50).

The lamellar structure may be considered as an interphase between oil and water. If sufficiently fluid, surface tension forces will produce a curvature according to the difference in tension between the hydrocarbon portion of the interphase molecules and the oil phase, and the hydrophilic portion of the interphase molecules and the water phase. The effect of the formation of a mixed interfacial film containing the water-soluble ionic surfactant and the oil-soluble amphipathic molecule is to lower the interfacial tension between the oil and water phases to only a fraction of a dyne/cm. A micro emulsion will form if the interface is liquid. In the case of emulsions of the type stabilized by a straight chain sulfate and cholesterol, the complex is too strongly condensed to be penetrated by oil molecules and the interfacial film cannot assume a large enough curvature to form droplets below about one micron diameter.

Preparation of Micro Emulsions (51)

In order to prepare micro emulsions it is necessary to obtain a negative interfacial tension. This can be achieved by the use of a sufficient concentration of an emulsifier combination that gives a mixed interfacial film. It is further required that the interfacial monolayer not be too highly condensed. Otherwise, the curvature necessary for very small droplets cannot be achieved. To obtain this disorder in the condensed monolayer it appears to be necessary for the molecules of the oil phase to interpenetrate or associate with the mixed interfacial film. Association may take place between the oil molecules and either component of the mixed interfacial film. Better micro emulsions are obtained if the oil molecules associate with both components.

If soap is used as one component of the mixed interfacial film, the use of large positive ions helps to produce disorder in the interfacial film. The cation 2-amino-2-methyl-1-propanol (AMP) is particularly effective in the formation of micro emulsions. Schulman and Montagne (49) showed that hydrogen bonding between the alcohol groups attached to the amino alkyl compound produces an open lattice structure that favors the penetration of the oil molecules into the mixed interfacial film.

In order to obtain a micro emulsion with a hydrocarbon oil, it is necessary that the chain length of the hydrocarbon oil not exceed the chain length of the components of the mixed interfacial film. The system cetyl alcohol and oleic acid neutralized to pH 10.5 will give micro emulsions with straight chain hydrocarbons varying in length from 7 to 18 carbon atoms. If the hydrocarbon chain length is greater than 18 carbon atoms no micro emulsions form. If the alcohol chain length is lengthened so that it is greater than that of the hydrocarbon, micro emulsions again form.

The above systems do not give micro emulsions with benzene. However, a micro emulsion will form if the straight chain alcohol is replaced by p-methylcyclohexanol.

Instead of using an alcohol in combination with soap, an acid soap may be used. At pH 8.8 there is a 1:1 ratio of free fatty acid to soap molecules. Above pH 10.5 kerosene will not form a micro emulsion with stearic acid or oleic acid and 2-amino-2-methyl-1-propanol, because of the absence of a second component necessary for the formation of a mixed interfacial film. A micro emulsion will form if stearyl alcohol is added, or if the pH is reduced to 8.7.

The technic of preparing micro emulsions is simple. For systems containing soap and a long chain alcohol, a crude emulsion may first be formed using the soap solution and the oil. This emulsion can then be titrated to transparency with the long chain alcohol. Alternatively, the soap, alcohol and the oil phase can be combined, and the mixture titrated to transparency with water under pH control. In the region of high soap concentration, where the diffuse double layer is suppressed, the system is a water-in-oil micro emulsion. As additional water is added, the soap becomes ionized, a high surface charge is developed, and the emulsion inverts to an oil-in-water micro emulsion. During this inversion the system becomes strongly viscoelastic and anisotropic, but again reverts to an isotropic fluid system when in the form of an oil-in-water micro emulsion.

The phase continuity of micro emulsions is governed by the same factors as macro emulsions. If the emulsifying agent is not ionized and preferentially soluble in the oil phase, the emulsion will be water-in-oil. If it is not ionized and preferentially soluble in water, the emulsion will be oil-in-water. If the interfacial film is charged, the emulsion will be oil-in-water. The addition of salt to remove the diffuse double layer can result in a phase reversal (49).

References

1. Bikerman, J. J., "Surface Chemistry", pp. 155–8, New York, Academic Press, 1958.
2. Harkins, W. D., "The Physical Chemistry of Surface Films", pp. 86–90, New York, Reinhold Publishing Corp., 1952.

3. Griffin, W. C., "Encyclopedia of Chemical Technology", edited by R. E. Kirk, and D. F. Othmer, Vol. 5, pp. 692–718, New York, Interscience Encyclopedia, 1950.
4. Clayton, W., "Theory of Emulsions", Philadelphia, The Blakiston Co., 1943.
5. Becher, P., "Emulsions", p. 46, New York, Reinhold Publishing Corp., 1957.
6. Bowcutt, J. E., and Schulman, J. H., *Z. Elektrochem.* **59**, 283 (1955).
7. American Cyanamid Company, "Aerosol 22 Surface-Active Agent", 1958.
8. Lawrence, A. S. C., and Mills, O. S., *Disc. Faraday Soc.* **18**, 98 (1954).
9. Kruyt, H. R., "Colloid Science", Vol. 1, p. 16, New York, Elsevier Publishing Company, 1952.
10. Einstein, A., *Ann. Physik* **19**, 371 (1906).
11. Von Smoluchowski, M., *Ann. Physik* **21**, 756 (1906).
12. Ref. 3, p. 695.
13. Gribnau, Fr. B., Kruyt, H. R., and Ornstein, L. S., *Kolloid Z.* **75**, 262 (1936).
14. Ref. 9, pp. 34–9.
15. Gumprecht, R. O., and Sliepcevich, C. M., *J. Phys. Chem.* **57**, 90 (1953); **57**, 95 (1953).
16. Mie, G., *Ann. Physik* **25**, 377 (1908).
17. Gumprecht, R. O., and Sliepcevich, C. M., "Tables of Light-Scattering Functions for Spherical Particles", University of Michigan, Engineering Research Institute, Special Publication: Tables; Ann Arbor, Michigan, 1951.
18. Penndorf, R. B., *J. Phys. Chem.* **62**, 1537 (1958).
19. Fradkina, E. M., *Zhur. Eksptl. Teoret. Fiz.* **20**, 1011 (1950).
20. Lifshits, S. G., and Teodorovich, V. P., *Energet. Byull*, **1947**, No. 8, 16.
21. Henry, D. C., *Proc. Roy. Soc.* (London), A, **133**, 106 (1931).
22. Powell, B. D., and Alexander, A. E., *Can. J. Chem.*, **30**, 1044 (1952).
23. Van der Minne, J. L., and Hermanie, P. H. J., *J. Colloid Sci.* **8**, 38 (1953).
24. Ref. 9, pp. 22–3.
25. Einstein, A., *Ann. Physik* (4) **19**, 289 (1906); **34**, 591 (1911).
26. Van der Waarden, M., *J. Colloid Sci.* **9**, 215 (1954).
27. Richardson, E. G., *J. Colloid Sci.* **5**, 404 (1950).
28. Richardson, E. G., *J. Colloid Sci.* **8**, 367 (1953).
29. Brown, G. L., and Myers, R. J., *Soap and Sanitary Chem.*, July.
30. Griffin, W. C., *J. Soc. Cosmetic Chem.* **1**, 311 (1949).
31. Davies, J. T., Proc. Intern. Congr. Surface Activity, 2nd, London, Butterworth, 1957.
32. Verwey, E. J. W., and Niessen, K. F., *Phil. Mag.* **28**, 435 (1939).
33. Van den Tempel, Rubber Stichting, Communication No. 225 (1953).
34. Sumner, C. G., *J. Appl. Chem.* **7**, 504 (1957).
35. Verwey, E. J. W., *Trans. Faraday Soc.* **36**, 192 (1940).
36. Powney, J., and Wood, L. J., *Trans. Faraday Soc.* **37**, 152, 220 (1941); 36, 57 (1940).
37. Mukerjee, P., Mysels, K. J., and Dulin, C. I., *J. Phys. Chem.* **62**, 1390 (1958); Mukerjee, P., ibid., 1397 (1958); Mukerjee, P., and Mysels, K. J., ibid., 1400 (1958); Mukerjee, P., ibid., 1404 (1958).
38. Fava, A., and Eyring, H., *J. Phys. Chem.* **60**, 890 (1956).
39. Verwey, E. J. W., and Overbeek, J., Th. G., "Theory of the Stability of Lyophobic Colloids", New York, Elsevier Publishing Co., 1948.
40. Ref. 9, p. 88.
41. Davies, J. T., and Rideal, E. K., *J. Colloid Sci., Suppl.* 1, 1 (1954).
42. Anderson, P. J., *Trans. Faraday Soc.* **55**, 1421 (1959).

43. Cockbain, E. G., and McRoberts, T. S., *J. Colloid Sci.* **8**, 440 (1953).
44. Lawrence, A. S. C., *Soap, Perfumery and Cosmetics* **33**, 1180 (1960).
45. Davies, J. T., and Mayers, G. R. A., *Trans. Faraday Soc.* **56**, 691 (1960).
46. Alberts, W., and Overbeek, J. Th. G., *J. Colloid Sci.* **14**, 501 (1959).
47. Blakey, B. C., and Lawrence, *Disc. Faraday Soc.* **18**, 268 (1954).
48. Nielson, L. C., Wall, R., and Adams, G., *J. Colloid Sci.* **13**, 441 (1958).
49. Schulman, J. H., and Montagne, J. B., *Ann. N.Y. Acad. Sci.*, **92**, Art. 2, 366 (1961).
50. Bowcott, J. E., and Schulman, J. H., *Z. Elektrochrm*, 59, **283** (1955).
51. Schulman, J. H., Stoeckenius, W., and Prince, L. M., *J. Phys. Chem.*, **63**, 1677 (1959).

CHAPTER 12

Foams

Foam consists of bubbles of gas whose walls are thin liquid films. Foams have rigidity as well as elasticity, and many foams can actually be cut with a knife, like a solid. They have a definite structure, with the arrangement of bubbles such that three films come together in one edge forming solid angles of 120° each, and not more than four edges form one corner. Depending upon the thickness of the liquid walls of the bubbles, the foam can be almost as dense as the liquid or almost as light as the gas comprising the interior of the bubbles. The diameter of foam bubbles can vary from several inches down to fractions of a micron (1).

Pure liquids do not foam and it is necessary to have at least two components present in a foaming liquid. While detergent solutions are ordinarily considered with regard to foaming systems, many other combinations are capable of producing stable foams. Aqueous solutions of proteins and other water-soluble polymers produce lasting foams. Even salt solutions foam. Non-aqueous solutions have been shown to produce very stable foams (1,2).

The formation of foam involves the expansion of the surface. The work required to produce foam is equal to the product of the foam surface-area and the surface tension. Here we are concerned with a dynamic rather than an equilibrium surface tension value, that is, the surface tension of a rapidly expanding surface. Thus, the lower the dynamic surface tension, the less work required to expand the surface. However, it does not follow that a lower surface-tension will result in a more voluminous foam. If the foam is very unstable, bubbles can collapse as rapidly as they are formed and no foam will be produced, as in the case of pure organic liquids of low surface-tension.

Single Soap Films (3)

Individual soap films formed from aqueous surfactant solutions have been studied extensively, initially to obtain information on the nature and ranges of molecular forces (4, 5), and more recently with the object of obtaining information on the behavior of films and emulsions (1, 2, 3).

Single soap films are readily formed, as by dipping a ring formed from glass or wire into a surfactant solution and withdrawing it partially. If protected from evaporation, the film can last for days or weeks. A broad source of light reflected on the film will show interference colors indicating the thickness of the film at each point. The light is reflected by the soap film at two interfaces, due to the difference in refractive index between the soap solution and air. The intensities of the two reflected beams are exactly equal, because the two changes in refractive index are equal, although in opposite directions. If the thickness of the soap film is a quarter of a wave length, the two beams are exactly in phase upon normal reflection, and the reflected light is of maximum intensity. A half wave-length thickness gives maximum destruction. A soap film of extreme thinness has a black appearance when viewed against a black background, in contrast to the multicolored reflections from the thicker films, and since it reflects no light of any color is called a black film. As the thickness of the film increases, the intensity of its reflection for each color increases to a maximum at its quarter wave-length and then decreases. When the thickness is about a quarter of the wave-length of the average color, the reflection is essentially white, and the film is called silvery.

Based on the observation of a large number of individual soap films, Mysels and co-workers (3) found considerable variations in their drainage properties, but considered three extreme types to be characteristic of a number of classes of soap films.

(1) Rigid films have surfaces offering high resistance to flow within the plane of the film, and drainage is very slow. These films apparently result from the formation of solid mixed surface layers. The film is initially very thick. Fringes and subsequently a black film slowly form along the border.

(2) The simple mobile film is frequently observed with commercial surfactants at concentrations above the cmc up to about 2 or 3 per cent. It drains in a rapid but orderly manner, with turbulent motion visible along the borders of the film. Horizontal smooth bands of color form in the center, with the boundary between black and silver colored films horizontal, except near the border.

(3) The irregular mobile films are observed at higher concentrations of commercial surfactants, or in more dilute solutions containing salt. These films appear as simple mobile films until the appearance of the black film. Its boundary with the colored part becomes highly irregular, and the film

gains the appearance of peacock feathers, due to the formation of streamers of black film. Thinning to a black film is especially rapid.

The mechanism of thinning of rigid films is much less complex than the thinning of mobile films. In general, the thinning of rigid films can be adequately explained by evaporation and by the downward flow of liquid between surfaces that are rigid in the plane of the film. Thinning by viscous flow results in a film having a parabolic cross section, with a black film at the top, and the maximum thickness at the lower border.

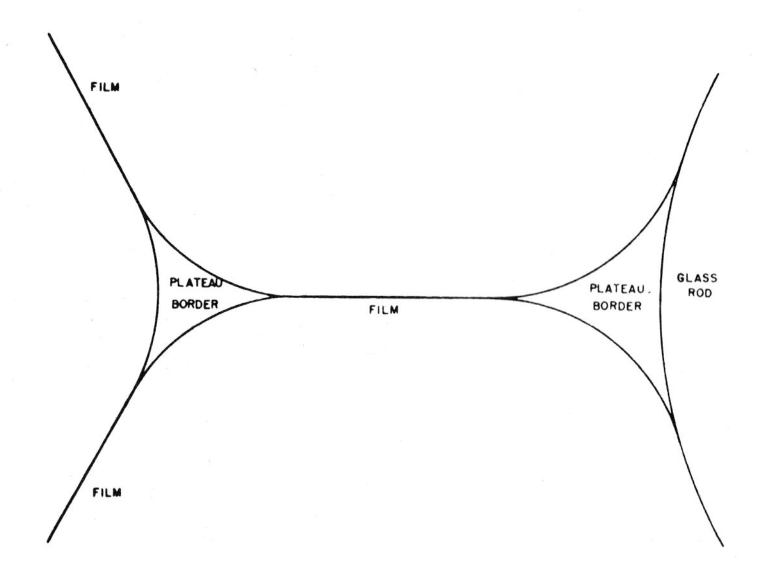

Figure 12.1. Cross section of a segment of film hanging between two Plateau borders. Three films meet at the border on the left while that on the right is supported by a glass rod of a film frame (5).

The major mechanism responsible for the thinning of mobile films is a constant regeneration of the film at the borders. The border is a region of sharp curvature of the surface, concave outward. This curvature corresponds to a pressure within the liquid that is less than the adjacent atmospheric pressure. The border region is called the "Gibbs ring" or "plateau border", and it occurs with both single films and films collected together as foam, as shown in Figure 12.1. The negative pressure is due to the force of gravity acting on the liquid. The suction at the Plateau border draws liquid out of the film, which thins at the border. The diminished thickness at the edges causes a rapid upward current on each side, while the central portion slowly descends. Thus, the thin elements that originate at the border rise under the action of gravity until they reach elements corresponding to their thickness in the film. Elements of the film with different thicknesses are two-dimensional systems, with the thinner elements being less dense.

According to Mysels, thick regions at a border are not merely drained by the border, but the whole film is sucked into the border. Since the total surface area remains constant, other regions must be created to take their place. Thus, thin films may be sucked out of the border. Careful observations appear to confirm that thick films are sucked into the border and the thinner film is pulled out in adjacent sections.

The plausibility of marginal regeneration is suggested by the fact that a film element, because of its thinness is very stable and little flow can occur within the walls of the film. The negative pressure at the border could more readily suck in the entire film than it can the fluid between immobile walls. If films of different thickness are in contact with the same border, the thick film would be subjected to the greater force and would tend to disappear into the border. In the transition region between the plateau border and the bulk of the film, there is a localized pressure drop which acts as a "wringer", so that film segments drawn from the border are quite thin.

Foam Stability

Foams are thermodynamically unstable, since their collapse is accompanied by a decrease in total free-energy. However, certain foams will persist for long periods while others break immediately after they are formed. Foams collapse as the result of drainage of liquid in the bubble walls until a portion of the film reaches a thickness of about 50 to 150Å, when the random motion of molecules is sufficient to cause the sudden breakdown of the film. However, the rate of drainage is not the only factor affecting stability. Fast-draining foams can be quite as stable as those which drain slowly. The important factor which determines the stability of a film Gibbs (6) called the elasticity of the film, and defined as

$$E = 2A\,(d\gamma/dA)_{T,N_1,N_2} \qquad (12.1)$$

where E, A, and γ are the elasticity, area and surface tension, respectively. T is the temperature, and N_1 and N_2 refer to the components of the system.

The elasticity is the tendency of a film to resist deformation. For a film containing an adsorbed surfactant, stretching of the film will decrease the concentration of foaming agent at the surface and increase the surface tension, thus increasing the work required for further enlargement of the surface area. Similarly, contraction of the film will decrease the surface tension by increasing the surface excess, which will oppose further contraction.

For pure liquids, the surface tension does not change with a change in area and the elasticity is zero. This is the theoretical basis for the observation that pure liquids do not foam. The elasticity of Gibbs depends on the equilibrium surface-tension of the stretched film. He noted that the restoring force may be greater than the elasticity under non-equilibrium conditions.

There are three causes for the thinning of bubble walls. (1) Liquid drains from the bubble walls owing to gravity. (2) There is a suction effect at the periphery of the film due to the high curvature of the film. This effect can be more pronounced than the effect of gravity, resulting in a pronounced thinning adjacent to the concave edge of the film. These thin portions of the film situated between thicker patches are unstable and rise in the film until they reach a stable position in a portion of the film of the same thickness. (3) Where evaporation of the solvent occurs, thinning of the film is accentuated.

Perhaps the most important concept, since Gibbs, in elucidating the mechanism of foam stability is the surface-transport theory of Ewers and Sutherland (7). According to this theory, the surface of a film will flow from a region of low surface-tension to one of high surface-tension. In doing so, it will drag underlying water molecules with it. A region of low surface-tension is one in which there is a relatively high concentration of adsorbed surfactant as compared to a high surface-tension area. Just as a gas distributes itself uniformly throughout a container, adsorbed surfactant will tend to distribute itself uniformly over a surface. Any movement of one layer of molecules will cause movement in the same direction, but to a lesser extent, on the part of underlying molecules.

The relationship between film elasticity and surface transport is evident. If a stretched film has a positive elasticity value, there will be an increase in surface tension at the stretched, or thinned, area. Surface transport will tend to restore the film to its original thickness.

However, there are two competing processes to consider. One is the rate of spreading of the surfactant from the region of low to that of high surface-tension. The other is the rate of adsorption of the surfactant from the bulk of the solution. If the rate of adsorption is rapid, the deficiency of surfactant molecules in the stretched surface may be made up largely by adsorption from the bulk of the film. This will destroy the surface tension gradient and the amount of solution returned by surface transport will not be sufficient to restore the original thickness of the film. The low stability of froth of alcohol-water solutions can be accounted for by the comparatively high rate of adsorption of alcohol. This can be contrasted with aqueous solutions of surfactants which are adsorbed at much lower rates below their cmc value. At higher concentrations, they are adsorbed more rapidly and show a corresponding decrease in foam stability.

Surface transport is consistent with the principles of hydrodynamics and has been observed by other workers in the field. In combination with film elasticity and dynamic surface-tension it provides a treatment that agrees, at least qualitatively, with experience.

Foams are subject to various types of disturbances which influence their longevity. Isolated foams are under the influence of gravity. Foams normally exposed to the atmosphere encounter mechanical and thermal shock, evaporation, and absorption of carbon dioxide. In washing, water-soluble and water-insoluble components are introduced into the foaming solution. In addition, the foaming agent is partially depleted by adsorption on the substrate and on suspended soil-components.

Regardless of the nature of the stress acting on a foam, the mechanism of collapse is always the same. Liquid in the bubble walls drains until a portion of the film becomes sufficiently thin for the ordinary motion of the molecules to cause breakdown. Several of the factors which influence the rate of drainage follow (7):

(1) High viscosity of the liquid in the bubble walls will reduce the rate of thinning. It will also tend to dissipate shock, thus increasing the persistence of the film.

(2) High viscosity of surface films will act as a viscous drag on neighboring molecules, reducing the rate of drainage.

(3) Gas permeability is an important factor in foam stability. A condensed surface-film will constitute a greater barrier to air diffusion than a gaseous surface-film, thus tending to preserve the foam. If a gas that is soluble in the foaming medium is used to form the foam, the bubbles will shrink rapidly, with correspondingly rapid liquid drainage and foam collapse.

(4) Electrostatic repulsion between similarly charged film surfaces reduces the rate of drainage. When the film has thinned to an extent that there is interference between the electrical double-layers associated with the surfaces, further thinning will be opposed by the electrostatic repulsion between the surfaces. Depending upon the ionic strength of the solution, the film thickness will be reduced to about 1000 Å before electrostatic repulsion is effective in retarding further thinning.

Viscous Surface Films

Aqueous solutions of pure surfactants in water frequently produce foams which drain rapidly and are unstable. The addition of a third component will sometimes greatly increase the stability of the foam. Frequently, the foam is observed to have become slow draining and the surface film viscous. Plastic flow may also be observed.

The system sodium lauryl sulfate-lauryl alcohol-water has been investigated most thoroughly. The addition of a few per cent of lauryl alcohol, based on the sulfate, changes the unstable foam to a stable slow-draining foam. For this and similar three-component systems there is a sharp transition temperature, above which slow-draining foams drain rapidly (8).

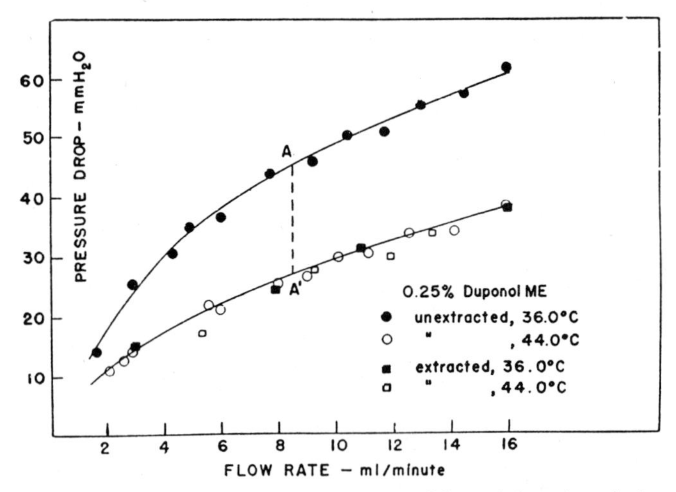

Figure 12.2. Pressure drop *versus* rate of flow. A is a slow-draining foam, while A[1] is a fast-draining foam (8).

Figure 12.2 illustrates the difference between fast- and slow-draining foams with a 0.25 per cent aqueous solution of Duponol ME, a commercial material which contained 88.1 per cent of mixed straight-chain alcohol sulfates—largely sodium lauryl sulfate—6.1 per cent of alcohol insolubles—sodium sulfate—5.1 per cent of ether-soluble materials—largely unsulfated alcohols—and 0.2 per cent of moisture. In the figure, a large pressure-drop corresponds to a large quantity of solution in the foam, or slow drainage. The Duponol ME solution is slow draining at 36.0°C but fast draining at 44.0°C, which is above the transition temperature of 41.5°C. After removal of the alcohols by extraction, the solution gives fast-draining foams at both temperatures.

The fact that the transition temperature can be estimated with considerable precision is shown in Figure 12.3. As the temperature of the foam is varied about the transition temperature, sharp changes in pressure drop and output occur. The output refers to the rate of flow of foaming solution in the column in which the foam is produced. Transition from fast to slow drainage corresponds to a decrease in output. The electrical resistance of the foam can also be used to determine the transition temperature (8).

For pure sodium lauryl sulfate and pure lauryl alcohol, the transition temperature varies with the concentrations of both solutes. Even below the cmc—0.234 per cent for pure sodium lauryl sulfate—transition temperatures have been found. Isotherms of film-drainage transition-temperatures are

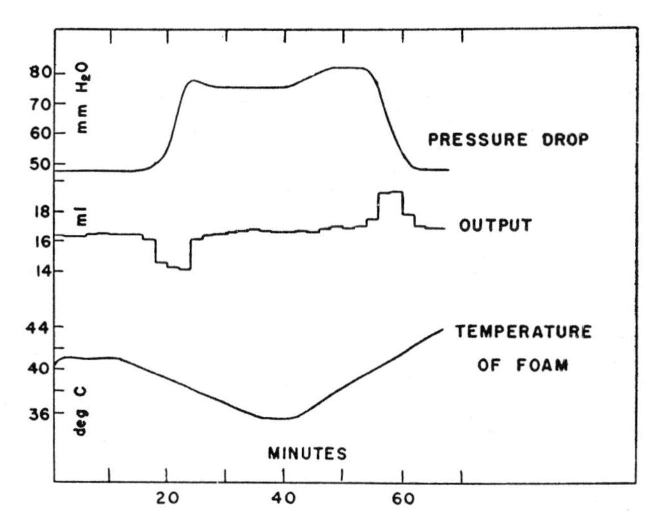

Figure 12.3. Foam transitions with 0.25 per cent Duponol ME, not extracted (8).

shown in Figure 12.4. A sharp break occurs near the cmc of the pure alcohol sulfate and there is an almost linear relationship between alcohol sulfate and alcohol concentration above the break. For concentrations above the cmc, all the data for a three-component system can be reduced to a single curve by plotting the transition temperatures against Z, the "apparent micellar mole fraction of alcohol". This mole fraction term was computed from the moles of alcohol sulfate diminished by the cmc, and the moles of alcohol in the system. Four plots are shown in Figure 12.5. They suggest that above the cmc, the transition temperature is a function of the composition of the micelles. At the lower end the curves appear to be bound by the Krafft temperature, below which micelles cease to exist for lack of solubility of single ions. At the upper end there is a limiting value for the transition temperatures and further additions of alcohol have no effect. In some instances, higher transition-temperatures are found in the region below the cmc. The break in the log Z curve appears to correspond to the first appearance of birefringence and turbidity in the solution (9).

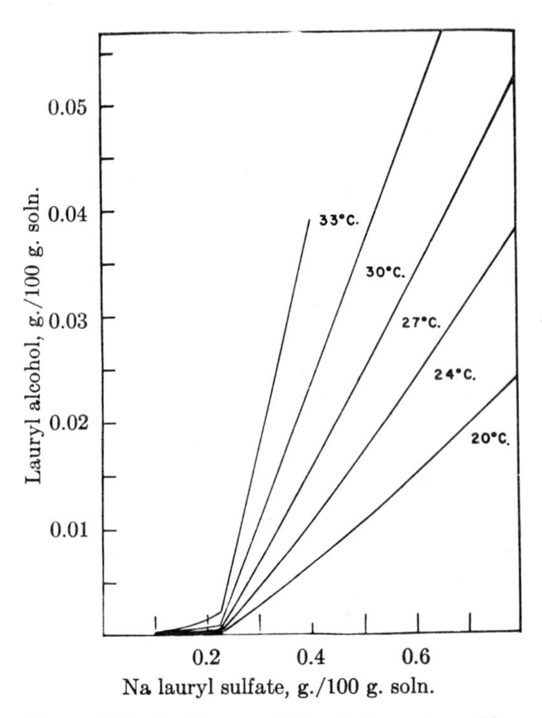

Figure 12.4. Isotherms of film drainage transition temperature (9).

The effect of sodium chloride on film drainage transition temperatures of solutions of sodium lauryl sulfate and lauryl alcohol is shown in Figures 12.6 and 12.7. The transition temperatures fall with increasing salt concentration. At each sodium chloride concentration, solutions which have the same calculated micelle compositions have similar transition temperatures. In all instances, the number of micelles present appears not to be a factor (10).

Crystals were obtained by cooling the solutions used in the foaming experiments. Analysis showed the crystals to be adducts which approached two compositions, a mole ratio of sodium lauryl sulfate to lauryl alcohol of 1:1

—39.3 per cent alcohol—and a mole ratio of 2:1—24.5 per cent alcohol. The composition of the adducts is shown in Table 12.1.

Epstein suggested that over the micellar region, constant transition temperatures mean constant adsorption of alcohol sulfate and alcohol corresponding to an essentially constant composition of the non-micellar medium.

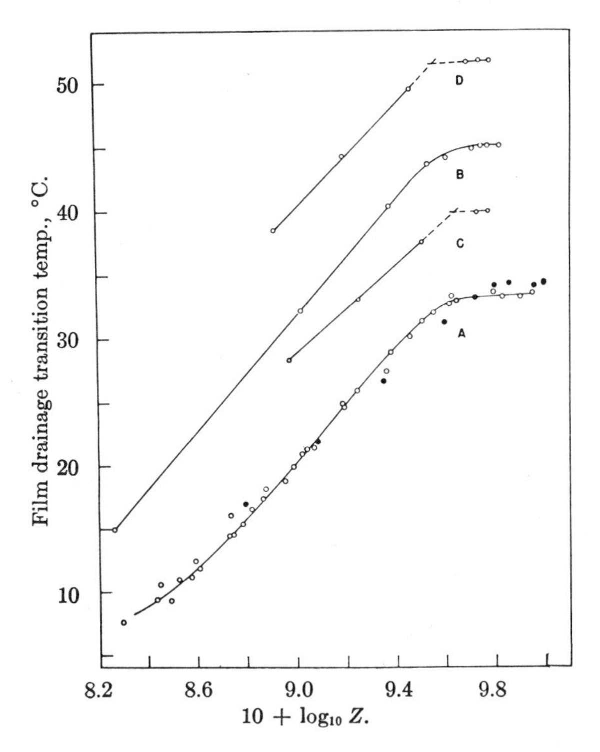

Figure 12.5. Film drainage transition temperature *vs.* log *Z*: A, sodium lauryl sulfate + lauryl alcohol (solid circles, 0.25% sodium lauryl sulfate; open circles, higher concn.). B, 0.6% sodium lauryl sulfate + myristyl alcohol. C, 0.2% sodium myristyl sulfate + lauryl alcohol. D, 0.2% sodium myristyl sulfate + myristyl alcohol (9).

The micelles act as reservoirs to maintain uniform the composition in the non-micellar medium. The phenomenon of slow drainage is believed due to an ordered surface-structure of considerable rigidity. It is not suggested that the composition of the surface film is the same as that of the crystalline adducts (10).

Several workers have confirmed that surface viscosity and foam life are

TABLE 12.1. COMPOSITION OF ADDUCTS (10).

NaCl, (moles/l)	Initial Solution Sodium Lauryl Sulfate (g./100 g. soln.)	Alcohol, (g./100 g. soln.)	Alcohol in Adduct (% by wt.)
nil	0.40	0.032	25.80
	.80	.070	36.07
	.80	.126	23.84
	.80	.151	24.47
	.81	.191	38.71
0.02	.80	0.057	23.15
	.80	.087	23.65[a]
	.81	.206	35.95[a]
	.81	.418	38.64
	.80	.539	38.95

[a] Examined for chloride.

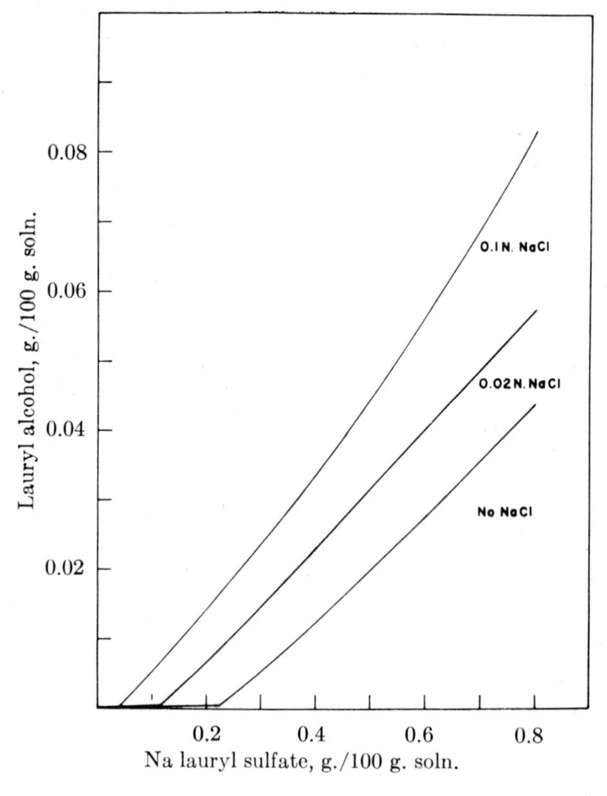

Figure 12.6. 25° Isotherms of film drainage transition temperatures (10).

related. Brown, Thuman, and McBain (11) employed a rotational viscometer to measure surface viscosity. Typical curves relating the rate of rotation of the cup to the deflection of the bob are shown in Figures 12.8, 12.9 and 12.10. With pure sodium lauryl sulfate, the surface viscosity is essentially the same as water. A small amount of lauryl alcohol greatly increases surface viscosity. A greater ratio of lauryl alcohol is required to produce an appreciable viscosity, when the sodium lauryl sulfate concentration is above the cmc than at lower concentrations.

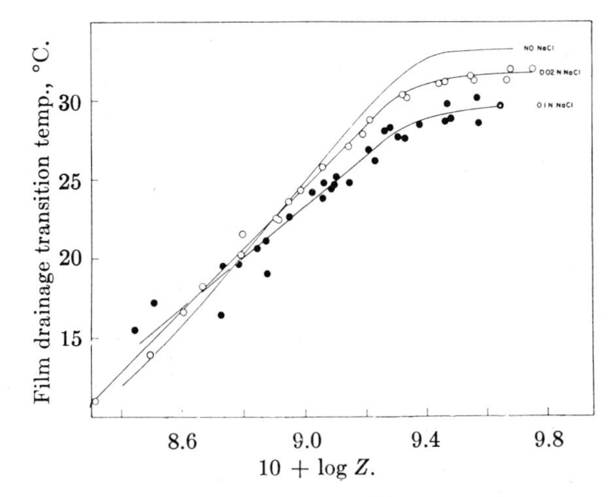

Figure 12.7. Film drainage transition temperatures *vs.* log *Z* (10).

Foam-life curves obtained with the solutions employed for the surface viscosity measurements are shown in Figures 12.11 and 12.12. It will be observed that in a general way, the solutions which showed the highest surface viscosities also yielded the most stable foams. In some instances, especially with the quaternary ammonium compounds, Quaternary O and E607L, the foams persisted though the surface viscosities of the solutions were low. However, these foams rapidly became thin and tenuous.

Davies (12) employed a viscous-traction surface viscometer. It will be observed from Figures 12.13 and 12.14 that both foam life and surface viscosity increase when lauryl alcohol or lauric-isopropanolamide are added to 0.1 per cent sodium laurate at pH 10.

It might be expected that an increase in surface viscosity, corresponding to a more compact surface-layer, would have the effect of decreasing the permeability of gas through the bubble walls. Brown, Thuman, and McBain

(13) studied the rate of decrease of the radius of individual bubbles blown with air. Tables 12.2, 12.3 and 12.4 compare air permeability, surface viscosity, and foam life. With sodium lauryl sulfate there is a narrow range of lauryl alcohol concentration through which the permeability of the solution changes sharply from a relatively high to a relatively low value. Increase in surface viscosity and decrease in air permeability first occur at different lauryl alcohol concentrations, possibly because surface viscosity involves

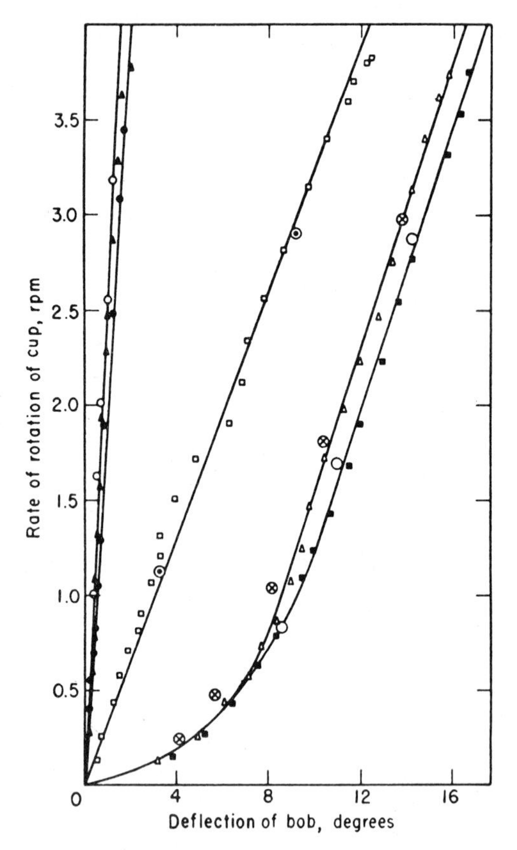

Figure 12.8. Surface viscosity curves for solutions of mixtures of pure sodium lauryl sulfate and lauryl alcohol. Small circles indicate curve for water alone. The remaining curves are for solutions 0.1% in pure sodium lauryl sulfate. Lauryl alcohol concentrations are: ▲ none; ● 0.001%; □ 0.003%; △ 0.005%; ■ 0.008%. Large circles show points obtained on decreasing speed (11).

TABLE 12.2. FOAM LIFE, SURFACE VISCOSITY, AND PERMEABILITY DATA FOR 0.1 PER CENT SOLUTIONS OF PURE SODIUM LAURYL SULFATE CONTAINING ADDED LAURYL ALCOHOL (13).

Lauryl Alcohol Conc. (g./100 ml.)	Permeability		Surface Viscosity		Foam Life
	dr/dt (cm./hr.) × 10^2	*k* (cm./sec.) × 10^2	n_s surface poises × 10^3	f_s (dynes/cm.) × 10^3	*Lf* minutes
0	2.84	0.99	2	0	69
0.00025	0.85	0.51	—	—	—
0.00025	0.56	0.30	—	—	—
0.00025	0.48	0.24	—	—	—
0.0005	0.47	0.27	—	—	—
0.001	—	—	2	0	825
0.002	0.43	0.39	—	—	—
0.003	—	—	31	0	1260
0.005	—	—	32	54	1380
0.008	0.43	0.47	32	64	1590

TABLE 12.3. FOAM LIFE, SURFACE VISCOSITY, AND PERMEABILITY DATA FOR 0.5 PER CENT SOLUTIONS OF PURE SODIUM LAURYL SULFATE CONTAINING ADDED LAURYL ALCOHOL (13).

Lauryl Alcohol Conc. (g./100 ml.)	Permeability		Surface Viscosity		Foam Life
	$-dr/dt$ (cm./hr.) × 10^2	*k* (cm./sec.) × 10^2	n_s surface poises × 10^3	f_s (dynes/cm.) × 10^3	*Lf* minutes
0	2.37	0.92	4.0	0	295
0.005	—	—	2.5	0	960
0.010	2.40	1.12	—	—	—
0.015	—	—	2.5	0	1100
0.020	2.50	1.30	—	—	—
0.025	0.51	0.32	24.5	29	1220
0.030	0.41	0.22	—	—	—

TABLE 12.4. FOAM LIFE, SURFACE VISCOSITY, AND PERMEABILITY DATA FOR 0.1 PER CENT SOLUTIONS OF SOME COMMERCIAL DETERGENTS (13).

Material	Surface Tension (dynes/cm.)	Permeability		Surface Viscosity		Foam Life
		$-dr/dt$ (cm./hr.) × 10^2	*k* (cm./sec.) × 10^2	n_s surface poises × 10^3	f_s (dynes/cm.) × 10^3	*Lf* minutes
Triton X-100	30.5	3.7	1.5	—	—	60
Santomerse 3	32.5	2.0	1.3	3	0	440
Quaternary O	37.0	1.4	0.65	1	0	1750
E 607 L	25.6	1.5	0.91	4	0	1650
Potassium laurate	35.0	0.28	0.14	39	59	2200
Sodium lauryl sulfate (comm.)	23.5	0.30	0.37	55	118	6100

only one adsorbed film while permeability involves two surface films with a solution layer between. Low permeability and high surface-viscosity relate similarly to foam stability. In general, solutions giving low permeability values give foams of high stability. However, where permeability values are intermediate, foam stability is high.

Interaction of Polar Additives in Foams and Micelles

Schick and Fowkes (14) discovered that polar additives which appreciably lower the cmc of anionic surfactants stabilize foam in the presence of anti-

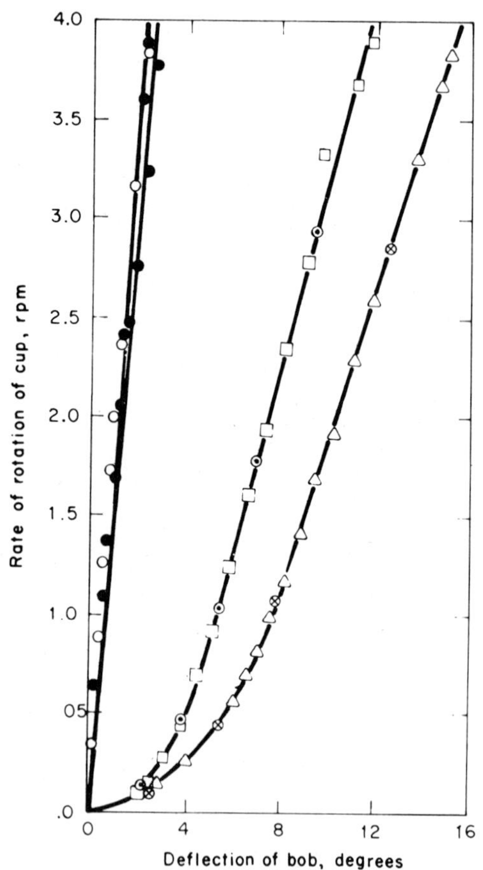

Figure 12.9. Surface viscosity curves for solutions of mixtures of pure sodium lauryl sulfate and lauryl alcohol. All solutions 0.5% in sodium lauryl sulfate. Lauryl alcohol concentrations are: ○ none; ▲ 0.005%; ● 0.015%; □ 0.025%; △ 0.040%. Points in large circles obtained on decreasing speed (11).

Figure 12.10. Surface viscosity curves of some commercial detergents. △ Santomerse 3; ○ Quaternary O; □ E 607 L; ■ potassium laurate; ● commercial sodium lauryl sulfate. Circled points obtained on decreasing speed (11).

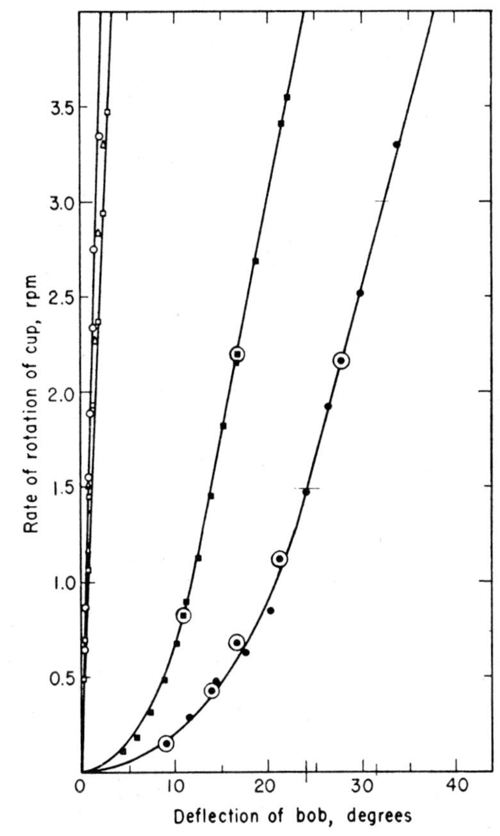

Figure 12.11. Foam life curves for solutions of mixtures of pure sodium lauryl sulfate (NaLS) and lauryl alcohol. Solid points are for solutions 0.1%; open points are for solutions 0.5% in sodium lauryl sulfate. Lauryl alcohol concentrations are: ● 0.005%; ○ 0.025%; ■ 0.008%; □ 0.040%. Dotted lines indicate changes in time scale (11).

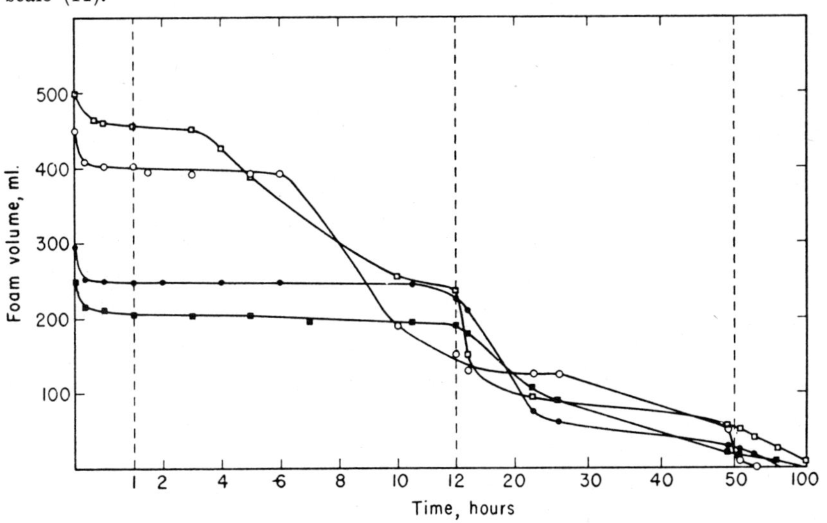

foaming agents. These investigators were interested in foam stabilizers for heavy-duty detergent compositions. Foam studies were generally conducted in the presence of General Dyestuff Corporation "Soiled Cloth" in hard water, equivalent to 350 ppm of calcium carbonate, containing 0.4 per cent of detergent. The built detergent consisted of 18 per cent of anionic surfactant, 27 per cent of sodium sulfate, 51 per cent of sodium tripolyphosphate, and 4 per cent of polar additive. Cmc values were determined in the absence of inorganic salts using pinacyanol dye. The dye method is known to give cmc values that are less reliable than those obtained by the conductivity method.

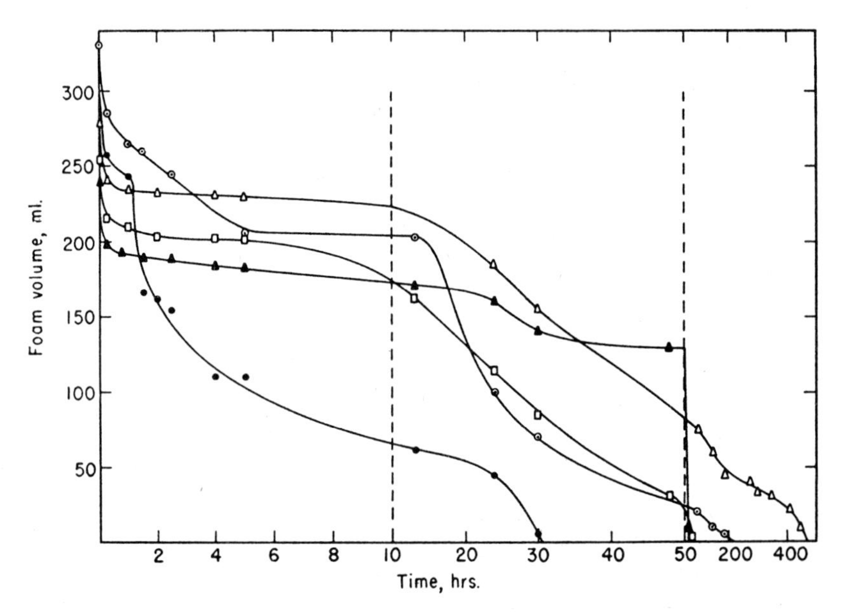

Figure 12.12. Foam life curves for 0.1% solutions of some commercial detergents. ● Santomerse 3; ▲ potassium laurate; □ Quaternary O; ○ E 607 L; △ commercial sodium lauryl sulfate. Dotted lines indicate changes in time scale (11).

The polar additives studied and their effect on the cmc of 2-*n*-dodecylbenzene sodium sulfonate is shown in Table 12.5. These results have been correlated with foaming tests carried out with propylene tetramer benzene sodium sulfonate (PTBS) in Table 12.6. Comparisons of cmc lowering by additives with foam performance of the built detergents are shown in Tables 12.7 and 12.8.

The results show that maximum lowering of the cmc occurs when the additive possesses a straight hydrocarbon-chain and the length of the chain is about the same as the straight hydrocarbon-chain of the detergent. Additives with bulky hydrocarbon chains decrease the cmc by only a small amount.

The effectiveness of the additive in reducing the cmc of sulfate- and sulfonate-type detergents is favored by hydrogen-bonding groups. Additives

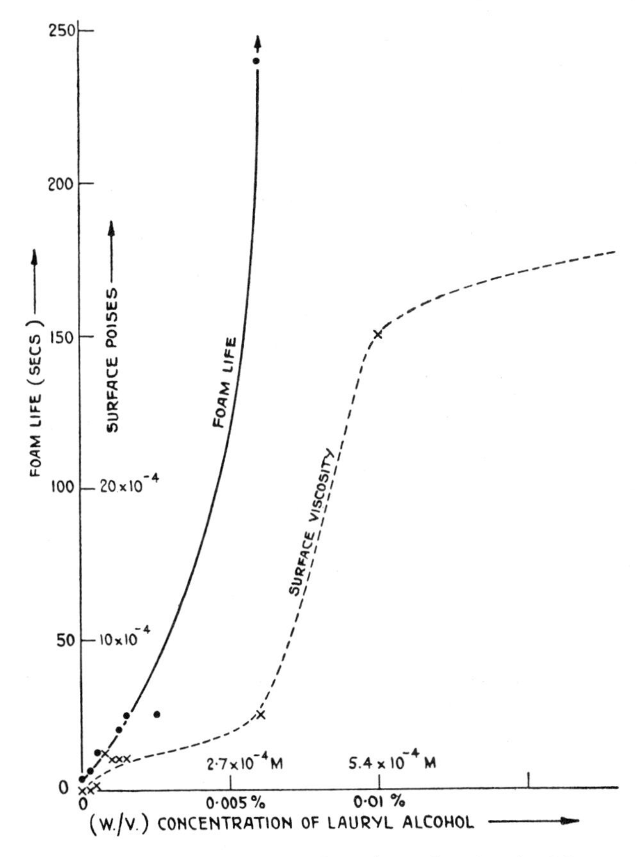

Figure 12.13. Foam life and surface viscosity in 0.1 per cent sodium laurate at pH 10 with added lauryl alcohol. The foam life rises to 6.5 min with 0.01 per cent lauryl alcohol, and 30 min with 0.025 per cent (12).

with many hydroxyl, primary or secondary amide, or sulfolanyl groups, or combinations of these groups cause the greatest reduction in the cmc of detergent solutions.

Comparison of foam stability data with cmc lowering shows that additives that have a pronounced foam-stabilizing effect lower the cmc to a great extent, 30 per cent or more. Additives with a minor effect on foam stability lower the cmc to only a small extent. In Table 12.9 are presented results of turbidity studies which show that the antifoaming agent for these systems, oleic acid, is solubilized by detergent solutions containing foam stabilizers.

That polar additives lower the cmc of ionic surfactants, while hydrocarbons do not, is well known. It is generally believed that polar additives are solubilized in the palisade layer of micelles while hydrocarbons are solubilized in the micelle interior. Orientation of the polar additive in the palisade layer is favored by hydrogen bonding.

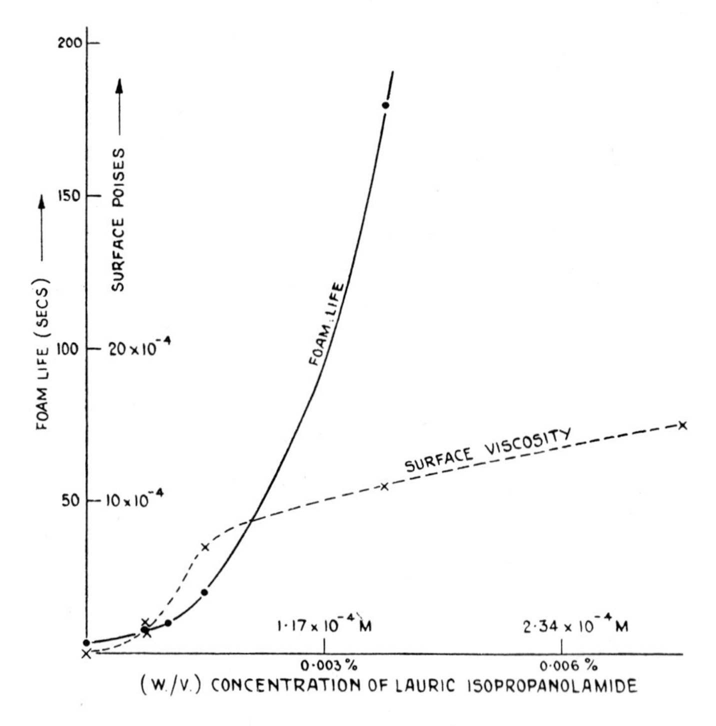

Figure 12.14. Foam life and surface viscosity in 0.1 per cent sodium laurate at pH 10 with added lauric-isopropanolamide. The foam life rises to 40 min with 0.0075 per cent additive (12).

Klevens (15) showed that the solubilizing action of detergent micelles is enhanced severalfold when additives are present in the palisade layer. This would provide one explanation for the relationship between cmc lowering and foam stabilizing. Another explanation is based on the similarity between the palisade layer and the surface film. An additive present in the

TABLE 12.5. EFFECT OF THE POLAR GROUPS OF ADDITIVES ON THE CMC OF 2-*n*-DODECYLBENZENE SODIUM SULFONATE (14).

Additive	Molecular Structure	% Lowering of cmc
Lauryl glycerol ether	$n\text{-}C_{12}H_{25}\text{—O—}CH_2\text{—CHOH—}CH_2OH$	51
Laurylethanolamide	$n\text{-}C_{11}H_{23}\text{—C(=O)—NH—}CH_2\text{—}CH_2OH$	48
Laurylsulfolanylamide	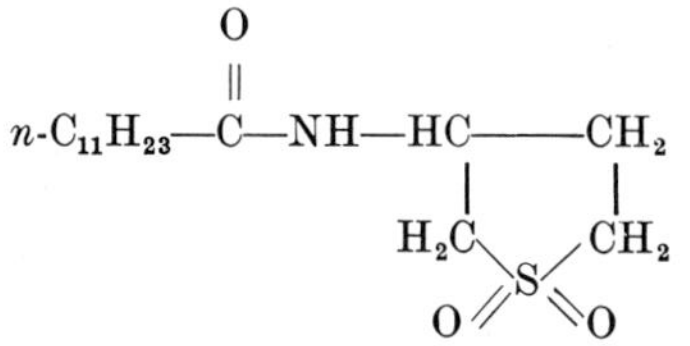 $n\text{-}C_{11}H_{23}\text{—C(=O)—NH—HC—}CH_2$; ring: H_2C, CH_2, $S(=O)_2$	41
Lauryl chlorohydrin glycerol ether	$n\text{-}C_{12}H_{25}\text{—O—}CH_2\text{—CH(Cl)—}CH_2\text{—OH}$	3
Tetradecanol-2	$n\text{-}C_{12}H_{25}\text{—C(OH)(H)—}CH_3$	
n-Decyl glycerol ether	$n\text{-}C_{10}H_{21}\text{—O—}CH_2\text{—CHOH—}CH_2OH$	44
n-Decyl-3-sulfolanyl ether	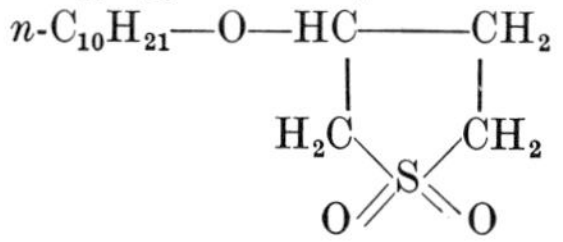 $n\text{-}C_{10}H_{21}\text{—O—HC—}CH_2$; ring: H_2C, CH_2, $S(=O)_2$	41
n-Decyl alcohol	$n\text{-}C_{10}H_{21}\text{—OH}$	30
n-Octyl glycerol ether	$n\text{-}C_7H_{15}\text{—}CH_2\text{—O—}CH_2\text{—CHOH—}CH_2OH$	39
Caprylamide	$n\text{-}C_{17}H_{15}\text{—}CONH_2$	15

palisade layer would also be expected in the surface film, resulting in tighter packing and lower surface tension (14).

Sawyer and Fowkes (16) calculated from surface-tension data that the polar foam-stabilizer is in fact present in the surface film to a considerable extent. With sodium 1-*n*-dodecyl sulfate plus 1-*n*-dodecanol, they found the dodecanol content of the monolayer to be about 90 per cent. This was the largest proportion of additive in the surface film for all of the systems studied. The important results of their foam stability studies (16) may be summarized as follows:

(1) The higher the surface tension of the surfactant solution, without additive, above its cmc value, the greater its susceptibility to foam promotion. Thus, unbranched alkylsulfates and alkylbenzene sulfonates have higher

surface-tensions and are more susceptible to foam promotion than branched-chain surfactants of the same types.

(2) The effectiveness of polar additives as foam stabilizers increases in the order primary alcohols $<$ glyceryl ethers $<$ sulfolanyl ethers $<$ amides $<$ N-polar substituted amides. This is generally the order of increasing surface activity and cmc-depressing activity.

TABLE 12.6. COMPARISON OF THE LOWERING OF THE CMC BY ADDITIVES IN 2-*n*-DODECYLBENZENE SODIUM SULFONATE SOLUTIONS WITH THE FOAM PERFORMANCE OF PROPYLENETETRAMER BENZENE SODIUM SULFONATE IN BUILT-DETERGENT COMPOSITIONS (14).

Additive	cmc g/l	% Decrease of cmc	Foam Stability: Foam vol. at 20 min. (ml.)
None	0.59	—	18
Lauryl glycerol ether	.29	51	32
Laurylethanolamide	.31	48	50
n-Decyl glycerol ether	.33	44	34
Laurylsulfolanylamide	.35	41	40
n-Decyl-3-sulfolanyl ether	.35	41	28
n-Octyl glycerol ether	.36	39	32
n-Decyl alcohol	.41	31	26
N-(3-Sulfolanyl)-oleylamide)	.48	19	23
Caprylamide	.50	15	17
Lauryl chlorohydrin glycerol ether	.57	3	22
Tetradecanol-2	.60	0	12

TABLE 12.7. COMPARISON OF THE LOWERING OF THE CMC BY ADDITIVES WITH THE FOAM PERFORMANCE OF THE SAME DETERGENTS (14).

Detergent	Additive	cmc, g./l.	% Lowering of cmc	Foam Stability: (20 min. foam vol.), ml.
Sodium lauryl sulfate	None	2.26	—	8
	Lauryl alcohol	0.76	66	54
Sodium 2-*n*-octylbenzene sulfonate	None	5.56	—	0
	Lauryl ethanolamide	3.43	38	49
	n-Decyl glycerol ether	3.47	38	27
Sodium 2-*n*-dodecylbenzene sulfonate	None	0.59	—	33
	Lauryl ethanolamide	.31	47	64
	n-Decyl glycerol ether	.33	44	38
	Lauryl sulfolanylamide	.35	41	58
Sodium tetradecane-2-sulfonate	None	1.13	—	16
	m-Decyl glycerol ether	0.59	39	64

(3) Increasing foam stability corresponds to an increasing mole fraction of additive in the surface film. The most stable foams were found with detergent-additive pairs having 60–90 per cent of additive in the adsorbed monolayers.

Table 12.10 verifies conclusions relating increasing susceptibility to foam promotion to high γ_1^0, the surface tension of the detergent solution without additive at concentrations in excess of the cmc, and the effectiveness of

TABLE 12.8. COMPARISON OF THE LOWERING OF THE CMC BY ADDITIVES WITH THE FOAM PERFORMANCE OF SODIUM LAURYL SULFATE (14).

Additive	% Lowering cmc	Foam Vol. at 20 min., ml.[a]
Lauryl alcohol	70	30–39
n-Decyl glycerol ether	64	39
(3,5-Dimethyl) phenyl glycerol ether	23	9
C_{15}-alcohol (highly branched)	20	9

TABLE 12.9. OPTICAL DENSITY OF 0.4 PER CENT SOLUTIONS OF BUILT SODIUM DODECYLBENZENE SULFONATE AT 25°C CONTAINING ADDED OLEIC ACID (14).

% Oleic Acid	None	Foam Additive LSA[a]	LEA[b]	DGE[c]
0	0	0.010	0.009	0.102
0.0015	0	.007	0	.0002
.0030	0.008	—	—	.009
.0075	.014		.015	
.015	.079	.088	.081	.028
.030	.180	.146	.142	.073
.045	.344	—	.177	.132
.060	—	.287	.307	—
.075	—	.365	—	—

[a] LSA, N-(3-sulfolanyl)-lauramide. [b] LEA, laurylethanolamide. [c] DGE, glycerol-α-*n*-decyl ether.

additives with decreasing values of γ_2^0 (satd.), the surface tension of saturated solutions of additives without detergent. Thus, while most of the additives tested were effective in stabilizing the foam of sodium dodecyl sulfate, very few were effective foam stabilizers for sodium propylene tetramer benzene sulfonate.

Ross and Bramfitt (17) confirmed the results of Schick and Fowkes (14) concerning the lowering of the cmc by foam stabilizers. However, they supplemented these findings with the discovery that antifoaming agents hinder the formation of micelles. Figure 12.15 is a generalized summary of the effect of additives on the relation between specific conductance and the surfactant concentration. With a foam stabilizer present, the cmc is lowered and the specific conductance is decreased at concentrations above the cmc. An antifoam raises the cmc and increases the specific conductance at concentrations above the cmc.

TABLE 12.10. COMPOSITIONS OF MIXED MONOLAYERS FORMED ON 0.4 PER CENT SOLUTIONS OF BUILT DETERGENTS AT 54°C (16).

Detergent[b]	Additive[b]	ΔCMC, %	γ_1^0	γ_2^0 (satd.)
PTBS	CPOH	−30[a]	31	40.7
	DOH	−30	31.0	35.4
	LOH	−30[a]	31	31.8
	DSE	−41	32.0	23.1
	DGE	−44	32.6	21.9
	LEA	−51	33.7	22.8
DDS	DOH	−65[a]	44	35.4
	LOH	−66	44.3	31.8
	DGE	−64	43.8	21.9
	DSE	−30[a]	38	23.1
DDS (25°)	LOH	−70	48.4	24.3

[a] Estimated values. [b] Detergents: PTBS = sodium propylene tetramer benzene sulfonate, DDS = sodium 1-n-dodecyl sulfate. Additives: CPOH = cyclohexylpentanol, DOH = decanol, LOH = dodecanol, DSE = decyl 3-sulfolanyl ether, DGE = α-(n-decyl) glycerol ether, LEA = N-(2-hydroxethyl) lauramide, LSA = N-(3-sulfolanyl) lauramide.

The surfactants used in this study were not pure, and it is not known whether antifoaming agents would raise the conductance—decrease the micellization—of pure surfactants. The effects reported may be the result of displacement of one additive for another in the micelles. Nonetheless, certain observations appear to be of general validity. Thus, addition of an antifoam to a surfactant solution at a concentration in excess of the cmc, increased the conductance to a constant value. The foam-inhibiting agent achieved its maximum effect at the concentration at which the specific conductance reached a constant value, as shown in Figure 12.16. Further, it was observed that the concentration of antifoam corresponding to the limit of foam-inhibiting action was marked by the appearance of a haze, indicating the limit of solubilization of the antifoam. This suggests that the limit of accommodation of the antifoam in the micelles corresponds to the maximum concentration of the antifoam that can be accommodated in the surface monolayer. The analogous effect produced by foam stabilizers of reducing the specific conductance is shown in Figure 12.17.

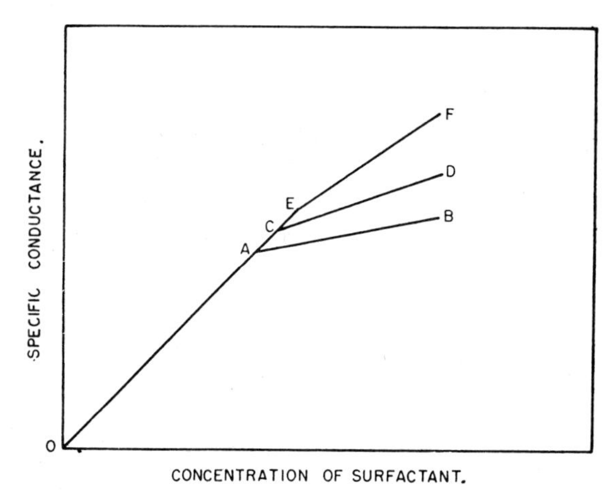

Figure 12.15. The effects of foam stabilizers and foam inhibitors on the relation between the specific conductance and the concentration of surfactant: Surfactant alone, OCD; plus foam stabilizer, OAB; plus foam inhibitor, OEF (17).

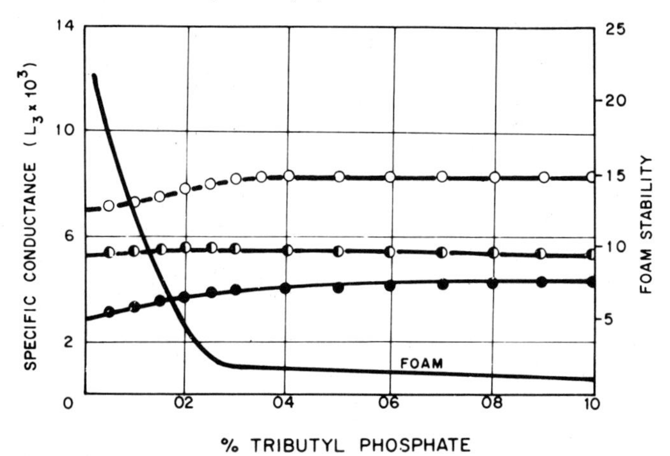

Figure 12.16. The effects of added tributyl phosphate (foam-inhibitor) on the specific conductance at 24.7° of colloidal electrolytes at fixed concentration: ○ = 0.4% sodium lauryl sulfate; ◐ = 0.40% Hyamine 1622; ● = 0.20% sodium oleate. The foam stabilities refer to 0.40% Hyamine 1622 plus tributyl phosphate at 25°, as measured by the Ross and Miles method (height of foam in cm. after five minutes) (17).

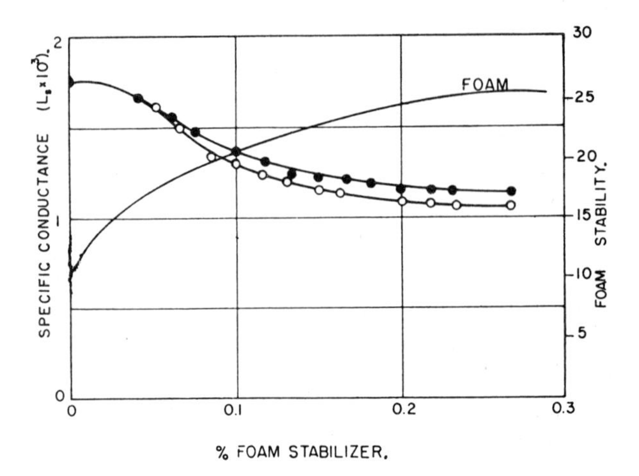

Figure 12.17. The effects of added foam stabilizers on the specific conductance at 24.7° of 0.10% sodium oleate: ● = addition of lauryl alcohol; O = addition of n-decyl alcohol. The foam stabilities refer to 0.10% sodium oleate with additions of n-decyl alcohol, as measured by the Ross and Miles method (height of foam in cm. after five minutes) (17).

Foam Inhibition

Foam breakers may destroy foam by reacting chemically with the foamer. Thus, an acid or a calcium salt can be used to break a soap foam. However, the foam breakers that are of greatest industrial importance are those that function by spreading, and thinning the bubble wall by transport of the fluid underlying the surface film.

The liquid acting as a foam breaker either spreads as a monolayer or as a lens (7). It is assumed, in either case, that the spreading liquid sweeps before it the film that is stabilizing the foam. Thus, the spreading film is initially in contact with a fresh surface that is identical in composition to the bulk liquid.

The condition that the foam breaker B will spread as a monolayer over the solution A is that the initial surface tension of the monolayer on the swept surface, $\gamma_{B(S)}$, shall be less than the surface tension of the foaming solution γ_A, or

$$S_{mi} = \gamma_A - \gamma_{B(S)} > 0 \tag{12.2}$$

where S_{mi} is the initial spreading-coefficient for a monolayer.

The condition that B will spread as a lens over the solution A is that the sum of γ_{SB} the initial interfacial tension of spreader over the swept solution and γ_B the surface tension of the spreader as liquid in bulk shall be less than γ_A the surface tension of the foaming solution, or

$$S_{li} = \gamma_A - \gamma_{SB} - \gamma_B > 0$$

In this, S_{li} is the initial spreading-coefficient for a lens. Thus, one criterion for a foam breaker is that S_{mi} or S_{li} should be positive.

Ewers and Sutherland (7) observed that the interfacial tension γ_{SB} as well as the surface tension $\gamma_{B(S)}$ may decrease with time if a mixed film is formed between the spreader and the molecules or ions of the foaming agent which reach the interface from the bulk of the solution. This reduced surface-energy will oppose the spreading of a further addition of foam breaker on the surface.

The rate and extent of spreading of the foam breaker will depend upon the character of the adsorbed layer of foaming agent. If it desorbs readily into solution there is no opposition to the spreader. However, if it desorbs slowly, excess surface-pressure will develop which will oppose spreading. Compression of the frother film will decrease γ_A and the spreading coefficients will decrease accordingly. This will also depend upon the area of the foamed surface. If it is large, the foam breaker will spread to a substantial extent before significant compression of the frother film occurs. In the case of saponin and various protein films, the viscosity of the films retards the rate of spreading. However, the low collapse-pressure of these films prevents the build up of a force resisting spreading. The bubble walls are slowly thinned by movement of the spreader until they burst.

The above discussion applies to the initial introduction of a small quantity of antifoaming agent into a foaming solution. An effective antifoam will generally cause the immediate collapse of foam and will retard additional foam formation for some time afterwards. However, after several hours it will frequently be found that the foam inhibitor has lost its effectiveness and additional foam breaker must be added.

Ross and Hack (18) found that the behavior of a foam inhibitor depends upon whether the cmc of the foaming agent in the solution is exceeded. Below the cmc of the solute, the antifoam is effective only if it is present as insoluble droplets. Above the cmc of the solute, antifoam solubilized in micelles is partially effective in reducing foam. Addition of the antifoam beyond the limits of solubilization causes a further increase in antifoaming action.

Figure 12.18 shows the effect of tributyl phosphate on the foam height of a

0.20 per cent solution of sodium oleate. This is above the cmc of this solute. The limit of solubilization of tributyl phosphate in this solution is 0.3 per cent. The antifoam is effective at low concentrations in the fresh solution, which contains dispersed droplets of the antifoam. However, the tributyl phosphate is slowly solubilized as the solution ages and its effectiveness is reduced. Further addition of antifoam until insoluble droplets form results in an additional abrupt drop of the foam stability of the aged system.

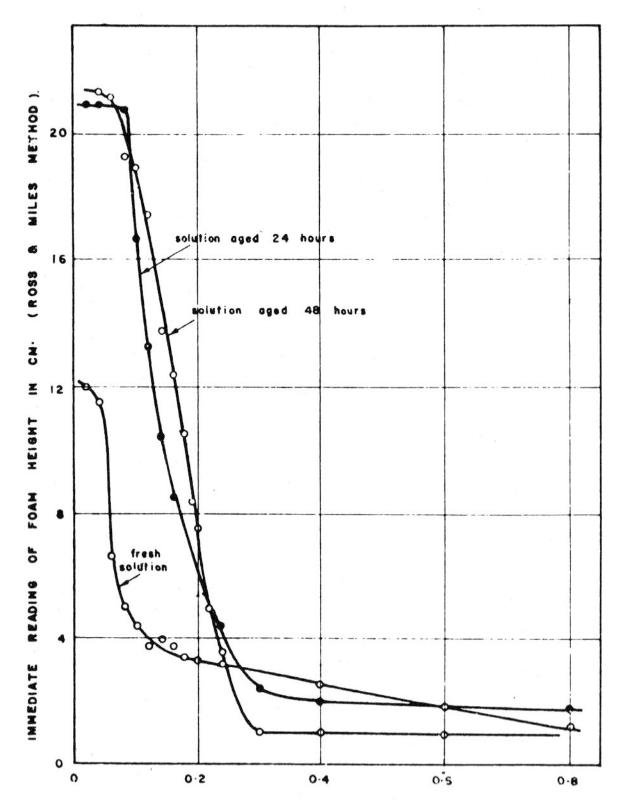

Figure 12.18. % Tributyl phosphate in 0.20% sodium oleate solution (18).

While the effectiveness of dispersed droplets of antifoam is due to the positive spreading-coefficient, another mechanism is ascribed to the less effective action of the solubilized antifoam. Ross and Hack (18) have shown that at concentrations above the cmc of the foamer, the rate of surface-tension lowering is greater with antifoam present in solubilized form than in the absence of antifoam. Foam stabilizers on the other hand appear to decrease the rate of surface-tension lowering. These results are shown in Figures 12.19 and 12.20. Burcik and Newman (19) also found a decrease in the rate of surface-tension lowering upon addition of the foam stabilizer, dodecanol-1, to sodium lauryl sulfate, as shown in Figure 12.21.

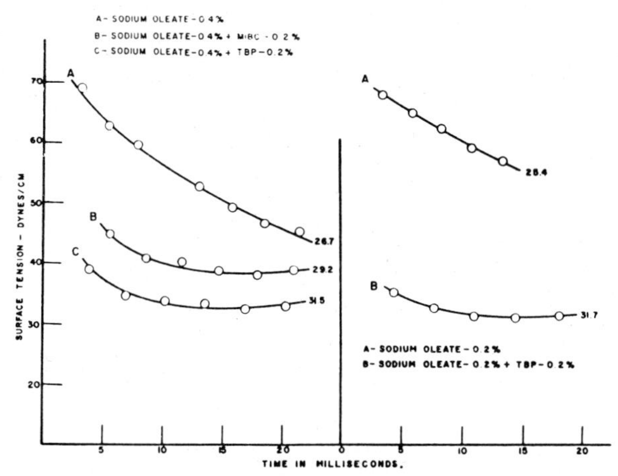

Figure 12.19. Dynamic surface tensions of sodium oleate solutions with and without added foam inhibiting agents: TBP = tributyl phosphate; MIBC = methylisobutylcarbinol. The numbers after the curves refer to the static surface tension of each solution (18).

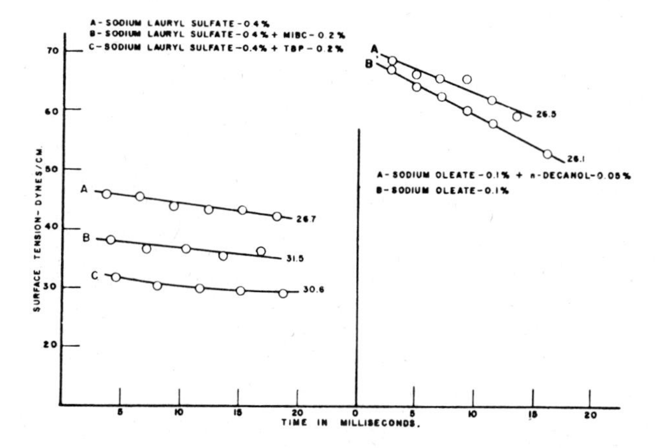

Figure 12.20. (a) Dynamic surface tensions of sodium lauryl sulfate solution with and without added foam inhibiting agents: TBP = tributyl phosphate; MIBC = methylisobutylcarbinol. (b) Dynamic surface tension of sodium oleate solution with and without added foam stabilizing agent, namely, *n*-decanol (18).

It is evident that a rapid lowering of surface tension resulting from a high adsorption rate will reduce the surface-tension gradient due to the stretched surface quite as effectively as the diffusion of the surface layer, but without restoration of the film thickness. Ross and his associates suggest that in the presence of antifoaming agents the micellar size is reduced and the micellar concentration is increased. Both factors favor a faster rate of attaining surface-tension equilibrium in solutions containing foam inhibitors.

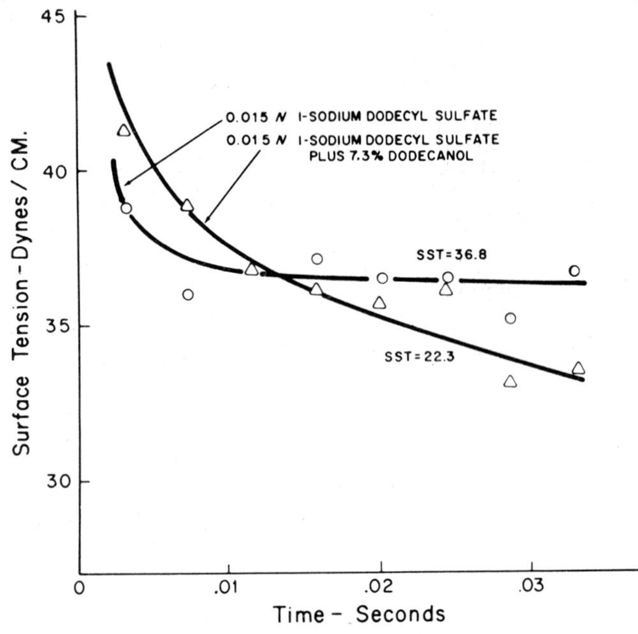

Figure 12.21. Surface tension *versus* time for 0.015*N* 1-sodium dodecyl sulfate plus dodecanol at 25°C. SST refers to the static surface-tension values (19).

Studies by Shearer and Akers (20) emphasize that foam inhibition in aqueous and organic systems is governed by the same principles. The foaming properties of four lube oils over a range of temperatures were examined in the presence and absence of silicone oils. In all instances, the silicone oils acted as profoamants up to the limit of their solubility in the lube oils. At higher concentrations they were foam inhibitors. It was observed that the silicones were effective antifoamers only when present as a dispersed phase having particles less than 100 microns in diameter. Other organic liquids failed to function as inhibitors. Only the silicone oils, of the various liquids tested, had a surface tension less than the lube oils and exhibited a positive spreading coefficient. The requirement of very low solubility and a lower surface-tension excludes almost all components except the silicones as anti-foaming agents for hydrocarbon oil systems.

Antifoaming Agents

Ross (21) has reviewed the various commercial antifoaming agents. These are classified according to chemical types as follows:

(1) *Alcohols.* Branched-chain higher alcohols are insoluble in water and have low surface-tension. One of the earliest chemical defoamers is 2-ethyl-hexanol. It is widely used for this purpose in beet sugar and paper production, textile printing, and glue spreading. Effective concentrations vary from 0.005 to 1.0 per cent, depending on the solution to be defoamed. Other competing alcohols are amyl alcohol, caprylic alcohol, 2, 6, 8-trimethyl nonanol-4, tetradecanol, and trimethylcyclohexanol. An effective defoaming agent is di-isobutyl carbinol (nonyl alcohol), which is used to defoam soap solutions in the wire-drawing industry, in the paper industry, and in the manufacture of printing inks and glues.

Certain polyalkylene glycols and derivatives are effective antifoams for a number of commercial applications. These products are sold by Union Carbide under the trade name of "Ucon".

(2) *Fatty Acids and Fatty-acid Esters.* A variety of water-insoluble fatty acids and their esters are employed for foam inhibition. Many of the esters are non-toxic and suitable for food applications. The sorbitan esters, produced by Atlas Powder Company, are widely used in this field. Span 20 (sorbitan monolaurate) is used to suppress foam during the evaporation of aqueous milk sugar solutions, in the concentration of molasses, and in the drying of egg white. Span 85 (sorbitan trioleate) effectively reduces foam during the evaporation of yeast slurries and casein glues.

Fatty acids are used to prevent foam in many fermentation processes. Glyceryl esters are used in pulp and paper making. Isoamyl stearate and diglycol laurate are useful for reducing the stability of bubbles of methane in drilling muds during the drilling of deep wells.

(3) *Amides.* High-molecular-weight substances containing one or more amide groups are used in the treatment of boiler waters to prevent priming or frothing. Substances containing at least 36 carbon atoms are more effective than simple fatty acid amides, such as stearamide. Distearoylethylenediamine is one of the more effective antifoams used in the treatment of boiler water.

(4) *Multiple Polar Groups.* A number of other highly effective foam inhibitors contain two or more polar groups. Wyandotte Chemicals Corpora-

tion markets Foamicide L and Foamicide A, both of which contain di-tertiary amyl phenoxyethanol as the active foam-inhibiting agent. These Foamicides are useful in washing operations employing soaps and alkaline detergents, sewage disposal plants, tanneries, etc.

Polymerized triethanolamine esterified with mixtures of high fatty acids are effective antifoams. Commercial Solvents Corporation markets an oxazoline from oleic acid and 2-amino-2-methyl-1, 3-pentanediol under the name Alkaterge O. A mixture of substituted oxazolines is sold as Alkaterge C. These are effective in controlling foam during fermentation. The Spans and Ucons mentioned earlier contain multiple polar groups.

(5) *Phosphate Esters.* Tributyl phosphate is sold primarily for use as an antifoam for rubber latex, textile sizes, inks, adhesives and many other products. When applied from a water-soluble solvent, such as ethanol or acetone, so that it is thoroughly dispersed, it is effective at concentrations of 0.01 to 0.1 per cent. Trioctyl phosphate is used to suppress foam in alkaline cleaning baths. Sodium octyl phosphate is effective in industrial cleaners. Several organic phosphates are claimed to control foam in petroleum oils. These include dimethylaniline isoamyl octyl phosphate and potassium tri-octylethylene diphosphate.

(6) *Metallic Soap of Fatty Acids.* Water-insoluble metallic soaps, such as aluminum, calcium, and magnesium stearates and palmitates, are frequently effective defoamers. They are ordinarily used by first dissolving them in an organic solvent or thoroughly dispersing them in water with a wetting agent. Unless the metallic soaps are thoroughly dispersed in the solution to be defoamed, effectiveness is low. They are sometimes used as defoamants for non-aqueous systems.

(7) *Organic Silicon Compounds.* Many organic silicon compounds are outstanding antifoamers for both aqueous and non-aqueous systems. The polysiloxanes (silicone oils) are prepared by condensation of dimethylsilanediol. When dispersed as particles smaller than 2 microns in diameter they are effective defoamers in hydrocarbon oils at concentrations of only a few parts per million. The polyalkoxysiloxanes are also effective antifoams for hydrocarbon oils.

These silicones are also useful antifoams for aqueous systems. Apparently, if the methyl groups are replaced by large alkyl groups, such as cetyl groups, the effectiveness in aqueous systems is enhanced. Organic silicon antifoaming agents are available as Antifoam 81066 from General Electric and DC Antifoam A from Dow-Corning Corporation. DC200 Fluid is used for hydrocarbon oil foams.

Action of Foam Stabilizers and Antifoams

The experimental results that were discussed in previous sections show that the behavior of foam stabilizers and antifoams are opposite in many respects. Foam stabilizers generally increase surface viscosity, and in so doing, reduce drainage and gas permeability. Antifoams tend to produce an expanded surface film and to reduce surface plasticity (22). Foam stabilizers lower the concentration for micelle formation and decrease the rate of surface-tension lowering. Antifoams have opposite effects.

There are superficial similarities. Both foam stabilizers and foam inhibitors are generally poorly soluble in the solution and both further lower the surface tension of the foaming solution. However, the mechanisms are different. Antifoams are most effective at concentrations in excess of their solubility in the solution, while foam stabilizers are effective when they are dissolved. At higher concentrations foam stabilizers can act as antifoams.

Antifoams sweep the surface, draining the bubble walls by surface transport. If they spread as a lens, film elasticity becomes zero. Otherwise, they may produce an expanded surface-film which offers little hindrance to further drainage. Surface-tension gradients are overcome by rapid adsorption or desorption, instead of by surface transport.

Foam stabilizers increase the solubility of antifoams. By lowering surface tension, they reduce the likelihood of the spreading of antifoams. Foam stabilizers are straight chain with the polar groups at one end of the molecule. The more effective antifoams minimize van der Waals attraction between hydrocarbon chains and tend to lie flat when oriented at an air-water interface. This is achieved by chain branching, placing the polar group toward the center of the molecule, scattering polar groups along the length of the molecule, including bulky groups, and using the smallest number of methylene groups necessary for insolubility in the system for which it was designed. Both foam stabilizers and antifoams are more effective in foaming solutions with relatively high surface-tensions.

References

1. Bikerman, J. J., "Surface Chemistry: Theory and Applications", pp 100–116, New York, Academic Press Inc., 1958.
2. Adam, N. K., "The Physics and Chemistry of Surfaces", pp. 142–146, London, Oxford University Press, 1941.
3. Mysels, K. J., Shinoda, K., and Frankel, S., "Soap Films", New York, Pergamon Press, 1959.
4. Plateau, J., "Statique Experimentale et Theorique des Liquides soumis aux Seules Forces Moleculaires", Gauthier-Villars, Paris, 1873.

5. Boys, C. V., "Soap Bubbles and the Forces Which Mould Them", Soc. for Promoting Christian Knowledge, London, E. & J. B. Young, New York, 1890. Reprinted Anchor Books, Doubleday Anchor Books, New York, 1959.
6. Gibbs, J. W., "Collected Works", Vol. 1, pp. 300, New Haven, Yale University Press, 1948.
7. Ewers, W. E., and Sutherland, K. L., *Australian J. Sci. Res.* **5**, 697 (1952).
8. Epstein, M. B., Ross, J., and Jakob, C. W., *J. Colloid Sci.* **9**, 50 (1954).
9. Epstein, M. B., Wilson, A., Jakob, C. W., Conroy, L. E., and Ross, J., *J. Phys. Chem.* **58**, 860 (1954).
10. Epstein, M. B., Wilson, A., Gershman, J., and Ross, J., *J. Phys. Chem.* **60**, 1051 (1956).
11. Brown, A. G., Thuman, W. C., and McBain, J. W., *J. Colloid Sci.* **8**, 491 (1953).
12. Davies, J. T., Proc. Intern. Congr. Surface Activity, 2nd, London, Butterworth, 1957.
13. Brown, A. G., Thuman, W. C., and McBain, J. W., *J. Colloid Sci.* **8**, 508 (1953).
14. Shick, M. J., and Fowkes, F. M., *J. Phys. Chem.* **61**, 1062 (1957).
15. Klevens, H. B., *J. Am. Oil Chemists' Soc.* **30**, 74 (1953).
16. Sawyer, W. M., and Fowkes, F. M., *J. Phys. Chem.* **62**, 159 (1958).
17. Ross, S., and Bramfitt, T. H., *J. Phys. Chem.* **61**, 1261 (1957).
18. Ross, S., and Haak, R. M., *J. Phys. Chem.* **62**, 1260 (1958).
19. Burcik, E. J., and Newman, R. C., United States Department of Commerce Report PB111420 (1954); *J. Colloid Sci.* **9**, 498 (1954).
20. Shearer, L. T., and Akers, W. W., *J. Phys. Chem.* **62**, 1264, 1269 (1958).
21. Ross, S., Rensselaer Polytechnic Institute Bulletin, Eng. and Sci. Series No. 63 (1950).
22. Ross, S., and Butler, J. N., *J. Phys. Chem.* **60**, 1255 (1956).

CHAPTER 13

Detergency

Detergency is economically the most important single application of surface-active agents. It consumes practically all the soap and more than one-half the volume of synthetic surfactants produced (1). Study of the detergency processes is of considerable theoretical interest. Essentially all of those properties of long-chain polar molecules which distinguish them from other materials are involved in the mechanism of cleaning.

Practical detergency is a particularly difficult subject, encompassing complex mixtures of soils bonded to a wide variety of substrates. The efficiency of different cleaning compositions depends upon both the soil and the substrate. Fineman and Kline (2) showed that metallic and non-metallic substrates behaved differently when coated with a carbon black-mineral oil soil and immersed in nonionic-detergent solutions. Metals produced rapid removal of the soil, which came off in finely dispersed form. The soil was removed slowly and in fairly large clumps from the various non-metallic substrates. When sulphur, which is also nonpolar, was substituted for carbon, the soil was not dispersed. With polar titanium dioxide in place of carbon, the soil dispersed from all substrates, even in water.

The efficiency of different surfactants frequently depends upon the nature of the substrate. Cationic surfactants promote soiling on glass. Anionic surfactants are effective in removing soil from glass and ceramic tile surfaces, but under some conditions cause the deposition of soil on metal surfaces (3). Reich and Snell (4) demonstrated the conditions under which the detergent could promote soiling rather than cleaning.

Textile fabrics are complex and differ in their properties, depending upon the weave, the nature of the fabric, and its prior history. Cotton fabric is predominantly hydrophobic, due to a waxy coating (5). However, it is permeable and can swell up to 40 per cent in water, predominantly in the cross-section of the fiber (6, 7). Cotton fabrics possess considerable cation exchange capacity (8, 9). Apparently, this is due to acidic carboxyl groups associated with noncellulosic constituents, such as pectins (10, 11). Almost raw cotton showed no adsorption of sodium carboxymethyl cellulose, while scoured and bleached cotton did adsorb this water-soluble polymer (12). Cotton fibers adsorb alkali, and silica to a lesser extent, from solutions of sodium metasilicate (12).

TABLE 13.1. ADSORPTION OF SODIUM LAURYL SULFATE AND CETYL TRIMETHYL AMMONIUM BROMIDE FROM ALKALINE SOLUTION AT 50° C. ON TEXTILE FIBERS (17).

	Sorption (millimoles/10⁴g fiber)	
Fiber	Sodium lauryl sulfate (0.086%)	Cetyl trimethyl ammonium bromide (0.08%)
Cotton	8	121
Viscose	1	318
Acetate	56	122
Nylon	68	54
Wool	139	520

Wool, silk, and nylon have some degree of proteinaceous character, with nylon the least (14). About one per cent of the wool surface is basic (15). In acid media, alkyl sulfate ion is strongly adsorbed on wool and will not desorb by washing (16). Differences in the properties of various fibers are evident from the adsorption data of Weatherburn and Bayley (17) reproduced in Table 13.1. The data were obtained by the analysis of the solutions used to contact the fibers.

Nature of the Soil

The nature of the soil to be removed in the detergent operation varies enormously, depending on the conditions of soiling. Thus, the soil on dishes will differ from that found on metals after stamping and polishing, or on textiles. Soil on fabrics will depend upon whether the clothing is worn close to the body or exposed to external sources of contamination. Soil on the outer garments of the office worker, coal miner or agricultural employee will differ considerably. Snell and coworkers (18) reviewed data by Brown (19) concerning the composition of soil found on soiled shirts, pillowcases and tea towels. They found up to about one per cent of fatty material in the soil. This comprised about one-third each of fatty acids, neutral fats, and unsaponifiables. The latter includes cholesterol and other unsaponifiable lipides present in perspiration.

Sanders and Lambert (20) analyzed street dirt from six different cities and found the soil to be quite similar. They formulated a synthetic soil, based on their analyses, with the composition: humus, 35 per cent; cement, 15 per cent; silica, 15 per cent; clay, 15 per cent; salt, 5 per cent; gelatin, 3.5 per cent; carbon black, 1.5 per cent; red iron oxide, 0.25 per cent; oils, 9.75 per cent. Powe (21) concluded that clay minerals that average about 0.1 micron in diameter are a major cause of soil build-up and redeposition on cotton fibers. Organic soil built up on cotton after repeated washing and use consists primarily of combined fatty acids in the form of lime soaps and triglycerides (22).

Substrate-Soil Bonds

The major factors responsible for the attachment of soil to fabric have been reported to be the following:

(1) *Mechanical.* Dirt can penetrate the interyarn capillary system of the fabric. A fabric having a loose weave or a rough or knobby weave will hold more dirt mechanically than a tight weave or a smoothly woven fabric (18). Goette (23) demonstrated that the more dispersed the dirt on the fabric, the more difficult it is to remove. When the dimensions of the dirt particles are reduced to the order of 0.1 micron, they cannot be removed from cotton with the best detergents.

(2) *Oil bonding.* Solid soils may adhere to fabric by oil bonding, because of the hydrophobic nature of both the oil and the fiber. Oily soils adhere to fiber in this manner. This cannot be the major factor responsible for the adhesion of soil to fabric, since solvent extraction of soiled shirts failed to render them sufficiently clean to be acceptable (19). Utermohlen and associates (24) found that pigment soil applied with an oily binder, with a water-soluble binder, and with no binder were removed to the same extent by detergent solutions. It has been suggested that soil becomes more difficult to remove as the soiled cloth ages, due to the polymerization of unsaturated fatty materials (18).

(3) *Electrical Forces.* Cellulosic and protein fibers ordinarily carry a negative charge in neutral and alkaline solutions. Similarly, most soils are negatively charged. However, under some conditions certain soils, such as carbon (25) and iron oxide (26), will carry a positive charge and will adhere to the fabric by attraction of unlike charges. Polyvalent cations such as calcium, magnesium, iron, and aluminum will adsorb on fabric and soil to produce positively charged surfaces. The bond between soil and fabric through this polyvalent cation "bridge" is said to be of major importance in the adhesion of soil to fabric (27, 28, 29, 30).

(4) *Hydrogen bonding.* Clays and other polar soils can adsorb hydroxyl or hydrogen ions and form hydrogen bonds with the hydroxyl groups in

cellulose. Similarly, fatty acids and proteins containing labile hydrogen atoms can hydrogen-bond to cotton (18, 29).

Detergent Compositions

A large variety of detergent compositions are employed, depending upon soil, substrate, cleaning conditions, and other factors not directly related to detergency. Thus, organic surfactants are generally not used in machine dishwashing. Salts of lauryl sulfate are preferably used in shampoos because of the quality and lubricity of the foam. In liquid compositions, water solubility over a broad temperature range is emphasized. This discussion is concerned primarily with the composition of cotton detergents.

Before the synthetic detergents became popular, fatty acid soaps were primarily used for cotton detergency. In general, palmitate, stearate, and oleate soaps have better detergent action than soaps of the smaller-chain fatty acids, particularly at higher temperatures. Packaged soaps frequently contain about 88 per cent of soap and 12 per cent of moisture. Of the soap content, 80 to 85 per cent is the sodium soap of tallow fatty acids and 15 to 20 per cent is the sodium soap of coconut-oil fatty acids. The latter is added for foam and to lower the solution temperature of the longer-chain soaps. Rosin soaps are poor detergents, but they have been included in packaged soap products to reduce cost. It has been claimed that up to 50 per cent of rosin soap in the composition will not seriously impair the detergency of the fatty acid soaps (30).

The addition of mild alkalis aids the detergent action of soap by repressing hydrolysis. Tomlinson (31) showed that the cleaning action of soap is at a minimum under conditions where hydrolysis is at a maximum. Snell (32, 33) and others (34) showed that soaps give their greatest detergency in the pH range of 10.5 to 11. Alkalis used in combination with soap to provide and maintain this optimum pH include trisodium phosphate, sodium carbonate, and sodium silicates.

In addition to the alkaline salts, various complex phosphates are used to soften the water and prevent the precipitation of calcium and magnesium soaps. These include sodium hexametaphosphate, $Na_6P_6O_{18}$, tetrasodium pyrophosphate, $Na_4P_2O_7$, and sodium tripolyphosphate, $Na_5P_3O_{10}$. The latter is most commonly used. There is considerable evidence that the complex phosphates and the silicates act as suspending agents, in the presence of soap, to decrease soil redeposition (35, 36).

Synthetic detergents, without builders, are effective for the cleaning of most fibers. However, they will not wash cotton white. The important discoveries leading to the large-scale use of the synthetics in cotton detergency were the observation that the combination of complex phosphates, particularly sodium tripolyphosphate, with synthetic anionic surfactants, greatly improves soil removal, and the introduction of sodium carboxymethyl cellulose to retard soil redeposition. To these can be added the production of low-cost sodium dodecylbenzene sulfonate.

Packaged cotton detergents for the retail market generally fall within the composition range:

Organic surfactant	15 to 40%
Sodium tripolyphosphate Tetrasodium pyrophosphate	30 to 50%
Sodium sulfate	5 to 15%
Sodium silicate	4 to 8%
Sodium carboxymethyl cellulose	0.5 to 1.5%
Optical bleach	trace

The organic surfactant used in the more widely sold high-foaming detergents is predominantly sodium dodecylbenzene sulfonate. In addition, from 1 to 5 per cent of lauryl monoethanolamide or lauryl monoisopropanolamide is added to maintain the foaming action and enhance detergency. At the higher amide concentrations, less active agent—organic surfactant—is required for high detergent action. The addition of C-16 and C-18 sodium alkyl sulfates further improves detergent action (37) and even less active agent is required in the composition.

The low-foaming or "controlled-suds" cotton detergents frequently employ nonionic in place of anionic surfactants, principally, polyoxyethylene derivatives of tall oil fatty acids or of tridecyl alcohol. One leading manufacturer uses a mixture of soap and sodium dodecylbenzene sulfonate, since the combination foams less than either agent alone (38).

The complex phosphates are the most important builders. While sodium tripolyphosphate is preferred, partial replacement of this phosphate by tetrasodium pyrophosphate facilitates manufacture. The built detergent composition is prepared as a slurry in water, which is then spray dried. A lower viscosity slurry is obtained by using a combination of the two phosphates. During spray drying, the sodium tripolyphosphate is partially hydrolyzed to tetrasodium pyrophosphate and disodium hydrogen phosphate.

Sodium sulfate is present as a by-product in the manufacture of the anionic surfactants. In compositions containing nonionic surfactants, sodium sulfate or sodium carbonate may be added as low-cost extenders. Sodium silicate serves primarily as a corrosion inhibitor to protect the washing machine and the plumbing.

Sodium carboxymethyl cellulose acts as an antigreying agent, to prevent the redeposition of soil initially removed from the cotton fabrics. Optical bleaches, colorless dyes that fluoresce blue in daylight, are used to counteract the yellowing of fabrics.

Evaluation of Cotton Detergency

Harris (39) has reviewed the various testing procedures used for the evaluation of detergents for a variety of cleaning operations. Only cotton detergency is considered here. Because practical washing tests are laborious and expensive for the evaluation of large numbers of compounded detergents, a number of laboratory tests have been developed for the screening of detergents. However, it is necessary to resort to practical washing tests on naturally soiled garments, before a final decision is made concerning the marketing of a particular detergent composition.

Several types of laboratory machines are available for detergency testing. These machines permit the testing of detergents under uniform conditions of temperature and agitation. Probably the first machine to be widely used was the Launder-Ometer, manufactured by Atlas Electric Devices Corporation, Chicago, Illinois, and developed by the American Association of Textile Chemists and Colorists. The standard machine can operate with twenty pint jars per loading. To each jar is added the detergent solution, test swatches, and rubber or steel balls to increase the mechanical energy applied in washing the fabrics.

The Terg-O-Tometer is distributed by the United States Testing Company, Hoboken, New Jersey. This machine consists of a gang of four small agitator washers in two-liter beakers. Both the speed of rotation and the angle through which the agitators can be oscillated are adjustable. The Deter-Meter is manufactured by the American Conditioning House, Boston, Massachusetts. The important feature of this machine is that it permits accurate variation and control of the mechanical force applied to the washing operation, and thus makes it possible to correlate mechanical force with cleaning action. The test fabric is placed in a circular metal cylinder, with ends fabricated of bronze screen. The fabric floats free between these ends. During operation, the cylinder containing the fabric is immersed in a four-liter beaker of detergent solution.

The preparation of standardized soiled test fabrics is difficult and requires considerable attention to be certain that different batches will give repro-

to be applicable to metals and possibly other valence bond crystals. However, for molecular solids, $\Delta\gamma$ should be about 1/8 to 1/3 $(\Delta h_f/A_w)$, depending upon the crystal structure.

At temperatures substantially below the melting point, equilibrium in

TABLE 10.20. SURFACE TENSION OF SOLIDS DETERMINED BY THE CREEP METHOD AND CALCULATED FROM THE SURFACE TENSION OF THE LIQUID AND EQUATION 10.50 (45).

Metal	Temperature (°C)	Surface tension of metal, by creep method	Crystal Plane	Surface tension calculated by Eq. 10.50	Increase in surface tension on solidification, calculated by Eq. 10.50
Silver	900	1140 ± 90	100	1150	220
			111	1180	250
Gold	1300	1400 ± 65	100	1396	256
		1510 ± 100	111	1438	298
Copper	1050	1430	100	1509	329
		1670	111	1560	380
γ-Iron	1400		100	2070	370
			111	2127	427
Tin	150	704	100	765	128
			001	672	35

TABLE 10.21. SURFACE TENSION OF LIQUID METALS (45).

Metal	Temperature (°C)	Surface Tension of Liquid Metal (dynes/cm)
Aluminum	700	900
Antimony	635	383
Bismuth	300	376
Cadmium	370	608
Copper	1140	1120
Gallium	30	735
Gold	1120	1128
Iron	1530	1700
Steel		950–1220
Lead	350	442
Magnesium	700	542
Mercury	20	476.1
Potassium	64	119
Selenium	220	105.5
Silver	995	923
Sodium	100	206.4
Thallium	313	(446)
Tin	700	538
Zinc	700	750

son (45) demonstrated that the globules are mainly water. He further showed that the primary mechanism by which the droplets form is osmosis. Thus, wool containing a small amount of residual oil is naturally saline. When this wool was soiled with mineral oil and immersed in distilled water, droplets grew rapidly at first under the oil film. Often the droplets coalesced to form larger droplets. When the water was replaced by a saturated sodium chloride solution, the droplets collapsed completely in 1 to 2 minutes. Upon replacing the brine with water, the droplets again grew. Similar observations were made using a variety of fibers, provided that they were soaked in brine overnight and dried before soiling. The rate of passage of water was more rapid on wool, silk, kapok, and cellulose acetate than on nylon, Terylene or glass, suggesting that with the first five fibers water travels to and from the droplets mainly through the fiber. Droplets grew faster on glass fibers soiled with castor oil than on those soiled with mineral oil. The water could pass through the fiber or the oil, or along the fiber-oil interface, depending upon which was more permeable.

Droplets have also been observed to grow on salt-free wool fibers soiled with mineral oil and immersed in certain electrolyte solutions. Alkalis were the most effective for the "secondary" droplet formation. The effect of other electrolytes corresponded with the lyotropic series for the swelling of gels:

$$\text{Alkali} > KCNS > KI > KNO_3 > KBr > KCl$$

While acids were without effect, certain polyvalent electrolytes had the effect of shrinking the droplets previously formed on naturally saline wool fiber. Again, the order agreed with the lyotropic series, the power of shrinking the droplets being found to decrease in the order:

$$MgSO_4 > \text{Na citrate} > CaCl_2 > Na_2SO_4 > BaCl_2$$

The formation of water droplets at the oil-fiber interface can play a part in detergency. Soiled fabrics will ordinarily contain salts derived from perspiration as well as traces from previous washings. The water droplets may assist the detachment of the oil from the fiber by decreasing the area of contact between oil and fiber. Similarly, solid soil may be forced away from the fiber.

Microscopic Study of Soiled Fabrics in Detergent Solutions

Where detergent solutions act on oiled fibers to remove the oil film a number of different phenomena have been observed. Adam (46) first demonstrated the manner in which oil drops roll up into spheres due to modification of the contact angle. Kling (47) and others (48, 49, 50) have studied this process, Emulsification, including the spontaneous formation of emulsified oil droplets, has been observed under the microscope (50). Solubilization has been deduced, but appears to play only a minor role in the removal of nonpolar soils (51).

One of the more interesting observations is the formation of myelinic figures. This occurs when long-chain polar compounds interact with detergent solutions. If the concentration of the water-insoluble polar compound is too great for complete solubilization, the mixed micelles become insoluble and separate in long cylindrical forms, as anisotropic droplets or as a coacervate (52). These forms are generally viscous or plastic in nature. The term "myelinic figures", which refers to the cylindrical forms, is derived from the myelin of the nerve sheath, which swells in this manner (50).

The interaction of long-chain polar compounds with detergents at interfaces, in micelles, or in myelinic figures is sometimes referred to as molecular association. Stevenson (50) soiled wool fibers as well as glass with various polar compounds and immersed them in detergent solutions. The microscopic changes that were observed are summarized in Table 13.2. In most instances, he recorded the formation of myelinic forms or anisotropic drops, emulsification, complete solution, or the formation of halos. The halos round the insoluble material indicated molecular association, since these halos frequently dissolved or broke into clusters of myelinic figures.

Stevenson also examined the complex suspensions in test tubes. Where there was considerable interaction, as with straight-chain detergents and polar compounds of similar chain length, a thick, gelatinous layer was observed upon centrifuging. This is in agreement with the rule (53) that viscous, birefringent gelatinous systems form readily in solubilized systems containing straight-chain compounds.

Spontaneous emulsification is frequently observed where the oil phase contains a water-soluble emulsifier, or the emulsifier is formed at the interface (54), or the oil phase contains a water-insoluble, long-chain polar compound (55). Spontaneous emulsification due to soap formation at the interface occurs when a fatty-acid soil is immersed in an alkaline solution. Emulsification due to molecular association was observed when a mineral oil soil containing 1 per cent cholesterol was immersed in 0.1 per cent sodium cetyl sulfate solution (50).

The action of 0.02 per cent solutions of detergents on soiled wool fibers is summarized in Table 13.3. Oildag is a colloidal suspension of graphite in mineral oil containing only a minimal amount of surfactant as the dispersing

TABLE 13.2. INTERACTION BETWEEN POLAR COMPOUNDS AND SURFACTANTS.
B – BIREFRINGENCE; W – SOIL ON WOOL FIBERS; G – SOIL ON GLASS (50)

Polar Compound	Sodium Oleate	Sodium Laurate	*sec.* Alkyl Sulphate	Non-ionic Detergent	Alkylaryl Sulphonate
Oleic Acid	(0.2 and 1.0%) W Coarse myelinic halo	(0.5) W BB Coarse myelinic halo—two zones	(0.2%) W Fine halo	(0.2%) W Diffuse halo Complete solution (1.5%) W B Spherical drops to myelinic forms	(1%) G Dense fine halo, not very coherent
Lauric Acid	(0.5%) W Fine halo not very coherent	(0.5%) W BB Coarse myelinic halo	(1%) W (B) Fine halo	(1.5%) W B Spherical droplets some myelinic forms	(1%) G Fine halo
Cetyl Alcohol	(1%) W B Poor halo not coherent	(0.5%) W B Coarse myelinic halo	(1%) W Poor halo not coherent	(1.5%) W B Poor halo action slow	(1%) W No apparent action
n-Decyl Alcohol	(1%) G BB Fine well developed two-zoned halo	(1%) G BB Fine well developed two-zoned halo	(1%) G BB Fine well developed two-zoned halo	(1.5%) G B Emulsification, some anisotropic drops	(1%) G B Fine well developed halo
Cholesterol	(1.0, 0.1 and 0.02%) G BB Long myelinic thread and clusters	(1%) G BB Long myelinic threads, drops, spirals and clusters	(1%) G B long myelinic threads	(1.5%) G Very little action	(1%) G Very little action
sec.–Heptyl Alcohol	(1%) G Complete solution, slow	(1%) G Complete solution	(1%) G Complete solution	(1.5%) G Solution into spherical drops with a violent jumping action	(1%) G Complete solution (0.1%) Little action
sec.–Dodecyl Alcohol	(1%) G Slight emulsification, mainly solution	(1%) G Complete solution into visible complex spherical droplets	(1%) G Complete solution	(1.5%) G Steady solution, visible streamers	(1%) G Complete solution (0.1%) Little action

TABLE 13.3. THE ACTION OF 0.02 PER CENT DETERGENT SOLUTIONS ON SOILED WOOL (50)

		Water	Sodium Oleate	*sec.* Alkyl Sulphate	Non-ionic Detergent
Nujol	Saline Wool	Primary droplets	Thick emulsion coating, mainly W/O	Primary droplets	Primary droplets
	Non-Saline Wool	No action	Slight action (as above)	No action	No action
Nujol + Oleic Acid (3 : 1)	Saline Wool	Primary droplets	Thick emulsion coating, mainly W/O	Moderate emulsion coating	Primary droplets
	Non-Saline Wool	No action	Slight action	No action	No action
Oleic Acid	Saline Wool	Temporary W/O emulsion coating, stable 1 hr.	Thick emulsion coating, mainly O/W held by invisible gel	Thick coarse O/W emulsion coating	Slight rolling up, O/W emulsion, solubilisation
	Non-Saline Wool	No action	Very slight action	No action	No action
Oildag	Saline Wool	Primary droplets	Thick emulsion coating, mainly W/O	Primary droplets	Primary droplets
	Non-Saline Wool	No action	Slight action	No action	No action
Oildag + Oleic Acid (3 : 1)	Saline Wool	Slight transient W/O emulsion	Thick emulsion coating, mainly W/O	Thick intermediate type emulsion coating	Primary droplets
	Non-Saline Wool	No action	Slight action	Slight action	No action

TABLE 13.4. REMOVAL OF OILY SOILS FROM WOOL FIBER (50)

	0.2% Sodium Oleate	0.2% *sec.* Alkyl Sulphate	0.2% Non-ionic Detergent	1.5% Non-ionic Detergent	0.2% Cetyl Trimethyl Ammonium Bromide	0.1% Na_2CO_3
Nujol	Normal rolling up, complete removal	No action	No action	No action	No action	No action
Nujol + Oleic Acid (3 : 1)	Rolling up accompanied by some spontaneous emulsification	Normal rolling up, complete removal	Some rolling up. Small drops disappear into complex, large drops give pear-shaped drops which leave fibre	Rapid and complete solution into complex	Complex formation, slight emulsification. Little removed	Temporary rolling up, spontaneous emulsification. Much removed
Oildag	Slow rolling up, slight complex formation and dispersion	No action	No action	No action	No action	No action
Oildag + Oleic Acid (3 : 1)	Slight rolling up, general complex formation, emulsification and dispersion	Rolling up, complete removal	Complex formation slow rolling up, dispersion of oil and solid	Large myelinic forms, complete sorption of oil into complex, dispersion of solid	Halo formed, slight dispersion, very little removed	Temporary rolling up, partial emulsification and dispersion of solid
Natural Soil	Partial spontaneous emulsification, emulsion held near oil drops. Complete soil removal	Rolling up, complete removal	Complete sorption of soil into complex, larger drops leave fibre as pear-shaped drops	Rapid and complete sorption into complex	Partial spontaneous emulsification, much soil removed	Almost instantaneous removal and emulsification

agent. On non-saline wool there was little if any action. Otherwise, activity associated with soil removal required the presence of a polar component in the soil. In the case of sodium oleate acting on saline wool soiled with white mineral oil, diffusion of the salt out from the fiber at pH 6.1 in contact with water was presumed to cause hydrolysis of the soap in the neighborhood of the fiber, with the result that a layer of acid soap formed around the fiber.

Results obtained by Stevenson using higher concentrations of detergents are shown in Table 13.4. The natural soil was obtained by extracting soiled pillowslips. At these higher concentrations, the rolling up of the soil droplets is more evident. Again, action is associated with the presence of a long-chain polar constituent in the soil.

Cryoscopic Theory of Detergency

Lawrence (55) has developed a theory of detergency, applicable to polar soils, that is based on cryoscopic forces. It explains many of the observations of Stevenson (50) and Harkins (56).

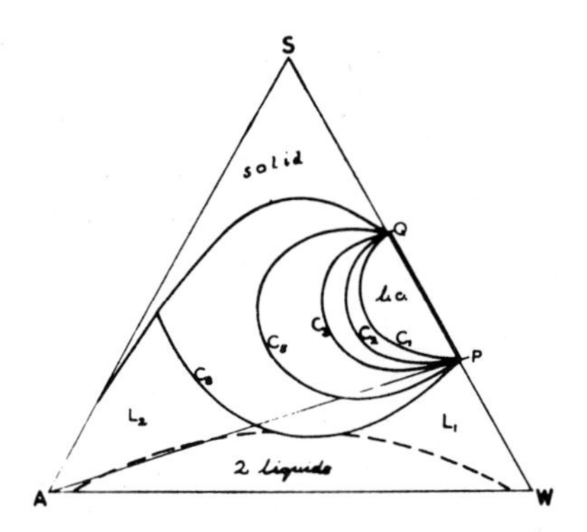

Figure 13.1. Generalized phase diagram for ternary systems (55).

Ternary systems composed of an ionic detergent, water, and a water-insoluble polar compound exhibit phase equilibria that are remarkably consistent, regardless of the nature of the detergent and the polar compound. Triangular diagrams show essentially five areas, as depicted in Figure 13.1. At the top of the diagram, at high soap concentrations, solid soap exists. At the bottom of the diagram, below about 10 per cent of soap, there are two liquid phases. The two phases may be present as separate layers, or as a single layer consisting of a colloidal dispersion of an aqueous isotropic phase in a liquid-crystalline phase. At the water-rich corner, L_1 is an isotropic solution corresponding to the solubilization of the polar compound in the soap micelles. At the corner containing a high content of organic compound, L_2 is an isotropic solution with water solubilized in soap micelles dispersed in the organic liquid. The final area in the triangular diagram is LC, the liquid-crystalline or smectic phase, flanked on either side by an area of LC + L_1 or LC + L_2. This area in the triangular diagram increases with the chain length of the water-insoluble polar compound.

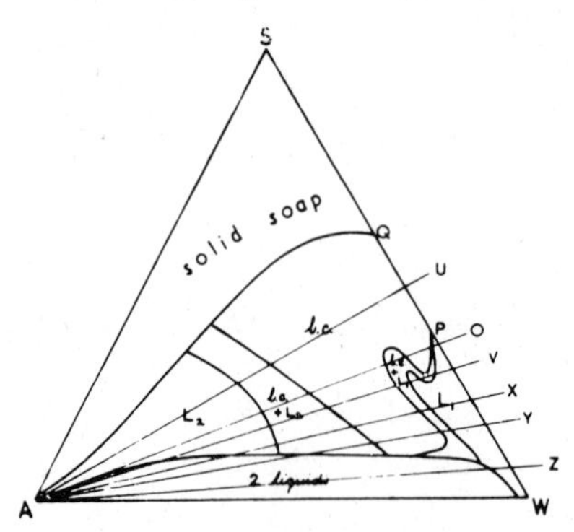

Figure 13.2. Phase diagram for the system: sodium dodecyl sulfate-water-caproic acid (55).

Figure 13.2 shows tie lines starting at point A, 100 per cent of organic compound. Following the line AO, the initial addition of soap solution results in a "soluble oil" phase L_2. Upon further addition of soap solution the anisotropic LC phase is formed. With continued addition of soap solution, the isotropic phase L_1 is developed, with the polar compound solubilized in micelles. The liquid-crystalline LC phase would give rise to the myelin forms observed by Stevenson (50).

According to Lawrence (55), the triangular diagrams may be viewed as soap-water-dirt phase-diagrams. He argued that the objection that the process requires a higher soap concentration than is normally present in detergent usage is not valid. At the point of attack, there will be a close-packed adsorbed layer of soap at the solution-dirt interface, and thus, a high concentration of soap in the environment of the soil. According to Lawrence, the viscous LC phase is always separated from the dirt by a fluid L_2 phase, which is the reason all washing requires agitation.

The theory can be further extended by replacing the polar compound at point A with a mixture of a polar and a nonpolar compound, such as oleic acid dissolved in mineral oil. Addition of soap solution would result in an

L_2 phase, followed by one or more emulsion areas in the diagram, in place of the LC phase, and finally an L_1 phase. Depending upon the composition of the soil and the soap solution, the gradual addition of the soap solution to the L_2 phase could result in a W/O emulsion which could become increasingly more viscous and then invert to an O/W emulsion, or it could go directly to an O/W emulsion.

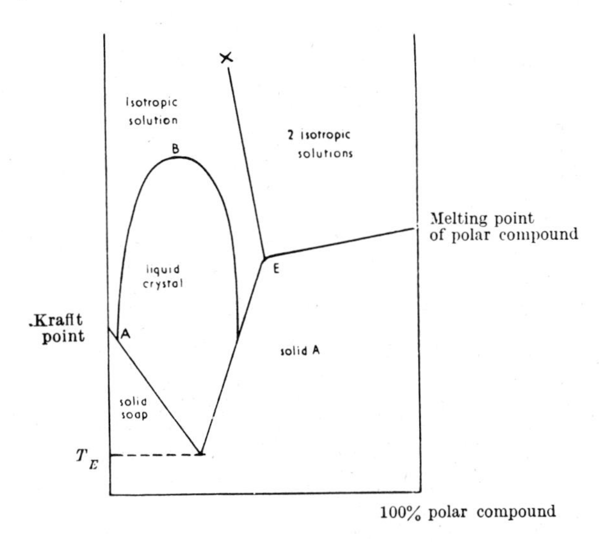

Figure 13.3. Generalized temperature-composition diagram for ionic surfactant-water-polar compound systems (55).

Thus, these phase diagrams can explain myelin formation or the formation of W/O or O/W emulsions during soil removal, as observed in microscopic studies. However, it cannot explain the rolling up of oil droplets or the phenomenon of soil redeposition.

In Figure 13.3 from Lawrence, is drawn a generalized temperature-composition diagram, for any selected soap solution above the minimum required to avoid the two-liquid-layer area, up to the maximum soap concentration at which the soap solution becomes liquid-crystalline. The melting point of the polar compound is shown at the extreme right side, at 100 per cent of the organic compound. The Krafft point of the pure soap solution, on the left, is taken as a melting point. At more or less equal ratios of soap solution and polar compound, there is a large depression of the freezing point, with T_E taken as the eutectic temperature. At a temperature below the melting point of the polar compound, mixing of the soap solution with the polar compound will result in a liquid-crystalline phase. This will not occur if the temperature is below the eutectic, T_E.

Rolling-up Process in Detergency

In addition to myelin formation and spontaneous emulsification, Stevenson (50) observed a rolling up of the oil film covering the wool fibers. This process was demonstrated with wool soiled with mineral oil and immersed in 0.2 per cent sodium oleate. It was also observed with wool soiled with the natural soil and immersed in the alkyl sulfate solution. In some instances, the rolling-up process was partially obscured by spontaneous emulsification.

The tendency for an oil film to roll up into a droplet can be expressed by the equation of Young

$$\cos\theta = \frac{\gamma_{ws} - \gamma_{os}}{\gamma_{ow}} \tag{13.1}$$

where θ is the contact angle measured in the oil, and γ_{ws}, γ_{os}, and γ_{ow} are the water-solid, oil-solid, and oil-water interfacial tensions, respectively.

Spontaneous removal of oil from the fiber requires that θ must be 180°. If γ_{ws} is less than γ_{os}, θ will tend toward 180° as γ_{ow} decreases. For a hydrophobic fiber, such as wool, adsorption of the detergent at the water-solid interface is required to lower γ_{ws}. If the polar compound is capable of forming a molecular association with the detergent at this interface, this will tend to produce a lower value of γ_{ws}. The same argument applies to γ_{ow}. If a mixed interfacial film does not form, removal of an oil film from a hydrophobic surface is improbable, since γ_{os} is necessarily low.

If the substrate is polar, preferential wetting by aqueous or oil phase will depend upon the orientation of the adsorbed surfactant-film. A cationic surfactant will adsorb on cellulosic fibers carrying a negative charge with the hydrocarbon chain directed away from the fiber. This orientation will promote retention of the oil film on the fiber.

Soil Removal and Surfactant-monomer Concentration

For white cotton muslin soiled with a lampblack-vaseline mixture, Preston (51) has shown the interesting relationship of Figure 13.4. Like many other properties of surfactant solutions, a plot of detergency defined as soil removal *versus* detergent concentration showed a change in slope at the cmc of the detergent. Below the cmc, detergent action increased rapidly with an increase in surfactant concentration. At higher concentrations, detergent action remained reasonably constant. This suggested to Preston that washing is proportional to the concentration of surfactant anions, and is, in fact, caused by the surfactant anions.

This is further shown in Figures 13.5, 13.6, 13.7 and 13.8. In Figure 13.5 for a homologous series of sodium alkyl sulfates, each curve exhibits a break at the cmc of the surfactant. The longer the chain, the lower the concentration required for washing. A similar pattern was found for the carboxylate

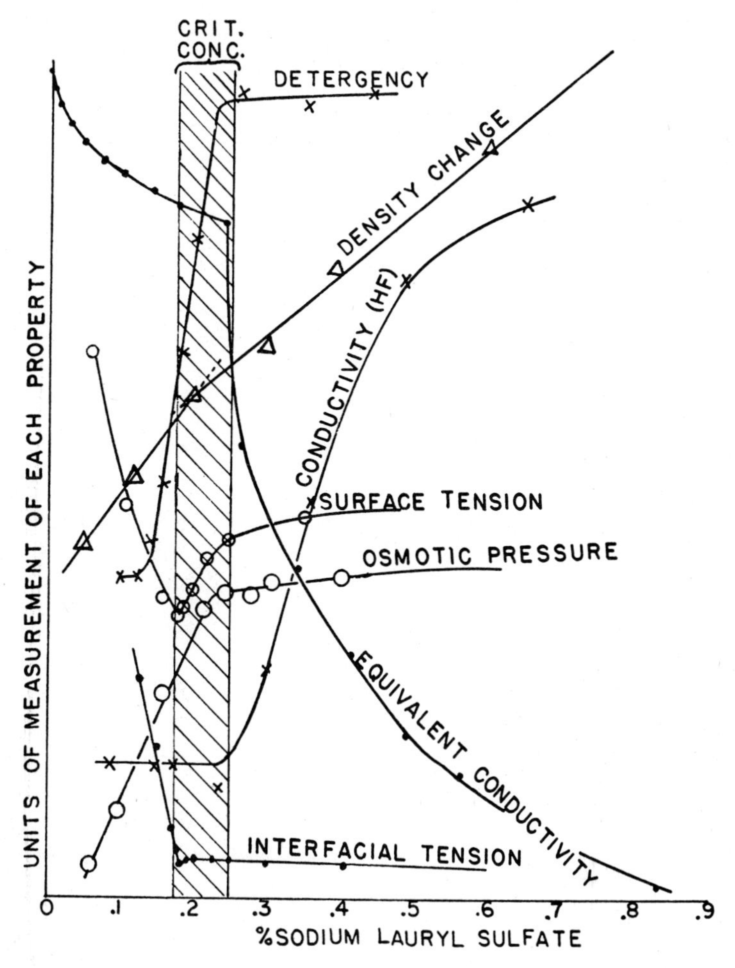

Figure 13.4. Properties of sodium lauryl sulfate at 25–38° C (51).

soaps, as shown in Figures 13.6, 13.7, and 13.8, except that at 71°C, the break in the curve for sodium palmitate occurred at a lower concentration than that of sodium stearate.

At 71°C, the greatest whiteness gain of the washed cloth was obtained with sodium stearate. At 55°C, sodium palmitate showed the most detergent action, while at 38°C, both the stearate and palmitate soaps were less effective than those of shorter chain length. The explanation for this behavior can be found in the solubility curves of Figure 13.9. If the wash temperature

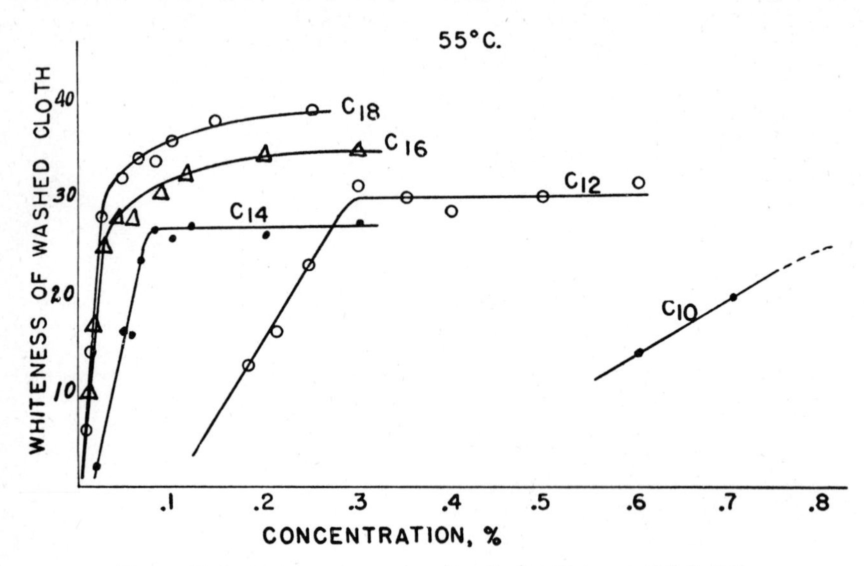

Figure 13.5. Detergency curves for alkyl sulfates at 55° C (51).

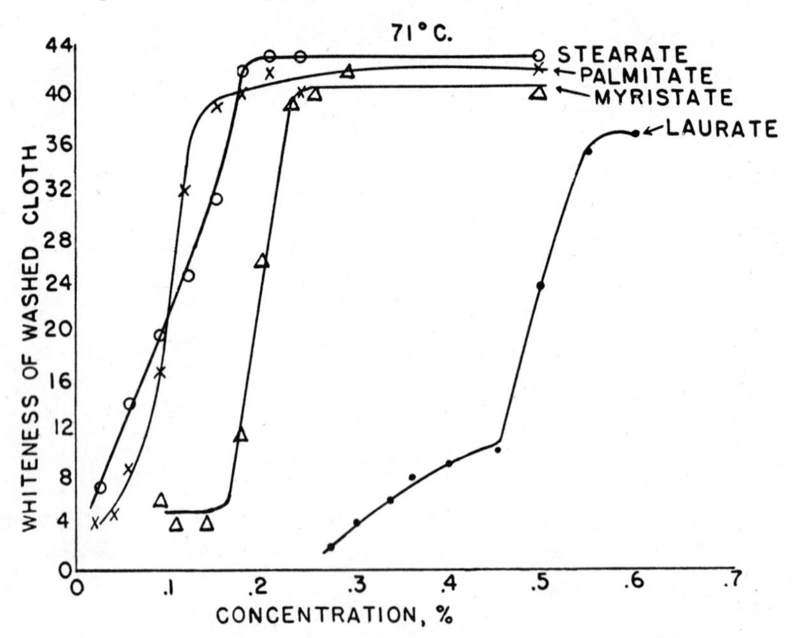

Figure 13.6. Detergency curves for sodium soaps at 71° C (51).

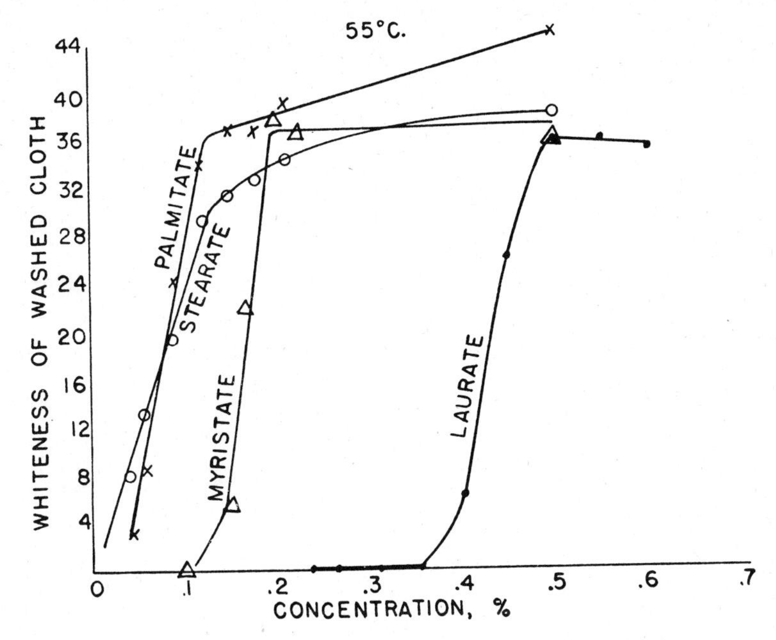

Figure 13.7. Detergency curves for sodium soaps at 55° C (51)

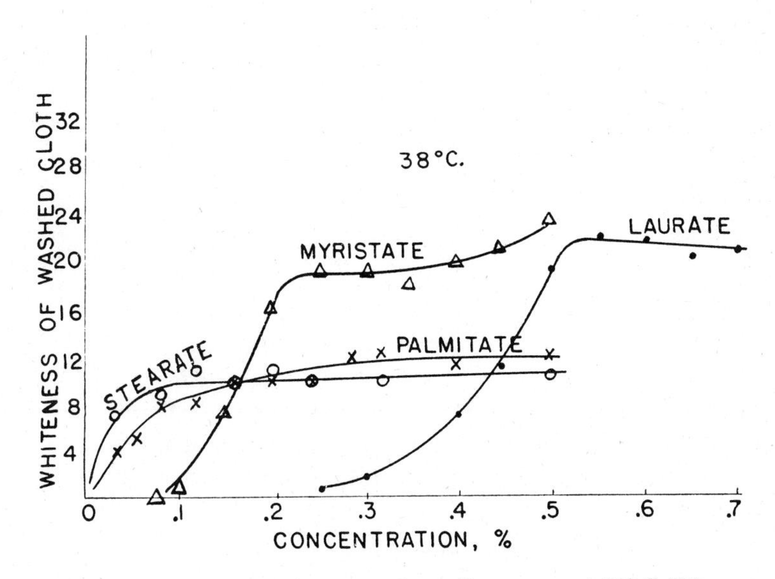

Figure 13.8. Detergency curves for sodium soaps at 38° C (51)

is below the Krafft temperature of the surfactant, the concentration of surfactant anions is less than at the cmc and detergency cannot attain its maximum value. It should be noted that with other soil-detergent systems, the detergency maximum occurs at higher concentration than the cmc.

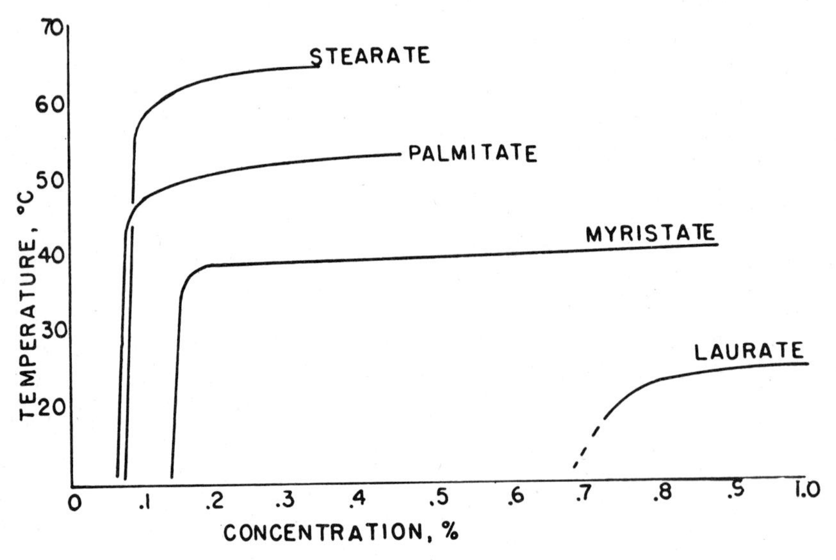

Figure 13.9. Solubility curves for sodium soaps, exclusive of hydrolysis products (51).

Adsorption

The adsorption of the surfactant at the soil-water and substrate-water interfaces is essential for all of the soil removal processes thus far observed. However, the adsorption isotherm is not a simple type. That shown in Figure 13.10 (57) appears to typify the adsorption of surfactants from aqueous solution onto various solid surfaces (57, 58, 59, 60). The adsorption isotherm shows a change in slope at point A, corresponding to the cmc of the surfactant. Adsorption reaches a maximum value at a somewhat higher concentration, and then decreases with further increase in the total concentration of the surfactant in solution. At the adsorption maximum, the amount of surfactant adsorbed is less than a close-packed monolayer (60).

Attempts to explain this adsorption behavior have not been wholly successful. If the unassociated surfactant anions are the only adsorbable surfactant species, adsorption should reach a maximum at the cmc and then decrease slowly with increasing surfactant concentration, as the concentration of the surfactant monomer decreases (58). However, the concentration at the adsorption maximum is substantially above the cmc.

Meader and Fries (59) suggested that at the cmc micellar adsorption commences. At still higher concentrations, the effect of the decrease in concentration of the single surfactant-ions predominates over the further slow increase in concentration of micelles, to account for the presence of a maximum in the adsorption isotherm. Adsorption of micelles had previously

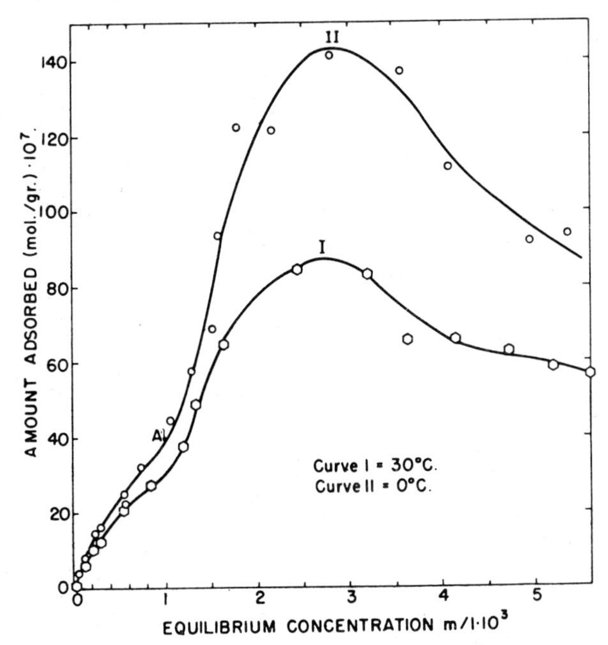

Figure 13.10. Adsorption isotherms of dodecylbenzene sodium sulfonate on cotton (57).

been postulated by Cockbain (61) to explain the stability of benzene-water emulsions. Vold and Phansalker (60) showed that this explanation cannot account for the rapid decrease in adsorption beyond the maximum. Further, since the micelles are necessarily a much less adsorbable species than the surfactant ions, this could not explain the increased rate of adsorption at concentrations above the cmc.

Fava and Eyring (57) suggested that the adsorption of small aggregates could explain increased adsorption at higher concentrations than the cmc. Thus, if in a surfactant solution, monomer is in equilibrium with micelles containing n single surfactant-ions, all aggregates of intermediate size from 2 to n-1 must necessarily be present. Mycelles and Mukerju (62) have demonstrated that dimers are present at concentrations considerably below the cmc. At the cmc and slightly above, when micelles are beginning to form, aggregates of intermediate size will probably be present to a greater extent than at other surfactant concentrations. Since these small aggregates have a substantial hydrocarbon surface exposed to water molecules, they can be considered to be adsorbable species.

This reasoning implicitly assumes that the surfactant is being adsorbed onto a low-energy surface. However, cotton fabrics contain exposed ionic groups. Polyvalent-cation bridges bonding fabric to micelles is not less plausible than the bonding of negatively-charged clay particles to fabric by this mechanism.

Regardless of whether monomer or aggregates are adsorbed, rate studies show that once on the surface the molecules are indistinguishable (57). The rate of adsorption of sodium dodecylbenzene sulfonate on cotton can be expressed by a reduced curve, with the ratio ϕ of the amount adsorbed to that adsorbed at equilibrium plotted against time. This behavior is shown in Figure 13.11. The quantity adsorbed as a function of time is shown in Figure 13.12 (57).

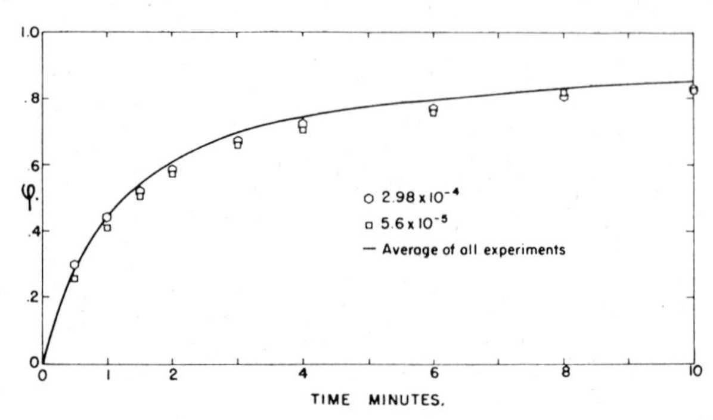

Figure 13.11. Adsorption of dodecylbenzene sodium sulfonate on cotton at 29.5° C (57).

The following kinetic equation is applicable to both adsorption and desorption at all concentrations:

$$-\frac{d\phi}{dt} = 2k'\phi \sinh b\,\phi \tag{13.2}$$

where ϕ is now defined as

$$\phi = \frac{a - a_f}{a_i - a_f} \tag{13.3}$$

with a_i, a, and a_f the amounts adsorbed at zero, t, and ∞ time, respectively.

The constant b has the same numerical value, 1.26, for both adsorption and desorption, and is temperature independent. The constant k' differs for adsorption and desorption. The two constants k' are temperature dependent and can be used to calculate the heat of activation for adsorption, $\Delta H_a^{\ddagger}$, and for desorption, $\Delta H_d^{\ddagger}$. These are 2075 calories for $\Delta H_a^{\ddagger}$ and 2720 calories for $\Delta H_d^{\ddagger}$. The difference of 645 calories is in reasonable agreement with a value of 778 calories for the heat of adsorption determined from equilibrium data. Application of the absolute reaction-rate theory gives

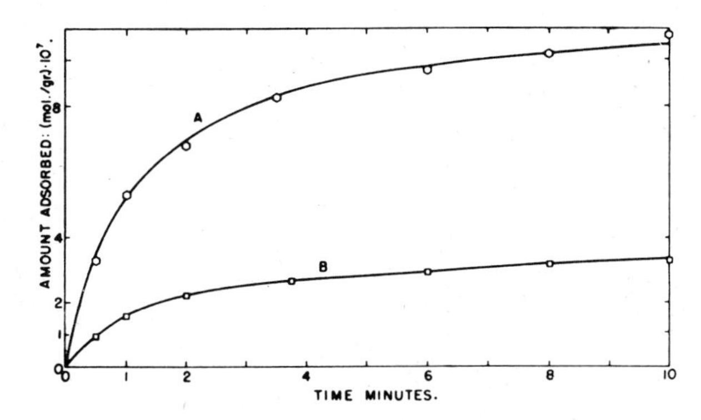

Figure 13.12. Adsorption of dodecylbenzene sodium sulfonate on cotton at 29.5° C; curve A, 2.98×10^{-4} mole/liter; curve B, 5.6×10^{-5} (57).

$\Delta S_a^{\ddagger} = -62.3$ e.u. and $\Delta S_d^{\ddagger} = -62.9$ e.u. for the entropies of activation for adsorption and desorption, respectively (57).

Effect of Nature of Surfactant and Substrate on Adsorption

Harris (14) has reviewed the adsorption of detergents. The area per molecule occupied by a large variety of surface active agents, adsorbed as a close-packed monolayer is summarized in Table 13.5. Nonionics exhibit greater surface coverage per molecule than the ionic surfactants. For the latter, surface coverage is greater when the ionic group is near the center of the molecule than when it is in a terminal position. Salts of oleic acid occupy a larger area per molecule than those of the trans isomer, elaidic acid, or the saturated acids.

Since surfactants adsorbed on cloth occupy somewhat less than the available surface area, the area per molecule for a close-packed film is only of casual interest. Selected data on the adsorption of surfactants on cotton and wool are shown in Table 13.6.

Figure 13.13 is a comparison of adsorption on different fibers of a homologous series of sodium alkyl sulfates. Measurements were made at 50°C for all but the C-18 compound. This alkyl sulfate was studied at 70°C, since it

TABLE 13.5. SELECTED VALUES OF SURFACE AREA PER MOLECULE REPORTED BY DIFFERENT INVESTIGATORS (14).

Compound	Area/Molecule Å²	Ref.
Na myristate	28	(1)
Na stearate	21	(1)
Na oleate	47	(1)
Mg stearate	50–51	(2)
Mg oleate	50–51	(2)
Ca stearate	70	(2)
Ca oleate	70	(2)
Na dodecyl sulfate	30–40	(3, 4)
Na hexadecyl sulfate	20–23	(5)
Na alkylbenzene sulfonate	20	(6)
Dihexyl Na sulfosuccinate	84–89	(4)
Nonylphenol, 9.5 ethylene oxide	78–92	(7)
Nonylphenol, 10.5 ethylene oxide	75–90	(7)
Nonylphenol, 15 ethylene oxide	110–130	(7)
Nonylphenol, 20 ethylene oxide	135–175	(7)
Nonylphenol, 100 ethylene oxide	1000	(7)

(1) Willson, E. A., Miller, J. R., and Rowe, E. H., *J. Phys. Chem.* **53**, 357 (1949).
(2) Solov'eva, L. R., *Colloid J. (U.S.S.R.)* **3**, 303 (1937).
(3) Pethica, B. A., *Trans. Faraday Soc.* **50**, 413 (1954).
(4) Lange, H., *Kolloid-Z.* **153**, 155 (1957).
(5) Edwards, G. R., and Ewers, W. E., *Australian J. Sci. Res.* **A4**, 627 (1951).
(6) Meader, A. L., Jr., and Criddle, D. W., *J. Colloid Sci.* **8**, 170 (1953).
(7) Hsiao, L., Dunning, H. N., and Lorenz, P. B., *J. Phys. Chem.* **60**, 657 (1956).

TABLE 13.6. ADSORPTION OF SURFACTANTS ON COTTON AND WOOL (14).

Surfactant	Adsorption, mg/g fabric Cotton		Wool	
Na laurate	0.3	(1)	3.0	(1)
Na myristate	1.1	(1)	2.4	(1)
Na palmitate	0.9–1.0	(1, 2)	7.8	(1)
Na stearate	0.9	(1)	3.4	(1)
Na oleate	1.2	(1)	11.8	(1)
Na dodecyl sulfate	0.09'	(3)	4.0	(3)
Na tetradecyl sulfate	0.5	(3)	1.4	(3)
Na hexadecyl sulfate	7.0	(3)		
Na octadecyl sulfate	0.07	(3)		
Na dodecylbenzene sulfonate	0.7	(3)	28.0–45.0	(4)
	5.0–7.0	(4)		
Cetyl trimethyl ammonium bromide	3.6	(3)		
Cetyl trimethyl ammonium chloride			15.0	(3)

(1) Weatherburn, A. S., Rose, G. R. F., and Bayley, C. H., *Canadian J. Res.* **F28**, 51 (1950).
(2) Meader, A. L., Jr., and Fries, B. A., *Ind. Eng. Chem.* **44**, 1636 (1952).
(3) Weatherburn, A. S., and Bayley, C. H., *Textile Research Journal* **22**, 797 (1952).
(4) Flett, L. H., Hoyt, L. F., and Walter, J. E., *J. Am. Oil Chemists' Soc.* **32**, 166 (1954)

was insufficiently soluble at 50°C. The data were obtained with the sodium alkyl sulfates at an initial concentration of 3.0 millimols per liter. The right hand side of Figure 13.13 was based on a similar series of experiments, with added sodium sulfate equal to 1 1/2 times the weight of sodium alkyl sulfate. Considerable differences in the adsorption on different fibers are evident. The least adsorption per unit weight of fiber occurred on cotton and the most on wool. Maximum adsorption on cotton and viscose occurred

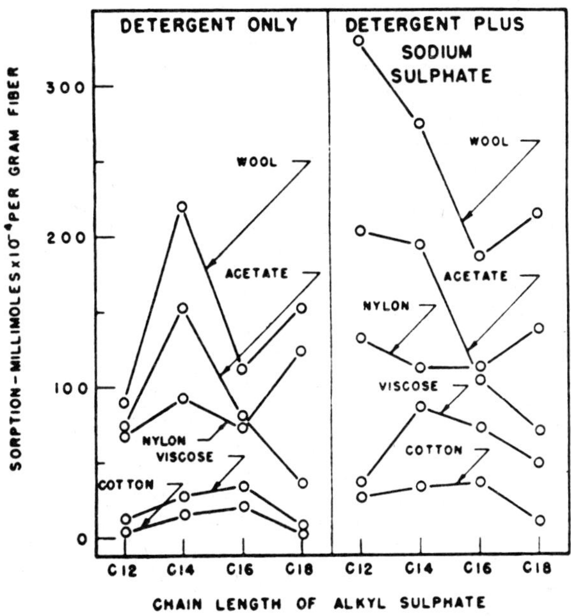

Figure 13.13. Adsorption of sodium alkyl sulfates by various fibers (17).

with the C-16 compound, while the C-14 compound was adsorbed to a greater extent than the C-16 compound on wool, acetate, and nylon. The effects of added electrolyte were to increase the degree of adsorption and to shift the adsorption maximum to sodium alkyl sulfates of shorter chain length.

Effect of Builders on Surfactant Adsorption (59)

The effect of sodium sulfate on the adsorption of sodium dodecylbenzene sulfonate on cotton cloth is shown in Figure 13.14. As previously observed, the adsorption maximum occurred at a higher concentration than the cmc. As determined by the pinacyanol method, the cmc values were 12, 7, 6, and $5 \times 10^{-4}M$ for solutions containing 100, 60, 40, and 20 per cent of active agent, respectively, with the balance sodium sulfate. In all instances, at

21°C the extent of adsorption increased with an increase in the sodium sulfate content.

Data on the adsorption of sodium alkylaryl sulfonate at 21°C in the presence of tetrasodium pyrophosphate are plotted in Figure 13.15. Increased

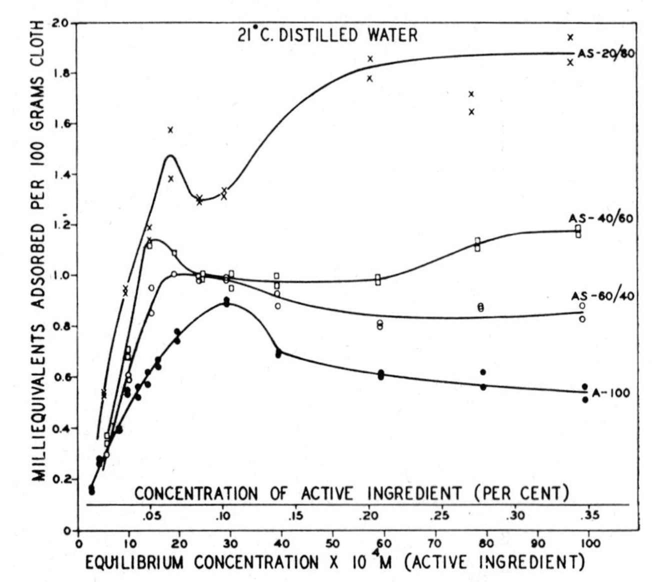

Figure 13.14. Adsorption of S^{35}–labeled detergent on cotton cloth. A, alkylaryl sodium sulfonate; S, sodium sulfate (59).

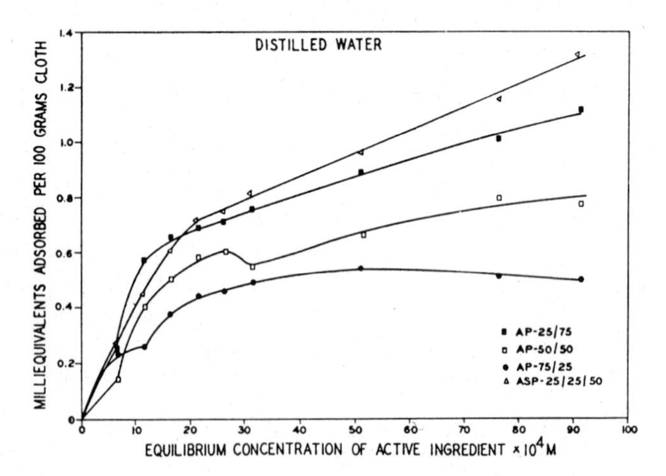

Figure 13.15. Adsorption on cotton cloth at 21° C. A, alkylaryl sodium sulfonate; P, tetrasodium pyrophosphate; S, sodium sulfate (59).

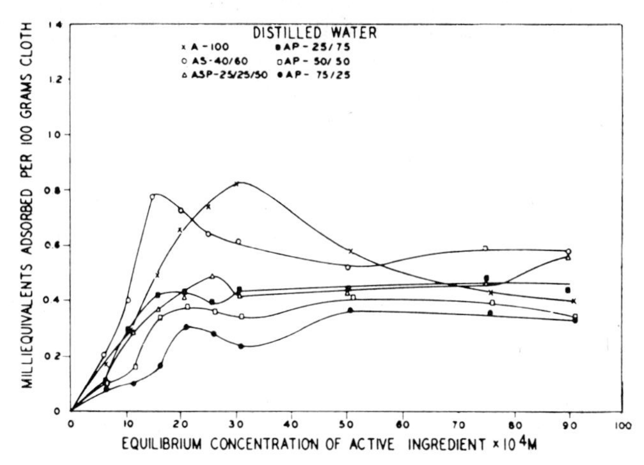

Figure 13.16. Adsorption on cotton cloth at 60° C. A, alkylaryl sodium sulfonate; P, tetrasodium pyrophosphate; S, sodium sulfate (59).

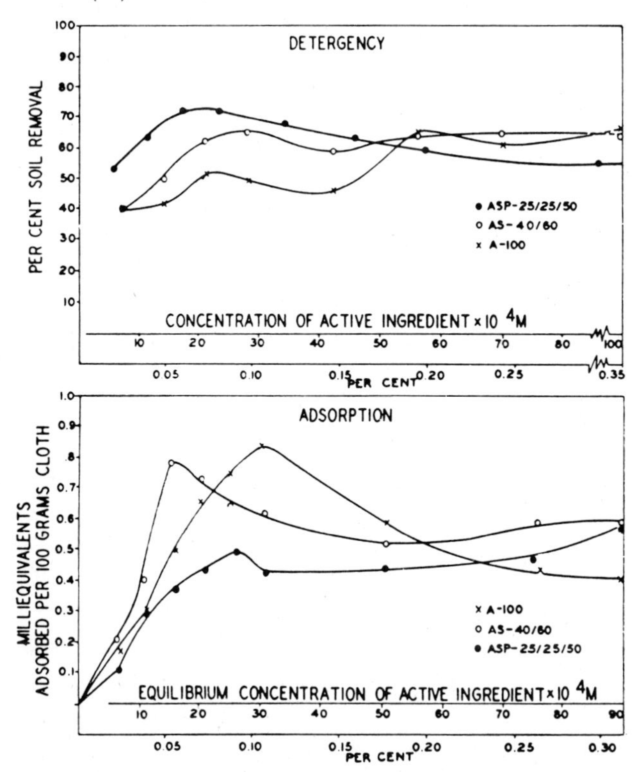

Figure 13.17. Comparison of adsorption and detergency at 60° C in distilled water (59).

phosphate content resulted in increased adsorption. However, at the same percentage of active agent, compositions containing tetrasodium pyrophosphate showed lower adsorption than those containing sodium sulfate. In general, the phosphate series did not show a pronounced adsorption maximum. At 60°C, adsorption in the presence of phosphate and sulfate was greatly diminished, while the 100 per cent active compound showed little if any reduction in adsorption, as plotted in Figure 13.16.

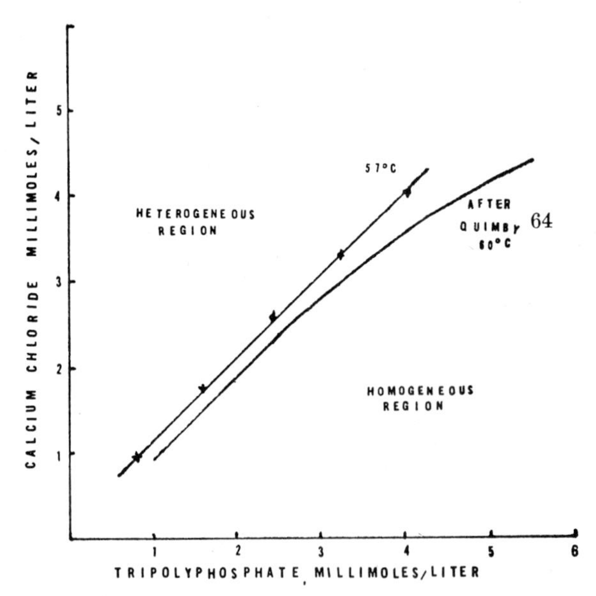

Figure 13.18. Phase diagram for $Na_5P_3O_{10}$-$CaCl_2$-H_2O system (63).

A comparison of detergency and adsorption is shown in Figure 13.17, using cotton cloth soiled with an oil-carbon black soil. While the adsorption and detergency maxima occurred at about the same concentration range, other similarities are absent. Thus, the composition containing tetrasodium pyrophosphate had the highest detergent action and the least adsorption of active agent, at concentrations below about 0.15 per cent of active agent.

Calcium ions normally present in hard water cause a substantial increase in the adsorption of sodium dodecylbenzene sulfonate on cotton cloth at 59°C, as shown in Figure 13.18 (63). The complex phosphates possess the capacity to sequester calcium ions and thus reduce their availability for adsorption on cotton or for interaction with the surfactant. The phase diagram in Figure 13.19 (63) shows the sequestering power of sodium tripolyphosphate for calcium ions. In the homogeneous region, the calcium ions are effectively sequestered. At higher ratios of calcium chloride to sodium tripolyphosphate, insoluble calcium tripolyphosphate is formed (63).

The effect of both calcium chloride and sodium tripolyphosphate on the adsorption of sodium dodecylbene sulfonate at 59°C on cotton sheeting is plotted in Figures 13.20 and 13.21. The tripolyphosphate reduced detergent adsorption at all calcium ion concentrations.

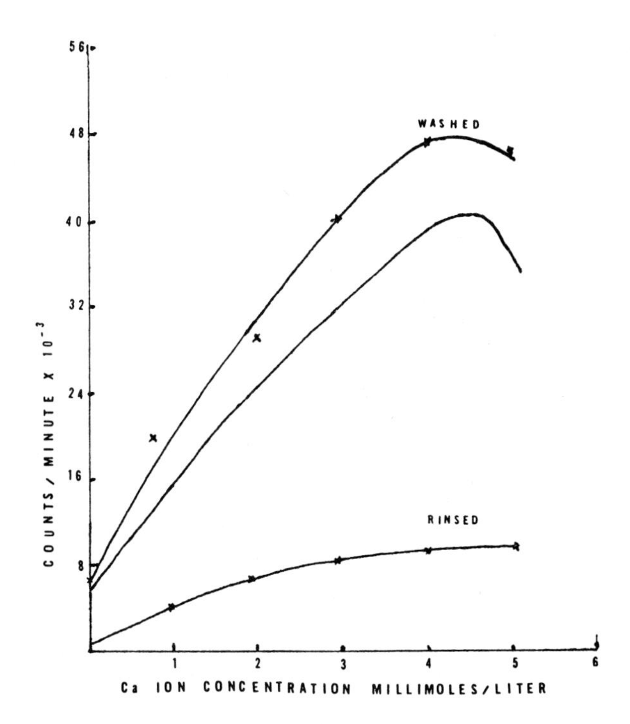

Figure 13.19. Adsorption of sodium dodecylbenzene sulfonate on cotton cloth in the presence of Ca^{++}; upper curve, after washing; lower curve, after rinsing; middle curve, difference in adsorption between washed and rinsed cloth (63).

Soil Redeposition

In the study of detergency, it is common practice to divide the problem of obtaining a clean wash into two parts. One is the actual removal of soil from the cloth, and the other is the redeposition of suspended solids onto the cloth. It is evident that the cloth will not appear white, if a portion of the soil is redeposited on the cloth during some phase of the washing cycle.

One aspect of the problem is the deposition of lime soap on cotton fabrics washed with soap in hard water. Figure 13.22 shows how the lime soap deposit builds up with the number of wash cycles. The data for this plot were obtained by washing cotton swatches in 360 ppm hard water with 0.5 per

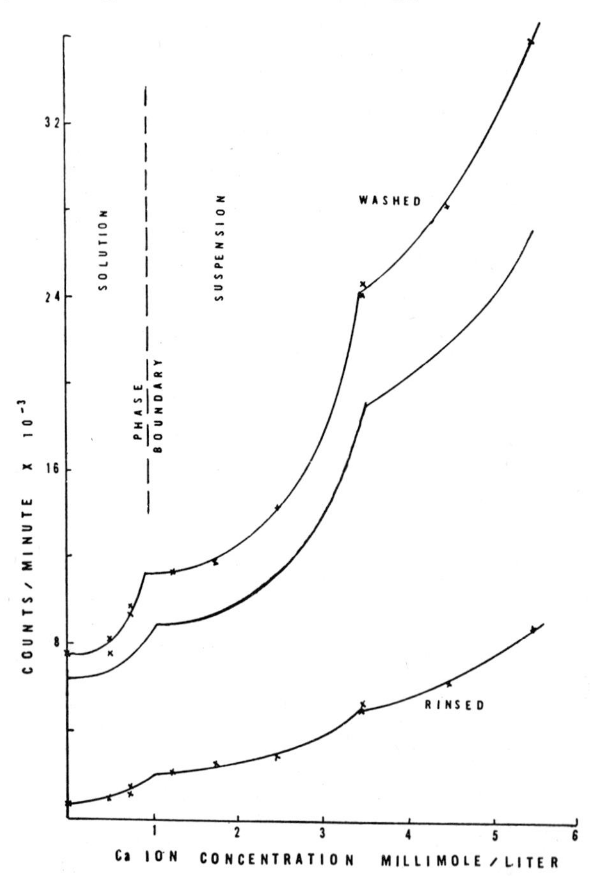

Figure 13.20. Adsorption of sodium dodecylbenzene sulfonate on cotton cloth in the presence of Ca^{++}, with 0.8 mmole/l of tripolyphosphate also present; upper curve, after washing; lower curve, after rinsing; middle curve, difference in adsorption between washed and rinsed cloth (63).

cent of a low titer sodium soap. In theory, 0.22 per cent of soap will precipitate all of the hardness from 360 ppm hard water (65).

When the cloth was washed in this level of hard water and then rinsed in distilled water, lime soap deposition reached a maximum at about 0.2 per cent of soap, as shown in Figure 13.23. In the range zero to 0.2 per cent of soap, all of the sodium soap is converted to lime soap. At higher soap concentrations, the sodium soap acts as a dispersing agent for the lime soap. At 0.4 per cent of total soap, the sodium soap remaining is sufficient to prevent any lime soap from depositing on the fabric. While these results show that sodium soap is an effective dispersing agent for lime soap, the situation

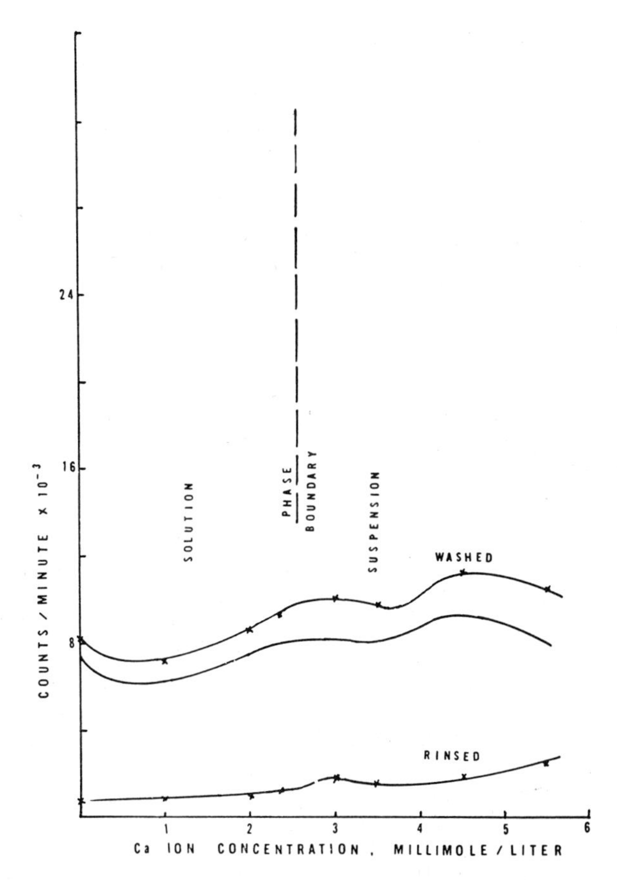

Figure 13.21. Like figure 13.20, but with 2.4 mmoles/l of tripolyphosphate (63).

is unrealistic, since it is not common practice to wash with hard water and rinse with distilled water.

The effect of a distilled water wash, followed by a hard water rinse, is shown in Figure 13.24. This is a more realistic set of conditions, since a softened water is sometimes used for the washing composition, with tap water for rinsing. The quantity of lime soap deposited after five cycles is approximately the same as that obtained using hard water for both washing and rinsing, as shown in Figure 13.22. During the hard water rinse, the sodium soap remaining on the fabric and in the occluded water is converted to the lime soap (65).

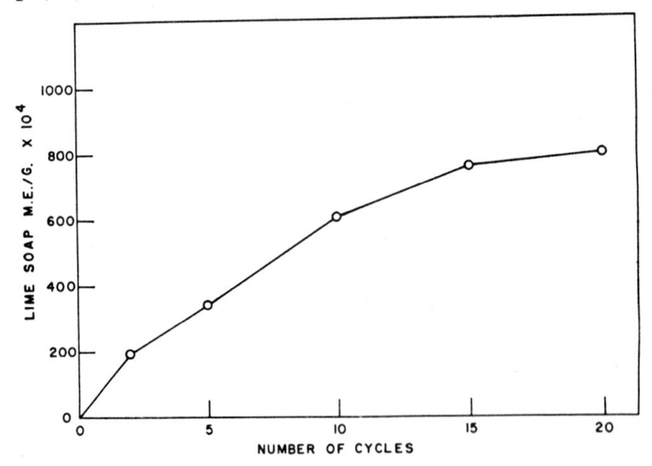

Figure 13.22. Deposition of lime soap (65).

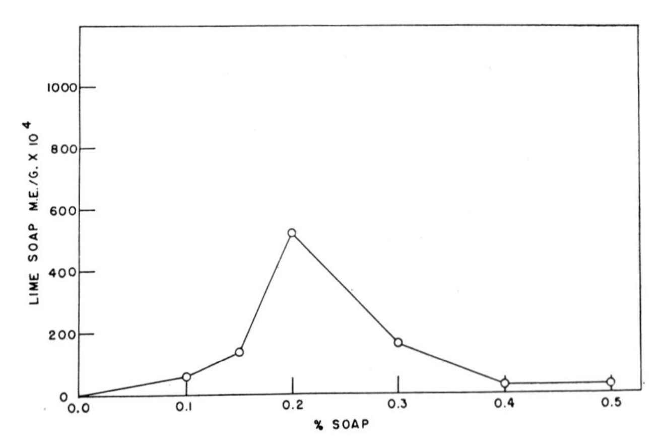

Figure 13.23. Deposition of lime soap *versus* soap concentration after 10 cycles (65).

Soaps are excellent dispersing agents, and they are effective in preventing soil redeposition in both soft and hard water. The synthetic surfactants do not have the disadvantage that soap has of forming adherent deposits on cotton in the presence of hard water. However, in the absence of builders, they do not wash cotton as clean as soap. Possibly because they do not remove calcium ions from solution and from the fabric and soil. They are particularly poor in permitting soil to redeposit on cotton. The sodium sulfate normally present with anionic surfactants further increases the amount of soil deposited. Figure 13.25 (66) is typical of soil redeposition with

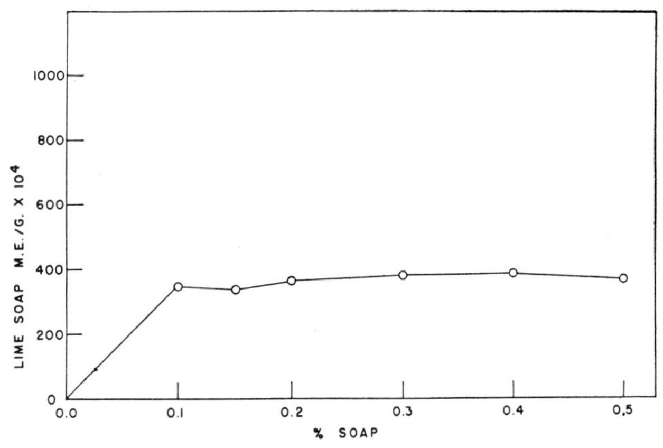

Figure 13.24. Deposition of lime soap *versus* soap concentration after 5 cycles (65).

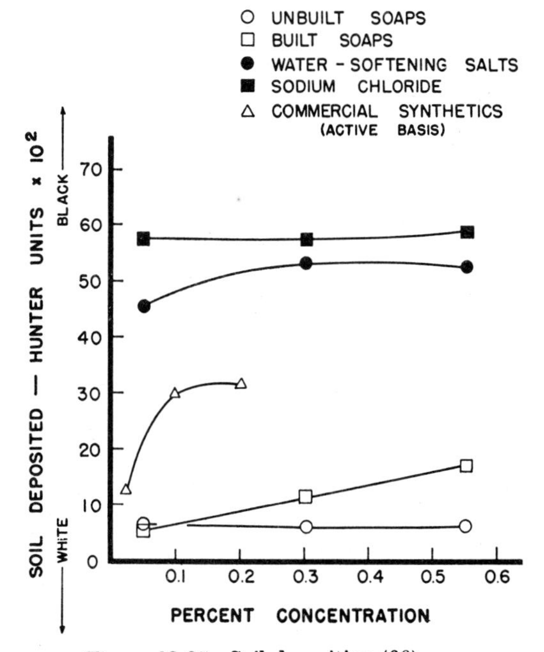

Figure 13.25. Soil deposition (66).

surfactants as compared with soap. The surfactants used contained 35 to 40 per cent of active agent and 60 to 65 per cent of sodium sulfate. Similar results were obtained with Nacconol NR, Igepon T and the sodium salts of sulfated coconut monoglyceride and sulfated coconut-tallow monoglycerides. The presence of builders with soap increased soil deposition. These results were obtained in distilled water at 110°F. "Aquadag", a colloidal dispersion of graphite in water, was used as the soil for deposition on cotton (66).

The presence of electrolytes, whether due to water hardness or added as builders, promotes soil deposition. In accordance with the Schultz-Hardy rule, the higher the valence of the electrolyte cation the greater the soil redeposition. The effect of mono-, di-, and tri-valent cations is demonstrated in Figure 13.26 (67).

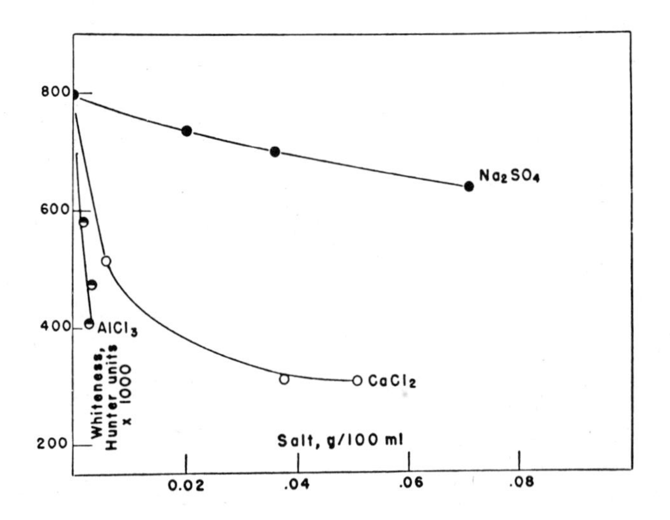

Figure 13.26. Deposition at 120° F with mono-, di-, and trivalent cations. 0.035% sodium alkyl sulfate, 0.03% Aquadag (67).

Fig. 13.27 is a plot of whiteness of cotton cloth *versus* concentration of sodium lauryl sulfate in the absence of other electrolytes. Whiteness increased to a maximum value at a lower concentration than the cmc of the surfactant and then decreased with a further increase in surfactant concentration (67). This decrease in anti-graying action with increasing surfactant concentration, which is generally found to be the case with both soaps and synthetic surfactants (66), can be ascribed to the increased electrolyte content of the solution.

The presence of sodium carboxymethyl cellulose appears to inhibit the effect of electrolytes on soil deposition. This is shown for hard water in Figure 13.28 and sodium sulfate in Figure 13.29 (67).

Once soil removal has occurred, both substrate and displaced soil become covered with an adsorbed film of surfactant which must be capable of maintaining the materials apart through both the remainder of the washing operation and the subsequent rinses. In the case of ionic surfactants, repulsion

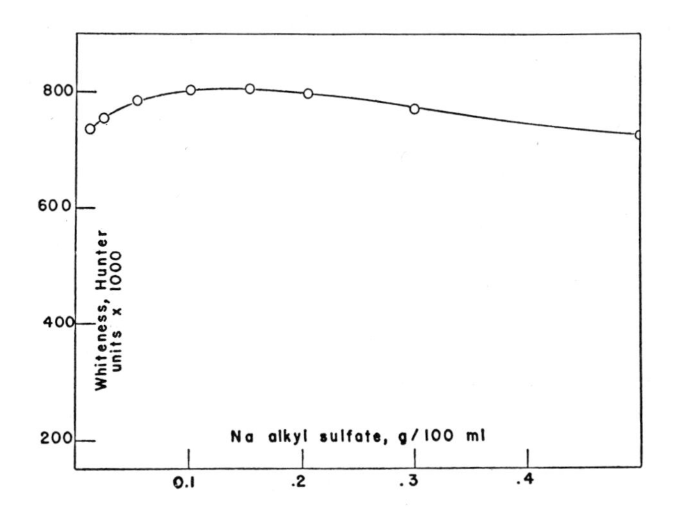

Figure 13.27. Deposition of 0.3% Aquadag with varying concentration of sodium alkyl sulfate at 120° F (67).

between surfaces of like charge discourages the redeposition of the soil on the substrate. A hydrated sheath surrounding an adsorbed film of nonionic surfactant may be even more effective than electrostatic repulsion in preventing soil redeposition, since many nonionic detergents are superior to the ionic surfactants in this respect.

Adsorbed surfactant ions are not firmly bonded to the surfaces. When fabric and soil bearing the same charge are brought together, through mechanical action, the electrical forces will cause adsorbed ions to leave the surface, thus decreasing the repulsion between the bulk materials. Van der Waals forces of attraction will tend to maintain the surfactants on these surfaces, and thus the longer hydrocarbon chain surfactants will desorb less readily than those with shorter hydrocarbon chains. Undoubtedly, sodium carboxymethyl cellulose is an effective antigraying agent because it does not desorb readily. The large size of the molecule is conducive to strong van der Waals interaction.

Role of Foam in Detergency

When soap is used for washing, the presence of foam provides evidence that sufficient soap is present for cleaning. In the absence of foam, it could be presumed that the soap had been precipitated as calcium or other insoluble soaps. When the synthetic detergents were first introduced for household

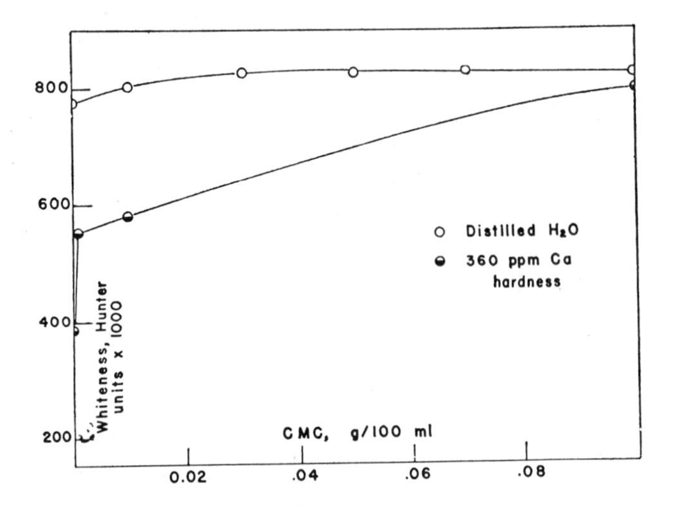

Figure 13.28. Deposition at 120° F. Effect of sodium carboxymethyl cellulose in distilled and 360 p.p.m. hard water. 0.035% sodium alkyl sulfate, 0.03% Aquadag (67).

use, and for many years thereafter, substantial foaming action was required to obtain consumer acceptance. However, foam has its drawbacks in automatic washing machines. It is not uncommon for the detergent solution to foam out of the machine, if excessive detergent is used. Particularly in the machines with vertical rotation, foam acts as a cushion to repress the mechanical action of the machine. Consequently, low-foaming detergents have been increasing in popularity.

In some detergent applications, foam can assist in the cleaning action. Stevenson (68) applied Oildag, a colloidal dispersion of graphite in mineral oil, to glass slides and allowed detergent solution to foam into contact with the slides. His photographs show the oil drawn into the thin liquid lamellar separating the air bubbles, and concentrating at the liquid junctions where two or more lamellar join. Stevenson suggested that this was probably due to the greater curvature of the air-liquid interface at these junctions, with consequent reduction in the pressure in the liquid at these junctions. Sodium oleate, a secondary alkyl sulfate, and a nonionic used in these tests were capable of detaching the oil from the glass slides, and the oil was sucked into the foam. Cetyl trimethyl ammonium bromide solutions do not displace mineral oil from glass. With this cationic surfactant, a slight amount of oil passed into the foam lamellar, but the oil was not removed from the glass.

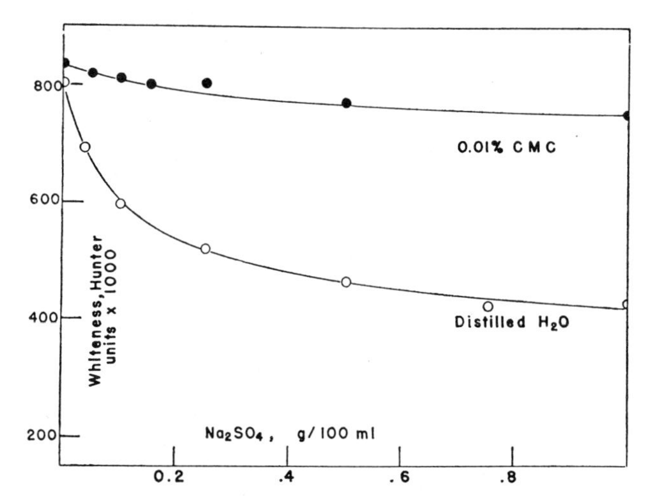

Figure 13.29. Deposition at 120°F. Effect of sodium sulfate in distilled water and in presence of 0.01% CMC. 0.035% sodium alkyl sulfate and 0.03% Aquadag (67).

While foam probably contributes insignificantly to soil removal in the washing machine, it can play an important role in some detergent applications, such as the removal of sebum from whiskers by shaving lather.

Mechanisms of Detergency

The first requirement for cleaning is the displacement of air and the thorough wetting of the soiled substrate by the detergent solution. The subsequent removal of soil can occur by a number of processes. The soil can be rolled-up and displaced from the substrate by the detergent solution. It can be removed in layers by emulsification. Alternatively, the surfactant can penetrate the soil to form inverse micelles, followed by the formation of liquid crystals and solubilization or emulsification. Finally, polyvalent cation bridges and hydrogen bonds connecting soil to substrate can be removed by ion-exchange and by the sequestering of the polyvalent cation. The final step is to maintain the soil in suspension, so that it does not redeposit during wash and rinse cycles.

The conditions for thorough wetting and air displacement have been discussed in the chapter on wetting. A low surface-tension is required for rapid wetting. Adsorption of the surfactant on the substrate can result in depletion of the surfactant, with a consequential rise in surface tension and increased wetting time as the process proceeds. However, in the absence of inorganic electrolytes, the adsorption maximum frequently corresponds with the detergency maximum (69).

Removal of an oil film by preferential wetting, cryoscopic effects, or emulsification are all favored by the formation of a mixed molecular film. Such mixed monomolecular films are most readily formed if there is strong interaction between both the hydrocarbon chains and the polar groups of the water-soluble surfactant and the polar oil-soluble components, either present in the soil or added with the detergent. Long, straight-chain alkyl groups are favorable for van der Waals interaction.

Electrolytes have a favorable effect on detergency in lowering the cmc of the surfactant. However, at concentrations of surfactant greater than the cmc, the effect of electrolytes is adverse, since they promote soil redeposition by reducing the effective surface charge on both soil and substrate. Charge neutralization can also result in exhaustion of the surfactant by adsorption. Polyvalent cations bond soil to substrate, as well as having a much greater effect on soil redeposition and surfactant exhaustion than monovalent cations. Complex phosphates sequester polyvalent cations and thus counteract their effect.

References

1. Pacifico, C., and Ionescu, M. E., *J. Am. Oil Chemists' Soc.* **34**, 11 (1957).
2. Fineman, M. N., and Kline, P. J., *J. Colloid Sci.* **8**, 288 (1953).
3. Fineman, M. N., *Soap and Sanitary Chemicals* **29** (2) 46, (3) 50 (1953).
4. Reich, I., and Snell, F. D., *Ind. Eng. Chem.* **40**, 1233 (1948).
5. Fowkes, F. M., *J. Phys. Chem.* **57**, 98 (1953).
6. Schwartz, A. M., *J. Am. Oil Chemists' Soc.* **26**, 212 (1949).
7. Niven, W. W., Jr., "Fundamentals of Detergency", New York Reinhold Publishing Corp., 1950.
8. McLean, D. A., and Wooten, L. A., *Ind. Eng. Chem.* **31**, 1138 (1939).
9. McLean, D. A., *Ind. Eng. Chem.* **32**, 209 (1940).
10. Snooke, A. M., and Harris, M., *J. Research Natl. Bur. Standards* **25**, 47 (1940).
11. Snooke, A. M., Fugitt, C. H., and Steinhardt, J., *J. Research Natl. Bur. Standards* **25**, 61 (1941).

12. Nieumenhuis, K. J., and Ton, K. H., Proc. Intern. Congr. Surface Activity. Vol. 4, p. 12, London, Butterworth, 1957.
13. Merrill, R. C., and Spencer, R. W., *Ind. Eng. Chem.* **42**, 744 (1950).
14. Harris, J. C., Soap and Chemical Specialties, Nov., Dec. 1958; Jan., Feb. 1959.
15. Myer, K. H., and Fikentscher, H., *Melliand Textilber* **7**, 605 (1926).
16. Aicken, R. G., *J. Soc. Dyers Colourists* **60**, 60, 170 (1944).
17. Weatherburn, A. S., and Bayley, C. H., *Textile Research Journal* **22**, 797 (1952).
18. Snell, F. D., Snell, C. T., and Reich, I., *J. Am. Oil Chemists' Soc.* **27**, 62 (1950).
19. Brown, C. B., *Research* (London) **1**, 46 (1947).
20. Sanders, H. L., and Lambert, J. M., *J. Am. Oil Chemists' Soc.* **27**, 153 (1950).
21. Powe, W. C., *Textile Research Journal* **29**, 879 (1959).
22. Powe, W. C., and Marple, W. L., *J. Am. Oil Chemists' Soc.* **37**, 136 (1960).
23. Goette, E. K., *J. Colloid Sci.* **4**, 459 (1949).
24. Utermohlen, W. P., Jr., Fischer, E. K., Ryan, M. E., and Campbell, G. H., *Textile Research Journal* **19**, 489 (1949).
25. Spring, W., *Z. Chem. Ind. Kolloid* **4**, 161 (1908).
26. Gotte, E., and Kling, W., *Kolloid-Z.* **64**, 227, 327, 331 (1933).
27. Kirk, R. E., and Othmer, D. F., "Encyclopedia of Chemical Technology", Vol. 4 p. 938, New York Interscience Publishers, Inc., 1949.
28. Rideal, E. K., *Chemistry and Industry*, June 26, 1948, p. 403.
29. Stayner, R. D., "Mechanism of Cotton Detergency", Oronite Chemical Company Bulletin, 1958.
30. Van Zile, B. S., and Borglin, , *Oil and Soap* **21**, (1944).
31. Tomlinson, K., *J. Soc. Dyers Colourists* **63**, 107 (1947).
32. Snell, F. D., *Ind. Eng. Chem.* **25**, 1240 (1933).
33. Snell, F. D., *et al., Soap and Sanitary Chemicals* **19**, (11) 63 (1943).
34. Rhodes, F. H., and Bascom, C. H., *Ind. Eng. Chem.* **23**, 778 (1931).
35. Hatch, G. B., and Rice, O., *Ind. Eng. Chem.* **31**, 51 (1939).
36. Kling, W., and Schmidt, Seifensieder-Ztg. **66**, 626 (1939).
37. Osipow, L., Marra, D., Snell, C. T., and Snell, F. D., *Ind. Eng. Chem.* **47**, 492 (1955).
38. Morrisroe, J. J., and Newhall, R. G., *Ind. Eng. Chem.* **41**, 423 (1949).
39. Harris, J. C., "Detergency Evaluation and Testing", New York Interscience Publishers, Inc., 1954.
40. Martin, A. R., and Davis, R. C., *Soap and Sanitary Chemicals*, May, June 1960.
41. Vaughn, T. H., and Suter, H. R., *J. Am. Oil Chemists' Soc.* **27**, 249 (1950).
42. Reich, I., Snell, F. D., and Osipow, L., *Ind. Eng, Chem.* **45**, 137 (1953).
43. Hensley, J. W., Kramer, M. G., Ring, R. D., and Suter, H. R., *J. Am. Oil Chemists' Soc.* **32**, 138 (1955).
44. Powney, J., *J. Textile Institute* **40**, T519 (1949).
45. Stevenson, D. G., *J. Textile Institute* **42**, T194 (1951).
46. Adam, N. K., *J. Soc. Dyers Colourists* **53**, 121 (1937).
47. Kling, W., *Melliand Textilber* **28**, 197, 427 (1947); **30**, 23, 412 (1949).
48. Kling, W., Langer, E., and Haussner, I., *Melliand Textilber* **25**, 198 (1944); **26**, 12, 56 (1945).
49. Reumuth, H., *Wascherie technik u. chemie* No. **4**, 21 (1950); No. **1**, 34 (1951).
50. Stevenson, D. G., *J. Textile Institute* **44**, T12 (1953).
51. Preston, W. C., *J. Phys. and Colloid Sci.* **52**, 84 (1948).
52. Dervichian, D. G., *Trans. Faraday Soc.* **42B**, 180 (1946).
53. Klevens, H. B., *Chem. Revs.* **47**, 1–74 (1950).
54. Winsor, P. A., *Trans. Faraday Soc.* **44**, 376 (1948).
55. Lawrence, A. S. C., *Nature* **183**, 1491 (1959).

56. Harkins, W. D., *J. Textile Institute* **50**, 189 (1959).
57. Favo, A., and Eyring, H., *J. Phys. Chem.* **60**, 890 (1956).
58. Corrin, M. L., Lind, E. L., Roginsky, A., and Harkins, W. D., *J. Colloid Sci.* **4**, 485 (1949).
59. Meader, A. L., and Fries, B. A., *Ind. Eng. Chem.* **44**, 1636 (1952).
60. Vold, R. D., and Phansalkar, A. K., 1st World Congr. on Surface Activity, Vol. 1, p. 137 (1954).
61. Cockbain, E. G., *Trans. Faraday Soc.* **48**, 185 (1952).
62. Mukerjee, P., Mysels, K. J., and Dubin, C. I., *J. Phys. Chem.* **62**, 1390 (1958).
63. Diamond, W. J., and Grove, J. E., *Textile Research Journal*, **29**, 863 (1959).
64. Quimby, O. T., *J. Phys. Chem.* **58**, 603 (1954).
65. Knowles, D. C., Jr., Berch, J., and Schwartz, A. M., *J. Am. Oil Chemists' Soc.* **29**, 158 (1952).
66. Vitale, P. T., *J. Am. Oil Chemists' Soc.* **31**, 341 (1954).
67. Ross, J., Vitale, P. T., and Schwartz, A. M., *J. Am. Oil Chemists' Soc.* **32**, 200 (1955).
68. Stevenson, D. G., *J. Soc. Dyers Colourists* **68**, 57 (1952).
69. Tachibana, T., Yabe, A., and Tsubomura, M., *J. Colloid Sci.* **15**, 278 (1960).

CHAPTER 14

Ore Flotation

Metal-bearing minerals are most commonly found as small particles embedded in a matrix of rock, clay, sand and other mineral matter of small commercial value. With most ores, it is necessary to separate the mineral from the worthless gangue before smelting. This process of ore dressing or concentrating may be accomplished by various mechanical separators. However, the most common method is that of flotation. Over 100 million tons of ore are said to be processed annually by flotation (1).

Before flotation, the ore is crushed to detach the mineral crystals from the gangue. The particle size must be sufficiently small for the particles to be floated by attached air bubbles. This will vary from 30 mesh down to smaller than 200 mesh. The crushed ore is fed to a flotation cell where it is mixed with water and the flotation reagents. Air is introduced by means of an agitator or an air-blowing device, to keep the pulp agitated and to form a froth. The mineral-laden foam is skimmed off as it is formed. Ordinarily, the desired mineral is removed with the froth, leaving the gangue behind. In some operations, this practice is reversed. In either event, the mineral is separated from the gangue (2).

A mineral particle will not become attached to an air bubble, unless it has a hydrophobic surface. That is, the water-solid interface must be at least partially replaced by an air-solid interface. The receding air-water-solid contact angle, measured through the water phase, must have a finite value. Since most minerals as they occur in the natural state possess a hydrophilic surface, it is necessary to add a reagent which will convert the surface to a hydrophobic state. This reagent, termed a collector, must be capable of imparting a hydrophobic film to the mineral which is to be floated, without altering the hydrophilic character of the surface of the gangue or the mineral to be left in the liquor.

In addition to the collector, secondary additives known as modifiers or conditioners may be added. A conditioner which promotes collection is known as an activator, while one which inhibits collection is known as a depressor. The pH of the liquor may also be adjusted by the addition of acid or alkali to an optimum value for the particular collector employed.

Collectors and conditioners modify the mineral surfaces for attachment to air bubbles. The second class of flotation reagents are the frothing agents. There are specific froth characteristics required. The froth must be sufficiently stable to resist disruption by the circulating mineral particles. However, it should not be too copious, and it should break down readily after removal from the flotation cell. The size of the individual bubbles should be suitable for the particle size and density of the mineral.

Flotation Reagents

The most commonly used frothing agents are pine oil, cresols, and the alcohols of medium molecular weight, usually C_5 to C_8. These frothers are effective at extremely low concentrations, well within their solubility limit. A typical pulp may contain 2 tons of water, 1 ton of ore, and only 0.1 pound of frothing agent (2). At higher concentrations, above their solubility limit, these frothing agents would be expected to act as antifoams.

The more typical foaming surfactants, such as the salts of the long-chain alkyl sulfates and amines, and the alkyl aryl sulfonates are employed as frothers to a lesser extent. In many systems they function both as frothers and collectors.

Broadly, collecting agents are classified into two types: thio compounds and non-thio compounds. The former include the xanthates, dithiophosphates and dithiocarbamates (3):

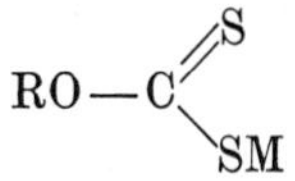

$$\mathrm{RO{-}C({=}S){-}SM} \qquad \mathrm{(RO)_2P({=}S){-}SM} \qquad \mathrm{R_2N{-}C({=}S){-}SM}$$

Xanthate 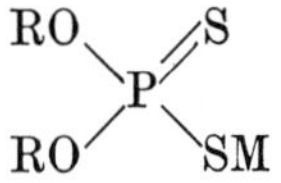 *Dithiophosphate* 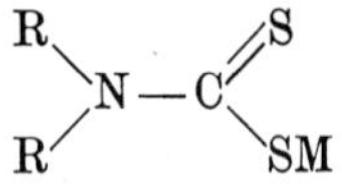 *Dithiocarbamate*

where R is an alkyl group and M is an alkali metal or ammonium ion. The alkyl group is frequently of small molecular weight, usually C_2 to C_5. The best known collectors are the alkyl xanthates.

Non-thio compounds used as collectors include fatty acids and fatty acid soaps, long-chain alkyl amines and their salts, quaternary ammonium salts, tall oil soaps, and sulfated fatty esters.

Differences Between Collector Types (4)

The alkyl thio compounds, such as the xanthates, mercaptans, dithiophosphates, and dithiocarbamates differ in many respects from the non-thio collectors which include the fatty acids, fatty amines, and their salts. Thus, the latter can be spread as monolayers while alkyl xanthates of the same chain length will not spread as monolayers. Instead they are oxidized to dicompounds, followed by association with the original species to form hydrophobic aggregates.

The alkyl thio compounds show negligible surface-tension lowering and little tendency to froth. Adsorption isotherms do not show detectable monolayer plateaus, but continue imperceptibly into multilayer adsorption. Contact angles, measured through the water phase at the air-water-solid interface, gradually increase to a maximum which is maintained at high concentrations of the thio compound.

Alkyl thio compounds adsorb initially as small patches of monolayer. However, these patches grow in thickness as well as laterally. The attainment of the maximum contact-angle is interpreted as the joining together of the isolated multilayer patches to form a continuous coating. The number of layers does not affect flotation.

The alkyl non-thio compounds adsorb initially to form isolated patches of monolayer. These grow laterally, as the concentration of the agent is increased, until a monolayer is completed. A second layer will start to build-up during the development of the monolayer or after its completion. Flotation is possible during the region of adsorption up to the starting point of the second layer.

The non-thio compounds are stable towards oxidation at the air-water interface, foam readily, and reduce the surface tension of water considerably. The adsorption isotherms show plateaus corresponding to 1 or 2 layers. The contact angles rise to a maximum with an increase in concentration. This is followed by a fairly sudden drop to zero contact-angle when the formation of a second layer prevents adhesion. These differences between the two types of collectors are shown in Figure 14.1.

Mixed Interfacial Films

Leja and Schulman (5) introduced a theory to the effect that frothers become effective only when there is molecular interaction between collector and frother molecules at the air-water and solid-water interfaces. This theory is based on earlier work by Schulman and coworkers (6, 7, 8, 9) which established the existence of molecular interaction between certain types of surfactants. It was shown that an insoluble monolayer, such as a film of a long-chain alcohol, could be penetrated by a soluble agent, such as sodium alkyl sulfate, injected into the aqueous solution under the monolayer.

Molecular interaction was determined by changes in the surface pressure, surface potential, and viscosity of the resulting film. Interaction takes place between the polar groups as well as the hydrophobic groups of the participating long-chain polar molecules. It is improbable that this interaction corresponds to the formation of true chemical compounds.

It has been demonstrated that monolayers of long-chain alcohols are readily penetrated by monolayers of long-chain xanthates. Thus, a mono-

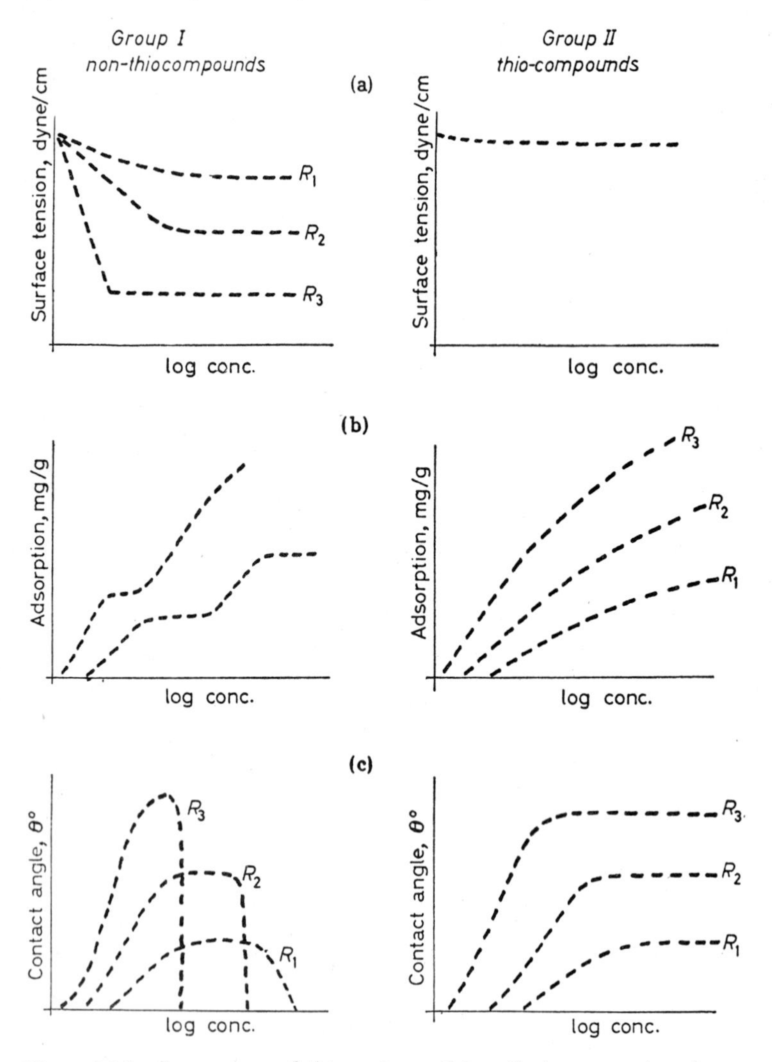

Figure 14.1. Comparison of thio and non-thio collector reagents, where the alkyl groups $R_1 < R_2 < R_3$ (4).

layer of cetyl alcohol was spread on $0.01N$ KCl solution at pH 11 and held at a constant area and at a surface pressure of 10 dynes/cm. Injection of 0.5 mg of potassium lauryl xanthate into the underlying 400 ml of solution led to an immediate rise in surface pressure to 52 dynes/cm. The potassium lauryl xanthate solution itself showed an initial surface pressure of 5 dynes/cm which rapidly decayed to zero pressure owing to oxidation of the xanthate into dixanthogen molecules, which form unstable films on their own. The mixed monolayer alcohol-xanthate remained stable at 52 dynes/cm, indicating that the penetrated monolayer stabilizes the xanthate against oxidation.

Weak penetration was observed with amyl xanthate into cetyl alcohol monolayers. These results were confirmed by Bowcott (10), who also found strong penetration by lauryl xanthate of films of cetyl amine and palmitic acid. In both instances, the decay in surface pressure was rapid, indicating that the association did not stabilize the xanthate. The xanthate had little tendency to form a mixed film with an alkyl sulfate.

Leja and Schulman (5) were unable to study adsorption with the xanthate-alcohol system, because decomposition of the xanthate interfered with the analyses. However, with other systems, adsorption of both collector and frother appeared necessary for flotation of the solids. For example, at pH 5.5 to 9.0 alkyl sulfonate gave excellent flotation of copper powder, acting as both collector and frother. At higher pH levels, the sulfonate desorbed, leaving the copper hydrophilic and non-floating. With potassium amyl xanthate present at pH 12.5, the copper floats and sulfonate is removed from solution by the floated copper.

Contact-angle measurements are applicable to flotation problems. The larger the contact angle measured through the water phase, the greater the probability that the solid will be efficiently floated. Measurements were made using air bubbles and also drops of various oils. The effects of concentration of potassium ethyl xanthate, time of conditioning, and the presence of an alcohol frother in the oil phase on the development of contact angles on electrolytically polished copper are shown in Figure 14.2. The air-water-solid contact angle can first be detected as a pronounced stick of the air bubble. For a conditioning time of five minutes this occurred at 0.5×10^{-3} M potassium ethyl xanthate, and the contact angle progressively increased to a maximum of 60° with the rise in concentration. For a conditioning time of 20 minutes, the stick of the air bubble occurred at a lower xanthate concentration and the maximum contact angle of 60° was reached sooner.

With a drop of pure benzene, the oil-water-solid contact angle could be detected at much lower concentrations of xanthate than those required for the development of air-water-solid contact angles, and the contact angles quickly rose to a maximum of 180°. At this contact angle, the drop of benzene spread almost completely on the xanthate-covered solid surface. With 0.01 *M* lauryl alcohol in the benzene, higher contact-angles were developed than with benzene, at the same xanthate concentration.

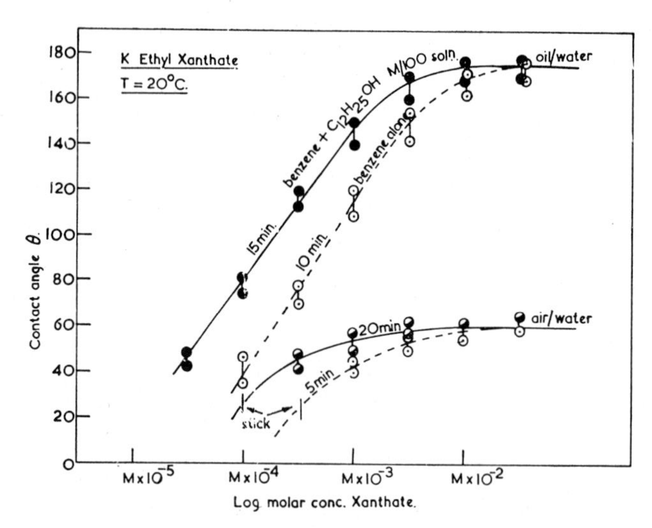

Figure 14.2. Contact angle *versus* concentration of ethyl xanthate on electrically-polished copper surfaces at pH 8.6, 20° C (5).

These and similar curves obtained with other xanthates show that the contact angle depends upon the concentration of xanthate and the time of conditioning. These factors determine the density of xanthate adsorbed on the surface. Other evidence shows that xanthate adsorbs in multilayers. The effect of the alcohol as a complexing agent is evident at lower concentrations of xanthate than that required to saturate the surface.

Surfactants capable of adsorption as orientated molecular-films at the air-water interface and the solid surfaces function as both collectors and frothers. Figure 14.3 shows the relationship between contact angle on $BaSO_4$ crystal faces and the concentration of sodium oleate, sodium dodecyl sulfate and sodium octyl sulfate. Initially, raising the concentration led to an increase in contact angle. At higher concentrations, the surfactant molecules associate into micelles, and coincident with micelle formation, the adsorbed layer takes on a hydrophilic character. The immediate effect is the nonadherence of air bubbles to the solid surface. That the surfactant is still present on the surface at these concentrations is shown by the high contact angles at the oil-water-solid interfaces.

Activators and Depressors

Collectors that do not ordinarily induce flotation of ore may become effective in the presence of a low concentration of some heavy-metal salt. The salt is said to activate the ore. It is generally accepted that in the activation process, the metal ions on the mineral surface are exchanged for

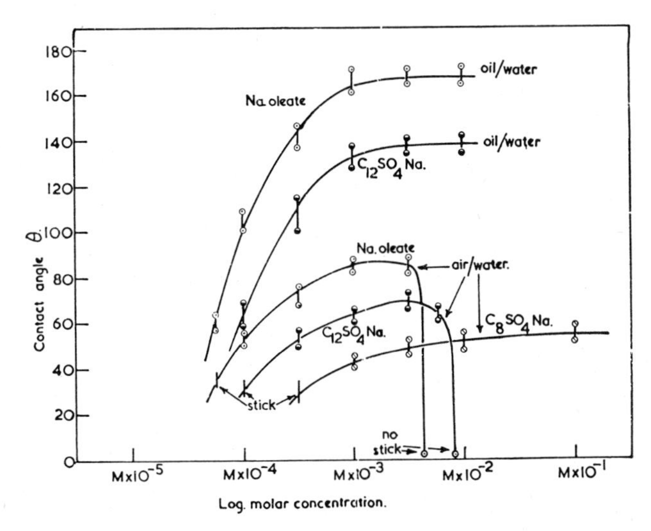

Figure 14.3. Contact angle *versus* concentration of carboxylic soaps and alkyl sulfates on cleaved $BaSO_4$ surfaces at pH 12.0, 20° C, after 5 to 10 minutes conditioning time (5).

the metal ions in the solution. In the activation of sphalerite—ZnS—activation is facilitated in accordance with the inverse solubilities of the metal sulfides. Thus, the sulfides of mercury, silver, and copper are much less soluble than zinc sulfide and ions of these metals are effective as activators. Electron diffraction studies by Sato (11) support this viewpoint. However, the salt found was not always the sulfide. When well-defined crystallites of the surface product were found, they were always oriented epitaxially. Sulfide minerals were activated with copper salts and then exposed to anionic collectors. The only products revealed by electron diffraction were the salts of the conditioning metal and the collector (11).

The activation of ores by metal ions has also been studied by the monolayer technique, using insoluble homologs of the soluble flotation-reagents spread over substrates containing the metal ion in question. Under suitable conditions, solidified films are formed.

According to Leja (4), the conditions which result in the formation of a solidified monolayer with an insoluble homolog of the collector are the same as those required for flotation. Thus, from the data of Wolstenholme (12) reproduced in Table 14.1, Leja showed that the pH conditions under which a myristic acid monolayer is solidified correspond to those required for the flotation of ores with caprylic acid. The ores studied by flotation contributed the same heavy-metal ions to the solution as were used in the monolayer experiments. In the absence of the metal ions, the spread monolayers were of the liquid-expanded type.

TABLE 14.1. COMPARISON OF THE pH RANGE REQUIRED FOR FLOTATION WITH THAT REQUIRED FOR THE SOLIDIFICATION OF THE MONOLAYER OF A HIGHER HOMOLOG (4).

Flotation		Metal-monolayer interactions	
Mineral	pH range of flotation M/1000 n-C_8 acid	Metal ions in the substrate	pH range of solidification with n-C_{14} acid
Pyrite, FeS_2	2–6.5	Fe^{3+}	2.0–5.5
Chalcocite, Cu_2S	4–7.5	Cu^{2+}	4.2–8.5
Alumina, Al_2O_3	3.5–8.5	Al^{3+}	3.7–8.0
Chalcopyrite, $CuFeS_2$; Bornite, Cu_3FeS_3	2.0–7.5	Cu^{2+}	4.2–8.5
		Fe^{3+}	2.0–5.5
Carrollite, $CuS.Co_2S_3$	4.8	Cu^{2+}	4.2–8.5
		Co^{2+}	4.6–9.0

The tendency to form a solidified monolayer depends upon van der Waals forces and polar interaction. These are influenced by steric factors and by pH. The latter determines the degree of ionization of the molecules in the monolayer, as well as the formation of basic-metal ions and other complex ions. Steric factors include: A_m the cross sectional area of the metal ion, A_h the cross sectional area of the hydrophobic portion of the molecules in the monolayer, and A_p the cross sectional area of the polar group of the monolayer molecules (4).

Some metal ions form basic-metal ions, such as $Cu(OH)^+$ at pH $\geqslant 4.2$, $Fe(OH)_2^+$ at pH $\geqslant 2.1$ In the presence of ions such as bicarbonate $(HCO_3)^-$, certain metal ions will form complex ions, such as $Cu(HCO_3)^+$ or $Ca(HCO_3)^+$.

The structure obtained from an interaction of a complex- or basic-metal ion with the monolayer molecules is portrayed in Figure 14.4a. If the ions are present as basic-metal ions, they may be held together by hydrogen-bonding, beneath the monolayer of surfactant molecules, with their array of counter-charges or dipoles. Hydrogen-bonding of the basic-metal ions would only be possible if A_m were larger than A_h and A_p. In that event, a two-dimensional solidified film of a mono-soap would form.

With metal ions such as calcium and magnesium, which do not form basic-metal ions, a di-soap will form. Since lateral interaction between the polar groups is comparatively insignificant, the film will not solidify unless multiple van der Waals forces can develop between the hydrocarbon chains of the monolayer. This is only possible when A_m and A_p are not larger than A_h. Two films solidified in this manner are shown in Figure 14.4b. In the case of fatty acid monolayers at low pH, most of the molecules are undissociated acid. The repulsion between the dipoles tends to keep the molecules apart. At higher pH levels, more di-soap molecules are formed and solidification becomes more pronounced.

When the hydrophobic portion of the molecules in the monolayer possesses a large cross sectional area, neither the development of multiple van der Waals forces nor lateral cross-linkages between the metal ions or complex-metal ions is likely, and the monolayer remains in the liquid-expanded form. This is represented in Figure 14.4c.

Thus carrollite—CuS . Co_2S_3—can be floated with 0.001 *M* caprylic acid, but not with the branched-chain 2-ethyl hexoic acid at the same concentration. Alkyl sulfate di-soaps do not solidify as readily as the di-soaps of the carboxylic acids, because the cross sectional area of the SO_4^- group is greater than that of the hydrocarbon chain.

Thus, activators encourage the formation of a condensed monolayer. Materials which are used as depressants retard polar interaction between the substrate and the collector. Depressor reagents include cyanides, ethylenediamine tetraacetic acid and tetrasodium pyrophosphate. These reagents can form complex anions with metal cations.

Selective Flotation of Salts

Crystals of potassium chloride are commercially separated from sodium chloride crystals, suspended in their saturated solution, by the use of alkyl amines. Selective flotation of the potassium chloride crystals has been explained on the basis of a lock-and-key fit of the corresponding size and shape of the adsorbing molecule, called epitaxis (13, 14). Thus, the $-NH_3^+$ ion is about the same size as the potassium ion, which it can displace in the crystal lattice. However, it is too large to fit in the sodium chloride lattice. Consequently, preferential adsorption occurs on potassium chloride.

More recently, Rogers and Schulman (15) presented a theory of more general applicability, which is based on the interaction of water molecules with the ions of the crystal-lattice surface. The relative attraction of water molecules to the crystal surfaces determines the selective adsorption of the surfactant. The rule is that adsorption of the surfactant ion is possible if the soluble salt has a sufficiently negative heat of solution. If the heat of solution is positive, water molecules will adsorb in preference to the flotation reagent and flotation will not occur.

The heat of solution ΔH_s is the sum of two energy terms, the lattice energy ΔH_e which is the energy required to separate the ions of the crystal lattice and the energy of hydration of the ions ΔH_h.

$$\Delta H_s = \Delta H_h + \Delta H_e \tag{14.1}$$

If the heat of hydration is greater than the lattice energy, the heat of solution is positive and the salt dissolves with evolution of heat. Otherwise, the process of solution is endothermic. When the water molecules interact strongly with the ions in the crystal surface, they prevent adsorption of the flotation reagent. If interaction is less strong, the surfactant ions will adsorb opposite their counter-ions on the crystal surface, aided by van der Waals forces available from the association of hydrocarbon chains.

Tables 14.2, 14.3 and 14.4 show the general relation between the energy terms and flotation. The scheme for the flotation of alkali halides is given in Table 14.2, with cations arranged in the order of decreasing hydration from Li^+ to NH_4^+ and anions arranged with hydration decreasing from F^- to I^-. The heats of solution (16) at infinite dilution for each salt are given in the table. The salts that float are to the right and below the dotted line. They have a negative heat of solution. The salts that do not float have a positive or a small negative heat of solution.

A similar scheme for the alkaline earth halides is given in Table 14.3. No

TABLE 14.2. HEATS OF SOLUTION IN KCAL/MOLE OF ALKALI HALIDES AT 25°C AND INFINITE DILUTION (15).

	Li^+	Na^+	K^+	Rb^+	Cs^+	NH_4^+
F^-	−0.7	0.2	4.2	6.3	9.0	−1.5
Cl^-	8.9	−0.9	−4.1	−4.4	−4.3	−3.8
Br^-	11.3	0.2	−5.1	−6.1	−6.2	−4.5
I^-	15.1	1.8	−4.9	−6.2	−7.9	−3.6
	no flotation			flotation		

flotation of these anhydrous salts has been observed, except for calcium fluoride, which has a negative heat of solution and is the only insoluble salt of the series.

Table 14.4 is similar to Table 14.2 with the values for free energy of hydration given is place of heats of solution. Salts which float have free energies of hydration below about 160 kcal/mole. The fact that lithium and sodium

TABLE 14.3. HEATS OF SOLUTION IN KCAL/MOLE OF ALKALINE-EARTH HALIDES AT 25°C AND INFINITE DILUTION (15).

	Mg^{2+}	Ca^{2+}	Sr^{2+}	Ba^{2+}
F^-		−3.2		
Cl^-	37.0	19.8	12.4	3.2
Br^-	44.5	24.9	16.4	6.1
I^-	51.2	28.7	21.6	11.4

TABLE 14.4. FREE ENERGIES OF HYDRATION IN KCAL/MOLE OF ALKALI HALIDES AT 25°C (15).

	Li^+	Na^+	K^+	Rb^+	Cs^+
F^-	228.5	203.6	187.9	181.4	174.7
Cl^-	198.8	173.9	157.7	151.7	145.0
Br^-	192.6	167.7	151.5	145.5	138.8
I^-	184.6	159.7	143.5	137.5	130.8
	no flotation			flotation	

fluorides have high hydration energies but yield small negative heats of solution is due to the high lattice energy or heat of separation of the ions. This lattice energy is larger for the smaller ions and offsets their greater heats of hydration. The high lattice energy is also related to the relatively low solubility of lithium and sodium fluorides. The lattice energy is still greater for calcium fluoride, so that it is less soluble and gives a more negative heat of solution.

Hydrated salts will frequently float while the anhydrous salt will not float in its saturated solution. This is the case with $MgSO_4$, which requires seven molecules of water of hydration to float. Similarly, $BaCl_2 \cdot 2H_2O$ has a negative heat of solution and the crystals in saturated solution will float with octyl sulfate and less readily with dodecylamine hydrochloride. Above 90°C, where anhydrous $BaCl_2$ is in equilibrium with its saturated solution, the heat of solution is positive and floating ceases. Rogers and Schulman (15) note that hydration will increase the cation-anion spacing in the crystal lattice. Thus, the ability of the polar group of the flotation reagent to fit into the lattice at the surface of the crystal is still retained as a factor in flotation.

Sodium octyl sulfate and dodecylamine hydrochloride are used as collectors in commercial flotation of soluble salts. They are sufficiently soluble in the saturated solutions to adsorb on the crystal surfaces, but not to reach the higher concentrations, related to the critical micelle concentration, at which flotation is prevented.

Native Floatability

Floatability is generally associated with a hydrocarbon surface, and non-floating solids are made to float by the adsorption of hydrocarbon-bearing species. However, there are a large number of solids that have native floatability. Gaudin, Miaw, and Spedden (18) showed that the tendency for pure solids to float depends upon the intensity of forces at the fracture or cleavage surfaces. Thus, if the surfaces offer a preponderance of ionic bonds to the surrounding liquid, the solid is non-floatable. However, native floatability results if fracture or cleavage surfaces form without rupture of interatomic bonds other than residual bonds.

In forming solid surfaces, whether by fracture or cleavage, pre-existing bonds are severed. When the severed bonds are residual intermolecular bonds, a non-ionic surface is obtained which is practically indifferent to the choice between water and air. The contact angle of water on the solid is greater than zero, the magnitude depending upon the slight preference for water or air. In contrast, solids which form ionic-fracture or cleavage surfaces are high-energy surfaces, completely wetted by water. Thus, solids that form molecular crystals, regardless of whether they are organic, should show native floatability.

Naphthalene and compounds of the paraffin class form molecular crystals and are floatable. Similarly, hexachlorobenzene, hexachloroethane, and carbon tetrabromide form molecular crystals, with distances between molecules much greater than distances between atoms. They float readily. Sulfur crystallizes in several forms, the most common consisting of arrays of molecules, each in the form of a puckered ring of eight sulfur atoms (19). Distances between atomic centers within a molecule are 2.10Å, while the nearest approach between atoms in different molecules is 3.30Å. This difference in interatomic distances shows that bond energies between molecules are small, and that sulfur forms molecular crystals. Sulfur is readily floated. Iodine also forms molecular crystals and is floatable. The atomic distance in the diatomic molecule is 2.70Å, as compared with least intermolecular distance between atoms of 3.54Å in some directions and 4.35Å in others (20, 21).

Talc and pyrophyllite are silicates which display native floatability. On the other hand, micas, chlorites, and clays, all of which are foliated silicates with some resemblance to talc and pyrophyllite, are non-floating. Crystal surfaces of these latter minerals are clearly ionic. Other solids exhibiting native floatability are realgar, arsenious oxide, boric acid, graphite, and boron nitride.

Hydrogen-bonded organic compounds which form dimers, linear structures, or sheet structures are floatable. Examples of floatable dimeric hydrogen-bonded substances are benzoic acid and salicylic acid. Linear structures are represented by formic acid, nicotinic acid, isocyanic acid, and hexamethylenediamine. Sheet structures include maleic acid, succinamide, oxamide, and stearic acid. All those that have been tested have been found to float. In contrast, if the hydrogen bonds connect the component parts by a three-dimensional network, the cleaved surfaces do not float. Examples of such non-floating substances include tartaric acid, sucrose, glucose, racemic acid, starch, cellulose, and rayon (18).

Mechanism of Ore Flotation

The requirements for flotation are that the particles to be floated become attached to air bubbles which convey them to the surface where they are retained by a froth of suitable stability. Solids that are completely wetted by water will not float unless a collector reagent is present in the pulp that is capable of imparting hydrophobic characteristics to the solid. For this, it is necessary that the collector reagent be adsorbed with the non-polar hydrocarbon chain directed towards the water phase.

In order for adsorption of the collector to occur, it is necessary that initially adsorbed water be displaced. The forces of attraction between the reagent and the ionic species at the solid surface must be stronger than the bonding of water molecules, H^+, OH^-, and other competitive ions to the solid.

Adsorption to provide a hydrophobic film is a necessary, but not a sufficient, condition for flotation. The film should also have the properties of a condensed two-dimensional phase. Hydrogen-bonding of the polar groups of the adsorbed species, as well as van der Waals interaction of hydrocarbon chains, contribute to the formation of a condensed monolayer. This is the more likely explanation of flotation resulting from collector-flother interaction at the solid-liquid or solid-air interface. A solidified film can also result from the formation of a di-soap or by hydrogen bonding of complex metal-ions at the solid-water interface, beneath the monolayer of amphipathic molecules.

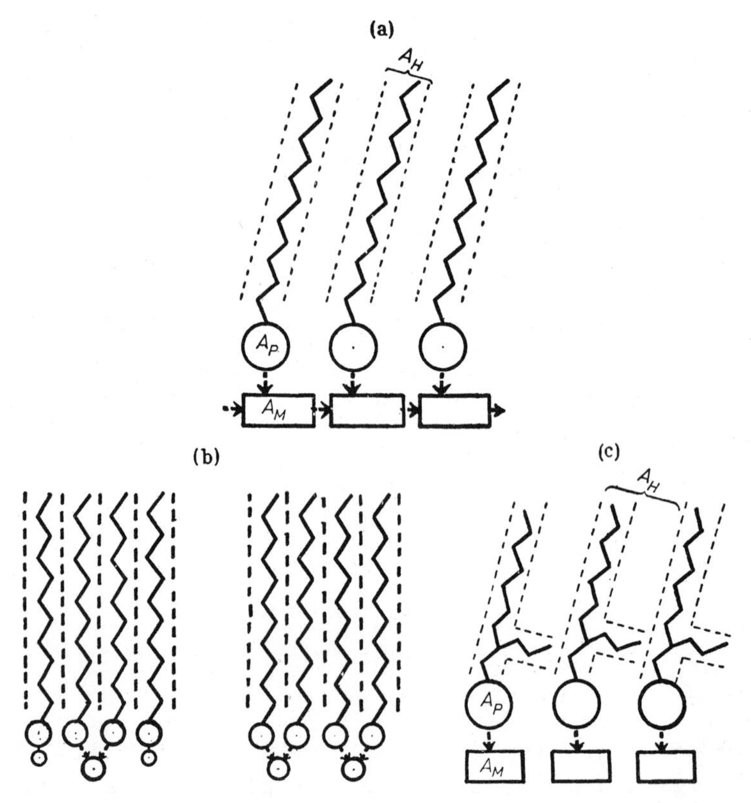

Figure 14.4. Types of films in metal-monolayer interactions; (a) solidified monolayer of a basic-metal mono-soap (or a complex ion mono-soap); (b) left: solidified monolayer of an acid-soap; right: solidified monolayer of a di-soap; (c) liquid-expanded monolayer of a soap (4).

Leja (4) has emphasized the importance of relatively strong adhesion of the solidified film to the particles, otherwise the film can be torn off by attrition and transferred to the air-water interface after contact with an air bubble. Thus, at pH $\leqslant$ 5.5 stearic acid and copper powder combine to form a solidified copper distearate film with an oil-water-solid contact angle of 150°. The adhesion of this film is negligible, as shown by lubrication tests, as well as by flotation experiments with copper powder and caprylic acid. However, at higher pH levels where basic copper-ions form, good lubrication and flotation result, even though the contact angle is lower, 105°. The greater adhesion of the film has been ascribed to the added contribution of hydrogen bonds within and beneath the basic-copper monolayer film.

Xanthates and dithiophosphates become immobilized after adsorption on sulfide minerals, because of the strong cohesional forces within the lattice. With progressive oxidation of the sulfide mineral, adsorption of the thio compound increases. However, cohesional forces within the layer of metal oxide are reduced and flotation is impaired. Reduced cohesion is shown with non-sulfide minerals as well. Thus, xanthates and dithiophosphates are strongly adsorbed on malachite—$CuCO_3$. $Cu(OH)_2$—but flotation is difficult. The metal-collector coating appears to be removed by attrition.

Quartz activated by metal ions is readily floated by fatty acids, but not by xanthates or dithiophosphates. With the thio compounds and the metal ions there is a stronger bond than with the fatty acids. However, the lateral cohesion between the long alkyl-chains is missing, and over-all adhesion is less.

References

1. Rideal, E. K., *Chemistry and Industry* **49**, 1528 (1959).
2. Schwartz, A. M., and Perry, J. W., "Surface Active Agents: Their Chemistry and Technology", Vol. 1, New York, Interscience Publishing Co., 1949.
3. Hagihara, H., Uchikoshi, H., and Yamashita, H., Proc. Intern. Congr. Surface Activity, 2nd, London, Butterworth, 1957.
4. Leja, J., Ibid.
5. Leja, J., and Schulman, J. H., *Min. Eng.*, N. Y. **6**, 221 (1954).
6. Schulman, J. H., and Rideal, E. K., *Proc. Royal Soc.* (London) **B122**, 29 (1937).
7. Schulman, J. H., and Rideal, E. K., *Nature* **144**, 100 (1939).
8. Schulman, J. H., and Cockbain, E. G., *Trans. Faraday Soc.* **35**, 1 (1939).
9. Schulman, J. H., and Hughes, A. H., *J. Biochem.* **29**, 1236, 1243 (1935).
10. Bowcott, J. E. L., Proc. Intern. Congr. Surface Activity, 2nd, London, Butterworth, 1957.
11. Sato, R., Ibid.
12. Wolstenholme, G. A., Ph.D. Thesis 1950, Ernest Oppenheimer Lab., Dept. Colloid Sci, Univ. Cambridge.
13. Bachmann, R., Z. *Erzbergh. Metallhuttenw.* 8, B109 (1955).
14. Fuerstenau, D. W., and Fuerstenau, M. C., *Min. Eng.*, N. Y. **8**, 302 (1956).
15. Rogers, J., and Schulman, J. H., Proc. Intern. Congr. Surface Activity, 2nd, London, Butterworth, 1957.
16. Bichowsky, F. R., and Rossini, F. D., "Thermochemistry of the Chemical Substances", New York, Reinhold Publishing Corp., 1936.
17. Latimer, W. M., Pitzer, K. S., and Slansky, C. M., *J. Chem. Phys.* **7**, 108 (1939).
18. Gaudin, A. M., Miaw, H. L., and Spedden, H. R., Proc. Intern. Congr. Surface Activity, 2nd, London, Butterworth, 1957.
19. Warren, B. E., and Burwell, J. T., *J. Chem. Phys.* **3**, 6 (1935).
20. Preston, M. H., Mack, E. Jr., and Blake, F. C., *J. Am. Chem. Soc.* **50**, 1583 (1928).
21. Straumanis, M., and Sauka, J., *Z. phys. Chem.* **B. 53**, 320 (1943).

CHAPTER 15

Lubrication

The field of lubrication can be conveniently divided into three categories. The first is referred to as hydrodynamic lubrication and applies to those situation where the moving surfaces are completely separated by a fluid layer. The principles of fluid mechanics are applicable to this type of lubrication. The second is called boundary lubrication. Here there is metal to metal contact, but the conditions are sufficiently mild that adsorbed films are capable of preventing the welding which occurs when surface asperities come into contact. The third category is similar to the second, except that sliding conditions are severe and a lubricant must be employed that will react chemically with at least one of the surfaces to provide a solid layer of low shear strength that is sufficiently durable to prevent welding and subsequent surface damage and wear. This type is termed extreme-boundary lubrication. Both of the latter types of lubrication are surface-chemistry phenomena.

When two solid surfaces are rubbed together, a resistance to sliding is experienced. This is expressed as a coefficient of friction μ which is related to the frictional force F and the applied load W. Thus

$$\mu = F/W \tag{15.1}$$

Effect of Adsorbed Gases on Friction

Solids that are cleaned in air are still covered by a thin film of oxide, water vapor, and other adsorbed impurities. This contaminant film, which is at least several molecules thick, has a profound effect on the friction. When metals are cleaned in the ordinary way, the coefficient of friction is of the order of $\mu = 0.5$ to 1. However, if the surfaces are cleaned and degassed by heating in a high vacuum, the coefficient of friction can rise to 100 or more.

Bowden (1) described experiments conducted by Young (2), who used an apparatus made of fused silica to facilitate degassing of the surfaces. The upper surface was dragged slowly over the lower surface, under an applied load of 15 to 20 grams. The area of contact was between a small curved protrusion and a flat surface. The metals were heated by high frequency induction. They were then allowed to cool nearly to room temperature, placed in contact and slid together, without disturbing the vacuum. When most metals were heated to 300°C, the effect of contamination was sufficiently reduced that complete seizure occurred on contact. This was observed with nickel, pure iron, and platinum. At room temperature, the contaminant film on iron was sufficient to reduce the friction to about $\mu = 3.5$. The sliding process

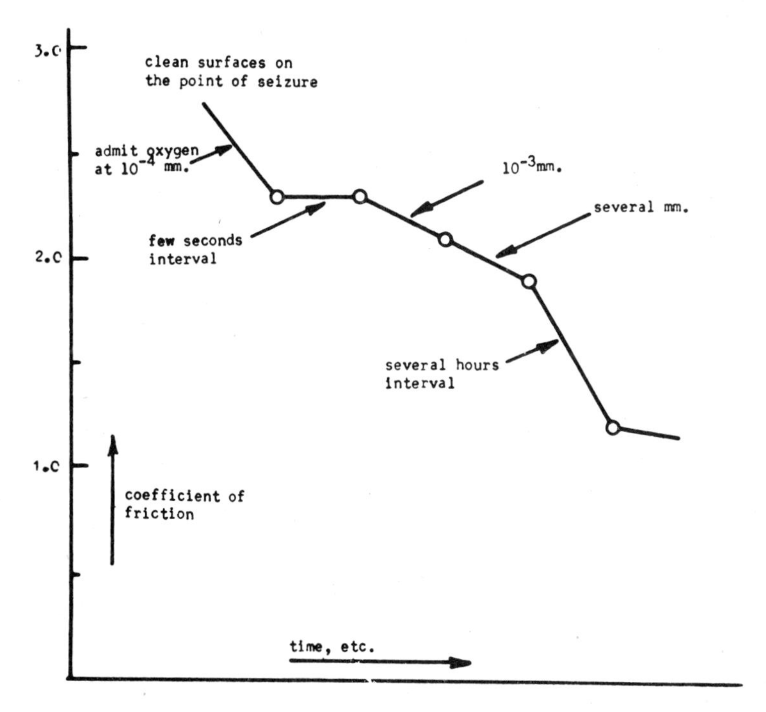

Figure 15.1. The effect of oxygen on the friction between outgassed iron surfaces (1).

itself led to a great increase in the true area of contact, with the apparent coefficient of friction increasing to 100 or more. When seizure occurred, the contact had the bulk strength of the metal itself.

The effect of adsorbed gases and vapors on the coefficient of friction of iron surfaces is shown in Figures 15.1 through 15.5. The presence of a trace of oxygen prevents seizure, and after prolonged exposure to oxygen, the coefficient of friction drops almost to 1 as shown in Figure 15.1. A similar decrease with water vapor, that is partially reversible, can be seen from Figure 15.2. An adsorbed layer of caproic acid reduces the coefficient of friction to about 2.5, as shown in Figure 15.3. After prolonged attack and the admission of oxygen, a further decrease is observed.

Figures 15.4 and 15.5 show the effects of chlorine and hydrogen sulfide, respectively. The reduction in friction is not reversible. The surfaces must

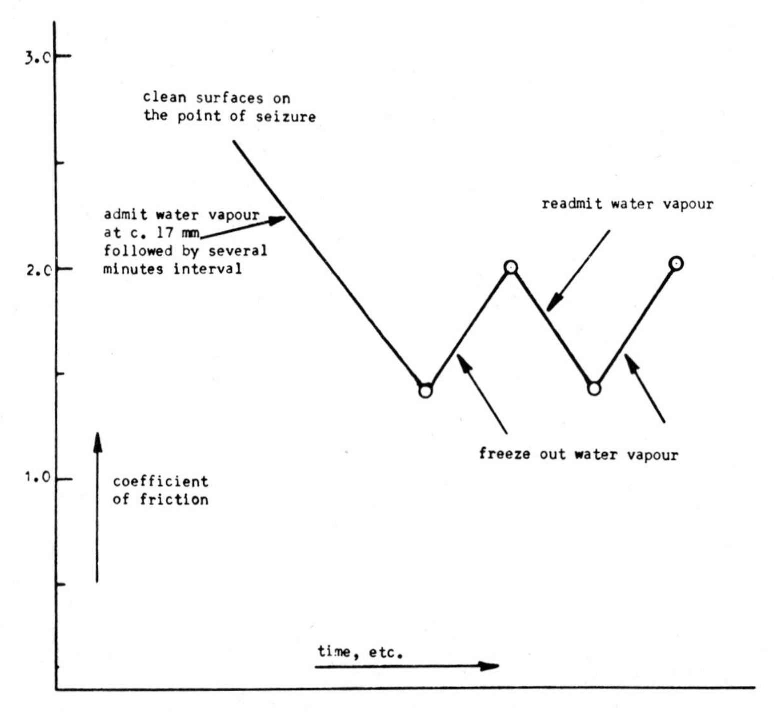

Figure 15.2. The effect of water vapor on the friction of outgassed iron surfaces (1).

be heated to 300 to 400°C before the chlorine is driven off and the friction again increases. The hydrogen sulfide reduces the friction, though not to the same extent as chlorine. However, the film is more stable and it is necessary to heat to over 800°C before the friction rises.

According to Amonton's law, the coefficient of friction is independent of the applied load. The ratio F/W is a constant. This law holds for a number of metals covered with their appropriate oxide-surfaces. These include zinc, cadmium, pure iron, and magnesium. When these surfaces are rubbed together, metallic welding occurs down to the lightest loads. However, with copper the rule is not applicable. At loads above about 10 grams, the coefficient is constant at about $\mu = 1.5$, but at lighter loads the friction drops to about $\mu = 0.4$. Metallic welding occurs only at the higher loads. Bowden (1) suggested that this behavior depends upon the relative mechanical properties of the oxide layer and the metal. With aluminum, the oxide layer is much harder than the metal. It may be broken through at the lightest load

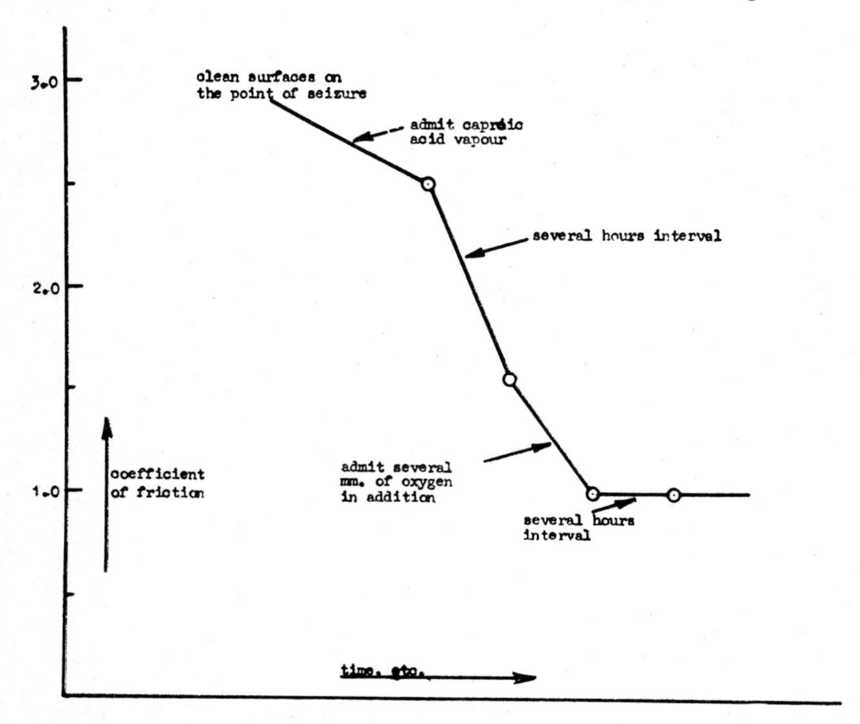

Figure 15.3. The effect of caproic acid vapor, alone and with oxygen, on the friction of outgassed iron surfaces (1)

and thus be unable to protect the metal. Copper and its oxide layer have a similar hardness and both are deformed together.

Gwathmey (3) prepared single crystals of metals in the form of spheres and polished the surfaces electrolytically in order to remove the distorted metal and expose the basic crystal. Static friction between bare crystals of copper was measured after heating in purified hydrogen at 500°C in order to remove oxide films. The friction measurements were carried out at room temperature with hydrogen passing through the apparatus. Profound differences in friction were observed for different faces of the copper crystals. For the (111) faces, the coefficient of friction ranged from 10 to 20, and for

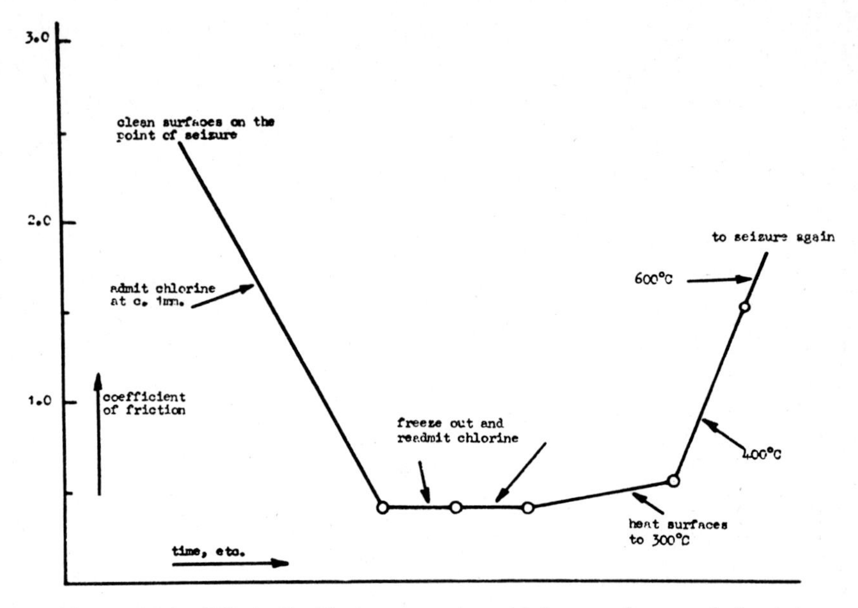

Figure 15.4. Effect of chlorine on outgassed iron surfaces and the influence of temperature (1).

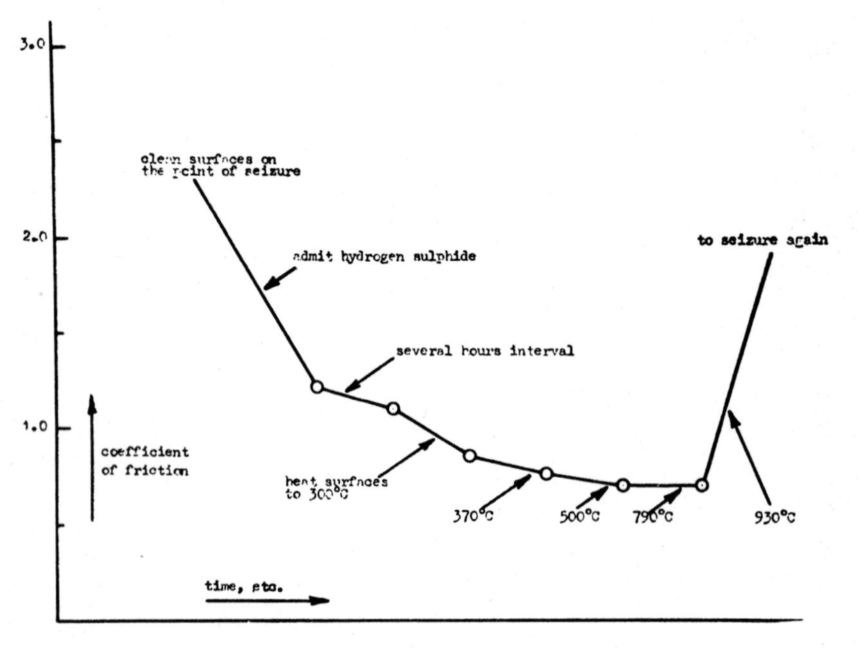

Figure 15.5. Influence of temperature on the frictional behavior of a sulphide film on outgassed iron surfaces (1).

the (100) faces from 62 to 100, the limit of the apparatus. Intermediate values were obtained with the (100) face pressed against the (111) face. Slip lines were visible in all instances, but were more pronounced on the (100) faces than on the (111). With pairs of (111) faces, there was only a slight tendency to dig into each other and the upper crystal slid smoothly over the lower one. With pairs of (100) faces, the upper crystal dug into and removed sizable chunks from the lower crystal, leaving pits as deep as 0.20 mm.

The effect of oxygen on wear was observed (3) by pressing a poly-crystalline copper sphere against a rotating copper-plated steel shaft, with a thrust of 15 grams. In a hydrogen atmosphere, the steel was exposed in only five seconds. In an atmosphere consisting of 500 parts of nitrogen to one part of oxygen, it required 6000 seconds to wear through the copper plating and expose the steel.

Effect of Films of Long-chain Polar Compounds on Friction

Hardy (4) and numerous other investigators have shown that films of polar organic compounds lower the coefficient of friction of rubbing solids. Langmuir (5) demonstrated that a monomolecular film of fatty acid deposited on glass could decrease the coefficient of friction to 0.1. Several years later, he (6) showed that a multilayer of calcium or barium stearate did not reduce the coefficient of friction significantly more than did a monolayer. However, the multilayer was much more durable and greatly reduced the rate of wear of the protected solid.

Zisman and his co-workers (7) concluded from their studies that the Langmuir-Blodgett multilayers (8) are artifacts and do not occur in lubrication or in any technological practice. However, monomolecular layers are readily formed under normal conditions. Zisman showed that three methods of preparing monolayers led to identical results. The withdrawal or retraction method consists simply of placing a solid with a clean, smooth surface in a solution of the polar organic compound and then slowly withdrawing the solid. Any polar-nonpolar organic compound may be used as the solute, and any liquid may be used as the solvent, provided that it is less adsorptive than the solute and it has a surface tension sufficiently greater than the critical surface-tension of the monomolecular film.

A second method of obtaining an adsorbed monolayer is to expose the clean solid to the vapor of the polar compound. This is done at a temperature sufficiently high to obtain an adequate vapor pressure but low enough to permit the molecules to adsorb as a close-packed film. After the solid has been allowed to cool to room temperature, excess condensed material above that in the monolayer is readily removed by vigorously wiping the surface with clean, grease-free tissue paper or adsorbent cotton. In the melt or thermal-gradient method (9), a small amount of the pure compound is placed on the solid, which is held with its plane slightly inclined from the horizontal. Upon the continued application of gentle heat to the upper edge of the solid, the organic material first melts and wets the surface, then suddenly recedes to the lower, cooler edge of the solid, exposing a dry band immediately adjacent to the liquid. Contact angle and coefficient of friction measurements

TABLE 15.1. WETTABILITY AND FRICTION DATA FOR MONOLAYERS OF FATTY AMINES AND THEIR DERIVATIVES AT 25°C, UNLESS OTHERWISE STATED (7).

Compound	Method of Prepn. of Film	Contact Angle θ max (degrees)	Coefficient of Friction μk
n-$C_{22}H_{45}NH_3Cl$	Retraction from aqueous soln.; concn. $\cong 1 \times 10^{-4}$ moles/l.; acidic pH	69	0.06
n-$C_{18}H_{37}NH_3Cl$	Retraction from aqueous soln.; concn. $\cong 1 \times 10^{-4}$ moles/l.; acidic pH	69	.06
n-$C_{17}H_{35}NH_3Cl$	Retraction from aqueous soln.; concn. $\cong 1 \times 10^{-4}$ moles/l.; natural pH	69	.04
n-$C_{16}H_{33}NH_3Cl$	Retraction from aqueous soln.; concn. $\cong 1 \times 10^{-4}$ moles/l.; natural pH	69	.06
n-$C_{12}H_{25}NH_3Cl$	Retraction from aqueous soln.; concn. $\cong 5 \times 10^{-4}$ moles/l.; natural pH	64[a]	.07
n-$C_{22}H_{45}NH_2$	Retraction from melt at 70°	69	.04
n-$C_{18}H_{37}NH_2$	1, vapor phase adsorption at 85°	69	.04
n-$C_{18}H_{37}NH_2$	2, retraction from n-cetane soln.; concn. $\cong 3 \times 10^{-4}$ moles/l.	70	.05
n-$C_{18}H_{37}NH_2$	3, retraction from nitromethane soln. (satd.)	69	.05
n-$C_{18}H_{37}NH_2$	4, retraction from melt at 70°	69	.05
n-$C_{16}H_{33}NH_2$	Retraction from melt at 55°	68–69	.06
n-$C_{14}H_{29}NH_2$	Retraction from melt at 50°	68	.04
n-$C_{12}H_{25}NH_2$	Retraction from melt at 45°	65[a]	.06
n-$C_{11}H_{23}NH_2$	Retraction from melt at 40°	62[a]	.06
n-$C_{10}H_{21}NH_2$	Retraction from melt at 35°	58[a]	.06
n-$C_8H_{17}NH_2$	Retraction from melt at 30°	53[a]	.10
n-$C_{16}H_{33}N(CH_3)_2$	Retraction from melt at 30°	66	.05
n-$C_{18}H_{37}N(CH_3)_3Cl$	Retraction from aqueous soln.; concn. $\cong 1 \times 10^{-4}$ moles/l.; natural pH	59	0.05
n-$C_{16}H_{33}N(CH_3)_3Br$	Retraction from aqueous soln.; concn. $\cong 1 \times 10^{-4}$ moles/l.; natural pH	59	0.06
n-$C_{12}H_{25}N(CH_3)_3Cl$	Retraction from aqueous soln.; concn. $\cong 5 \times 10^{-4}$ moles/l.; natural pH	58	0.06–0.07

[a] Test drop consisted of a dilute solution of the bulk film material in methylene iodide.

gave identical values in the dry band as on monolayers obtained by retraction from solution and by vapor adsorption. This is shown in Table 15.1. Contact-angle measurements were obtained with sessile drops of methylene iodide, with θ_{max} corresponding to the contact angle obtained with maximum packing of the adsorbed film. The coefficient of friction μ was determined by sliding a clean steel ball at a speed of 0.01 cm/sec over coated glass microscope slides. The frictional force F was measured while pressing the steel ball against the glass plate under a total load W which varied from a few grams to 9000 grams. The ratio F/W, which equals μ, was constant showing that Amonton's law was obeyed.

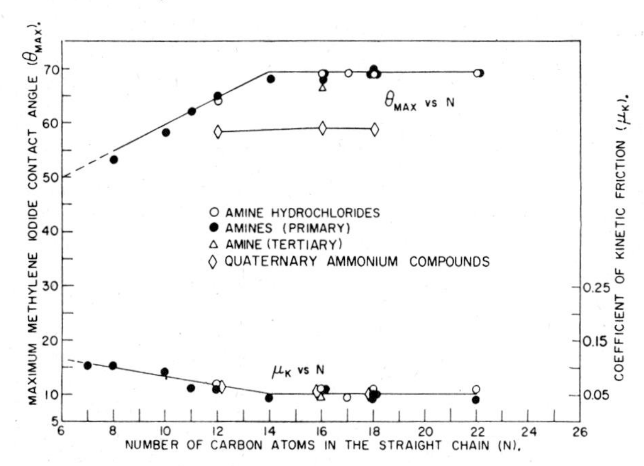

Figure 15.6. Friction and wetting properties of monolayers of fatty amines and their derivatives (7).

In Figure 15.6 are graphs of μ and θ_{max}, obtained with condensed monolayers of an homologous series of five *n*-alkyl primary amines, as a function of the number N of carbon atoms in the principal chain of the molecule. The monolayers were prepared under the conditions summarized in Table 15.1. A condensed monolayer of primary *n*-octadecylamine on glass caused μ to decrease from 1.1 on clean glass to 0.05. A clean ball under a 200 gram load left a wear scar on glass after one traverse. With the condensed monolayer a wear track was not visible even under a load of 5000 grams.

Essentially the same μ values were obtained for amine hydrochlorides, primary amines, tertiary amines, and quaternary ammonium compounds containing the same number of carbon atoms in the principal chain of the molecule, with μ reaching a limiting minimum value of 0.05 with $N \geqslant 14$. All of the amine derivatives, except the quaternary ammonium compounds, gave data falling on the same graph of θ_{max} *versus* N. Lower contact angles obtained with methylene iodide on monolayers of the quaternary ammonium compounds are probably due to the larger cross-sectional area of the polar heads of these compounds resulting in greater separation of the terminal methyl groups. The maximum contact angle θ_{max} on the amine monolayers reached its highest value with $N \geqslant 14$.

Similar studies were conducted (7) with other homologous series of polar compounds. With both fatty amines and fatty acids, abrupt transitions in the curves of μ *versus* N and θ_{max} *versus* N are well defined at $N = 14$. The fatty alcohols exhibit abrupt transitions in the μ *versus* N graph at $N = 18$ and in the θ_{max} *versus* N graph at $N = 15$. Perfluoro-alkanoic acids having the general formula $F_3C(CF_2)_xCOOH$, as well as acids and alcohols belonging to the homologous series $HF_2C(CF_2)_xCOOH$ and $HF_2C(CF_2)_xCH_2OH$ exhibit a sharp break in the μ *versus* N curve at $N = 11$, and a gradual transition in the θ_{max} *versus* N curve at about $N = 12$.

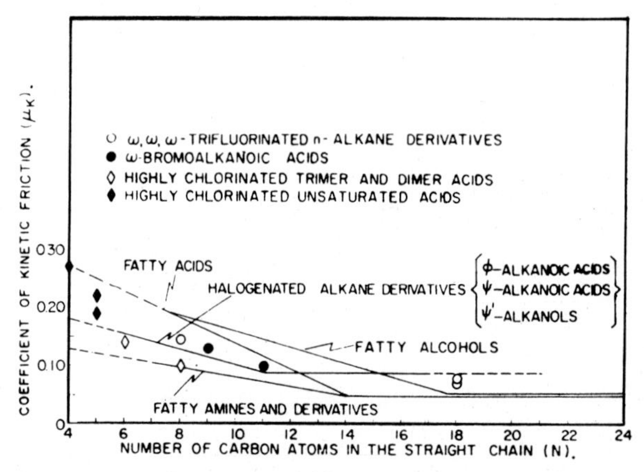

Figure 15.7. Comparison of frictional properties of monolayers of alkane and halogenated alkane derivatives (7).

Zisman deduced that the change in slope in any of the μ *versus* N curves on going to higher values of N is due to a transition from a condensed liquid film to a solid film, and that a two-dimensional solid film is necessary for maximum lowering of friction by a physically-adsorbed monolayer. The value of N corresponding to the abrupt transition in the μ *versus* N graphs is an indication of the adhesion of the polar group to the substrate and of the intermolecular cohesion of the film. Thus, the lower value of $N = 14$ for fatty acids and fatty amines, as compared with $N = 18$ for fatty alcohols shows the greater adhesion of acid and amine for glass, as compared with the hydroxyl group. Similarly the halogenated compounds examined exhibited greater intermolecular cohesion than the paraffin chains.

This is shown in Figure 15.7 for the liquid condensed region. At $N \leqslant 14$, the fatty amine curve of μ *versus* N is below the fatty-acid curve, which is displaced below the fatty alcohol curve. For the n-alkane derivatives, the energy of adsorption of the polar group to glass in the liquid condensed region increases in the order: alcohol $\leqslant$ acid $\leqslant$ amine $\leqslant$ quaternary ammonium compound.

It will also be observed from Figure 15.7 that the halogenated alkanoic acids are better boundary lubricants than the corresponding hydrocarbon acids at $N \leqslant 11$. However, asymptotic minimum values of $\mu = 0.05$ were obtained for the paraffin polar-derivatives, as compared with $\mu = 0.09$ for the highly fluorinated paraffin-derivatives. Thus, while fluorination has a strong effect on the wettability properties of the films, it does not reduce the friction of solid monolayers.

Durability of Surface Films

In continuing their studies on the lubricating properties of monolayers, Levine and Zisman (10) measured the effect of multiple traverses of a clean steel ball on the durability of monolayers on glass microscope slides. Differences in lubricating properties were observed that were not discernable by the single traverse of a steel ball under an applied load. Thus, while the same μ value was obtained in one traverse with monolayers of n-hexadecylamine hydrochloride and n-hexadecyltrimethylammonium bromide on glass, after 30 successive traverses under a 5000 gram load μ remained essentially unchanged with the primary amine. However, as shown in Figure 15.8, with the quaternary ammonium compound μ increased from 0.05 to 0.40 after only five traverses under the same load.

Condensed monolayers of octadecylamine prepared by retraction from nitromethane solution and from hexadecane solution showed the same initial μ values. However, the films prepared by retraction from hexadecane solution were considerably less durable because of the formation of mixed monolayers comprising the amine and hexadecane. The nitromethane molecule, with its almost spherical shape, is unable to adlineate between the adsorbed molecules of amine and form mixed metastable films. The decreased durability of these mixed films results from the lesser number of polar molecules in the monolayer, with consequent loss of adhesional energy between the glass and the polar groups. These mixed films were obtained after a two-thirds hour immersion of the glass slide in the solution. If 20 hours were allowed to elapse before retraction, the film was composed entirely of the amine. Thus, while the mixed film corresponds to a non-equilibrium condition, it can occur in practical lubrication conditions. These investigators also showed that the monolayer can be desorbed if the steel ball is rotated while sliding, so that a fresh surface is constantly being exposed. However, if sliding occurs in a pool of the polar organic compound, the monolayer will be replenished.

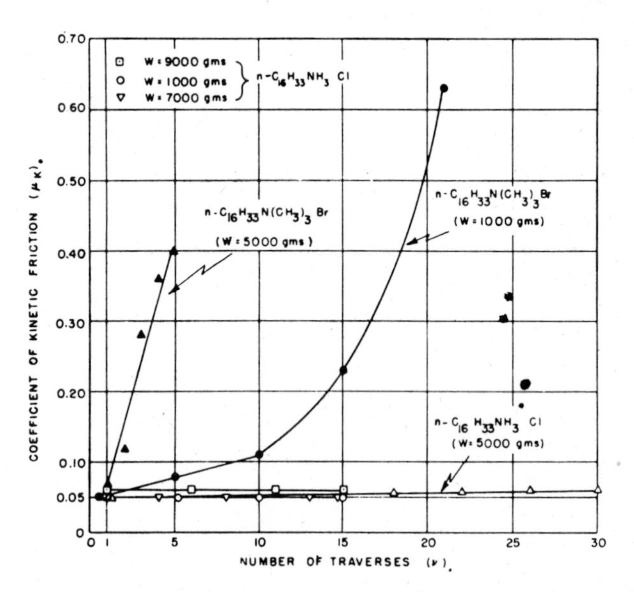

Figure 15.8. Durability of monolayers of amino compounds (10).

The effect on durability of varying the length of the principal chain of the molecule for a homologous series of fatty acids is shown in Figures 15.9 and 15.10. Only monolayers with 13 or more carbon atoms per molecule were able to endure 10 or more traverses under a 1000-gram load. Below 14 carbon atoms, there was a rapid decrease in durability with decreasing number of carbon atoms per molecule. Monolayers behave like liquids when $8 \leqslant N \leqslant 12$, amorphous or plastic solids when $13 \leqslant N \leqslant 15$ and crystalline solids when $N \geqslant 16$.

Condensed monolayers of perfluorolauric acid are considerably more durable than lauric acid. Replacing one of the terminal fluorine atoms with hydrogen reduces the durability slightly. Condensed monolayers of ω, ω, ω-trifluorostearic acid behaved similarly to stearic acid. Substitution of a bromine atom for an ω-hydrogen atom of an 11-carbon fatty acid, as esti-

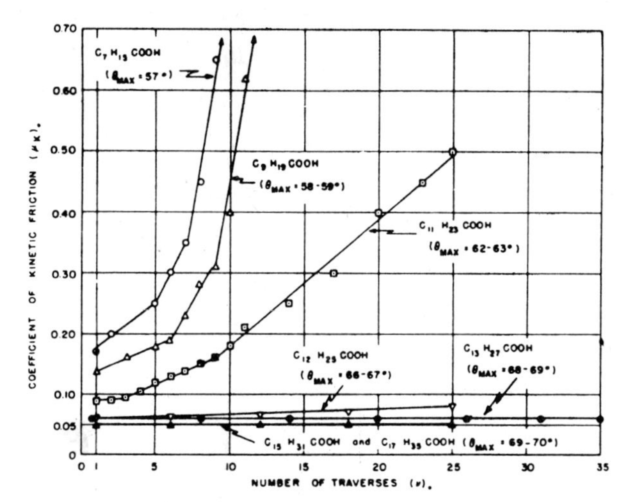

Figure 15.9. Durability of monolayers of fatty acids under a load of 1000 g. (10).

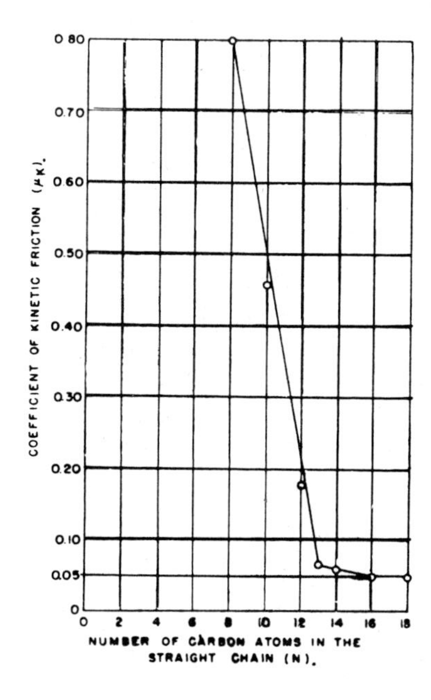

Figure 15.10. Coefficient of kinetic friction after 10 traverses under a load of 1000 g. for monolayers of fatty acids (10).

mated by interpolation, produced a more durable monolayer. Zisman presumed this to be due to an increase in the intermolecular cohesion.

When monolayers are applied to polished, stainless-steel plates instead of glass slides, upon which stainless steel balls are slid, the durability of the films is greatly reduced (11). The solid curves show multiple traverses on monolayers on steel with an 800 gram load as compared with the dashed curves for monolayers on glass with a 1000-gram load. Even the shortest acid studied appeared five times more durable on glass than on steel, and the differences in durability were more pronounced for the longer-chain acids. In contrast to negligible wear scars obtained on glass, scoring marks were deep and numerous on steel and the tear damage became progressively worse with successive traverses. This was due in part at least, to the cumulative build-up of adhering metallic fragments on the surface of the ball, each particle acting as a cutting tool.

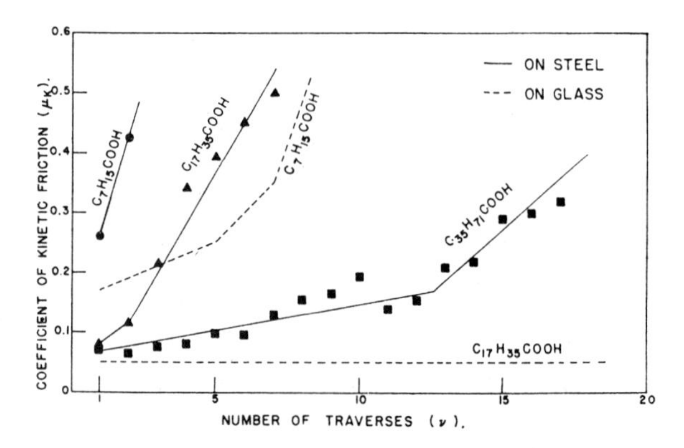

Figure 15.11. The durability of fatty acid monolayers (11).

For single traverses on monolayers of a homologous series of fatty acids on stainless steel a minimum value of $\mu=0.07$ was found, as compared with $\mu=0.05$ on glass. The graphs are shown in Figure 15.11. This difference in μ has been ascribed to either greater adhesion of the acid molecules to glass than to stainless steel or to the greater shear strength of the adhering asperity joints of steel to steel than of steel to glass causing a slight increase in the friction (11).

With fatty amines on steel the limiting value of $\mu=0.08$ was found, and the amine films were less durable than the corresponding acid-films. Thus, while the primary amines were found to be more strongly adsorbed on glass than the acids, they appear to be less strongly adsorbed on steel. The aliphatic alcohols were even less durable. With perfluorinated alkanoic acids, a minimum value of $\mu=0.15$ was obtained, showing that they were inferior as lubricants to the long-chain fatty acids. Trifluoromethyl-substituted acids were close to the fatty acids in general properties.

Bowden and Tabor (12) have demonstrated that adhesion occurs when asperities meet in the course of sliding. The extent of adhesion depends upon the freedom from organic contaminants, oxides, and other inorganic coatings. Wear, material transfer, and scoring result from the tearing of these adhesive joints. Ernst and Merchant (13) and Coffin (14) have proved that surface damage resulting from boundary friction is greater when pairs of metals capable of forming solid solutions or intermetallic compounds are rubbed together, as compared with metals of low affinity. This explains the differences observed upon sliding steel on glass, as compared with steel on steel. There is low mutual-affinity between steel and glass and relatively weak adhesive joints are formed.

From the work discussed above, certain conclusions (11) are evident concerning the molecular structure required for minimum friction between rubbing surfaces. First, a solid monolayer is required. The adsorbed molecules must be able to adlineate in order to attain a sufficiently large value of the intermolecular cohesional-energy to form the most impenetrable solid film possible. A long chain of methylene groups, or covalently bonded chlorine or bromine atoms substituted along the chain are advantageous structural features. Next, the polar group of the monolayer should adhere strongly to the substrate. Third, the outermost terminal groups of the adsorbed molecules should comprise a surface having the lowest possible free surface-energy. The outermost terminal groups are preferably $-CF_3$, $-CF_2H$ and $-CH_3$. Finally, the conditions for the application of the monolayer should be such as to avoid the possibility of forming a mixed monolayer containing solvent molecules which reduce the cohesional or adhesional energy. Thus, the function of a boundary lubricant appears to be the replacement of areas of contact between metals by contact between coatings having low adhesion to each other (4, 12). Preservation of the monolayer requires a high energy of adhesion of the polar group to the substrate and a high energy of intermolecular cohesion in the adsorbed organic monolayer.

Extreme Boundary Lubrication

The cutting of metals by a single point tool is a convenient means of studying the frictional properties of metals in sliding contact (15). The chip cut from the metal by the tool slides along the surface of the tool. In the absence of an efficient lubricant, the surface peaks periodically weld and separate. There is metal transfer, and a built-up edge can be observed at the point of the tool. The freshly produced metal chip is in a highly reactive stage, since it is clean and free of oxide films, and also contains a large amount of internal stress. The temperature at the chip-tool interface frequently exceeds 1000°F. These factors are conducive to chemical reaction and the formation of chemisorbed films. Thus, the cutting process is a useful means of testing extreme-pressure lubricants.

Shaw (15) was able to measure the coefficient of friction of a milling tool cutting aluminum by means of a two-component force dynamometer. The tool used was an 18-4-1 high-speed steel having a rake angle of 15° and a clearance angle of 5°. The cutting edge was normal to the direction of cut. The work material was annealed, commercially-pure aluminum having a Rockwell hardness of H-50.

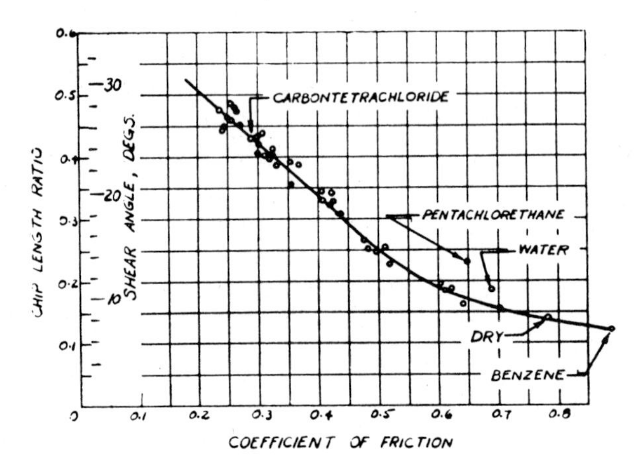

Figure 15.12. Variation of chip length ratio and shear angle with coefficient of friction for aluminum sliding on steel (15).

From measurements conducted in the presence of a number of fluids, used as lubricants, it was found that the coefficient of friction is related to the chip-length ratio, and therefore, to the shear angle. This is shown in Figure 15.12. The chip-length ratio is defined as the ratio of the length of the chip to the corresponding length of cut. The shear angle ϕ can be computed from the chip length ratio r_e and the rake angle a by the equation

$$\tan \phi = \frac{r_e \cos a}{1 - r_e \sin a} \qquad (15.2)$$

TABLE 15.2. FRICTION DATA FOR ORGANIC COMPOUNDS WITH ALUMINUM SLIDING ON HIGH-SPEED STEEL AT 5.5 IN/MIN (15).

Fluid	Formula	Chip Length Ratio	Coefficient of Friction
Chlorinated Hydrocarbons:			
Carbon tetrachloride	CCl_4	0.427	0.288
Chloroform	$CHCl_3$	0.401	0.312
Dichloromethane	CH_2Cl_2	0.265	0.477
Pentachlorethane	$Cl_2HC—CCl_3$	0.230	0.648
S. tetrachlorethane	$Cl_2HC—CHCl_2$	0.320	0.419
β Trichlorethane	$ClH_2C—CHCl_2$	0.344	0.407
Methyl chloroform	$Cl_3C—CH_3$	0.340	0.422
Ethylene dichloride	$ClH_2C—CH_2Cl$	0.250	0.481
Ethylidene dichloride	$Cl_2HC—CH_3$	0.330	0.406
Ethyl chloride	$H_3C—CH_2Cl$	0.255	0.511
Tetrachloroethylene	$Cl_2C=CCl_2$	0.195	0.603
Trichloroethylene	$Cl_2C=CHCl$	0.188	0.623
T. dichloroethylene	$ClHC=CHCl$	0.185	0.612
Hexyl chloride	$C_6H_{13}Cl$	0.308	0.440
Lauryl chloride	$C_{12}H_{25}Cl$	0.385	0.369
Esters:			
Ethyl acetate	$H_3C—C(=O)—O—C_2H_5$	0.227	0.518
Hexyl acetate	$H_{11}C_5—C(=O)—O—C_2H_5$	0.403	0.299
Lauryl acetate	$H_{23}C_{11}—C(=O)—O—C_2H_5$	0.440	0.240
Ethyl caproate	$H_3C—C(=O)—O—C_6H_{13}$	0.391	0.353
Ethyl laurate	$H_3C—C(=O)—O—C_{12}H_{25}$	0.447	0.244

continued

TABLE 15.2—CONTINUED

Fluid	Formula	Chip Length Ratio	Coefficient of Friction
Mercaptans:			
Ethyl mercaptan	H_5C_2—SH	0.437	0.308
Propyl mercaptan	H_7C_3—SH	0.440	0.276
Butyl mercaptan	H_9C_4—SH	0.473	0.267
Amyl mercaptan	$H_{11}C_5$—SH	0.473	0.265
Hexyl mercaptan	$H_{13}C_6$—SH	0.477	0.262
Heptyl mercaptan	$H_{15}C_7$—SH	0.483	0.255
Lauryl mercaptan	$H_{25}C_{12}$—SH	0.473	0.237
Disulfides:			
Methyl disulfide	H_3C—S—S—CH_3	0.417	0.302
Ethyl disulfide	H_5C_2—S—S—C_2H_5	0.448	0.271
Propyl disulfide	H_7C_3—S—S—C_3H_7	0.457	0.256
Butyl disulfide	H_9C_4—S—S—C_4H_9	0.457	0.257
Amyl disulfide	$H_{11}C_5$—S—S—C_5H_{11}	0.462	0.250
Alcohols:			
Ethyl alcohol	H_5C_2—OH	0.327	0.425
Hexyl alcohol	$H_{13}C_6$—OH	0.396	0.320
Decyl alcohol	$H_{21}C_{10}$—OH	0.400	0.325
Undecyl alcohol	$H_{23}C_{11}$—OH	0.412	0.325
Lauryl alcohol	$H_{25}C_{12}$—OH	0.385	0.331
Hydrocarbons:			
Hexane	C_6H_{14}	0.163	0.639
Dodecane	$C_{12}H_{26}$	0.156	0.703
Miscellaneous:			
Benzene	C_6H_6	0.123	0.892
Acetic acid	H_3C—C(=O)—OH	0.247	0.496
Hexanoic acid	$H_{11}C_5$—C(=O)—OH	0.349	0.353
Heptaldehyde	$H_{13}C_6$—C(=O)—H	0.430	0.298
Water	H_2O	0.187	0.690
Air		0.140	0.785

Data obtained with various fluids are reproduced from Shaw in Table 15.2. The disulfides, mercaptans, and long-chain-length esters were particularly effective in reducing the coefficient of friction. The lowest value, 0.237, was obtained with *n*-lauryl mercaptan. With each homologous series, the coefficient of friction was found to decrease with increased chain length.

With the chlorinated hydrocarbons, neither chlorine content nor chain length could be correlated with the coefficient of friction. Carbon tetrachloride gave the lowest coefficient of friction of the chlorinated hydrocarbons tested, while the unsaturated chlorinated hydrocarbons gave particularly high values.

By refluxing the chlorinated hydrocarbons in the presence of aluminum, Shaw showed that the fluids which reacted readily with aluminum produced less friction than those which did not react. In Table 15.3 the reflux conditions and time for reaction are compared with the coefficients of friction.

TABLE 15.3. COMPARISON OF REACTIVITY TOWARD ALUMINUM AND PERFORMANCE AS CUTTING FLUIDS FOR CHLORINATED HYDROCARBONS (15).

Fluid	Formula	Boiling Point (°C.)	Time to React (min.)	Coefficient of Friction	Chip Shape
Carbon tetrachloride	CCl_4	76.7	8.5	0.288	
Chloroform	$CHCl_3$	61.2	135	0.312	
Dichloromethane	CH_2Cl_2	39.8	360	0.477	
Pentachlorethane	$Cl_3C—CCl_2H$	161.9/180 mm.	6/125 mm.	0.648	
S. tetrachlorethane	$HCl_2C—CCl_2H$	146.5/360 mm.	2/200 mm.	0.419	
β Trichlorethane	$H_2ClC—CCl_2H$	113.5/700 mm.	1.5/500 mm.	0.407	
Methyl chloroform	$H_3C—CCl_3$	74.1	2	0.422	
Ethylene dichloride	$H_2ClC—CClH_2$	83.7	not in 24 hr.	0.481	
Ethylidene dichloride	$H_3C—CCl_2H$	57.3	not in 24 hr.	0.406	
Ethyl chloride	$H_3C—CClH_2$	12.5	not in 24 hr.	0.511	
Tetrachloroethylene	$Cl_2C=CCl_2$	120.8	not in 24 hr.	0.603	
Trichloroethylene	$HClC=CCl_2$	86.7	not in 24 hr.	0.623	
T. dichloroethylene	$HClC=CClH$	48.4	not in 24 hr.	0.612	

References

1. Bowden, F. P., *Ann. N. Y. Acad. Sci.* **53**, 805 (1951).
2. Bowden, F. P., and Young, J. E., *Nature* **164**, 1089 (1949).
3. Gwathmey, A. T., *Ann. N. Y. Acad. Sci.* **53**, 987 (1951).
4. Hardy, W. B., "Collected Scientific Papers", p. 867 Cambridge University Press, 1936.

5. Langmuir, I., *Trans. Faraday Soc.* **15**, 62 (1920).
6. Langmuir, I., *J. Franklin Inst.* **218**, 143 (1934).
7. Levine, O., and Zisman, W. A., *J. Phys. Chem.* **61**, 1068 (1957).
8. Blodgett, K., *J. Am. Chem. Soc.* **57**, 1007 (1935).
9. Baker, H. R., Shafrin, E. G., and Zisman, W. A., *J. Phys. Chem.* **56**, 405 (1952).
10. Levine, O., and Zisman, W. A., *J. Phys. Chem.* **61**, 1188 (1957).
11. Cottington, R. L., Shafrin, E. G., and Zisman, W. A., *J. Phys. Chem.* **62**, 513 (1958).
12. Bowden, F. P., and Tabor, D., "The Friction and Lubrication of Solids", London, Oxford University Press, 1950.
13. Ernst, H., and Merchant, M. E., "Conference on Friction and Surface Finish", Cambridge, Mass., Mass. Inst. Tech., 1940.
14. Coffin, L. F., Jr., *Lubrication Eng.* **12**, 50 (1956).
15. Shaw, M. C., *Ann. N. Y. Acad. Sci.* **53**, 962 (1951).

CHAPTER 16

Corrosion Inhibition

Rusting takes a tremendous annual toll of factories, ships, machines and buildings. The cost of corrosion has been estimated at $6 billion per year (1). Consequently, efforts are constantly being made to check the ravaging of corrosion. Methods used include the application of a barrier film between the metal and its environment, such as a paint or varnish. Suitable adsorbed films are capable of retarding corrosion. Another approach is the use of metals that are not attacked by the environment in which they are exposed. The passivity of metals and corrosion inhibition by adsorbed films are surface phenomena.

Corrosion Inhibition with Monomolecular Films

Corrosion is an electrochemical phenomenon and invariably requires the presence of moisture. A film capable of preventing direct contact of moisture with the metal will prevent corrosion. Pure nonpolar liquids, or weakly associated liquids, such as the hydrocarbons, ethers, esters, silicones, and halogenated hydrocarbons will not prevent corrosion, because they can be displaced from contact with clean metals by water. When long-chain polar molecules capable of adsorption are dissolved in these liquids, a hydrophobic film is formed on clean metals immersed in them, and water can no longer wet the metal. A rust inhibitor used in a petroleum oil must be soluble or dispersible in the oil, readily adsorbed by the metal and relatively insoluble in water. The latter requirement is necessary to prevent leaching of the monomolecular film.

An adsorbed film of hydrophobic character is not an impenetrable barrier. Oxygen and water molecules can diffuse between the long hydrocarbon chains to reach the metal surface. However, molecular movement is restricted and liquid water cannot contact the metal. Corrosion products which do form quickly clog the molecular pores and stop further attack. The less permeable the film, the more effective it is in reducing attack.

Reduced permeability can be obtained by increasing the size of the hydrocarbon chain and minimizing steric effects which prevent close packing, such as branching or bulky polar groups. An increase in temperature will decrease the closeness of packing. This can be prevented by increasing the concentration of the amphipathic compound, by increasing its chain length, or by selection of a polar group which is more firmly bonded to the metal. A limiting factor is the solubility of the compound in the oil.

Surfactants are also used in aqueous media to prevent corrosion by the media, as in acid-pickling. The same principles apply, the inhibitor must be soluble in the vehicle and capable of adsorption as a close-packed monolayer.

Inhibitors used in oils include the alcohols, amines and N-ring compounds, petroleum sulfonates, as well as fatty acids and their divalent soaps. The latter may function as reservoirs for the fatty acids. Compounds containing the C=S group and its modifications are commonly used in acid media. Vapor-phase inhibitors include organic nitrites that are sufficiently volatile to evaporate and condense on metal surfaces kept in closed containers. Long-chain amines and amides are used to prevent corrosion in boiler-water condensate lines. These compounds are steam distilled and then condense on the exposed metal.

Passivity and Chemisorption

Uhlig (1) has presented an informative discussion relating the passivity of metals to chemisorption. The definition of passivity stated by him is: "A metal active in the EMF Series, or an alloy composed of such metals, is considered passive when its electrochemical behavior becomes that of an appreciably less-active or noble metal." (2)

Since surface atoms have on the average only half the number of atoms as neighbors as those atoms underneath, there is a marked unbalance of forces at the surface. This situation accounts for the affinity of the surface for its environment. Adsorption supplies in effect additional neighbors, with chemisorbed atoms more completely satisfying surface forces than physically-adsorbed species. Since chemisorption renders a surface less attractive to its environment, it is more readily associated with the passivity of metals.

Langmuir (3, 4, 5) showed that chemisorbed oxygen atoms on tungsten become negatively charged and thereby drastically reduce the rate of electron emission from heated tungsten wires. The work function of a metal is reduced in accord with the equation,

$$\Delta \phi = 4 \pi n \mu \tag{16.1}$$

where $\Delta\phi$ is the change in the work function, n is the number of adsorbed oxygen atoms and μ is the dipole moment, which is equal to twice the product of the negative charge of the adsorbed atom and the distance from the center of gravity of the negative charge to the metal surface.

This change in potential at the surface of the metal accounts for its more noble potential as measured in aqueous media. The metal becomes more noble with an increase in the number of adsorbed atoms, and reaches its maximum value when the monolayer is completed.

According to the Langmuir equation for the adsorption of a gas on a surface as a function of pressure,

$$X = \frac{aX_m p}{1 + ap} \tag{16.2}$$

where X is the amount of gas adsorbed per unit area of surface at pressure p, X_m is the maximum adsorbate at high values of p, and a is a constant related to the heat of adsorption. The change in potential ΔE is proportional to the amount of the specie adsorbed per unit of surface area. Using Henry's Law, and replacing p by the concentration of solute C it follows that

$$\frac{C}{\Delta E} = \frac{C}{\Delta E_m} + \frac{1}{a'\Delta E_m} \tag{16.3}$$

where ΔE_m is the maximum potential change and a' is related to the heat of adsorption. When Langmuir's equation is applicable, a plot of $C/\Delta E$ *versus* C will produce a straight line.

Uhlig (6) showed the Langmuir adsorption isotherm to be applicable when iron is exposed to increasing concentrations of chromate and when stainless steel is exposed to increasing pressures of oxygen. The results are reproduced

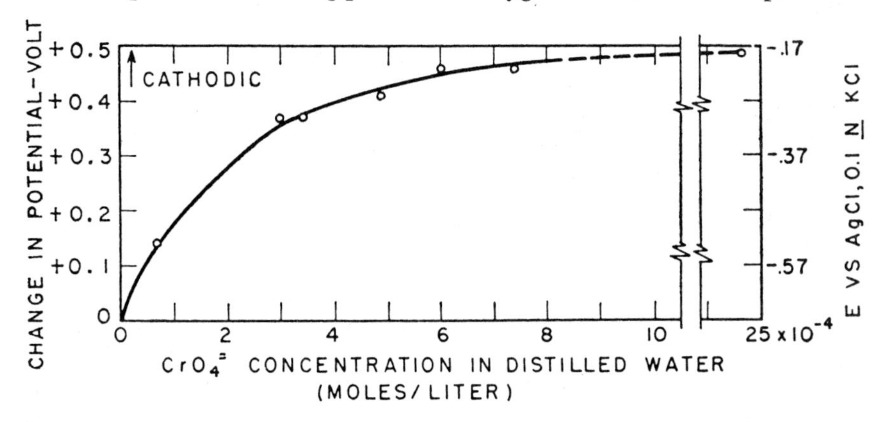

Figure 16.1. Effect of $CrO_4^{=}$ additions on the electrode potential of electrolytic iron in aerated distilled water at 25° C (1).

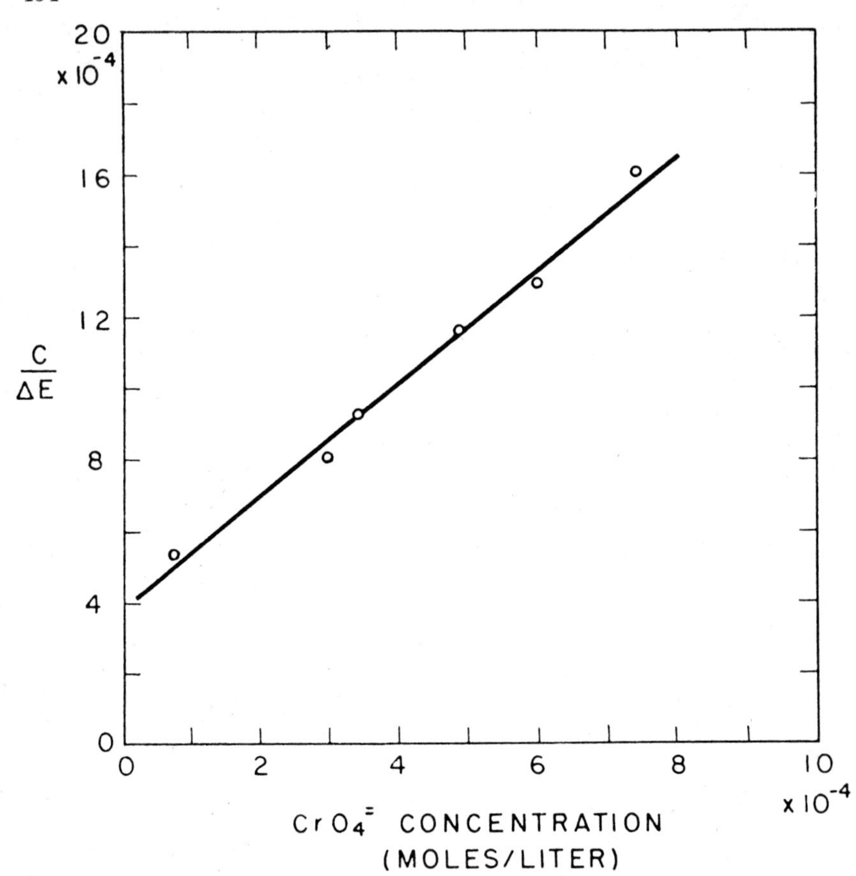

Figure 16.2. Langmuir adsorption plot for electrolytic iron in distilled water containing $CrO_4^=$ inhibitor at 25° C (1).

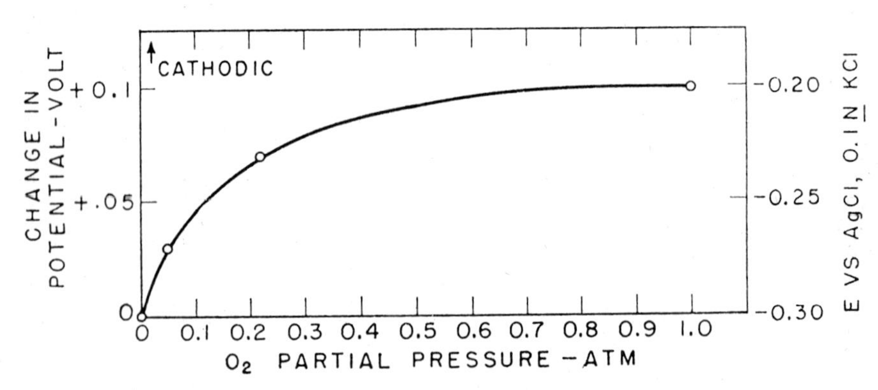

Figure 16.3. Effect of varying O_2 partial pressure on the electrode potential of 18–8 stainless steel in 4 per cent NaCl with 0.3*M* NaOH at 25° C (1).

in Figures 16.1 through 16.4. The minimum concentration of chromate required for optimum corrosion inhibition has been shown to be 10^{-3} molar (7). This is the approximate concentration for maximum potential change, corresponding to monolayer formation.

Since obedience to the Langmuir equation is frequently an indication of chemisorption, Uhlig concluded from this and other data that passivity

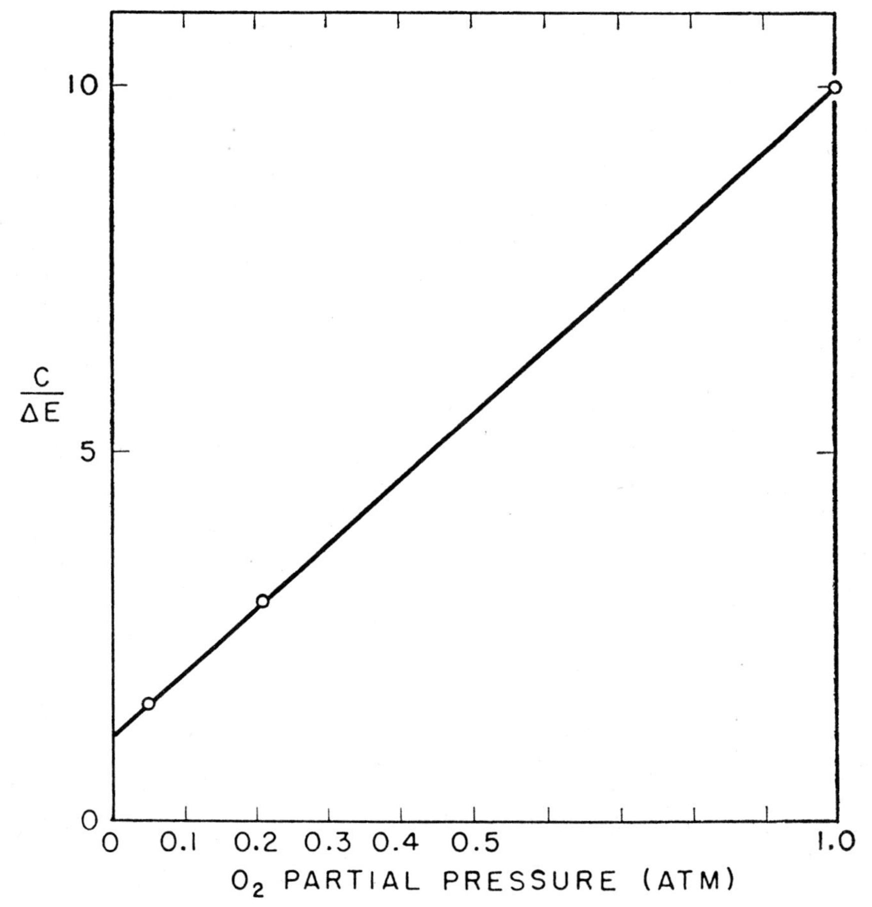

Figure 16.4. Langmuir adsorption plot for 18–8 stainless steel in 4 per cent NaCl with $0.3M$ NaOH at 25° C (1).

is related to chemisorption. The requirements for passivity are stated to be (1) a high affinity of the environment for the metal surface, and (2) a high activation energy for the adsorption. The effect of chlorides and chlorine in destroying passivity is said to be due to the low activation-energy of the adsorption reaction. It is difficult to understand this requirement for a high activation-energy, since this is not necessary for chemisorption.

Uhlig (1) relates the tendency for chemisorption and passivity with the properties of the metal and its electron configuration. Low work-function or easy surrender of an electron within the metal to its environment is a factor favorably disposing the metal to the formation of chemisorbed films. On the other hand, the heat of sublimation indicates the tendency of a metal atom to leave the crystal lattice. A low work-function and high heat of sublimation favor escape of an electron leading to chemisorption and passivity. If the work function is high and the heat of sublimation is low, the atom is more prone to escape, resulting in oxidation or corrosion. In Table 16.1, the work function ϕ and the sublimation energy ΔH_s, both in electron volts, is expressed as a ratio. If $\phi/\Delta H_s$ is less than one, passivity is favored, if greater than one we have oxidation and lack of passivity. With the transition metals of the Periodic Table, which are characterized by unfilled *d*-electron energy bands, the ratio suggests the tendency to become passive. This agrees with experience.

The transition metals with unfilled *d*-bands are among the most active

TABLE 16.1. RATIOS OF WORK FUNCTION TO HEAT OF SUBLIMATION FOR VARIOUS METALS (1).

	Heat of Sublimation, ΔH_S, 25°C, electron-volts	Work Function, electron-volts	Ratio, Work Function $/\Delta H_s$
K	0.93	2.24	2.4
Na	1.12	2.37	2.1
Ca	1.85	3.09	1.7
Mg	1.56	3.60	2.3
Zn	1.35	3.7–4.2	2.7–3.1
Hg	0.63	4.50	7.1
Pb	2.01	3.8	1.9
Al	2.92	3.6–4.4	1.2–1.5
Ag	3.0	3.9–4.7	1.3–1.6
Cu	3.54	4.80	1.36
Mn	3.02	3.76	1.25
Pt	5.40	6.3	1.17
Ni	4.26	4.1–5.0	0.96–1.2
Fe	4.20	4.2–4.7	1.0–1.1
Cr	3.87	4.35	1.1
W	8.8	4.52	0.51
Mo	6.75	4.12	0.61
C	8.6	4.82	0.56

catalysts. When a transition metal is alloyed with a metal with a filled *d*-band, the catalytic activity is reduced. Thus, palladium has 0.6 electron vacancies per atom in the *d*-band of energy levels, as determined by magnetic data. When alloyed with gold, which has a filled *d*-band, the electrons of gold tend to fill the partially empty *d*-band of palladium. The alloy remains relatively good as a catalyst as long as the *d*-band is partially empty. However, at 60 atoms per cent of gold or higher, when the *d*-band fills, catalytic activity deteriorates.

Since chemisorption is a requirement for surface catalysis, if passivity also requires a chemisorption, there should be a similarity in the behavior of alloys containing transition metals in both processes. This has been found

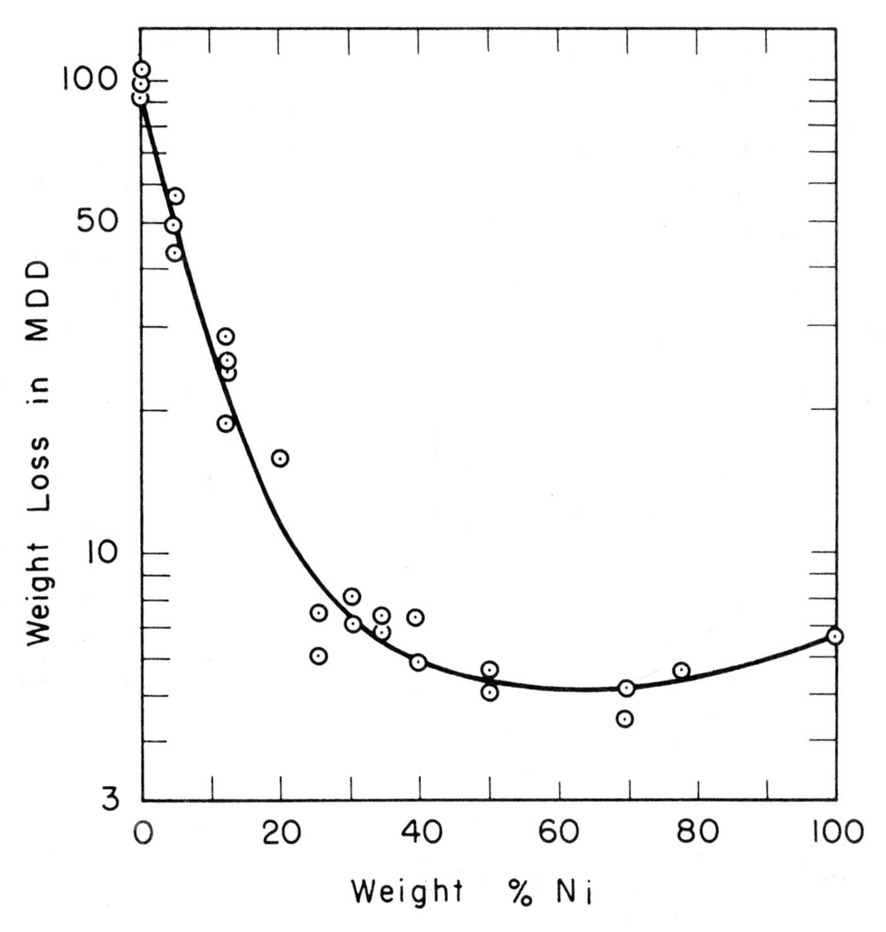

Figure 16.5. Corrosion rates of Cu-Ni alloys in aerated NaCl solution at 80°C (1).

to be the case. One example given by Uhlig (1) is the copper-nickel system. Nickel has 0.6 electron vacancy per atom in the *d*-band and is a passive metal. Copper with a full component of 10 *d*-electrons is not passive and corrodes in hot sodium chloride solution. Nickel alloyed with copper forms a solid-solution system throughout the composition range, with one electron from each copper atom available to fill the *d*-band of nickel. Above a certain critical level of nickel, the number of copper atoms is no longer sufficient to fill all of the *d*-band vacancies and the alloy behaves as a transition metal. Magnetic and specific-heat data show that this critical level is about 40 atom per cent of nickel. Corrosion data graphed in Figure 16.5 show that alloys containing more than 30 or 40 atom per cent of nickel corrode at lower rates than alloys of lower nickel content. Similar results are obtained with the familiar stainless steels. Above certain critical levels of chromium or nickel, there are electron vacancies remaining and the alloys behave like transition metals and are passive.

Protective Films and Passivity

There are two general theories concerning the passivity of metals. The one based on chemisorption of the corrosion inhibitor has been discussed above. The other is based on the formation and maintenance of a protective film, which is a chemical reaction product of the metal and its environment. This may be an oxide film, a precipitation film, or a mixture of both. While the chemisorbed film is essentially monomolecular, the chemical film is many molecules thick. It will continue to grow until direct contact between the metal and the corrosion medium has been eliminated. In accordance with this generalized film theory (8), only dense, impenetrable films are effective in preventing further corrosion.

In accordance with the chemisorption theory, the ability of chloride ion to destroy passivity was ascribed to the low or zero energy of activation for chemisorption of chloride ion and its capacity for displacing chemisorbed oxygen and other passivators. The chemical-film theory of passivity explains chloride ion corrosion on the basis of penetration of the oxide or precipitate film by the small chloride ions.

Much of the evidence in support of the generalized film-theory has been obtained by isolation of the passivating film. Several investigators have found that oxide films in the range of ten monolayers thick are closely connected with the development of passivity (8, 9, 10, 11, 12). Rhodin (13) passivated stainless steels by immersion in 5 per cent nitric acid -0.5 per cent potassium dichromate solution for 30 minutes and then removed the passive film using either 2 per cent bromine in anhydrous methanol, or by stripping with a high-purity pressure-sensitive synthetic rubber adhesive on "Mylar" tape. His data indicated multimolecular layer films composed of hydrous mixed metal oxides enriched in silica. Chromium, nickel, and molybdenum enrichment was relatively slight. The passive films resembled gel structures.

References

1. Uhlig, H. H., *Ann. N. Y. Acad. Sci.* **58**, 843 (1954).
2. Uhlig, H. H., "The Corrosion Handbook", p. 21, New York, John Wiley & Sons, Inc., 1948.
3. Langmuir, I., *J. Am. Chem. Soc.* **38**, 2221 (1916).
4. Langmuir, I., *Trans. Electrochem. Soc.* **29**, 260 (1916).
5. Langmuir, I., *J. Chem. Soc.* **1940**, 518.
6. Uhlig, H. H., and Geary, A., *J. Electrochem. Soc.* **101**, 215 (1954).
7. Robertson, W. D., *J. Electrochem. Soc.* **98**, 94 (1951).
8. Heumann, T., *Z. Elektochem.* **55**, 287 (1951).
9. Evans, V. R., and Stockdale, J., *J. Chem. Soc.* **1929**, 2651.
10. Vernon, W. H. J., Wormwell, F., and Nurse, T. J., *J. Iron Steel Inst.* **150**, 81 (1944).
11. Nurse, T. J., and Wormwell, F., *J. Applied Chem.* **2**, 550 (1952).
12. Mahla, E. M., and Nielsen, N. A., *Trans. Electrochem. Soc.* **93**, 1 (1948).
13. Rhodin, T. N., Jr., *Ann. N. Y. Acad. Sci.* **58**, 855 (1954).

Author Index

Subject Index